MULTI-SCALE
INTEGRATED
ANALYSIS
OF AGROECOSYSTEMS

Advances in Agroecology
Series Editor: Clive A. Edwards

MULTI-SCALE
INTEGRATED
ANALYSIS
OF AGROECOSYSTEMS

MARIO GIAMPIETRO

CRC Press
Taylor & Francis Group
Boca Raton London New York

CRC Press is an imprint of the
Taylor & Francis Group, an **informa** business

CRC Press
Taylor & Francis Group
6000 Broken Sound Parkway NW, Suite 300
Boca Raton, FL 33487-2742

First issued in paperback 2019

ISBN-13: 978-0-8493-1067-6 (hbk)
ISBN-13: 978-0-367-39481-3 (pbk)

Library of Congress Cataloging-in-Publication Data

Giampietro, M. (Mario)
    Multi-scale integrated analysis of agroecosystems / Mario Giampietro.
        p.   cm. (Advances in agroecology)
    Includes bibliographical references and index.
    ISBN 0-8493-1067-9 (alk. paper)
    1. Agricultural ecology. 2. Agricultural systems. I. Title II. Series.

S589.7.G43 2003
338.1—dc22
                                                2003059613

Library of Congress Card Number 2003059613

Visit the Taylor & Francis Web site at
http://www.taylorandfrancis.com

and the CRC Press Web site at
http://www.crcpress.com

If a student is not eager, I won't teach him;

If he is not struggling with the truth, I won't reveal it to him.

If I lift up one corner and he can't come back with the other three,

I won't do it again.

— The Analects, *Confucius*

Preface

Warning to the Potential Reader of This Book

Discussing the implications of a paradigm change in science, Allen et al. (2001) said: "A paradigm change modifies protocols, vocabulary or tacit agreements not to ask certain questions" (p. 480). If we agree with this brilliant definition, and therefore if we accept that a scientific paradigm is "a tacit agreement not to ask certain questions," the next step is to find out why certain questions are forbidden. In general, the questions that cannot be asked from within a scientific paradigm are those challenging the basic assumptions adopted in the foundations of the relative disciplinary scientific knowledge.

The enforcement of this tacit agreement is a must for two reasons. First, it is required to preserve the credibility of the established set of protocols proposed by the relative disciplinary field (what the students learn in university classes). Second, it makes it possible for the practitioners of a disciplinary field to focus all their attention and efforts only on how to properly run the established set of protocols, while forgetting about theoretical issues and controversies. In fact, the acceptance of a scientific paradigm prevents any questioning of the usefulness of the established set of protocols developed within a disciplinary field for dealing with the task faced by the analyst.

When dealing with a situation of crisis of an existing scientific paradigm — and many seem to believe that in relation to the issue of sustainability of human progress we are facing one — we should expect that such a tacit agreement will get us into trouble. Whenever the established set of protocols (e.g., analytical tool kits) available for making analysis within disciplinary fields is no longer useful, the number of people willing to ask forbidden questions reaches a critical size that overcomes the defenses provided by academic filters. After reaching that point, criticizing the obsolete paradigm is no longer a taboo. In fact, nowadays, several revolutionary statements that carry huge theoretical implications about the invalidity of the foundations of leading scientific disciplines are freely used in the scientific debate. For example, expressions like "the myth of the perpetual growth is no longer acceptable (why?)," "it is not possible to find an optimal solution when dealing with contrasting goals defined on different dimensions and scales (why?)" and "we cannot handle uncertainty and ignorance just by using bigger and better computers (why?)" in the 1970s and 1980s have been sanguinary battlefields between opposite academic disciplines defending the purity of their theoretical foundations. These expressions are now no longer contested. Actually, we can even find softened versions of these statements included in the presentation of innovative academic programs and in documents generated by United Nations agencies.

This situation of transition, however, generates a paradox. In spite of this growing deluge of unpleasant forbidden questions about the validity of the foundations of established disciplinary scientific fields, nothing is really happening to the teaching of protocols within the academic fields under pressure for change. In fact, at this point, the lock-in that is protecting obsolete academic fields no longer works against posing forbidden questions. Rather, it works by preventing the generation of answers to these forbidden questions. The mechanism generating this lock-in is simple and conspiracy-free. Academic filters associated with obsolescent disciplinary knowledge do their ordinary work by attacking every deviance (those who try to find new perspectives). This applies to both those who develop nontraditional empirical analyses (e.g., putting together data in a different way, especially when they obtain interesting results) and those who develop nontraditional theories (e.g., putting together ideas in a nonconventional way, especially when they obtain interesting results). The standard criticism in these cases is that "this is just empirical work without any sound theory supporting it" or that "this is just theoretical speculation without any empirical work supporting it." When innovative theories are developed to explain empirical results, the academic filter challenges every single assumption adopted in the new theory (even though it is totally neglecting to challenge even the most doubtful assumptions of its own discipline). Finally, whenever the academic filter is facing the unlikely event that (1) a new coherent theory is put forward, (2) this theory can be defended step by step starting from the foundations, (3) experimental data are used to validate such a theory and (4) this theory is useful for dealing with the tasks faced by the analysts, the unavoidable reaction is always the same: "This is not what our disciplinary field is about. Practitioners of our field would never be interested in going through all of this."

Obviously, the analysis of this mechanism of lock-in — very effective in preventing the discussion of possible answers to forbidden questions — has a lot to do with the story that led to the writing of this book. This is why I decided to begin with this preface warning potential buyers and readers. This book represents an honest effort to do something innovative in the field of the integrated analysis of sustainability of agricultural systems, that is, an honest effort to answer a few of the forbidden questions emerging in the debate about sustainability. This book reflects a lot of work and traveling to visit the most interesting groups that are doing innovative things related to this subject in various disciplinary and interdisciplinary fields. I wrote this book for those who are not happy with the analytical tools actually used to study and make models about the performance of farming systems, food systems and agroecosystems, and especially for those interested in considering various dimensions of sustainability (e.g., economic, ecological, social) simultaneously and willing to reflect in their models the nonequivalent perspectives of different agents operating at different scales.

The mechanism that generated the writing of this book is also simple. There is an old Chinese saying (quoted by Röling, 1996, p. 36) that puts it very plainly: "If you don't want to arrive where you are going, you need to change direction."

What does this mean for a scientist or practitioner changing direction? In my interpretation of the Chinese saying, this means going back to the foundations of the disciplinary knowledge that has been used to develop the analytical tools that are available and in use at the moment and trying to see whether it is possible to do things in an alternative way. When I started my journey many years ago, as a scientist willing to deal with the sustainability of agriculture, I had to swim in a sea of complaints about the inadequacy of reductionism, the lack of holism and the need of a paradigm shift in science. This ocean of complaints was linked to the acknowledgment of a never-ending list of failures of the applications of the conventional approach in relation to the sustainability of agriculture in both developed and developing countries. However, in spite of all of these complaints, when looking at scientific papers dealing with the sustainability of agriculture, in the vast majority of cases I found models that were based on the same old set of tools (e.g., statistical tests and differential equations). These models were applied to an incredible diversity of situations, always looking for the optimization of a function assumed to represent a valid (substantive) formal definition of performance for the system under investigation.

Since I was then and still am convinced that I am not smarter than the average researchers of this field, I was forced to realize that if I wanted to arrive in a different place, I had to change the path I was on. Otherwise, I would have joined the party of optimizers already jammed at the end of it. When you take a wrong path and want to get on another one, you must go back to the bifurcation where you made the bad turn. This is why I decided to go back to the theoretical foundations of the analytical tools I was using, to try to see if it were possible to develop an alternative set of tools useful to analyze in a different way the complex nature of agroecosystems. Then I found out that the new field of complex systems theory implied the rediscovery of old epistemological issues and new ways of addressing the challenge implied by modeling.

This book is an attempt to share with the reader what I learned during this long journey. The text is organized in three parts:

- **Part 1: Science for Governance: The Clash of Reductionism against the Complexity of Reality.** After acknowledging that there is a problem with reductionism when dealing with the sustainability of agroecosystems (in Chapter 1), the remaining four chapters provide new vocabulary, narratives and explanations for the epistemological predicament entailed by complexity. Chapter 2 starts by looking at the roots of that predicament, focusing on the neglected distinction between the perception and representation of reality. Additional concepts required to develop an alternative narrative are introduced and illustrated with practical examples in Chapter 3. The resulting challenge for science when used for governance in the face of uncertainty and legitimate contrasting values is debated in general terms in Chapter 4. Finally, an overview of the problems associated with the development of scientific procedures for participatory integrated assessment is discussed in Chapter 5.
- **Part 2: Complex Systems Thinking: Daring to Violate Basic Taboos of Reductionism.** This part introduces a set of innovative concepts derived from various applications of complex systems thinking. These concepts can be used to develop a tool kit useful for handling multi-scale integrated analysis of agroecosystems. In particular, three key concepts are introduced and elaborated on in the three chapters making up this second part:
  1. Chapter 6 — Multi-scale mosaic effect
  2. Chapter 7 — Impredicative loop analysis
  3. Chapter 8 — Unavoidable necessity of developing useful narratives to surf complex time
- **Part 3: Complex Systems Thinking in Action: Multi-Scale Integrated Analysis of Agroecosystems.** This part presents a tool kit based on the combined use of the previous three concepts to obtain a multi-scale integrated analysis of agroecosystems. This third part is organized into three chapters:
  1. Chapter 9 — Bridging disciplinary gaps across hierarchical levels
  2. Chapter 10 — Bridging changes in societal metabolism to the impact generated on the ecological context of agriculture
  3. Chapter 11 — Benchmarking and tailoring multi-objective integrated analysis across levels

After having put the cards on the table with this outline, I can now move to the warning for potential readers and buyers: Who would be interested in reading such a book? Why?

This is not a book for those concerned with being politically correct, at least according to the definitions adopted by existing academic filters. This book is weird according to any of the conventional standards adopted by reputable practitioners. This is scientific research in agriculture that is not aimed at producing *more* and *better*. Rather, this is research aimed at learning how to define what *better* means for a given group of interacting social actors within a given socioeconomic and ecological context. Within this frame, the real issue for scientists is that of looking for the most useful scientific problem structuring.

It should be noted that hard scientists who use models to individuate the best solution (a solution that produces *more* and *better* than the actual one) are operating under the bold assumption that it is always possible to have available: (1) a win–win solution, that is, that *more* does not imply any negative side effects and (2) a substantive formal definition of *better* that is agreed to by all social actors and that can be used without contestation as an input to the optimizing models. According to this bold assumption, the only problem for hard scientists is that of finding an output generated by the model that determines a maximum in improvement for the system.

If we were not experiencing the tragic situation we are living in (malnutrition, poverty and environmental collapse in many developing countries associated with bad nutrition, poverty and environmental collapse in many developed countries), this blind confidence in the validity of such a bold assumption would be laughable. After having worked for more than 20 years in the field of ecological economics, sustainable development and sustainable agriculture in both developed and developing countries, I no longer, unfortunately, find the blind confidence in the validity of this bold assumption amusing.

Sustainability, when dealing with humans, means the ability to deal in terms of action with the unavoidable existence of legitimate contrasting views about what should be considered an improvement. Winners are always coupled to losers. To make things more difficult, nobody can guess all the implications of a change. If this is the case, then how can this army of optimizers know that their definition of what is an improvement (the one they include in formal terms in their models as the function to be optimized) is the right one? How can it be decided by an algorithm that the perspectives and values of the winners should be considered more relevant than the perspectives and values of the losers?

Sustainability means dealing with the process of "becoming." If we want to avoid the accusation of working with an oxymoron (sustainable development), we should be able to explain what in our models remains the same when the system becomes something else (in a sustainable way). That is, we should be able to individuate in our models what remains the same when different variables, different boundaries and emerging relevant qualities will have to be considered to represent the issue of sustainability in the future. Optimizing models either maximize or minimize something within a formal (given and not changing in time) information space.

When dealing with a feasible trajectory of evolution, the challenge of sustainability is related to the ability to keep harmony among relevant paces of change for parts (that are becoming in time), which are making up a system (that is becoming in time), which is coevolving with its environment (that is becoming in time). This requires the simultaneous perception and representation of events over a variety of space–time scales. The various paces of becoming of parts, the system and the environment are quite different from each other. Can this cascade of processes of becoming and cross-relations be studied using reducible sets of differential equations and traditional statistical tests? A lot of people working in hierarchy theory and complex systems theory doubt it. This book discusses why this is not possible.

These fundamental questions should be taken seriously, especially by those who want to deal with sustainability in terms of hard scientific models (by searching for a local maximum of a mathematical function and for significance at the 0.01 level). It is well known that when dealing with life, hard science often tends to confuse formal rigor with rigor mortis. In this regard, the reductionist agenda is well known. To study living systems, we first have to kill them to prevent adjustment and changes during the process of measurement. The rigorous way, for the moment, provides only protocols that require reducing wholes into parts and then measuring the parts to characterize the whole. Is it possible to look at the relation of wholes and parts in a new way? Can we deal with chicken–egg paradoxes, when the identity of the parts determines the identity of the whole and the identity of the whole determines the identity of the parts? Obviously, this is possible. This is how life, languages and knowledge work. This book discusses why and how this can be done in multi-scale integrated analysis of agroecosystems.

Finally, there is another very interesting point to be made. Are these forbidden questions about science new questions? The obvious answer is not at all. These are among the oldest and most debated issues in human culture. Humans can represent in their scientific analyses only a shared perception about reality, not the actual reality. Models are simplified representations of a shared perception of reality. Therefore, by definition, they are all wrong, even though they can be very useful (Box, 1979). But to take advantage of their potential usefulness — in terms of a richer understanding of the reality — it is necessary to be aware of basic epistemological issues related to the building of models. The real tragedy is that activities aimed at developing this awareness are considered not interesting or even not "real science" by many practitioners in hard sciences. On the contrary, this is an issue that is considered very seriously in this book. From this perspective, complex systems theory has merit to have put back on the agenda of hard scientists a set of key epistemological issues debated in disciplines such as natural philosophy, logic and semiotics, which, until recently, were not viewed as hard enough.

It is time to reassure those potential readers who got scared by the outline and the ensuing discussion. What does all of this have to do with a multi-scale integrated analysis of agroecosystems? Well, the point I have been trying to make so far is that it has a lot to do with multi-scale integrated analysis of agroecosystems.

In the last 20 years, I have been generating a lot of numbers about the sustainability of agricultural systems by studying this problem from different perspectives (technical coefficients, farming systems, global biophysical constraints, ecological compatibility) and using various sets of variables (energy, money, water, demographics, sociality). In the beginning, this was done by following intuitions about how to do things in a different way. Later on, after learning about hierarchy theory, postnormal science and complex systems theory (especially because of the gigantic contributions of Robert Rosen), I realized that it was possible to back up these intuitions with a robust theory. This made possible the organization of the various pieces of the mosaic into an organic whole. This is what is presented in Part 2 of this book. Part 2 provides new approaches for organizing data and examples of applications of multi-scale integrated analysis of agroecosystems to real cases. The results presented in Part 3, in my view, justify the length and heterogeneity of issues presented in Parts 1 and 2. In spite of this, I understand that cruising Parts 1 and 2 is not easy, especially for someone not familiar with the various issues discussed in the first eight chapters. On the other hand, this can be an occasion for those not familiar with these topics to have a general overview of the state of the art and reference to the literature.

There is a standard predicament associated with scientific work that wants to be truly interdisciplinary. Experts of a particular scientific field will find the parts of the text dealing with their own field too simplistic and inaccurate (an uncomfortable feeling when reading about familiar subjects), whereas they will find the parts of the text dealing with less familiar topics obscure and too loaded with useless and irrelevant details (an uncomfortable feeling when reading about unfamiliar subjects). This explains why genuine transdisciplinary work is difficult to sell. As readers we are all bothered when forced to handle different types of narratives and disciplinary knowledge. Nobody can be a reputable scholar in many fields. To this end, however, I can recycle the apology written by Schrödinger (1944) about the unavoidable need of facing this predicament:

> A scientist is supposed to have a complete and thorough knowledge, at first hand, of some subjects and, therefore, is usually expected not to write on any topic of which he is not a life master. This is regarded as a matter of noblesse oblige. For the present purpose I beg to renounce the noblesse, if any, and to be the freed of the ensuing obligation. My excuse is as follows: We have inherited from our forefathers the keen longing for unified, all-embracing knowledge. The very name given to the highest institutions of learning reminds us that from antiquity to and throughout many centuries the universal aspect has been the only one to be given full credit. But the spread, both in width and depth, of the multifarious branches of knowledge during the last hundred odd years has confronted us with a queer dilemma. We feel clearly that we are only now beginning to acquire reliable material for welding together the sum total of all that is known into a whole; but, on the other hand, it has become next to impossible for a single mind fully to command more than a small specialized portion of it. I can see no other escape from this dilemma (lest our *true who aim* be lost for ever) than that some of us should venture to embark on a synthesis of facts and theories, albeit with second-hand and incomplete knowledge of some of them — and at the risk of making fools of ourselves.

To make the life of the reader easier, the text of the first eight chapters has been organized into two categories of sections:

1. **General sections** that introduce main concepts, new vocabulary and narratives using practical examples and metaphors taken from normal life situations
2. **Technical sections** that get into a more detailed explanation of concepts, using technical jargon and providing references to existing literature

The sections marked "technical" can be glanced through by those readers not interested in exploring details. In any case, the reader will always have the option to go back to the text of these sections later. In fact, when dealing with a proposal for moving to a new set of protocols, vocabulary and tacit agreements not to ask certain questions, one cannot expect to get everything in one cursory reading of a book. Actually, the goal of the first eight chapters is to familiarize the reader with new terms, new concepts and new narratives that will be used later on to propose innovative analytical tools. This means that the structure of this book implies a lot of redundancy. The same concepts are first introduced in a discursive way (Part 1), reexplored using technical language (Part 2) and then adopted in the development of procedures useful to perform practical applications of multi-scale integrated analysis of agroecosystems (Part 3). Because of this, the reader should not feel frustrated by the high density of the information faced when reading some of the chapters in Parts 1 and 2 for the first time.

References

Allen, T.F.H., Tainter, J.A., Pires J.C. and Hoekstra T.W., (2001), Dragnet ecology, "just the facts ma'am": The privilege of science in a post-modern world, *Bioscience*, 51, 475–485.

Box, G.E.P., 1979 Robustness is the strategy of scientific model building. In R.L. Launer and G.N. Wilkinson (Eds.) Robustness in Statistics. Academic Press, New York. pp. 201–236.

Röling, N., (1996), Toward an interactive agricultural science, *Eur. J. Agric. Educ. Ext.*, 2, 35–48.

Schrödinger, E., (1944), *What Is Life?* based on lectures delivered under the auspices of the Dublin Institute for Advanced Studies at Trinity College, Dublin, February 1943, available from The Book Page: http://home.att.net/~p.caimi/schrodinger.html.

Acknowledgments

As discussed in a convincing way by Aristotle, it is not easy to individuate a single direct cause of a given event — i.e. the writing of a book — since, in the real world, several causes (material, efficient, formal and final) are always at work in parallel. Because of this, it is not easy for me to start the list of people to be included in the acknowledgments from one given point. Crucial to the writing of this book was a vast array of people that is impossible to handle in a linear way. Therefore, I will start such a list from the category of efficient cause (those who were instrumental in generating the process). This dictates starting with two key names associated with my choice of dedicating my life to research in this field — David Pimentel and Gian-Tommaso Scarascia Mugnozza.

In the six years spent at Cornell University with Professor Pimentel, I learned how to sense the existence of hidden links, when considering biophysical, economic, social and ecological issues in agriculture simultaneously. I learned from him how to follow the prey (looking for hard data to prove the existence of these links), even when this requires putting together scattered clues and going for creative investigation. The lessons I got in this field were invaluable. But the most important lesson was in another dimension: the human side. That is, following his example, I understood that, to do this job, one has to work hard and forget about trying to be politically correct. Even when building up your career you must resist the sirens' song of *cosí fan tutte*. You must keep going your own way, no matter what.

Professor Scarascia Mugnozza not only pushed me into the world of agriculture, but made it possible for me to engage in a nomadic "learning path," stabilized now for more than a decade, by facilitating international contacts and supporting my applications for funding.

Next, I would like to acknowledge the vital input of Wageningen University. In particular, I recall the demiurgic intervention of Niels Röling, who came up (over an Italian dinner) with the idea of the writing of such a book. As we were in a restaurant, the recipe came out pretty clear: one third epistemology, one third complex system theory and one third examples of real applications to the sustainability of agriculture. During my first seminar at Wageningen, Herman van Keulen did the rest by posing the following question: "You seem to believe that it is possible to establish a link between the various changes in indicators defined across different scales. But how can you establish a bridge across nonequivalent descriptive domains?" This question has been very important to me for two reasons: (1) this was the first time in my life that I found someone who understood perfectly what I was talking about when discussing Multi-Scale Integrated Analysis and (2) this question made me aware that my firm belief in the possibility of establishing a link across nonreducible indicators (something I had done in the past just following intuition) was not at all obvious to other people. Actually, when confronted with such a direct question in public, I was not able to offer a systemic explanation of my approach. Finally, the last key element in Wageningen was the enthusiasm for complexity shown at that time by Hans Schiere. My brief visit there (5 months) was to explore the possibility of using new concepts derived from this field for improving analytical models of sustainable development. At that time, I had a few discussions with Schiere about the problem of boundary definition in modeling. On one of those occasions I was asked: "Can you prove that it is impossible to find and use a unique "substantive" boundary definition for a given system?" When I was finally able to answer such a question in a very simple and direct way I realized that this book was finished.

The third and last item under the efficient-cause category is the input I received from Wisconsin-University at Madison. Tim Allen and Bill Bland invited me to work on the application of complex system thinking to the development of analytical tools related to agroecology. It was during that period that the various pieces of the puzzle were put together into the first draft of this book.

Getting to the formal cause, I would like to begin the list of people who were instrumental in shaping my understanding of complexity with someone I never managed to meet: Robert Rosen. In my view, Rosen, who died in 1998, was one of the greatest scientists of the last century. Hopefully, he will get due recognition in this century. In this book, I tried to build on his deep understanding of the link between basic epistemological issues and basic principles of a theory of complex systems.

Continuing with those I was lucky enough to work with, I can organize the list according to topics:

- **Epistemology, Science for Governance, and Post-Normal Science** — Silvio Funtowicz (who I visited at the Joint Research Center of the European Commission in Ispra in relation to writing this book), Jerome Ravetz, Martin O'Connor and Niels Röling (who opened for me the doors of Soft-Systems Methodology developed by Checkland). I now consider all friends as well as mentors.

- **Multi-criteria Analysis, Societal Multi-criteria Evaluation applied to Ecological Economics** — Joan Martinez-Alier and Giuseppe Munda (with whom I visited for 2 years at the Universitat Autonoma de Barcelona in Spain). I learned from them many of the ideas expressed in Chapter 5. A few paragraphs of that chapter are based on a technical report written with Giuseppe in 2001.

- **Complex Systems Theory and Hierarchy Theory** — Tim Allen (the 5 months spent with him in Madison accelerated my brain more than if it had been placed in a Super Proton Synchrotron), James Kay, David Waltner-Toews and Gilberto Gallopin. Again, it is a honor for me to consider all these people friends (none of us will ever forget the first meeting of the "Dirk Gently group" of holistic investigators in 1995).

- **Energy Analysis and Thermodynamics Applied to Sustainability Analysis** —The list includes Kozo Mayumi (the co-author of Chapters 6, 7 and 8), who is another fraternal friend with whom I have been working now for a decade. Together we developed the concept of multi-scale integrated analysis of societal metabolism. In this category I have to mention again Martin O'Connor, then James Kay and Roydon Frazer, two exquisite theoreticians interpreting the concept of rigor in the correct way (avoiding sloppiness, but at the same time daring to violate taboos when needed). Bob Ulanowicz is another important pioneer in this field from whom I got the main idea of the four-angle model for the analysis across hierarchical levels of metabolic systems. Vaclav Smil, another guru of the analysis of energy and food security, proved to be a very amiable and collaborative person. The list continues with Joseph Tainter, one of the few nonhard scientists who is perfectly comfortable with handling these scientific concepts when dealing with the sustainability of human societies. Last but not least, Sergio Ulgiati, Bob Herendeen and Sylvie Faucheux, other friends or colleagues with whom I have been interacting in this field for many years now.

- **Multi-Scale Integrated Analysis of Agroecosystems** — This list begins with Tiziano Gomiero (co-author of Chapter 11), who a few years ago decided to do his Ph.D. on this approach, and since then has never stopped working on it. Gianni Pastore and Li Ji were very active in 1997 during the first development of the method, when processing a dataset gathered in a 4-year project in China. I had several discussions about theory and applications with Bill Bland of the Agroecology program at the University of Madison. Finally, I would like to acknowledge various researchers involved in a project in South-East Asia with whom I am collaborating now (and hopefully in the future): H. Schandl, C. Grünbühel, N. Schulz, S. Thongmanivong, B. Pathoumthong, C. Rapera and Le Trong Cuc.

Moving now to the material cause: Many people helped in different ways during the actual writing, preparation and correction of the manuscript. The list includes: Sandra Bukkens, Nicola Cataldi, Maurizio Di Felice, Stefan Hellstrand, Joan Martinez-Alier, Igor Matutinovic, Alfredo Mecozzi, David Pimentel, Stefania Sette, Sigrid Stagl and Sergio Ulgiati. Sylvia Wood, at CRC Press, also contributed.

Finally, there is an unwritten rule about the layout of acknowledgment sections: They all finish with a reference to family and friends. This is where the final cause enters into play. In this case, particular mention is really due to my wife, Sandra Bukkens, who has contributed both indirectly and directly in a substantial way to this book. Indirectly, she sustained the burden associated with the running of our household for the last 5 years, a period during which our family moved six times across four different countries. And more directly, before this nomadic madness, she contributed by co-authoring with me several published papers dealing with related topics. A few of these are quoted and used in this book as sources of tables and figures.

Author

Mario Giampietro's interdisciplinary background began as a chemical engineer. His undergraduate degree was in biological sciences and his Master's degree in food system economics. He received his Ph.D. from Wageningen University.

Dr. Giampietro is the director of the Unit of Technological Assessment at a governmental research Institute in Italy (INRAN — National Institute of Research on Food and Nutrition). He was Visiting Scholar (from 1987 to 1989 and from 1993 to 1994) and Visiting Professor (1995) at Cornell University; Visiting Fellow at Wageningen University (1997); Visiting Scientist at the Joint Research Center of the European Commission of Ispra, Italy (1998), Visiting Professor at the Ph.D. Program of Ecological Economics at the Universitat Autonoma Barcelona, Spain (1999 and 2000); Visiting Fellow at the University of Wisconsin-Madison (2002). He is one of the organizers of the Biennial International Workshop "Advances in Energy Studies" held in Portovenere (Italy) since 1996.

Dr. Giampietro serves on the editorial boards of *Agriculture Ecosystems and Environment* (Elsevier), *Population and Environment* (Kluwer), *Environment, Development and Sustainability* (Kluwer) and *International Journal of Water* (Interscience). He has published more than 100 papers and book chapters in the fields of ecological economics, energy analysis, sustainable agriculture, population and development, and complex systems theory applied to the process of decision making in view of sustainability.

Introduction

Science for governance — The new challenge for scientists dealing with the sustainability of agriculture in the new millennium.

The development of agriculture in the 21st century is confronting academic agricultural programs with the need for handling new typologies of trade-offs and social conflicts. These multiple trade-offs are associated nowadays with the concept of multifunctionality of land uses.

In fact, it should be noted that this is nothing new. Throughout the history of humankind the agricultural sector has always been multifunctional and at the basis of social conflicts. Because of this, until a recent past (say before 1900) (1) the perception of the agricultural sector (the criteria of performance), (2) the representation of the agricultural sector (the attributes of performance) and (3) the regulation of agricultural activities (selection and evaluation of policies and laws based on the selected set of criteria and attributes) have always been based on the simultaneous consideration of various perspectives and dimensions of analysis. In modern jargon we can say that in the past (e.g., in preindustrial times) the development of agriculture was always driven by policies that were selected and evaluated considering both long-term and short-term effects in relation to different dimensions of analysis (political, social, economic, ecological). Land was perceived as a source of food for survival, as well as a required asset to sustain soldiers. Depending on the location, land was also seen as crucial to controlling trade. In addition to that, land always had a sacred dimension to anchor cultural values (people tend to associate their cultural identity with familiar landscapes, their homeland). Finally, in relation to the ecological dimension, land was often confused with nature and therefore considered as the given context within which humans have to play their part in the larger process of life.

If this is true, how is it that academic agricultural programs perceive the concept of multifunctional land use and the relative need of addressing multiple trade-offs and dimensions to be new? To answer this question, it is important to realize the deep transformations that the period of colonies first and the massive process of industrialization later induced in the metabolism of social systems in Europe and in other developed countries. In these privileged spots, economic growth could dramatically expand, escaping, at least in the short term, local biophysical constraints. This special situation was able to change in a few decades the codified perception about the role of the agricultural sector. Fossil energy-based inputs and imports were used to offset bottlenecks in the natural supply of production inputs. In this situation, the choice of considering (perceiving, representing and regulating) agriculture as just a set of economic activities aimed at producing goods and profit — while neglecting other dimensions — was very rewarding.

This change in the perception of agriculture in Western academic programs in the past decades was associated with a rapid economic growth in developed countries and a rapid demographic growth in developing countries. During this rapid transition, those operating in the developed world learned that introducing major simplifications in the codified way of perceiving, representing and regulating agriculture could generate comparative advantages for their economies, at least in the short term. That is, by ignoring the constraints imposed by the old set of cultural values (e.g., the sacredness of land) and by ignoring ecological aspects (e.g., the necessity to maintain human exploitation within the limits required by eco-compatibility), farmers and those investing in farming could take out much more food from the same unit of land and at the same time could increase their operative profits. In this way, developed societies were able to support more "soldiers" per unit of land. It should be noted, however, that after the industrial revolution the social role of preindustrial soldiers was replaced by nonagricultural workers. That is, the fraction of the workforce invested in operating machines was able to achieve an economic return per hour of labor much higher than that generated by farmers. To make things tougher for agriculture, the large variety of economic activities expressed by industrial societies implied that the very same land could be invested in alternative and more profitable uses. Modern economic sectors (both in production and in consumption) are competing with old-style agricultural practices for the use of the available endowment of human activity and land. In this situation, workers invested in just producing food or land invested in just keeping low the ecological stress associated with food production (e.g., fallow) involves a high opportunity cost in developed countries. In other words, old-style agricultural practices were the losers in the definition of priorities when deciding the new development strategies for modern economies. As a consequence, for more than five decades now, technical progress in agriculture has been driven by two simple goals:

1. Maximizing biophysical productivity: reducing the number of human workers and the amount of space required to produce food, so that these valuable resources can become available to other economic activities that give higher returns.
2. Maximizing economic performance: the high opportunity cost of capital in developed countries requires reaching levels of return on investment comparable with those achieved in other sectors.

These two goals, when combined, tend to generate a mission-impossible syndrome. In fact, the goal of maximizing the biophysical productivity in terms of higher throughput per hectare and per worker translates into the need for massive investments of capital per worker. On the other hand, a large difference in the opportunity cost of production factors such as land and labor — required in large quantities in the agricultural sector compared with other economic sectors — translates into low competitiveness of developed farmers on the international market (in relation to farmers operating in developing countries). In developed countries with enough land (e.g., the United States or Canada), the second goal was still achievable, at least before the third millennium. On the contrary, in other developed countries with high population densities (e.g., European countries or Japan), the second goal soon became impossible without subsidies. As soon as the downhill slope of subsidies was taken, the definition of the goal of maximizing economic performance changed dramatically.

At that point, the goal of maximizing economic performance in developed and crowded countries became that of reducing the fraction of total available economic capital that must be invested in the agricultural sector. In developed countries, the capital has a high economic opportunity cost. This implies that the agricultural sector with a high requirement of capital per worker and a low economic return on investments is forced to continuously compress the number of workers to handle this double task. The solution to this dilemma can be obtained by (1) increasing the ratio of capital per worker to capital per unit of land, while at the same time (2) reducing both the number of workers and the land in production. Obviously, the pace of reduction of the number of workers and the area of land in production has to be faster than the pace of growth of the required amount of capital per worker and per unit of land.

After having taken such a suicidal path for the sustainability of agriculture, many in developed countries were forced to recognize the original capital sin. The effects of the drastic simplifications adopted to perceive, represent and regulate agriculture, seen simply as another economic sector that is just producing commodities, became crystal clear (at least to those willing to see it). The decision to adopt a mechanism of monitoring and control based mainly on money (e.g., the implementation of agricultural policies in the 1960s and 1970s based mainly on economic analysis) was reflecting such a hidden simplification. Basing the evaluation of policies mainly on economic terms resulted in missing for decades a lot of relevant information referring to additional dimensions of agriculture. These neglected dimensions (e.g., ecology — health of ecosystems and cultural and social dimension — health of communities) are now slashing back on those in charge of determining agricultural policies. Even worse is the situation of developing countries where the societal context of agriculture is completely different from that of the developed world. In these countries there is less capital available for agricultural activities in the face of a growing demand for services and investments in the development of other economic activities. Moreover, the meager capital left to agriculture has to be used to deal with a dramatic reduction of land per capita. Obviously, in this situation, the challenge of developing new technologies and new policies for agricultural development is becoming harder and harder to tackle. Because the context of agriculture in developing countries is totally different from that in developed countries, we should expect that the idea of transferring either technologies that were generated in developed countries (e.g., high-tech genetically modified organisms (GMOs)) or policy tools (e.g., full market regulation) to developing countries to tackle their problems is, in general, a recipe for failure.

The scale of the global transformation implied by the oil civilization has now reached a point at which a simplified perception of agriculture that involves ignoring important dimensions of sustainability can no longer be held without facing important negative consequences. The perception that humans have passed this critical threshold is indicated by the widespread use of the buzzword *globalization* to indicate that something new is happening. As observed by Waltner-Toews and Lang (2001), the scale of human activity on this planet has reached a point that no longer leaves room for externalizations (shortcuts providing temporary comparative advantage to those deciding to use them) to the global economy. In terms of pollution, the term *globalization* means that "what goes around comes around." In terms of international development, the term *globalization* means that increasing someone's profit because of favorable terms of trade implies impoverishing someone else. That someone else, sooner or later, will require assistance. Ignoring negative side effects on the environment in the long term (the key to the dramatic success of Western science and technology in the last century) no longer pays. The environment will sooner or later present the bill, and it will be a very high one. Put another way, the term *globalization* means acknowledging that sooner or later (the sooner the better) we will have to go back to the ancient practice of integrating the goal of economic growth with a set of additional goals such as equity, environmental compatibility and respect for diversity of cultures and values. This will require looking for wise solutions, rather than for optimal solutions.

This new situation, which is challenging the conventional ideological paradigms of perpetual growth, is generating an additional dose of stress for human societies. Social systems are facing a continuous need of fast adjustments of their established rules and "truths." Human societies all over the planet are forced to learn how to make tough calls to find the right compromise between too much and too little technical progress. This is the back door through which the concept of multifunctionality of agriculture was rediscovered by the high-tech society. Within the army of scientists fully dedicated to maximizing and optimizing, those who are meditating on the various dilemmas associated with the issue of sustainability are discovering that many additional goals have to be considered when dealing with the sustainability of human development. That is, the two goals of economic growth and technical progress have to be considered as members of a larger family of goals that include respect for ecological processes, more equity for the present generation, respect for the rights of future generations and protection of cultural diversity to arrive at deeper and more basic procedural issues such as learning how to define quality of life when operating in a multi-cultural setting. In spite of the fact that these goals are becoming more and more important in the choice of sound policies for agricultural development, the scientific capability of supplying useful representations and structuring of these sustainability dilemmas is far behind the demand.

Niels Röling (2001) characterizes the need of a total rethinking of the performance of agriculture as the need for stipulating a new social contract among the actors of the food system (farmers, consumers, industry, scientists, administrators and their constituencies). This new social contract should be about how to use and distribute common resources in relation to an agreed-upon (1) set of activities judged as needed and admissible in the food systems and (2) set of indicators of performance used for discussing and implementing what should be considered a desirable food system. This new social contract requires considering shared goals, legitimate contrasting views about positive results and negative side effects of human actions; discussing the validity of available analytical tools, which can be used to characterize the performance of the food system in relation to different attributes of performance and generating viable procedures able to guarantee quality in decision processes (quality has to do with competence, fairness, transparency and the ability to learn and adapt).

This sudden change in the terms of reference of agriculture is challenging the conventional codified knowledge associated with the production of food and fibers. Such knowledge, religiously preserved in the various departments of agricultural colleges, is nowadays just one of the many pieces of information required for solving the puzzle. The puzzle is the necessity of continuously updating both the definition and regulation of agricultural activities in a fast-changing social context. This updating is getting more and more difficult because of (1) the speed at which new social actors, social dynamics and technical processes emerge at different scales and (2) the increasing awareness of the crucial and growing role that the ecological dimension plays in a discussion about sustainability. For these reasons, the challenge of finding new analytical tools that can be used to deal with the sustainability of agricultural development is extremely important within both the developed and developing worlds.

The very idea of multifunctional land uses requires the adoption of the concept of multi-criteria analysis of performance. This, in turn, requires a previous definition, at the social level, of an agreed-upon problem structuring. A problem structuring can refer to the decision of how to represent the system under analysis (e.g., when dealing with a simple monitoring of its characteristics) or of what scenarios should be considered (e.g., when discussing potential policies). By a given problem structuring, I mean the individuation of:

1. A set of alternatives to be considered feasible and acceptable (the agreed-upon option space)
2. A set of indicators reflecting legitimate but contrasting perspectives found among the stake-holders (the relevant attributes of the system and the direction of change that should be considered an improvement or a worsening — a multi-criteria space)
3. A set of nonreducible models useful for understanding and simulating different types of causal relations (a multi-objective, multi-scale integrated representation of changes in relevant attributes) in relation to the set of alternatives and the set of indicators
4. The gathering of enough data to be able to run the models and discuss the pros and cons of different options in relation to the set of relevant criteria

In this new framework, the scientists are just another class of nonequivalent-observers, part of a given society. As such, they have to learn, together with the rest of the society, how to perceive and represent in a more effective way the performance of a multifunctional agriculture.

To face such a challenge, scientists have to learn how to put their old wine (sound reductionist analytical tools) into new bottles to address new types of problems. Their new goal is no longer that of finding optimal solutions — Optimal for whom? Optimal for how long? Optimal in relation to which criteria? Who is entitled to decide about these questions? Rather, scientists are asked to help different social actors negotiate satisfying compromises about how to use their land, human time, technology and financial resources in relation to noncomparable types of costs and benefits (e.g., social, economic, ecological, individual gain or stress) that are expected (but with large doses of uncertainty) to be associated with different policy choices.

In human affairs, to be able to solve a problem one has, first of all, to be willing to admit that such a problem exists in the first place. The second step is to try to understand the nature of the problem in a way that can help the finding of solutions. An evident sign of crisis in the conventional scientific paradigm, when dealing with sustainability, is represented by the fact that the necessity of a paradigm shift is much clearer for the general public than for the community of politicians and scientists giving them advice. Often the sustainability predicament currently experienced by humankind is ignored (or even denied) in the analyses provided by many conventional academic disciplines and in the strategic planning of large national and international institutions. Common people, on the contrary, are forced to watch, in their daily life and every night in the news, the growing and widespread crumbling of ecological and social fabrics all over the planet. In front of this emotional stress, they are not receiving convincing explanations that current trends of environmental deterioration and uncontrolled growth of either population or aspirations are not just the result of a temporary crisis, but the challenge for the stability of any political process in the next century. The implications of this in terms of science for governance are at least twofold:

1. The scientific capability of providing useful representations and structuring of these new sustainability problems
2. The political capability of providing adequate mechanisms of governance

This book deals with only the first of these two implications. However, the dual nature of this challenge implies that when dealing with the issue of sustainability, society is trapped in a chicken–egg paradox: (1) scientists cannot provide any useful input without interacting with the rest of society and (2) the rest of society cannot perform any sound decision making without interacting with the scientists. In general, these concerns have not been considered relevant by "hard" scientists in the past. Thus, the goal of improving the quality of a decision-making process was not considered to belong to the realm of scientific investigation. On the other hand, the new nature of the problems faced in this third millennium implies that very often, when deciding on facts that can have long-term consequences, we are confronting issues "where facts are uncertain, values in dispute, stakes high and decisions urgent" (Funtowicz and Ravetz, 1991; Ravetz and Funtowitz, 1999).

Funtowicz and Ravetz coined the expression "postnormal science" to indicate this new predicament for scientific activity. Whenever scientists are forced by stakeholders to tackle specific problems at a given point in space and time, they can face a mission impossible according to the terms of reference of normal science. There are problems and situations in which risk (defined as an assessment based on probabilities) cannot be assessed (e.g., potential environmental problems of large-scale application of GMOs are associated with uncertainty and ignorance). There are other situations (e.g., whenever they are told "to fix Chicago in 30 days") in which scientists are facing (1) events that do not make possible repetitions in experiments and (2) a flow of questions from the stakeholders that would require a flow of scientific answers at a rate not compatible with the development of a sound scientific understanding. When operating in a normal mode, scientists are used to having the privilege of picking up the best experimental setting for studying what they want to study, and in doing so, they can take all the time they need to work out robust answers.

In a situation of postnormal science, scientific rigor does not always coincide with sound science. On the contrary, using risk assessment (e.g., using frequencies or estimated probabilities to assess risks) in cases in which one deals with irreducible uncertainty and genuine ignorance should be considered sloppy science. That is, the use of sophisticated statistical tests providing a significance of 0.01 should not be confused with sound science when used in situations in which they do not make any sense (Giampietro, 2002). In this situation, those who refuse to sell fake rigorous science in exchange for power and academic recognition can find themselves marginalized in the debate over the future of our development. To make things worse, this situation enables the establishment of ideological filters based on pseudo-scientific rigor to avoid confronting unpleasant realities. The denial of the existence of a problem of global warming related to the accumulation in the atmosphere of greenhouse emissions is a well-known example of this fact. When dealing with a complex reality and large-scale problems (e.g., global warming) there is always some rigorous test that can be found to challenge the evidence supplied by the adverse side. But a broken clock indicating the exact time twice a day is much less useful for decision making than a clock that slows down a second every year and therefore never gives the exact time during any day for the following months. When dealing with large-scale issues, it is much better to have a sound understanding of the big picture, even if details are missing, than a very accurate picture of just one piece of the puzzle, which can only be studied rigorously when considered in pieces and held out of context.

This book wants to answer three questions crucial for scientists willing to be effective in the development of a science that can be more useful for governance in relation to the issue of sustainability of agriculture:

Part 1: What is the role that scientists working in the field of sustainability of agriculture should play in this process?

Part 2. Can we develop different scientific analyses using complex systems thinking?

Part 3. What alternative analytical tool kits can be developed for integrated analysis of agroeco-systems?

References

Funtowicz, S.O. and Ravetz, J.R., (1991), A new scientific methodology for global environmental issues. In: R. Costanzam (Ed.). *Ecological Economics*, Columbia, New York, pp. 137–152.

Giampietro, M., (2002), The precautionary principle and ecological hazards of genetically modified organisms, *AMBIO*, 31, 466–470.

Ravetz, J.R. and Funtowicz, S.O., Guest Eds., (1999), *Futures*, (Vol. 31). Special issue dedicated to postnormal science.

Röling, N., (2001), *Gateway to the Global Garden: Beta/Gamma Science for Dealing with Ecological Rationality*, Eighth Annual Hoper Lecture, 8 Centre for International Programs, University of Guelph, Ontario.

Waltner-Toews, D. and Lang, T., (2000), The emerging model of links between agriculture, food, health, environment and society, *Global Change Hum. Health*, 1, 116–130.

Contents

Part 2 Complex Systems Thinking: Daring to Violate Basic Taboos of Reductionism

Part 3   Complex Systems Thinking in Action: Multi-Scale Integrated Analysis of Agroecosystems

Part 1

Science for Governance:
The Clash of Reductionism
Against the Complexity of Reality

1

The Crash of Reductionism against the Complexity of Reality

This chapter presents two practical examples of the general impasse that reductionism is experiencing when attempting to deal with the issue of sustainability. The goal of this section is to provide a narrative and introduce basic issues; technical explanations and a more detailed analysis are provided in the next two chapters.

1.1   Example 1: In a Complex Reality It Is Unavoidable to Find Multiple Legitimate Views of the Same Problems

1.1.1   Contrasting but Legitimate Policy Suggestions for Sustainability

In 1996 I was invited to an international conference in Zurich to debate in front of the media the problem of food security for humankind in the 21st century (SAGUF, 1996). The conference was held to commemorate the 50th anniversary of the Swiss Academy of Science. To celebrate this special event, besides the work of the conference, the organizers invited a panel of very distinguished scholars to close the conference with reliable policy suggestions. A list of six of the suggestions given by different panelists is provided in Figure 1.1. It includes three pairs of contrasting suggestions referring to the following fields: (1) food policies within countries, (2) international trade policies and (3) social policies dealing with the role of women in guaranteeing food security. The rationale of these suggestions is briefly discussed below:

- **Food security within countries** — In the next century a growing fraction of the population in developing countries will be urbanized. Keeping food prices low will be a key policy to guarantee food security for those who will be relying heavily on the market for their food supply. This is the explanation and the policy advice obtained when reading the problem at a given point in space and time (assuming the *ceteris paribus* hypothesis as valid). In fact, this was the view expressed by the international expert on food policy. On the other hand, when looking at evolutionary trends, it will only be possible to feed such a growing urban population if the productivity of farmers is dramatically increased. Matching the additional food demand of the few billions of people arriving in the next decades, mainly in the South, will require a major increase in the flow of investments in agriculture. A dramatic increase in investments in farming will not happen without an adequate return on these investments. This can only be obtained by guaranteeing high prices paid to farmers. This was the view expressed by the professor of agricultural development. Framing the analysis of the future food security within an evolutionary context leads to a completely different policy.
- **The effects of world trade** — It is true that developed countries are using both nonrenewable and renewable resources taken from ecosystems of poor countries. In this way, people living in developed countries are reducing the amount of natural capital that can be used for devel-

NATIONAL POLICY	
Keep prices of food commodities LOW	I.F.P.R.I. — U.S. scientist
Keep prices of food commodities HIGH	Ag. Econ. — Prof. from Pakistan
INTERNATIONAL POLICY	
REDUCING imports from the South	Wuppertal Inst. — German scientist
INCREASING imports from the South	Ag. Dev. — Prof. from Ghana
SOCIAL POLICY	
PRESERVING local cultural heritage	NGO — Swiss feminist
FIGHTING local cultural heritage	Sociologist — Prof. from India

FIGURE 1.1 International Conference on World Food Security, SAGUF, Zurich, October 9–10, 1996.

opment by people living in developing countries. This was the view expressed by an institute of research focused on the sustainability of human development. On the other hand, many developing countries invested heavily in the intensification of their agricultural sectors. In this way, they are attempting to use comparative advantages (lower cost of labor, abundance of land) to boost their processes of economic development, attracting foreign currency. When considering this effort, they are today hampered by those policies of developed countries aimed at restricting food imports from the South. This was the view expressed by a professor of international development. Also in this case, both rationales used to develop policy suggestions are correct. They are generated simply by different formulations of the problem.

•   **Empowering women or preserving cultural heritage?** — It is obvious that the protection of the diversity of cultural identities and local historic heritages is a must for a sustainable global development. Such a long-term perspective was embraced by the representative of a Swiss nongovernmental organization (NGO). On the other hand, a demand for empowerment coming from women strongly abused by existing social habits cannot be dismissed just because of the need to preserve cultural diversity at any cost. In this case, the other speaker (also a woman) came from a region in India where wives are burned alive with their dead husbands. Her strong perception of the urgent need for change in cultural habits pushed her, in her role of analyst, to reduce the time horizon considered in evaluating the desirability and side effects of policies.

In this example, the experts asked to provide advice to humankind on how to improve the sustainability of food security were perfectly comfortable in making and defending their points. Moreover, they were all right. However, as soon as they were confronted with the fact that the information given by the panel was contradictory (a journalist actually made the obvious remark), they could not figure out why this was happening. They immediately started to defend their theses *against* the others and defensively showing their academic credentials ("I know what I am talking about … I am a well-known professor. … I have been working on this problem for decades …"). Under the pressure of the moment, nobody even considered the possibility that legitimate but contrasting scientific truths can coexist.

Another interesting point can be driven home from this example. Looking at the different conclusions reached by the scientists, it becomes immediately clear that scientists coming from different social contexts (e.g., developed countries vs. developing countries) were adopting for their analysis different preanalytical choices of their problem structuring (choice of relevant goals, variables, explanatory dynamics for the select explanatory model). Put another way, the differences in policy recommendations were not generated by differences in the accuracy or validity of models or equations. Rather, they were just reflecting basic differences in how the various scientists perceived and represented the problem to be tackled. Scientists operating in developed societies were suggesting policies aimed at preserving the current steady state (keep prices low, reduce trading, keep cultural diversity at any cost). Scientists coming from less developed countries were suggesting policies aimed at changing as fast as possible the current situation (boost the evolutionary rate of the system). Probably, this clear-cut division at that conference has been generated by chance by the particular combination of invited speakers and topics assigned. However, different perceptions of a given problem tend to reflect differences in the social context in which the scientist is operating. The possibility of multiple nonequivalent perceptions of the same situation is one of the typical characteristics of complexity, and it is further elaborated in the following section.

1.1.2   Looking at Nonequivalent Useful Pictures of a Person, Which One Is Right?

Before moving to the second example, this section discusses in more detail the impossibility of obtaining the "right" picture of a given situation when dealing with complex systems organized in a nested hierarchy. In this section I want to make the point that it is literally impossible to get the "right picture" of a given complex system. Even when talking of real pictures (those printed on a paper or shown on a monitor), the complexity of the reality entails the unavoidable existence of multiple identities that, to be represented, require the parallel use of nonequivalent pictures.

Imagine that we are requested to pick up a visiting scientist at the airport. We are given the name — Dr. X — but we do not know her or his face. The most obvious additional input needed to perform our task is a picture of Dr. X. Now imagine that we ask for a picture and what we get in the mail is the picture given in Figure 1.2a with a note saying: "Please find enclosed the picture of Dr. X that you requested." Such a picture is completely useless for our task, even though we cannot say that such a picture does not contain relevant information about Dr. X. This picture makes it possible to study how Dr. X digests nutrients to keep him or her alive. Therefore, this picture (which has been taken from an experimental nutrition lab in my institute) reflects a very important option available to us for looking at human beings. It should be considered a crucial piece of information to study human sustainability.

Getting back to our story, we ask for another picture of Dr. X, this time a picture taken at a larger scale. In response to our request we get another picture, that shown in Figure 1.2c, with a note saying: "Fulfilling your request, please find enclosed a larger-scale picture of Dr. X, who is the one indicated by the arrow." Also in this case, even if we cannot use this picture at the airport, this picture tells us

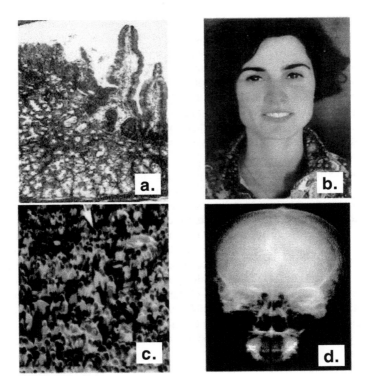

FIGURE 1.2 Nonequivalent views of the same person. (Photos by Andrea Ghiselli.)

useful information about Dr. X (her or his relations within her or his social context). In fact, after a close examination, we can say that Dr. X is a resident of Italy and he or she is concerned with the problems of the environment (this is a picture of a rally of the Italian Green Party). Using this picture, we got useful information about Dr. X, but still, this is not the information we needed for our task.

When writing for yet another picture, we specify this time that the picture of Dr. X has to include the whole head and only the head (we fix the scale at which the boundary is set for the representation, to get rid of possible nonequivalent representations generated by the arbitrariness of such a choice). The next picture (Figure 1.2d) matches our constraints, but it still leaves us disappointed. In the picture given in Figure 1.2d we see the "head" of Dr. X, but we cannot see her face. Now I use "her" since at this point, with the help of a physician, we can know that Dr. X is a woman. But the selection of mapping mechanism (pattern recognition based on x-rays rather than visible-length waves) still hides the information we need. The choice of using x-rays to take her picture prevented us from seeing the face (the pattern recognition that we need at the airport). However, the very same choice of using x-rays makes it possible for us to learn useful information about Dr. X (she has a sinusitis that should be taken care of).

The picture that enables us to recognize Dr. X among the people getting out from the gate at the airport is given in Figure 1.2b.

Three points from this discussion follow:

1. Systems that are organized over different hierarchical levels (such as humans made of organs, made of cells, made of molecules, made of atoms, etc., which at the same time are part of a household, part of a community, part of a country, part of a large whole) show different identities when looked at (and represented) on these different levels. No matter how many photos we make of a person with a microscope (pictures of the type shown in Figure 1.2a), we will never be able to see her face. The face is an emergent property that can be seen only in adopting an appropriate selection of descriptive domain, that is, (1) an appropriate space–time scale and (2) an appropriate system of encoding relevant qualities (the head of Dr. X seen using x-rays does not show her face).

2. Nonequivalent descriptions referring to nonequivalent descriptive domains are not reducible to each other. They catch and represent some aspects of the system (on a given scale) and hide other aspects of the system (either on the same scale or on other scales). Therefore, a description that makes us happy at a given point in space and time and in relation to a specific goal (picking up a given person at the airport) should not be considered as better than others. Someone with a different problem can find much more useful descriptions. Going back to our story, Dr. X, experiencing a continuous headache and knowing very well her own face, would be happier with the encoding of the x-rays telling her about her sinusitis (Figure 1.2d) than with the encoding used in Figure 1.2b (seen every day when looking in the mirror). This is why one pays to get x-rays.

3. Emerging properties generated when aggregating information at a different level can also generate emergence in behaviors in those that are using the information. For example, imagine that the person in charge of the "airport rescue mission" is a young single male scientist. After receiving the picture given in Figure 1.2b, he could expand in his mind the set of potential interactions with Dr. X by including a plan to go out to dinner with her. Actually, this enlargement of his option space can dramatically change his perception of the duty that has been assigned to him. What before was considered a chore (wasting valuable leisure time to get an unknown Dr. X at the airport) suddenly becomes a pleasant opportunity justifying his investment of leisure time. This dramatic change in perception would never have occurred if the information available was only that of Figure 1.2a, c and d. When dealing with reflexive systems (humans), sudden changes in the characteristics of the observers can become crucial in determining the validity of a given problem structuring.

The relation of the nonequivalent views presented in Figure 1.2 with the contrasting policy suggestions presented in Figure 1.1 will be discussed in Chapter 2. For the moment, what can be said is that reductionistic analysis based on the selection of a set of observable qualities encoded in measurable variables over a given descriptive domain represents just one of the possible pictures given in Figure 1.1. That is, by using formal models, we can look at only one of the possible identities of the system at a time. This is what generates the existence of legitimate but contrasting policy recommendations based on nonequivalent problem structuring.

1.2   Example 2: For Adaptive Systems *"Ceteris"* Are Never *"Paribus"* — Jevons' Paradox

1.2.1   Systemic Errors in Policy Suggestions for Sustainability

Jevons' paradox (Jevons, 1990) was first enunciated by William Stanley Jevons in his 1865 book *The Coal Question* (Jevons, 1865). Briefly, it states that an increase in efficiency in using a resource (defined as a better output/input ratio) leads, in the medium to long term, to an increased use of that resource (rather than to a reduction). At that time, Jevons was discussing possible trends of future consumption of coal and reacting to scenarios generated by advocates of technological improvements. Also at that time, some were urging that the efficiency of engines be dramatically increased to reduce the consumption of fossil energy (the preoccupation for nonrenewable fossil energy has a long history). Jevons' point was that more efficient engines would expand the possible uses of coal for powering human activities, and therefore, they would have boosted rather than reduced the rate of consumption of existing coal reserves.

Jevons' paradox proved to be true not only with regard to the demand for coal and other fossil energy resources but also with regard to the demand for food resources. Doubling the efficiency of food production per hectare over the last 50 years (the green revolution) did not solve the problem of hunger; it actually made it worse, since it increased the number of people requiring food and the absolute number of malnourished (Giampietro, 1994a). In the same way, doubling the area of roads did not solve the problem of traffic; instead, it made it worse, since it encouraged the use of personal vehicles (Newman, 1991). As more energy efficient automobiles were developed as a consequence of rising oil prices, American car owners increased their leisure driving (Cherfas, 1991). Not only the number of miles increased but also the expected performance of cars grew; U.S. residents are increasingly driving minivans, pickup trucks and four-wheel drives. More efficient refrigerators have become bigger (Khazzoom, 1987). A promotion of energy efficiency at the microlevel of economic agents tends to increase energy consumption at the macrolevel of the whole of society (Herring, 1999). In economic terms, we can describe these processes as increases in supply boosting the demand in the long term, much stronger than the so-called Say's law.

Jevons' paradox has different names and different applications; for example, it is called the "rebound effect" in energy literature and the "paradox of prevention" in relation to public health. In the latter case, the paradox consists of the fact that the amount of money saved by prevention of a few targeted diseases generates in the long term a dramatic increase in the overall bill of the health sector. Due to the fact that humans sooner or later must die (which is a fact that seems to be ignored by steady-state efficiency analysts), any increase in life span of a population directly results in an increase in health care expenses. Besides the higher fraction of retired in the population needing assistance, it is well known that the hospitalization of elderly is much more expensive than the hospitalization of adults.

This last example leads us to the heart of the paradox. Technological improvements in the efficiency of a process (e.g., increases in miles traveled per gallon of gasoline) represent improvements in *intensive variables*. That is, they can be defined as improvement per unit of something and under the *ceteris paribus* hypothesis that everything else remains the same. This means that an increase in efficiency could translate into savings only when the system does not evolve in time (when our steady-state representation of the system is satisfactory for the purposes of decision making). Unfortunately, complex systems, especially human systems, tend to adapt quite fast to changes. As soon as such technological improvements are introduced into a society, room is generated for either (1) the addition of new activities (e.g., new models of cars, including new features, or a change in the distribution of individual overage classes with an increase of elderly) or (2) a further expansion of current levels of activity (e.g., more people make more use of their old cars, or a larger population size). Former expansion refers to the qualitative aspects of the systems (an addition of possible options in the set of accessible states); latter expansion represents a change in *extensive variables*, that is, in the dimension of the process (but using the same set of options). In conclusion, increasing efficiency (that is, doing better according to the picture obtained at a particular point in space and time, in relation to a given associative context and a given set of goals) simply makes it possible for systems to consider new alternatives or to expand the set of criteria considered for defining the performance of an activity. In no way could this be associated with an improvement of sustainability. For example, car air conditioning was at the beginning a fancy option, then it became a reasonable option on mid-size and compact cars, and finally it became a relevant criteria considered by potential buyers at the moment of deciding what to buy. This change was made possible by the dramatic improvement in energy efficiency of the engine and more in general cars. The energy savings obtained in relation to the activity of just moving around was used to make possible another activity, that of moving around with a controlled and pleasant temperature. In terms of sustainability (e.g., human consumption of fossil energy), the dramatic increase in energy efficiency of engines and cars did not result in a reduction of consumption. This very same pattern will be discussed later on when discussing the so-called agricultural treadmill.

The important point for our discussion is how the system will expand and what consequences will be generated by this expansion, questions that cannot be answered (let alone predicted) by those studying the efficiency of a given process. For these reasons, sound decision making should be based on an assessment of the problem, which should be based on complementing views (e.g., both steady-state and evolutionary views (trend analysis)). Moreover, considering a relevant system's qualities one at a time tends to generate contrasting indications on the direction of changes (as in the problems of contrasting policy suggestions discussed in the previous example). Going back to the paradox of prevention, policies aimed at reducing health costs (e.g., smoking restrictions to prevent lung cancer) have the effect, in the long term, of increasing the cost of health care. However, paradox in the paradox, this is a good result for society. In fact, when assessing health care costs in an evolutionary perspective, we can easily recognize that the ability to afford a larger bill for health care is an indicator of development for a human society. However, this double paradox points to the existence of a systemic error in many analyses of development (more on this in the following two sections).

1.2.2 Systemic Errors in the Development of Strategies: The Blinding Paradigm

The issue of sustainability is determining a generalized critical appraisal of the current strategy of technical progress. As a matter of fact, whenever we enlarge the scale used to evaluate the results of previous innovations, we are able to appreciate the general validity of Jevons' paradox. Technologies developed to solve a particular problem often have the effect of making things worse in relation to that particular problem. The mechanism seems to be repeated endlessly in three steps:

Step 1: Humans face a specific problem and define simple terms of reference for the solution of it. They look for a fix of the problem according to a simplified description of it as seen at a particular point in space and time by a particular social group. This translates into (1) the choice of a single space–time scale on which dynamics are represented and (2) a representation of the problem over a finite performance space, based on a limited selection of relevant variables. This reflects a preanalytical value call on the relevance of the set of qualities to be monitored. Often this simplification of the representation of the reality leads to the adoption of a monocriteria mechanism of representation of the performance (e.g., cost–benefit analysis maximizing welfare or a maximization of the efficiency of a given transformation). At this point, it is possible to generate, using hard science, optimizing functions individuating the best possible courses of action. According to these simplified terms of reference, scientists look for quick technological fixes aimed at eliminating the perceived problem according to the above set of assumptions.

Step 2: Technological fixes able to move the system closer to the so defined optimal solution are found, developed and applied on a larger scale.

Step 3: The implementation of Step 2 induces in the long term unexpected effects; a few factors and aspects not considered in the simplified problem structuring and therefore neglected in the formulation of the policy are amplified in their relevance and get into the picture. In many cases this mechanism leads even to an amplification of the very same problem that generated Step 1. As noted earlier, the successful implementation of a fix on a large scale leads to change in the societal perception of what should be considered desirable, important or irrelevant.

What is disturbing in this pattern is its regular repetition in human affairs. A perfect example is represented by the fate of the green revolution. In the 1960s, decision makers and scientists of developed countries entered into Step 1. They were facing the problem of 200 to 300 million malnourished people located mainly in developing countries. The representation of the problem was as follows: the current demand of food is larger than the current supply (an analysis based on a quasi-steady-state representation). The obvious solution based on this problem structuring was to develop farming techniques able to produce more food (boost the supply). At this point, Step 2 was followed. The chosen technological fix was the green revolution. The name was chosen to stress the dramatic increase in yields obtained in food production thanks to technological fixes. The solution was applied to many areas of our planet, fulfilling the goal fixed in Step 1 to increase the supply, and the result was revolutionary indeed. The unavoidable arrival of Step 3 has been confirmed by the world conference organized by the Food and Agriculture Organization (FAO) in 1996. That is, 40 years after the first world conference, the aim of the second conference was the same: focusing world attention on the urgent need to fight hunger. The unexpected side effect of the technological fix developed in the 1960s has been a huge increase in population induced by the increase in food supply. That is, the solution of doubling yields put forward in the 1960s resulted into a more than proportional increase in the absolute number of malnourished (Srinivasan, 1986; Kates and Haarmann, 1992). It must be mentioned, however, that in these 40 years many reputable scientists — including some of the fathers of the green revolution — have been warning against confusing the symptoms with the disease. Boosting food yields with the green revolution represented a short-term technical patch capable of buying some time for the implementation of more structural changes in developing economies. For example, Norman Borlaug — who won the Nobel Prize for his contribution to the green revolution — in his 1970 Nobel lecture said:

> The green revolution has won a temporary success in man's war against hunger and deprivation; it has given man a breathing space. If fully implemented, the revolution can provide sufficient food for sustenance during the next three decades. But the frightening power of human reproduction must also be curbed; otherwise the success of the green revolution will be ephemeral only. (Lecture available at: http://www.nobel.se/peace/laureates/1970/borlaug-lecture.html)

However, such a message did not go through.

The last FAO summit of 1996 concluded with a pressing call for a new and convincing jump into a new Step 1. That is, decision makers and scientists gathered on this occasion adopted the same terms of references (problem structuring) as in the 1960s, which led to the same solution: developing another wave of silver bullets to increase food production, this time also resorting to genetically engineered crops (Giampietro, 1994a).

The blinding paradigm is quite clear. Improvements in efficiency of a certain activity can only be defined and assessed in terms of quantifiable increases in what is perceived to be at the moment a system quality to which top priority is given. This translates into a strategy for increasing short-term returns in relation to such a priority based on the *ceteris paribus* hypothesis. This is a snapshot of the situation based on a given mix of goals, perceived boundary conditions, available technical options, existing institutional settings, sets of relevant criteria selected and an agreed-upon definition of performance. The validity of this mix is assumed indefinitely into the future, even though this is never the case. Improvements in efficiency (which must refer necessarily to a specific and short-term view of improvement by those social groups in power) are in general paid for in terms of a lower sustainability of the improved societal activity in the long term (as noted before) or when considering a different set of viewpoints. That is, improvements in efficiency based on the consideration of a given set of criteria tend to induce

TABLE 1.1

Jevons' Paradox in Action: How Many Treadmills Are Out There Waiting to Be Found?

The Agricultural Treadmill (after Röling, 2002)

1. Many small farms all produce the same product.
2. Because not one of them can affect the price, all will produce as much as possible against the going price.
3. A new technology enables innovators to capture a windfall profit.
4. After some time, others follow (diffusion of innovations).
5. Increasing production and efficiency drives down prices.
6. Those who have not yet adopted the new technology must now do so lest they lose income (price squeeze).
7. Those who are too old, sick, poor or indebted to innovate eventually have to leave the scene. Their resources are absorbed by those who make the windfall profits (scale enlargement).

The Golfer Treadmill

1. Many amateur players all play with the same result (e.g., very high handicap).
2. Because none of them have enough time to dramatically improve their personal skills, they all can happily play against their own high score.
3. A new technology enables innovators to reduce their score (e.g., clubs made of new, expensive material capable of hitting the ball to a longer distance).
4. After some time, others follow (diffusion of innovations).
5. Increasing the average distance achieved using high-tech clubs induces an increase in the average length of par 4 and par 5 holes in golf courses.
6. Those who have not yet adopted the new technology must now do so; otherwise, they will not even be able to maintain their original high handicap (increase of the cost of the game).
7. Those who are too old or unable to innovate because of a low income eventually will see their status plummeting among the other golfers.

a worsening in relation to those criteria not considered in the original problem structuring. A winning solution at a certain hierarchical level tends to imply losers on other hierarchical levels. This fact can go undetected at the beginning, since these unpredicted phenomena tend to occur within different descriptive domains and on different space–time windows, often not considered by those in power (Giampietro, 1994b).

Side effects associated with the mismatch between the definition of improvement at a location-specific scale and the open definition of improvement referring to an evolutionary scale are at the root of the so-called agricultural treadmill (Cochrane, 1958). A quick view of the various steps is given in Table 1.1 (from Röling, 2002). Again, it is important to observe that the existence of bifurcations between (1) a local definition of improvement and (2) a larger-scale definition of improvement is a general feature to be expected when dealing with the mechanism of self-organization of adaptive systems. An innate tension among processes of evolution and adaptation operating in parallel on different scales has the effect of keeping high the level of selective pressure on the evolving system (thanks to the previous innovation, the system is forced to innovate more and at a faster pace). The negative side effects are represented by the fact that a faster pace of innovation translates into a larger level of stress for the components of the system. The trajectory of technical development in agriculture is just one of the fields of possible application. Table 1.1 shows the remarkable similarity between the agricultural treadmill and the golfer treadmill.

1.2.3 Systemic Errors in the Representation of Evolution: The Myth of Dematerialization of Developed Economies (Are Elephants Dematerialized Versions of Mice?)

As noted earlier, unless a comprehensive analysis of the changes induced by technological improvement is performed, including intensive and extensive changes and an adequate selection of descriptive domains considering changes on higher and lower hierarchical levels, it is easy to be misled by the counterintuitive behavior of evolving complex systems. The myth of dematerialization of developed economies can be used as a good practical example.

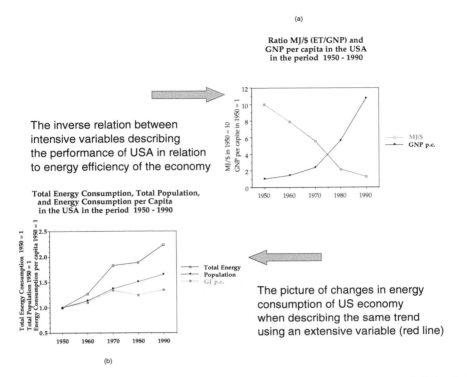

FIGURE 1.3 (a) The inverse relation between intensive variables describing the performance of the U.S. in relation to energy efficiency of the economy. (b) The picture of changes in energy consumption of the U.S. economy when describing the same trend using an extensive variable. (after Pastore et al. 2000. Societal metabolism and multiple-scales integrated assessment: Empirical validation and examples of application. *Popul. Environ.* 22 (2): 211–254.

When adopting an economic definition of energy efficiency (megajoules of energy consumed per dollar of gross national product (GNP) produced [MJ/\$–GNP]), one can get the false impression that technological progress is decreasing the dependence of modern economies on energy. In fact, many seem to be reassured by the fact that technological progress is associated with a decreasing ratio of MJ/\$-GNP in developed countries. However, this mapping of improvement relies on a variable that refers to an intensive variable, which is therefore not useful when dealing with the issue of sustainability. For example, in 1991 the U.S. operated at a much lower ratio of energy consumption per unit of GNP than, for instance, the People's Republic of China (12.03 vs. 69.82 MJ/\$, respectively); see Figure 1.3a. On the other hand, because of this greater efficiency, the U.S. managed to have a GNP per capita much higher than that of China (22,356 vs. 364 \$/year, respectively) (WRI, 1994). That is, if we change the mechanisms of mapping changes, moving to an extensive variable (by multiplying the energy consumption per unit of GNP by the GNP per capita), the picture is totally reversed. In spite of the significantly higher economic energy efficiency, the energy consumed per U.S. citizen is 11 times higher than that consumed by a Chinese citizen. Obviously, this is just one of the possible mechanisms of mapping; if we compare these two countries, also considering population size, we would get another set of assessments referring to another relevant dimension of analysis.

If we look in parallel at the trends of energy consumption per unit of GNP, GNP per capita, population size, total energy consumption and energy consumption per capita, as reported in Figure 1.3a and b for the U.S. in 1950–1990, we can clearly see the inverse relation between energy consumption per unit of GNP and the aggregate level of GNP per capita. In this way, the degree of dematerialization induced by technological progress in the U.S. economy can be checked by analyzing data of aggregate energy consumption published by USBC (1991) — see Figure 1.3. A reduction in the energy consumption per unit of gross domestic product (GDP) from 113.1 MJ/\$ in 1950 to 25.1 MJ/\$ in 1990 (a ratio of reduction of 4.5/1) had the effect of increasing the aggregate consumption of commercial energy in the U.S.

economy from 34.5 Terra Joules (TJ) in 1950 to 77.0 TJ in 1990 (more than doubling the total energy consumption).

As indicated by Figure 1.3b, the aggregate energy consumption of the U.S. increased not only because of an increase in consumption per capita but also because of an increase in population size. This phenomenon is explained not only by differences between fertility and mortality, but also by immigration, driven by the attractive economy. Actually, strong gradients in standard of living among countries — generated by gradients in efficiency — tend to drive labor from poorer to richer countries (Giampietro, 1998). For example, the dramatic improvement in energy efficiency that California has achieved in the past decade (in terms of the intensive variable useful energy/energy input) will not necessarily curb total energy consumption in that state. Present and future technological improvements are likely to be nullified by the dramatic increase in immigration, both from outside and inside the U.S., which makes California's population among the fastest growing in the world. Again, we find the systematic failure of accounting for the change in boundary conditions induced by the change in technology at the root of this counter-intuitive trend.

In relation to the myth of dematerialization of developed economies, it is time to mention the striking similarity in the pattern relating the two variables: (1) intensity of metabolism and (2) size found when comparing socioeconomic systems and biological organisms. In biology it is well known that animals with a smaller body size have a higher rate of energetic metabolism per kilogram of biomass. Actually, there is abundant literature on this phenomenon — a good overview of the literature can be obtained for empirical analyses in the book of Peters (1986) and for more recent theoretical applications in the book of Brown and West (2000). Using available data (organized in tables or in parameters for equations that can be applied to different typologies of animals), we can calculate, for example, the relation between size and metabolic intensity for mammals of different sizes. For example, a male mouse weighing 20 g — an extensive variable — has a metabolic rate of 0.06 W (joules/second). That is, male mice have a metabolic rate of 3 W/kg of body mass — an intensive variable. A male elephant weighing 6000 kg (an extensive variable) has a total metabolic rate of 2820 W, which is equivalent to a metabolic rate of 0.5 W/kg of body mass (an intensive variable) (Peters, 1986, p. 31).

If we apply the same reasoning used by some neoclassical economists to describe the process of dematerialization of modern economies (using an intensive variable assessing energy intensity per unit of GNP), we would find a quite bizarre result. When looking at animal biomass across the evolutionary ranking (using an intensive variable assessing the energy expenditures per unit of biomass), we would find a quite peculiar way of defining the process of dematerialization in mammals. Since 10,000 kg of elephants consume much less (4700 W) than 10,000 kg of mice (30,000 W), we have to conclude that elephants (with a low energy intensity per unit of biomass) should be considered a dematerialized version of mice (with a high energy intensity per unit of biomass). After all, this is exactly what we are told by neoclassical economists (and what is taught to students in the vast majority of colleges dealing with the issue of sustainable development). According to this perception, the process through which very poor countries (based on location-specific subsistence economies) are evolving into large developed countries (based on a global economy) is described as a process of dematerialization of world economy.

The fact that modern neoclassical economic analysis sees elephants as dematerialized versions of mice would be an amusing finding indeed, if this silly idea were not taught to students in almost every academic program dealing with the sustainability of human progress.

References

Brown, J.H. and West, G.B., Eds., (2000), *Scaling in Biology*, Oxford University Press, Oxford, 384 pp.

Cherfas, J., (1991), Skeptics and visionaries examine energy savings, *Science*, 251, 154–156.

Cochrane, W.W., (1958),The Agricultural Treadmill, in *Farm Prices, Myth and Reality*, Minneapolis, University of Minnesota Press ch. 5.

Giampietro, M., (1994a), Using hierarchy theory to explore the concept of sustainable development, *Futures*, 26, 616–625.

Giampietro, M., (1994b), Sustainability and technological development in agriculture: a critical appraisal of genetic engineering, *Bioscience*, 44, 677–689.

Giampietro, M., (1998), Energy budget and demographic changes in socioeconomic systems, in *Life Science Dimensions of Ecological Economics and Sustainable Use*, Ganslösser, U. and O'Connor, M., Eds., Filander Verlag, Fürth, pp. 327–354.

Herring, H., (1999), Does energy efficiency save energy? The debate and its consequences, *Appl. Energy.*, 63, 209–226.

Jevons, F., (1990), Greenhouse: a paradox, *Search*, 21, 171–172.

Jevons, W.S., [1865] (1965), *The Coal Question: An Inquiry Concerning the Progress of the Nation, and the Probable Exhaustion of Our Coal-Mines*, 3rd ed. rev., Flux, A.W., Ed., Augustus M. Kelley, New York.

Kates, R.W. and Haarmann, V., (1992), Where the poor live, *Environment*, 34, 4–11, 25–28.

Khazzoom, J.D., (1987), Energy saving resulting from the adoption of more efficient appliances, *Energy J.*, 8, 85–89.

Newman, P., (1991), Greenhouse, oil and cities, *Futures*, Vol. 23(3), 335–348.

Peters, R.H., (1986), *The Ecological Implications of Body Size*, Cambridge University Press, Cambridge, U.K.

Pastore, G., Giampietro, M. and Mayumi, K. 2000. Societal metabolism and multiple-scales integrated assessment: Empirical validation and examples of application. *Popul. Environ.* 22 (2): 211–254.

Röling, N., (2002), There is life after agricultural science? in *Proceedings Workshop on New Directions in Agroecology Research and Education*, Bland, W., Ed., May 29–31, 2002, Madison. University of Wisconsin.

SAGUF, (1996), Symposium for the 50th Anniversary of the Swiss National Council of Research: *"How Will the Future World Population Feed Itself ?"* October 9–10, 1996, Zurich.

Srinivasan, T.N., (1986), Undernutrition: extent and distribution of its incidence, in *Agriculture in a Turbulent World Economy*, Maunder, A.H. and Renborg, V.I., Eds., Gower, Aldershot, Hants, U.K., pp. 199–208.

USBC (U.S. Bureau of the Census), (1991), *Statistical Abstract of the United States 1991*, U.S. Department of Commerce, Washington, D.C.

WRI (World Resources Institute), (1994), *World Resources 1994–95*, Oxford University Press, New York.

2

The Epistemological Predicament Entailed by Complexity

This chapter has the goal of clarifying a misunderstanding that often affects the debate about how to handle, in scientific terms, the challenge implied by sustainable development. The misunderstanding is generated by confusion between the adjectives *complicated* and *complex*. Complicatedness is associated with the nature and degree of formalization obtained in the step of representation (the degree of syntactic entailments implied by the model). That is, *complicated* is an adjective that refers to models and not to natural systems. Making a model more complicated does not help when dealing with complexity. Complexity means that the set of relations that can be found when dealing with the representation of a shared perception is virtually infinite, open and expanding. That is, *complex* is an adjective that refers to the characteristics of a process of observation. Therefore, it requires addressing the characteristics of a complex observer–observed that is operating within a given context. Dealing with complexity implies acknowledging the distinction between perception and representation, that is, the need to consider not only the characteristics of the observed, but also the characteristics of the observer. Scientists are always inside any picture of the observer–observed complex and never acting from the outside. In scientific terms, this implies (1) addressing the semantic dimension of our choices about how to perceive the reality in relation to goals and scales; (2) acknowledging the existence of nonequivalent observers who are operating in different points in space and time (on different scales), using different detectors and different models and pursuing independent local goals; and (3) acknowledging that any representation of the reality on a given scale reflects just one of the possible shared perceptions found in the population of interacting nonequivalent observers. To make things more difficult, both observed systems and the observers are becoming in time, but at different paces.

2.1 Back to Basics: Can Science Obtain an Objective Knowledge of Reality?

The main point of this chapter is that understanding complexity entails going beyond the conventional distinction between epistemology and ontology in the building of a new science for sustainability. To introduce such a basic epistemological issue, I have listed quotes taken from the paper "Einstein and Tagore: Man, Nature and Mysticism" (Home and Robinson, 1995), which is about a famous discussion between Einstein and Tagore about science and realism.

- "In classical physics, the macroscopic world, that of our daily experience, is taken to exist independently of observers: the moon is there whether one looks at it or not, in the well known example of Einstein." … "The physical world has objectivity that transcends direct experience and that propositions are true or false independent of our ability to discern which they are." (pp. 172–173).
- "The laws of nature which we formulate mathematically in quantum theory deal no longer with the elementary particles themselves but with our knowledge of the particles." … "The nature of reality in the Copenhagen interpretation is therefore essentially epistemological, that is all meaningful statements about the physical world are based on knowledge derived from

observations. No elementary phenomenon is a phenomenon until it is a recorded phenomenon."
Einstein declared himself skeptical of quantum theory because it concerned "what we know
about nature," no longer "what nature really does." In science, said Einstein, "we ought to be
concerned solely with what nature does." Both Heisenberg and Bohr disagreed: in Bohr's view,
it was "wrong to think that the task of physics is to find out how nature *is*. Physics concerns
what we can say about nature" (p. 173).

- Quote of Tagore: "This world is a human world — the scientific view of it is also that of the
  scientific man. Therefore the world apart from us does not exist. It is a relative world, depending
  for its reality upon our consciousness" (p. 174).
- Quote of Einstein: "The mind acknowledges realities outside of it, independent of it. For
  instance nobody may be in this house, yet that table remains where it is" (p. 174).
- Quote of Tagore: "Yes, it remains outside the individual mind, but not the universal mind. The
  table is that which is perceptible by some kind of consciousness we possess. ... If there be
  any truths absolutely unrelated to humanity, then for us it is absolutely non-existing" (p. 175).

At the end of this paper, three positions related to the question "Does reality exist and can science obtain
an objective knowledge of it?" are summarized as follows:

1. Einstein's position — Science must study (and it can) what nature does. Entities do have well-
   defined objective properties, even in the absence of any measurement, and humans know what
   these objective properties are, even when they cannot measure them.
2. Bohr's position — Science can study starting from what we know about nature. Objective
   existence of nature has no meaning independent of the measurement process.
3. Tagore's position — Science is about learning how to organize our shared perceptions of our
   interaction with nature. Objective existence of nature has no meaning independent of the human
   preanalytical knowledge of typologies of objects to which a particular object must belong to
   be recognized as distinct from the background.

The first two positions can be used to point at the existence of a big misunderstanding that some
physicists have about the role of the observer in the process of scientific analysis. Quantum physics finally
was forced to admit that the observer does play a role in the definition of what is observed, but still, the
interference generated by the observer in quantum physics is only associated to the act of measurement.
Put another way, it is the interaction between the measuring device and the natural system (an interaction
required to obtain the measurement) that alters the natural state of the measured system. This is why
smart microscopic demons could get rid of this problem. According to this view, if it were possible to
look directly at individual molecules in some magic uninvasive way, one could get knowledge (measures)
while at the same time avoiding the problem of the recognized interference observer–observed system.

Unfortunately, things are not that easy. Epistemological problems implied by complexity (multiple
scales, multiple identities, and nonequivalent observers) are so deep that, even with the help of friendly
demons, it would not be possible to escape the relative basic epistemological impasse.

In any scientific analysis of complex natural systems, the step of measuring is not the only step in
which the observer affects the perception and representation of the investigated system. Another and
much more important interference of the observer is associated with the very definition of a formal
identity for the system to be studied. This is a type of interference that has been systematically overlooked
by hard scientists. The nature of this interference is introduced in the next section, again using a practical
example. A more detailed description of relative concepts is given in Section 2.2.

2.1.1   The Preanalytical Interference of the Observer

In a famous article, Mandelbrot (1967) makes the point that it is not possible to define the length of the
coastal line of Britain if we do not first define the scale of the map we will use for our calculations. The
smaller the scale (the more detailed the map), the longer will be the length of the same segment of coast.

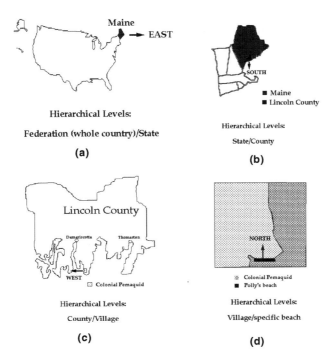

FIGURE 2.1 Orientation of the *coastal line*: nonequivalent perceptions.

This means that the length of a given segment of the coast — its numerical assessment — is affected not only by the intrinsic characteristics of the observed system (i.e., the profile of a given segment of coast), but also by a preliminary agreement about the meaning of what a segment of coast is (i.e., a preliminary agreement among interacting nonequivalent observers about the shared meaning of "a segment of coast"). Put another way, this implies reaching an agreement on how a given segment of coast should be perceived and how it should be represented. This means that such a number will unavoidably reflect an arbitrary choice made by the analyst when deciding which scale the system should use to be perceived and represented (before being measured). To better explore this point, let us use a practical example, provided in Figure 2.1, which is based on Mandelbrot's idea. The goal of this example is to explore the mechanism through which we can "see" different identities for the same natural system (in this case, a segment of coast) when observing (perceiving and representing) it in parallel on different scales. The arbitrary choice of deciding one of the possible scales by which the coast can be perceived, represented and observed will determine the particular identity taken by the system and its consequent measure.

Imagine that a group of scientists is asked to determine the orientation of the *coastal line* of Maine, providing scientific evidence backing up their assessment. Before getting into the problem of selecting an adequate experimental design for gathering the required data, scientists first have to agree on how to share the meaning given to the expression "orientation of the shore of Maine." Actually, it is at this very preanalytical step that the issue of multiple identities of a complex system enters into play. In fact, imagine that we give to this group of scientists the representation of the coast shown in Figure 2.1a. Looking at that map, the group of scientists can safely state (it will be easy to reach an agreement on the related perception) that Maine is located on the East Coast of the U.S. A sound statistical experiment can be easily set to confirm such a hypothesis. For example, the experiment could be carried out by calling from London and Los Angeles 500 Maine residents randomly selected from a phonebook during their daytime and asking them, What time is it? Using such input and the known differences in time zones between London and Los Angeles, it is possible to scientifically prove that Maine is on the East Coast of the U.S.

However, if we had given to the same group of scientists a map of Maine based on a smaller scale for the representation of the coast — for example, a map referring to the county level, as in Figure 2.1b — then the group of scientists would have organized their perceptions in a different way. Someone who is preparing computerized maps of Maine by using satellite images could have easily provided empirical evidence about the orientation of the *coastal line*. By coupling remote sensing images with a general geo-referential system, it can be "proved" that the orientation of the coast of Lincoln County is south.

What if we had asked another group of scientists to work on the same question, but had given them a smaller map of the coast of Maine from the beginning? For example, consider the map referring to the village of Colonial Pemaquid, in Lincoln County, Maine (Figure 2.1c). The scientists looking at that map would have shared yet another perception of the meaning to be assigned to the expression "orientation of a tract of shore of Maine." When operating from within this nonequivalent shared meaning assigned to this expression, they could have provided yet another contrasting statement about the orientation of this tract of *coastal line*. According to empirical analyses carried out at this scale, they could have easily concluded that the *coastal line* of Maine is actually facing west. Also, in this case, such a statement can be scientifically "proved." A random sample of 1000 trees can be used to provide solid statistical evidence, by looking at the differences of color on their trunks in relation to the sides facing north. In this way, this group of scientists could have reached a remarkable level of confidence in relation to such an assessment (e.g., $p = .01$). This new scientific inquiry performed by a different group of scientists operating within yet another distinct shared perception of the identity of the investigated system can only add confusion to the issue, rather than clarifying it.

The situation experienced in our mental example by the various groups of different scientific observers given different maps of Maine is very similar to that experienced by scientists dealing with sustainability from within different academic disciplines. Our hypothetical groups of scientists were given nonequivalent representations of the coastline of Maine, and this pushed them to agree on a particular perception of the meaning to be assigned to the label/entity "tract of *coastal line*." As will be discussed in more detail in the rest of this chapter, the existence of different legitimate formal identities for a natural system is generated by the possibility of having different associations between (1) a shared perception about the meaning of a label (in this case, "tract of *coastal line*") and (2) the corresponding agreed-upon representation (in this case, the nonequivalent maps shown in Figure 2.1). Differences about basic assumptions and organized perceptions are in fact at the basis of the problem of communication among disciplinary sciences. For example, a cell physiologist assumes that the biomass of wolves (seen as cells) is operating at a given temperature and a given level of humidity, whereas an ecologist considers temperature and humidity key parameters for determining the survival of a population of wolves (parameters determining the amount of wolf biomass). Neoclassical economists often assume the existence of perfect markets, whereas historians study the processes determining the chain of events that make imperfect actual markets.

The mechanism assigning an identity to geographic objects implies that we should expect (rather than be surprised) to find new identities whenever we change the scale used to look at them. Getting back to our example, it would be possible to ask yet another group of scientists to clarify the messy scientific empirical information about the orientation of the *coastal line* of Maine. We can suggest to this group that, to determine the "true" orientation of the *coastal line*, sophisticated experimental models should be abandoned, getting back to basic empiricism. Following this rationale, we can ask this last group of scientists to go on a particular beach in Colonial Pemaquid to gather more reliable data in a more direct way (they should use the "down to Earth" approach). The relative procedure is to put their feet into the water perpendicular to the waterfront while holding a compass. In this way, they can literally "see" what the "real" orientation is. If they would do so on Polly's Beach (Figure 2.1d), they would find that all the other groups are wrong. The "truth" is that Maine has its shore oriented toward the north. Such a shared perception of the reality, strongly backed by solid evidence (all the compasses used in the group standing on the same beach indicate the same direction), will be difficult to challenge.

The point to be driven home from this example is that different observers can make different preanalytical choices about how to define the meaning assigned to particular words, such as "a segment of coast," which will make them work with different identities for their investigated system. This will result in the coexistence of legitimate but contrasting scientific assessments. This example introduces a major problem for reductionism. Whenever different assessments are generated by the operation of nonequivalent measurement schemes, linked to a logically independent choice of a nonequivalent perception/representation of the same natural system, it becomes impossible to reduce the resulting set of numerical differences just by adopting a better or more accurate protocol of measurement or using a more powerful computer.

The four different views in Figure 2.1 show that there are several possible couplets of organized perceptions (the meaning assigned to the label "*coastal line*") and agreed-upon representations (types of map used to represent our perception of *coastal line*s) that can be used to plan scientific experiments aimed at answering the question "What is the orientation of a tract of *coastal line* of Maine?"

If we do not carefully acknowledge the implication of this fact, we can end up with scientifically "correct" (falsifiable through empirical experiments) but misleading assessments. For example, the assessment that Maine is on the East Coast (based on an identity of the *coastal line* given in Figure 2.1a and scientifically proved by a sound experiment of 500 phone calls) is misleading for a person interested in buying a house in Colonial Pemaquid with a porch facing the sun rising from the sea. For this goal, the useful identity (and the relative useful experiment) to be chosen is that shown in Figure 2.1d. At the same time, the information based on the identity of Figure 2.1a is the right one for the same person when she needs to determine the time difference between Los Angeles and Colonial Pemaquid to make a phone call at a given time in Los Angeles. So far, the story told through our mental example has shown the practical risk that honest and competent hard scientists can be fouled by donors who provide research funds to make them prove whatever should be proved (that the coast is oriented toward the north, south, east or west). Put another way, the existence of multiple potential identities entails the serious risk that smart and powerful lobbies can obtain the scientific input they need just by showing in parallel to honest and competent scientists a given map of the system to be investigated, together with a generous check of money for research.

The set of four different views (couplets of perceptions/representations) of the coastline given in Figure 2.1 obviously can be easily related to the example of the four different identities of the same natural system (in that case, a human being) given in Figure 1.2. The same natural system is observable (generating patterns on data stream) on different scales, and therefore it entails the coexistence of multiple identities. The message given by these two figures is clear. Whenever we are in a situation in which we can expect the existence of multiple identities for the investigated system (complex systems organized on nested hierarchies), we must be very careful when using indications derived from scientific models. That is, we cannot attach to the conclusions derived from models some substantive value of absolute truth. Any formal model is based on a single couplet of organized perception and agreed-upon representation at the time. Therefore, before using the resulting scientific input, it is important to understand the epistemological implications of having selected just one of the possible couplets (one of the possible identities) useful for defining the system. The quality check about how useful the model is has to be related to the meaning of the analysis in relation to the goal and not to the technical or formal aspects of the experimental settings (let alone the significance of statistical analysis checked through $p = .01$ tests). The soundness of the chain of choices referring to experimental setting (e.g., sampling procedure and measurement scheme) in relation to the statistical test used in the analysis can be totally irrelevant for determining whether the problem structuring was relevant or useful for the problem to be tackled. Rigor in the process generating formal representations of the reality (those used in hard science) is certainly indispensable, but rigor is a necessary but not sufficient condition when dealing with complexity. Actually, a blind confidence in formalizations and algorithmic protocols can become dangerous if we are not able to define first, in very clear terms, where we stand with our perception of the reality and how such a choice fits the goals of the analysis.

It is time to return to the original discussion about the "querelle" between Einstein and Tagore about science and realism. If we admit that the observer can interfere with the observed system even before getting into any action, during the preanalytical step, simply by deciding how to define the identity of the observed system, then it becomes necessary to discuss in more detail the steps and implications of this operation. The concept of identity will be discussed in detail in Section 2.2; for now it is enough to say that the definition of an identity coincides with the selection of a set of relevant qualities that makes it possible for the observer to perceive the investigated system as an entity (or individuality) distinct from its background and from other systems with which it is interacting. We can distinguish between semantic and formal definitions of identity; the former are sets of expected qualities associated with direct observations of a natural system (e.g., a fish). This definition still belongs to the realm of semantics since it is open (e.g., the list of relevant and expected qualities of a fish is open and will change depending on who we ask). Moreover, a semantic identity does not specify the procedure that will be used to make the observations (e.g., what signal detectors will be used to check the presence of fish or to establish a measurement scheme useful for representing it with a finite set of variables). For example, bees and humans see flower colors in different ways, even though they could reach an agreement about the existence of different colors. A semantic definition of identity, therefore, includes an open and expanding set of shared perceptions about a natural system (see the examples given in Figure 2.2). A semantic identity becomes a formal identity when it refers not only to a shared perception of a natural system, but also to an agreed-upon finite formal representation. That is, to represent a semantic identity in formal terms (e.g., to represent a fish in a model), we have to select a finite set of encoding variables (a set of observable qualities that can be encoded into proxy variables) that will be used to describe changes in the resulting state space (for more, see the theory about modeling relations developed by Rosen (1986)). This, however, requires selecting within the nonequivalent ways of perceiving a fish (illustrated in Figure 2.3) a subset of relevant attributes that will be included in the model.

In conclusion, we can make a distinction that will be used later in this book:

- *Semantic identity* = the open and expanding set of potentially useful shared perceptions about the characteristics of an equivalence class
- *Formal identity* = a closed and finite set of epistemic categories (observable qualities associated with proxies, e.g., variables) used to represent the expected characteristics of a member belonging to an equivalence class associated with a type

By using this definition of semantic identity, we can make an important point about the discussion between Einstein and Tagore. The preliminary definition of an identity for the observed systems (associated with an expected pattern to be recognized in the data stream, which makes possible the perception of the system in the first place) must be available to the observer before the actual interaction between observer and observed occurs. This applies either when detecting the existence of the system in a given place or when measuring some of its characteristics, let alone when we make models of that system. This means that any observation requires not only the operation of detectors gaining information about the investigated system through direct interaction (the problem implied by the operation of a measurement scheme, indicated by Bohr), but also the availability of a specified pattern recognition, which must be know *a priori* by the observer (the point made by Tagore). The measurement scheme has the only goal of making possible the detection of an expected pattern in a set of data that are associated with a set of observable qualities of natural systems. These observable qualities are assumed to be (because of the previous knowledge of the identity of the system) a reflection of the set of relevant characteristics expected in the investigated system.

An observer who does not know about the identity of a given system would never be able to make a distinction between (1) that system (when it is possible to recognize its presence in a given set of data in terms of an expected pattern associated with observable qualities of the system) and (2) its background (when the incoming data are considered just noise). The table in the room mentioned by Einstein in his discussion with Tagore can be there, but if the epistemic category associated with the equivalence class table is not in the mind of the observer — in the "universal mind," as suggested by Tagore, or in the "World 3 of human culture," as suggested by Popper (1993) — it is not possible to talk of tables in the first place, let alone check whether a table (or that table) is there.

The concepts of identity, multiple identity and different perceptions/representations on different scales are discussed in more detail in the following section. The main point of the discussion so far is that scientists can only measure specific representations (using proxies based on observable qualities) of their perceptions (definition of sets of relevant qualities associated with the choice of a formal identity to be used in the model) of a system. That is, even when adopting sophisticated experimental settings, scientists are measuring a set of characteristics of a type associated with an identity assigned to an equivalence class of real entities (e.g., cars, dogs, spheres). This has nothing to do with the assessment of characteristics of any individual natural systems.

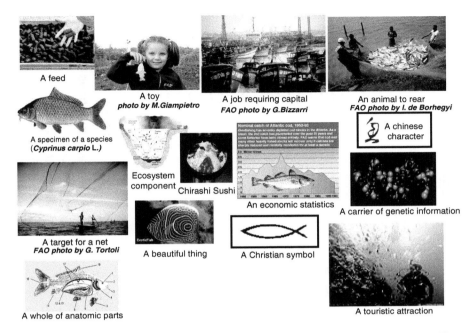

FIGURE 2.2 The open universe of semantic identities for a fish determined by goals and contexts. (Courtesy FAO Photo Library.)

In fact, it is well known that, when doing a scientific inquiry, any measurement referring to special qualities of a special individual is not relevant. For example, when asked to provide an assessment of the energy output of 1 h of human labor, we would be totally uninterested in assessing the special performance of Hercules during one of his mythical achievements or a world record established during the Olympic games. In science, miracles and unique events do not count. Coming to the assessment of the energy equivalent of 1 h of labor, we want to know average values (obtained through sound measurement schemes) referring to the energy output of 1-h of effort performed by a given typology of human worker (e.g., man, woman, average adult). This is why we need an adequate sample of human beings to be used in the test. Scientific assessments must come with appropriate error bars. Error bars and other quality checks based on statistical tests are required to guarantee that what is measured are observable qualities of an equivalence class (belonging to a given type, i.e., average adult human worker) and not characteristics of any of the particular individuals included in the sample.

Put another way, when doing experimental analyses we do not want to measure the characteristics of any real individual entity belonging to the class (of those included in the sample). We want to measure only the characteristics of simplified models of objects sharing a given template (which are describable using an identity). That is, we want to measure the characteristics of the type used to identify an equivalence class (the class to which the sampled entity belongs). This is why care is taken to eliminate the possibility that our measurement will be affected by special characteristics of individual objects (individual, special natural systems) interacting with the meter.

The previous paragraph points to a major paradox implied by science: (1) science has to be able to make a distinction between types and individuals belonging to the same typology (or between roles and incumbents, using sociological jargon, or essences and realization of essences, using philosophical jargon) when coming to the measurement step, but, at the same time, (2) science has to confuse individuals belonging to the same type when coming to the making of models, to gain predictive power and compression. This paradox will be discussed in detail in the rest of the book. This requires, however, the rediscovery of new concepts and ideas that have been developed for centuries in philosophy (for an overview, see Hospers (1997)) or in disciplines related to the process through which humans organize their perceptions to make sense of them — e.g., semiotic (for an overview, see Barthes (1964) or the

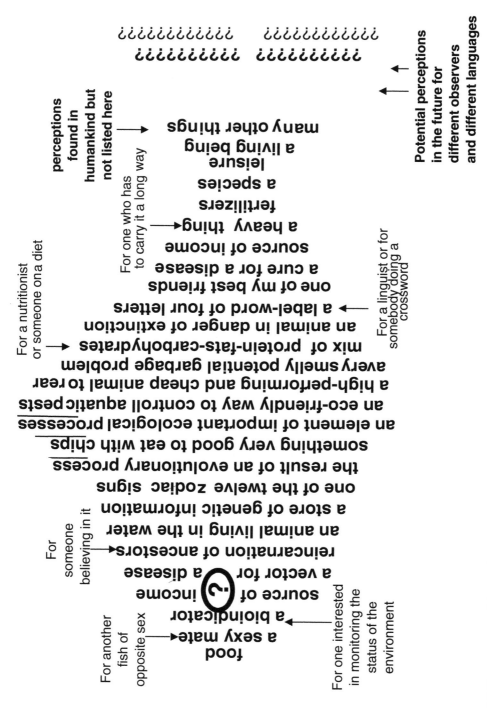

FIGURE 2.3 The open nature of the set of attributes making up the semantic identity of a fish. (After Gomiero, T. 2003 Multi-Objective Integrated Representation (Moir) As a Tool To Study and Monitor Farming System Development and Management. Ph.D. Thesis to be submitted in Environmental Science, Universitat Autonoma de Barcelona, Spain.)

work of Polany (1958, 1977) and Popper (1993)). This issue has been explored recently within the field of complex systems theory, especially in relation to the epistemological implications of hierarchy theory — Koestler (1968, 1978), Simon (1962), Allen and Starr (1982), Allen and Hoekstra (1992), O'Neill (1989), Ahl and Allen (1996). In the rest of this chapter I will just provide an introductory overview of these themes. The reader should not feel uncomfortable with the high density of concepts and terms found in this chapter. These concepts will be discussed again, in more detail, later on. The main goal now is to induce a first familiarization with new terms, especially for those who are seeing them for the first time.

2.1.2 The Take of Complex Systems Thinking on Science and Reality

The idea that the preanalytical selection of a set of encoding variables (deciding the formal identity that will be used as a model of the natural system) does affect what the observer will measure has huge theoretical implications. When using the equation of perfect gas ($PV = nRT$) we are adopting a model (a formal identity for the gas) that perceives and describes a gas only in terms of changes in pressure, volume, number of molecules and temperature, with R as a gas constant. Characteristics such as smell or color are not considered by this equation as relevant qualities of a gas to be mapped in such a formal identity. Therefore, this particular selection of relevant qualities of a gas has nothing to do with the intrinsic real characteristics of the system under investigation (a given gas in a given container). This does not mean, however, that a modeling relation based on this equation is not reflecting intrinsic characteristics of that particular gas kept in the container, and therefore that our model is wrong or not useful. It means only that what we are describing and measuring with that model, after having selected one of the possible formal identities for the investigated system (a perfect gas), is a simplified version of the real system (a real amount of molecules in a gaseous state).

Any numerical assessment coming out from a process of scientific modeling and then measurement is coming out of a process of abstraction from the reality. "The model shares certain properties with the original system [those belonging to the type], but other properties have been abstracted away [those that make the individual member special within that typology]" (Rosen, 1977, p. 230). The very concept of selecting a finite set of encoding variables to define a formal identity for the system (defining a state space to describe changes) means "replacing the thing measured [e.g., the natural system] by a limited set of numbers" (e.g., the values obtained through measurement for the selected variables used as encoding] (Rosen, 1991, p. 60).

According to Rosen, experimentalists should be defined as those scientists who base their assessments on procedures aiming at generating abstractions from reality. The ultimate goal of a measurement scheme is, in fact, to keep the set of qualities of the natural system, which are not included in the formal definition of system identity, from affecting the reading of the meters. Actually, when this happens, we describe the result of this event as a noise that is affecting the numerical assessment of the selected variable.

When assuming the existence of simple systems (e.g., elementary particles) that can be usefully characterized with a very simple definition of identity (e.g., position and speed), one can be easily fouled by the neutral role of the observer. In this situation one can come up with the idea that the only possible interference that an observer can induce on the observed system is due to the interaction associated to the measurement process. But this limited interference of the observer is simply due to the fact that simple systems and simple identities that are applicable to all types of natural systems are not very relevant when dealing with the learning of interacting nonequivalent observers (e.g., when dealing with life and complex adaptive systems). Simple systems, in fact, can be defined as those systems in which there is a full overlapping of semantic identity (the open set of potential relevant system qualities associated with the perception of the system) with formal identity (representation of the system based on a finite set of encoding variables). This assumes also that with the formal identity we are able to deal with all system qualities that are considered relevant by the population of nonequivalent observers: the potential users of the model.

This means that simple systems such as ideal particles and frictionless or adiabatic processes do not exist; rather, they are artifacts generated by the simplifications associated with a particular relationship between perception and representation of the reality. This particular forced full overlapping of formal and semantic identities of the investigated system has been imposed on scientists operating in these fields by the basic epistemological assumption of elementary mechanics. This explains why simple models of the behavior of simple systems are very useful when applied to real situations (e.g., movements of planets). In these models the typologies of mechanical systems are viewed as not becoming in time. Unfortunately, when this is true, the relative behaviors are not relevant to the issue of sustainability.

Whenever the preanalytical choices made by the observer when establishing a relation between the set of potential perceptions (the semantic identity) and the chosen representation (the formal identity used in the model) of a natural system cannot be ignored, we are dealing with complexity. Imagine, for example, that the task of the scientist is to perceive and represent her mother (which I hope reductionist scientists will accept to be a natural entity worth of attention). In any scientific representation of the behavior of someone's mother, the bias introduced by the process of measurement would be quite negligible when compared with the bias generated by the decision of what relevant characteristics and observable qualities of a mother should be included in the finite and limited set of variables adopted in the formal identity. Dealing with 1000 persons, it is much more difficult to reach an agreement about the right choice of the set of relevant qualities that have to be used in the definition of a mother, to describe with a model her changes in time, rather than to reach an agreement on the protocols to be used for measuring any set of agreed-upon encoding variables. On the other hand, without an initial definition of what are the relevant characteristics associated with the study of a mother, it would be impossible to work out a set of observable qualities used for numerical characterizations (no hard science is possible).

This problem becomes even more important when the future behavior of the observer toward the observed system is guided by the model that the observer used. The problem of self-fulfilling prophecies is in fact a standard predicament when discussing policy in reflexive systems (see Chapter 4 on postnormal science).

These basic epistemological issues, which have been systematically ignored by reductionist scientists, are finally being addressed by the emerging scientific paradigm associated with complex systems thinking (and not even by all those working in complexity). In fact, an intriguing definition of complexity, given by Rosen (1977, p. 229), can be used to introduce the topic of the rest of this chapter: "a complex system is one which allows us to discern many subsystems [a subsystem is the description of the system determined by a particular choice of mapping only a certain set of its qualities/properties] depending entirely on how we choose to interact with the system." The relation of this statement to the example of Figure 2.1 is evident.

Two important points in this quote are: (1) The concept of complexity is a property of the appraisal process rather than a property inherent to the system itself. That is, Rosen points at an epistemological dimension of the concept of complexity, which is related to the unavoidable existence of different relevant perspectives (choices of relevant attributes in the language of integrated assessment) that cannot all be mapped at the same time by a unique modeling relation. (2) Models can see only a part of the reality — the part the modeler is interested in. Put another way, any scientific representation of a complex system is reflecting only a subset of our possible relations (potential interactions) with it. "A stone can be a simple system for a person kicking it when walking in the road, but at the same time be an extremely complex system for a geologist examining it during an investigation of a mineral site" (Rosen, 1977, p. 229).

Going back to the example of the equation of perfect gas (PV = nRT), as noted earlier it does not say anything about how it smells. Smell can be a nonrelevant system quality (attribute) for an engineer calculating the range of stability of a container under pressure. On the other hand, it can be a very relevant system quality for a chemist doing an analysis or a household living close to a chemical plant. The unavoidable existence of nonequivalent views about what should be the set of relevant qualities to be considered when modeling a natural system is a crucial point in the discussion of science for sustainability.

2.1.3 Conclusion

Before closing this introductory section, I would like to explain why I embarked on such a deep epistemological discussion about the scientific process in the first place. There are subjects that are taboo in the scientific arena, especially for modelers operating in the so-called field of hard sciences. Examples of these taboos include avoiding acknowledging:

1. **The existence of impredicative loops** — Chicken–egg processes defining the identity of living systems require the consideration of self-entailing processes across levels and scales (what

Maturana and Varela (1980, 1998) call *processes of autopoiesis*). That is, there are situations in which identities of the parts are defining the identity of the whole and the identity of the whole is defining the identity of the parts in a mechanism that escapes conventional modeling.

2. **The coexistence of multiple identities** — We should expect to find different boundaries for the same system when looking at different relevant aspects of its behavior. Considering different relevant dynamics on different scales requires the adoption of a set of nonreducible assumptions about what should be considered as the system and the environment, and therefore, this requires the simultaneous use of nonreducible models.

3. **The existence of complex time** — Complex time implies acknowledging that (1) the observed system changes its identity in time, (2) the observed system has multiple identities on different scales that are changing in time but at different paces and (3) the observed system is not the only element of the process of observation that is changing its identity in time. Also, the observer does change in time. This entails, depending on the selection of a time horizon for the analysis we can observe, (1) multiple distinct causal relations among actors (e.g., the number of predators affecting the number of preys or vice versa) and (2) the obsolescence of our original problem structuring and relative selection of models (the set of formal identities adopted in the past in models no longer reflects the new semantic identity — the new shared perceptions — experienced in the social context of observation). That is, changes in (1) the structural organization of the observed system, (2) the context of the observed system, (3) the observer and (4) the context of the observer (e.g., goals of the analysis) can indicate the need to adopt a different problem structuring (an updated selection of formal identities), that is, a different meaningful relation between perception and representation of the problem.

Keeping these taboos within hard science implies condemning scientists operating within that paradigm to be irrelevant when dealing with topics such as life, ecology and sustainability. The challenges found when dealing with these three forbidden issues while keeping a serious scientific approach are discussed in Chapter 5. Alternative scientific approaches that can be developed by adopting complex systems thinking are discussed in Chapters 6, 7 and 8, and applications to the issue of multi-scale integrated analysis of agroecosystems are given in Chapters 9, 10 and 11. However, facing these challenges requires being serious about changing paradigms. This is why, before discussing potential solutions (in Parts 2 and 3), it is important to focus on the following points (the rest of Part 1):

1. Hard scientists must stop the denial. These problems do exist and cannot be ignored.

2. There is nothing mystical about complexity: current epistemological impasses experienced by reductionism can be explained without getting into deep spirituality or meditations (even though their understanding facilitates both).

3. These three taboos can no longer be tolerated: the development of analytical tools based on the acceptance of these three taboos is a capital sin that is torpedoing the efforts of a lot of bright students and becoming too expensive to afford.

To make things worse, many hard scientists are more and more getting into the business of saving the world, and they want to do so by increasing the sustainability of human progress. They tend to apply hard scientific techniques aimed at the development of optimal strategies. The problem is that they often individuate optimal solutions by adopting models that in the best-case scenario are irrelevant. Unfortunately, in the majority of cases, they use models based on the *ceteris paribus* hypothesis or single-scale representations that are not only irrelevant for the understanding of the problems, but also wrong and misleading.

To contain this growing flow of optimizing strategies supported by very complicated models, it is important to get back to basic epistemological issues that seem to be vastly ignored by this army of good-intentioned world savers. Moreover, in the field of sustainability, past validation has only limited relevance. Scientific tools that proved to be very useful in the past (e.g., reductionist analyses, which were able to send a few humans to the moon) are not necessarily adequate to provide all the answers to new concerns expressed by humankind today (e.g., how to sustain decent life for 10 billion humans on this planet). As noted in Chapter 1, humans are facing new challenges that require new tools.

Epistemological complexity is in play every time the interests of the observer (the goal of the mapping) are affecting what the observer sees (the formalization of a scientific problem and the resulting model — the choice of the map). That is, preanalytical steps — the choice of the space–time scale by which reality should be observed and the previous definition of a formal identity of what should be considered the system of interest (a given selection of encoding variables) — are affecting the resulting numerical representation of a system's qualities. If we agree with this definition, we have to face the obvious fact that, basically, any scientific analysis of sustainability is affected by such a predicament.

In spite of this basic problem, there are several applications of reductionist scientific analysis in which the problems implied by epistemological complexity can be ignored. This, however, requires acceptance without reservations from the various stakeholders who will use the scientific output of the reductionist problem structuring. Put another way, reductionist science works well in all cases in which power is effective for ignoring or suppressing legitimate but contrasting views on the validity of the preanalytical problem structuring within the population of users of scientific information (Jerome Ravetz, personal communication). Whenever we are not in this situation, we are dealing with postnormal science, discussed in Chapter 4.

2.2 Introducing Key Concepts: Equivalence Class, Epistemic Category and Identity (Technical Section)

To make sense of their perceptions of an external reality, humans organize their language-shared perceptions into epistemic categories (e.g., words able to convey a shared meaning). Obviously, I do not want to get into a detailed analysis of this mechanism. The study of how humans develop a common language is very old, and the relative literature is huge. This section elaborates rather on the concept of *identity*, which was already introduced in the previous section.

Before getting into a discussion of the concept of identity, however, we have to introduce another concept — *equivalence class*. An equivalence class can be defined as a group or set of elements sharing common qualities and attributes. The formal mathematical definition of an equivalence relation — a relation (as equality) between elements of a set that is symmetric, reflexive and transitive and for any two elements either holds or does not hold — is difficult to apply to real complex entities. In fact, as discussed in the example of the *coastal line*, nonequivalent observers adopting different couplets of shared perceptions and agreed-upon representation can perceive and represent the same entity as having different identities; therefore, they would describe that entity using different *epistemic categories*. As a result, the same segment of *coastal line* could belong simultaneously to different equivalence classes, depending on which nonequivalent observer we ask (e.g., *coastal line* segments oriented toward the south, *coastal line* segments oriented toward the north, etc.).

Imagine dealing with the problem of how to load a truck. Nonequivalent observers will adopt different relevant criteria to define the identity (set of relevant characteristics) that defines a load in terms of an equivalence class of items to be put on the truck. For example, a hired truck driver worried about not exceeding the maximum admissible weight of her or his truck will perceive/represent a relevant category for defining as equivalent the various items to be loaded — the weight of these items. With this choice, whatever mix of items can be loaded, as long as the total weight does not surpass a certain limit (e.g., 5 tons). The accountant of the same company, on the other hand, will deal with the mix of items loaded on the truck in terms of their economic value. This criterion will lead to the definition of a different equivalent class based on the economic value of items. For example, to justify a trip (the economic cost of investing in a truck and driver), the load must generate at least $500 of added value. Thus, whereas 100 kg of rocks and 100 kg of computers are seen as the same amount of load by the truck driver, according to her or his equivalence class based on weight, they will be considered differently by the accountant and her or his definition of equivalence class.

Any definition of an equivalence class used for categorizing physical entities is therefore associated with a previous definition of a semantic identity (a set of qualities that make it possible to perceive those

entities as distinct from their context in a goal-oriented observation). An equivalence class of physical entities is therefore the set of all physical entities that will generate the same typology of perception (will be recognized as determining the same pattern in the data stream used to perceive their existence) to the same observer. At the same time, the possibility of sharing the meaning given to a word (the name of the equivalence class) by a population of observers requires the existence of a common characterization of the expectations about a type (about the common pattern to be recognized) in the mind of the population of observers.

At this point the reader should have noticed that the series of definitions used so far for the concepts epistemic category, equivalence class and identity look circular. Actually, when dealing with this set of definitions, we are dealing with a clear impredicative loop (a chicken–egg paradox). That is, (1) you must know *a priori* the pattern recognition associated with an epistemic category to recognize a given entity as a legitimate member of the class (e.g., you have to know what dogs are to recognize one) and (2) you can learn the characteristics associated with the label of the class only by studying the characteristics of legitimate members of the relative class (e.g., you can learn about the class of dogs only by looking at individual dogs). A more detailed discussion of impredicative loops, and how to deal in a satisfactory way with the circularity of these self-entailing definitions, is given in Chapter 7. On the other hand, the reader should be aware that scientists are used to handling impredicative loops all the time without much discussion. For example, this is how statistical analysis works. You must know already that what is included in the sample as a specimen is a legitimate member of the equivalence class that you want to study. At that point you can study the characteristics of the class by applying statistical tests to the data extracted from the sample. Thus, you must already know the characteristics of a type (to judge what should be considered a valid specimen in the sample) to be able to study the characteristics of that type with statistical tests.

In spite of this circularity, the impredicative loop leading to the definition of identities works quite well in the development of human languages. In fact, it makes it possible for a population of nonequivalent observers to develop a language based on meaningful words about an organized shared perception of the reality. This translates into an important statement about the nature of reality. The ability to generate a convergence on the validity of the use of epistemic categories in a population of interacting nonequivalent observers points to the existence of a set of ontological properties shared by all the members of the equivalence classes. *Dog* is *perro* in Spanish, *chien* in French and *cane* in Italian. Different populations of nonequivalent observers developed different labels for the same entity (the image of the equivalence class associated with members belonging to the species *Canis familiaris*). This identity is so strong that we can use a dictionary (establishing a mapping among equivalent labels) to convey the related meaning across populations of nonequivalent observers speaking different languages. That is, the essence of a dog (the set of characteristics shared by the members of the equivalence class and expected to be found in individual members by those using the language) to which the different labels (*dog, perro, chien, cane*) refer must be the same. Also in this case, a discussion of the term *essence* and its possible interpretations and definitions within complex systems thinking will be discussed at length in Chapter 8.

This remarkable process of convergence of different populations of nonequivalent observers on the definition of the same set of semantic identities (associated with the words of different languages) can only be explained by the existence of ontological aspects of the reality that are able to guarantee the coherence in the perceived characteristics of the various members of equivalence classes associated with different identities (e.g., a dog) over a large space–time domain (over the planet, across languages). If various observers interacting with different individual realizations of members of the class (e.g., having distinct different experiences with individual dogs) are able to reach a convergence on a shared meaning assigned to the same set of epistemic categories (e.g., share a meaning when using the label "dog"), then the ontological properties of the equivalence class dog must be able to determine a recognized pattern on a space–time domain much larger than that of individual observers, individual dogs and even individual populations of interacting nonequivalent observers using a common language. Put another way, if all the observers perceiving the characteristics of a dog can agree on the usefulness and validity of the identity associated with such a label, we can infer that something "real" is responsible for the coherence of the validity of such a label. Such a real thing obviously is not an organism belonging to the species *Canis familiaris*. In fact, any organism can only generate local patterns in the data stream (those recognized by a few observers) on a very limited space–time domain. To generate coherence across languages, we must deal with an equivalence class of physical objects sharing the same pattern of organization and expressing similar behaviors on a quite large space–time domain. This class must exist and interact with several populations of nonequivalent observers to make possible the convergence of the use of a set of meaningful labels in a language. It is the shared meaning of different words in different languages that makes it possible to organize them in a dictionary.

The search for equivalence classes useful for organizing our knowledge of physical entities through labels is a quite common experience for humans. We are all familiar with the use of assigning names to human artifacts (e.g., a refrigerator or a model of a car such as the Volkswagen Golf). In different languages this implies establishing a correspondence between a given essence (a semantic identity in our mind expressed as an expected set of common characteristics of the class of objects that are considered to be a realization of that essence) and a label (the name used in the language for communicating — representing — such a perception). At this point it becomes possible to associate these labels with a mental representation of perceived essences — the most habitual images in our mind. The same mechanism applies in biology, where equivalence classes of organized structures are also very common, e.g., the individual organism (that dog) belonging to a given species (*Canis familiaris*).

I am arguing that this similarity between human-made artifacts organized in equivalence classes and biological structures organized in species is not due to coincidence but, on the contrary, is a key feature of autopoietic systems. The very essence of this class of self-organizing systems is their ability to guarantee the coherence between:

1. The ability to establish useful relational functions, which define the essence of their constituent elements. This coherence has to be obtained on a large scale.
2. The ability to guarantee coherence in the process of fabrication of the various elements of the corresponding equivalence class — e.g., using a common blueprint for the realization of a set of physical objects sharing the same template. This coherence has to be obtained at a local scale.

According to the terms introduced so far, we can say that elements belonging to the same equivalence class are different realizations of the same essence (they share the same semantic information about the common characteristics of the class) (Rosen, 2000).

The variability of the characteristics of different realizations belonging to the same equivalence class will depend on (1) the quality of the process of fabrication (how well the process of realization of the essence is protected from perturbations coming from the environment) and (2) the accuracy of the information stored, carried and expressed by the reading of the blueprint against gradients between the expected associative context of the type and the actual associative context of the realization.

At this point it should be noted again that any assessment of the characteristics of the template used to make an equivalence class or of the type used to represent members belonging to the class does not refer to the characteristics of any individual organized structure observed in the process of assessment. Rather, both measurements and assessments refer only to the relevant attributes used to define the equivalence class. Put another way, scientific assessments refer to the image of the class (the type) and not to special characteristics of realizations. The variability of individual realizations will only affect the size of error bars describing various characteristics of individual elements in relation to the average values for the class.

We have now accumulated enough concepts to attempt a more synthetic definition of identity.

The etymology of the term *identity* comes from the Latin *identidem*, which is a contraction of *idem et idem*, literally "same and same" (*Merriam-Webster Dictionary*). An identity implies using the same label for two tasks: (1) to identify mental entities (types representing the essence of the equivalence class as images in our mind) and (2) to identify physical entities perceived as members of the corresponding equivalence class (the set of all the specific realizations of that essence). As noted before, such mechanism of identification (obtained using a sort of stereo complementing mapping) must be useful to: (1) see distinct things as the same (to gain compression), e.g., all dogs are handled as if they were just dogs and (2) handle each real natural system one at a time (to gain anticipation) e.g., we can infer knowledge about this particular dog from our general knowledge about dogs..

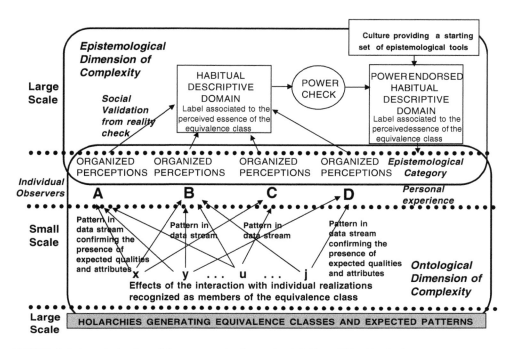

FIGURE 2.4 A synthetic view of the processes leading to the definition of identities for holons.

The concept of identity helps individual observers (at a given point in space and time) to handle their daily experience with natural systems. In fact, by using identities, an observer can either:

1. Identify an element of an equivalence class as an entity distinct from its context (e.g., to perceive the existence of an individual cow or the table in the room), since it is possible for the individual observer to distinguish the relative pattern recognition in the data stream coming from the reality. Depending on the type of detectors used to perceive the existence of that cow — sight, smell, touch — nonequivalent observers will adopt a different selection of relevant attributes (a different characterization of the type — selection of nonequivalent formal identities — for the same label). Obviously, a large carnivorous predator or an endoparasite will map the formal identity of a cow in different ways.

2. Infer information about the characteristics of a particular realization (any element of the class — again, a particular cow met in a particular point in space and time) obtained from the knowledge of the class and not by previous experience with the same physical entity. Put another way, an observer knowing about cows can know about characteristics of the essence, which are common to the members of the equivalence class and therefore can safely be assumed to be present also in that particular specimen. This implies that even when meeting a particular cow for the first time, an observer knowing about the characteristics of the type can infer that that particular cow has, inside her body, a pulsing heart even when it is 500 m away.

The ability to associate the "right" set of epistemic categories to an individual realization (physical entity) recognized as belonging to a given equivalence class (associated with a label) can provide a huge power of compression and anticipation. But at the same time, this can be a source of confusion. In fact, one must be always aware that every cow, as well as every farm, every farmer and every individual ecosystem, is special.

The validity of an identity requires two quality checks in parallel, as illustrated in Figure 2.4:

1. A congruence check (in relation to an external referent) over a small scale. This check is about the validity of the correspondence between the mental object (semantic identity associated with an epistemic category in the mind of the observer) and the physical object (the experienced characteristics expressed by a member of the corresponding equivalence class) in relation to the given label used to link the expectations associated with the mental objects to the experience associated with the interaction with the individual realization of the corresponding type. This validity check is related to a local space–time domain. That is, it requires that individual observers using the set of epistemic categories associated with the label be able to verify the congruence between expectations — how a cow is expected to look and behave — and the pattern found in the data stream obtained when interacting at a given point in space and time with a real-world entity that is assumed or recognized to be a member of the equivalence class — how that particular entity, identified as a cow, is actually looking and behaving when interacting with the observer.

2. A congruence check (in relation to an external referent) over a much larger scale on the congruence of the various identities assigned to different objects by a population of observers within a given language. It should be noted that the universe of words, semantic identities and epistemic categories is there before any individual human observer enters into play. That is, new human observers learn, when they are babies, how to name objects, use adjectives and locate events in space and time according to an established set of epistemic tools found in the culture within which they grow up. This mismatch between the space–time domain at which these tools are defined and the process of learning of individuals is at the basis of the perception that types and epistemic categories are out of time. The reader can recall here the world of ideas (e.g., Plato) out of time or, in more recent times, the World 3 of established concepts (e.g., Popper, 1993). Obviously, the universe of epistemic tools of a cultural system is there before any individual observer enters into play, and therefore it looks like it is given (as the laws of a country) to individuals. In fact, the mental image of object A has to be shared by nonequivalent observers. This is what makes it possible to reach an agreed-upon representation of that object at the social level. This is why newcomers to the culture and language have to learn how to converge on the set of categories usually associated with such a label — the habitual descriptive domain adopted by society. This would be, for example, the definition of an entity found in the dictionary (see Figure 2.4). Obviously, the official definition of an entity at the societal level does not match entirely the personal perceptions that different individuals have of dogs, cars or other entities. However, when operating at a very large scale (as implied by the interaction of a population of nonequivalent observers over a large period of time), the problem of congruence among official definitions in languages operates above the level of individual observers and users of the language. The agreed-upon definition (representation) of the semantic identity for individual objects (e.g., object A) must be compatible with the definition (representation) of other organized perceptions of other objects interacting with object A. Actually, this condition is a must. Humans are able to define a mental image only in relation to other mental images (Maturana and Varela, 1998). That is, as soon as we define other identities (e.g., for objects B, C and D, which are interacting with A) we have to socially organize (at a level higher than that of individual observers) the sharing of the meaning assigned to labels (the representation of organized shared perceptions) about other mental objects interacting with A. Put another way, the operation of a language requires reaching a socially validated habitual descriptive domain for each of the mental images of the objects (e.g., B, C and D) interacting with A; this is shown in the upper part of Figure 2.4. To do that, humans have to introduce additional concepts such as time and space, which are needed to make sense of their shared perceptions in relation to the various selections of identities. In fact, it is only within time (the relative representation of a rate of change compared to another rate of change used as a reference) and space (the relative position of an object compared to another one used as a reference) that we can represent the relation and interaction between different sets of mental objects. This is what generates the existence of multiple identities for systems organized on different hierarchical levels. In fact, the various definitions of identities sharing compatible categories tend to cluster

on different scales (meaningful relations between a perception and representation of reality — (Allen and Starr,1982). This is what generates the phenomenon discussed in Figure 1.2 and Figure 2.1 of emergence of different identities on different scales. The need to reach a mutual compatibility and coherence among the reciprocal definitions of the various epistemic tools used in the process requires the use of different clusters of epistemic categories to perceive and represent the same system at different levels of organization. Only in this way does it become possible to share the meaning of representations about perceptions in a common language. The coherence of a language entails a set of reciprocal constraints derived from the mutual information carried out by epistemic categories.

When we talk of a check on the reciprocal compatibility of the universe of epistemic categories used to handle the perception and representation of the reality at different scales and in relation to different typologies of relevant qualities, we deal with a validity check that does not refer to any specific interaction between observers and individual physical elements of equivalence classes. Rather, this is a validity check that refers to the emergent properties of the whole language (Maturana and Varela, 1998).

2.3 Key Concepts from Hierarchy Theory: Holons and Holarchies

2.3.1 Self-Organizing Systems Are Organized in Nested Hierarchies and Therefore Entail Nonequivalent Descriptive Domains

All natural systems of interest for sustainability (e.g., complex biogeochemical cycles on this planet, ecological systems and human systems when analyzed at different levels of organization and scales above the molecular one) are dissipative systems (Glansdorf and Prigogine, 1971; Nicolis and Prigogine, 1977; Prigogine and Stengers, 1981). That is, they are self-organizing, open systems, away from thermodynamic equilibrium. Because of this they are necessarily becoming systems (Prigogine, 1978); that in turn implies that they (1) are operating in parallel on several hierarchical levels (where patterns of self-organization can be detected only by adopting different space–time windows of observation) and (2) will change their identity in time. Put another way, the very concept of self-organization in dissipative systems (the essence of living and evolving systems) is deeply linked to the idea of (1) parallel levels of organization on different space–time scales, which entails the need of using multiple identities and (2) evolution, which implies that the identity of the state space, required to describe their behavior in a useful way, is changing in time. The two sets of examples are discussed in Chapter 1.

Actually, the idea of systems having multiple identities has been suggested as the very definition of hierarchical systems. A few definitions, in fact, say:

- "A dissipative system is hierarchical when it operates on multiple space–time scales — that is when different process rates are found in the system" (O' Neill, 1989).
- "Systems are hierarchical when they are analyzable into successive sets of subsystems" (Simon, 1962, p. 468) — in this case we can consider them as near decomposable.
- "A system is hierarchical when alternative methods of description exist for the same system" (Whyte et al., 1969).

As illustrated in the previous examples, the existence of different levels and scales at which a hierarchical system is operating implies the unavoidable existence of nonequivalent ways of describing it. Examples of nonequivalent descriptions of a human being (Figure 1.2) or geographic objects (Figure 2.1) have already been discussed. Human societies and ecosystems are generated by processes operating on several hierarchical levels over a cascade of different scales. Therefore, they are perfect examples of nested dissipative hierarchical systems that require a plurality of nonequivalent descriptions to be used in parallel to analyze their relevant features in relation to sustainability (Giampietro, 1994a, 1994b; Giampietro et al., 1997; Giampietro and Pastore, 2001). The definition of hierarchy theory suggested by Ahl and Allen is perfect for closing this section: "Hierarchy theory is a theory of the observer's role in any formal study of complex sysems" (Ahl and Allen, 1996, p. 29).

2.3.2   Holons and Holarchies

Holons and holarchies are a new class of hierarchical systems relevant for the study of biological and human systems. In fact, in the case of biological and human holarchies, this class is made up of self-organizing (dissipative) adaptive (learning) agents that are organized in nested elements. Gibson et al. (1998) suggest for these systems the term *constitutive hierarchies*, following the suggestion of Mayr (1982). Personally, I prefer the use of the term *holarchies*, to acknowledge the theoretical work already developed in this field (started in the 1960s), discussed below and at the end of Chapter 3.

Each component of a dissipative adaptive system organized in nested elements can be called a holon, a term introduced by Koestler (1968, 1969, 1978) to stress its double nature of whole and part (for a discussion of the concept, see also Allen and Starr (1982, pp. 8–16)). A holon is a whole made of smaller parts (e.g., a human being made of organs, tissues, cells, atoms), and at the same time it forms a part of a larger whole (an individual human being is a part of a household, a community, a country, the global economy). The choice of the term *holon* points explicitly to the obvious (in the perception of everyone, yet denied in the representation of reductionist science) fact that entities belonging to dissipative adaptive systems, which are organized in nested elements (say a dog or a human being), have an inherent duality.

Holons have to be considered in terms of their composite structure at the focal level (they represent emergent properties generated by the organization of their lower-level components) — a tiger as organism. We obtain this view when looking at the black box and inside it at the pieces that are making it work. In this way we can perceive and represent *how* they work.

Because of their interaction with the rest of the holarchy, holons perform functions that contribute to other emergent properties expressed at a higher level of analysis — functions that are useful for the higher-level holon to which they belong — e.g., the role of tigers in ecological systems. We obtain this view when looking at what the black box does within its larger context. In this way, we can perceive and represent *why* the black box makes sense in its context.

"A nested adaptive hierarchy of dissipative systems [a system made of holons] can be called a holarchy" (Koestler, 1969, p. 102). A crucial element to be clarified is that the very concept of holarchy — what represents the individuality of a holarchy — implies the ability of preserving a valid mapping between a class of organized structures (e.g., a population of individual organisms belonging to a species) and the associate functions (e.g., the set of functions related to the ecological role of the species). That is, the two nonequivalent views of a holon must be and remain consistent with each other in time. This means that a holarchy, to remain alive, must have the ability to coordinate across levels: (1) mechanisms generating the realization of a class of organized structures expressing the same set of characteristics (e.g., making similar organisms by using a common blueprint when making a population of tigers) at one level and (2) mechanisms guaranteeing the stability of the associative context within which the agency of these organized structures translates into the expression of useful functions (e.g., the preservation of a favorable habitat for the individuals belonging to the species *Panthera tigris*). The informed action of agents has to guarantee an admissible environment for the process of fabrication of organisms. Recalling the discussion about the concept of identity, the individuality of a holarchy can be associated with the ability to generate and preserve in time the validity of an integrated sets of viable identities (on different scales).

When dealing with holons and holarchies, we face a standard epistemological problem. The space–time domain that has to be adopted for characterizing their relational functions (when considering higher-level perceptions and descriptions of events) does not coincide with the space–time domain that has to be adopted for characterizing their organized structures (when considering lower-level perceptions and descriptions of events).

When using the word *dog*, we refer to any individual organism belonging to the species *Canis familiaris*. At the same time, the characterization of the holon dog refers both: (1) to a type characterized in terms of relational functions associated with the niche of that species; these functions are expressed by the members of the relative equivalence class (the organisms belonging to that species) in a given ecosystem and (2) to a type characterized in terms of structural organization; the same organization pattern is shared by any organism belonging to the equivalence class. This means that when using the word *dog*, we loosely refer both to the characteristics relevant in relation to the niche occupied by the species in the ecosystem (to "dogginess," so to speak) and to the characteristics of any individual organism belonging to it (the organization pattern expressed by realized dogs, that is, by individual organisms, including the dog of our neighbor). Every dog, in fact, belongs by definition to an equivalence class (e.g., the species *Canis familiaris*) even though each particular individual has some special characteristics (e.g., generated by stochastic events of its personal history) that make it unique.

That is, any particular organized structure (the dog of the neighbor) can be identified as different from other members of the same class, but at the same time, it must be a legitimate member of the class to be considered a dog.

Another example of holon, this time taken from social systems, could be the president of the U.S. In this case, President George W. Bush is the lower-level organized structure, that is, the incumbent in the role of president for now. Any individual human being (required to get a realization of the type) has a time closure within this social function — under the existing U.S. Constitution, a maximum of 8 years (two 4-year terms). The U.S. presidency as a social function, however, has a time horizon that can be estimated in the order of centuries. In spite of this, when we refer to the president of the U.S., we loosely address such a holon, without making a distinction between the role (social function) and the incumbent (organized structure) performing it. The confusion is increased by the fact that you cannot have an operational U.S. president without the joint existence of (1) a valid role (institutional settings) and (2) a valid incumbent (person with appropriate sociopolitical characteristics, verified in the election process). On the other hand, the existence and identity of President Bush as an organized structure (e.g., a human being) able to perform the specified function of U.S. president are totally, logically independent (when coming to the representation of its physiological characteristics as human being) from the existence and identity of the role of the presidency of the U.S. (when coming to the representation of its characteristics as a social institution) and vice versa. Human beings were present in America well before the writing of the U.S. Constitution.

The concept of holon as a constitutive complex element of self-organizing systems operating on nested hierarchical systems is crucial for understanding the standard epistemological predicaments faced by hard science. In fact, the dual nature of holons entails, even requires, the existence of an image of it in the epistemological tool kit used by humans to describe complex systems. This image refers not to the specific characteristics of an incumbent or individual realizations, but rather to the characteristics of the type itself. However, these characteristics must be found in a particular natural system recognized as belonging to that class. In the literature of complex systems there is a significant convergence on the point that to deal with complexity, scientists should look for a new mechanism of mapping based on the overlapping of two nonequivalent representations. This means using a sort of representation of the natural system "in stereo" based on the simultaneous use of two complementing views. Herbert Simon (1962) proposes the need to use in combination two concepts — organized structure and relational function — as a general way to describe elements of complex systems. Kenneth Bailey (1990) proposes the same approach, but using different terms — role and incumbent — when dealing with human societies. Salthe (1985) suggests a similar combination of mappings based on yet another selection of terms: individuals (as the equivalent of organized structures or incumbents) and types (as the equivalent of relational functions or roles). Finally, Rosen (2000) proposes, within a more general theory of modeling relation, a more drastic distinction that gets back to the old Greek philosophical tradition. He suggests making a distinction between individual realizations (which are always special and which cannot be fully described by any scientific representation due to their intrinsic complexity) and essences (associated with the typical characteristics of an equivalence class). The logical similarities between the various couplets of terms are quite evident.

As noted earlier, when developing his theory of modeling relation, Rosen (1986) suggests that scientists must always keep a clear distinction between natural systems (which are always special and which cannot be fully described by any scientific representation due to their intrinsic complexity) and representation of natural systems (which are based on the use of epistemic categories, based on the definition of a set of attributes required to define equivalence classes used to organize the perception and representation of elements of reality over types). The use of epistemic categories makes possible a compression in the demand of computational capability when representing the reality (e.g., say "dog," and you include them all). But this implies generating a loss of one-to-one mapping between representation and direct perception (this implies confusing the identities of the individual members of equivalence classes).

2.3.3   Near Decomposability of the Hierarchical System: Triadic Reading

To better understand the nature of the epistemological predicament faced when making models of holarchic systems, it is opportune to reflect on how it is possible to describe a part of them (or a given view of them) as a well-defined entity separated from the rest of the reality (as having an identity separated from the rest of the holarchy) in the first place. Put another way, if the holarchy represents continuous nested elements across levels and scales, how is it that we can define a given identity (the perception of a face, cell or crowd, as illustrated in Figure 1.2) for a part of it, as if it were separated from the rest? Any definition of a system, in fact, requires a previous definition of identity that makes it possible to individuate it as distinct from the background (that makes it possible to define a clear boundary between the system and its environment). However, when we apply this rationale to the representation of a holon, we have to include in our representation of the environment of a given holon the remaining part of its own holarchy. For example, humans have to be considered as the environment (given boundary conditions) of their own cells. At the same time, we have to admit that cells' behaviors (e.g., the insurgence of some disease) can affect directly those large-scale mechanisms guaranteeing the boundary conditions of the cell (the health of the individual to which the cells belong). In the same way, when considering humans as entities operating within a given ecosystem (their environment), it is well known that with their behavior humans can affect the stability of their own boundary conditions (e.g., pollution or greenhouse emissions). When dealing with the representation of a part of a holarchy as distinct from the environment, we must be aware of the fact that this is an artifact, since holarchic dissipative systems cannot be isolated from their context. Their very identity depends on the interaction across boundaries in cascade across levels. When dealing with dissipative holarchies, the clear distinction between system and environment becomes fuzzy and ambiguous, especially when we want to consider several dynamics on different levels (and scales) at the same time.

In spite of this general problem, the possibility to perceive and represent a part of a holarchy as a separated entity from the whole to which that part belongs is related to the concept of near decomposability — introduced by Simon (1962) in his seminal paper "The Architecture of Complexity." This principle refers more to the epistemological implications of hierarchy theory.

Hierarchy theory sees holarchies as entities organized through a system of filters operating in a cascade — a consequence of the existence of different process rates in the activity of self-organization (Allen and Starr, 1982). For example, a human makes decisions and changes her or his daily behavior based on a timescale that relates to her or his individual life span. In the same way, the society to which he or she belongs also makes decisions and continuously changes its rules and behavior. Differences in the pace of becoming generate constraints within the holarchy: "Slaves were accepted in the United States in 1850, but would be unthinkable today. However, society, being a higher level in the hierarchy than individual human beings, operates on a larger spatio-temporal scale" (Giampietro, 1994b). The lower frequency of changes in the behavior of the society is perceived as laws (filters or constraints) when read from the timescale by which individuals are operating. That is, individual behavior is affected by societal behavior in the form of a set of constraints ("this is the law") defining what individuals can or cannot do on their own timescale.

Getting into hierarchy theory jargon, the higher level, because of its lower frequency, acts as a filter constraining the higher-frequency activities of the components of the lower level into some emergent property. For more, see Allen and Starr (1982). Additional useful references on hierarchy theory are Salthe (1985, 1993), Ahl and Allen, (1996), Allen and Hoekstra (1992), Grene, (1969), Pattee (1973) and O'Neill et al. (1986). Obviously, processes occurring at the lower hierarchical levels also matter. In fact, this is where structural stability is guaranteed. This means that there are different dynamics and mechanisms operating in parallel on different levels that are actually affecting each other. The deep epistemological punch of hierarchy theory is that it is not possible to recognize (perceive) and describe (represent) a system organized in nested hiearchies by adopting a single validated model (a model able to make valid predictions on the basis of the congruence of simulated inferences on the values taken by variables with a stream of data coming from the reality). In fact, such a validation must necessarily refer to a single scale — or a single descriptive domain.

We all know the popular line within the community of dynamical modelers that stocks are just flows that go extremely slowly and that flows are just fast-going stocks. In this example, the decision to call something either a stock or a flow will depend on the choice, made by the modeler, when selecting a given time differential for the model. Put another way, the possibility of associating a label (either stock or flow) with a recognized pattern in the reality (to assign an identity to a process in our representation or our perception of it) is determined by the speed at which such an identity is perceived to change in time compared with the rate of perceived changes in its context.

If we adopt this view, it is clear that we must expect the existence of different perceptions (and therefore representations) of the same reality made by nonequivalent observers (e.g., a human being with a life span of several decades and a drosophila with a life span of a few days) even when dealing with the same natural systems — again, see Figure 2.1. For example, even though humans do change their aspects during their lifetime, the pace of such a process is slow enough to make possible the neglecting of the perception of this change on a daily base. In fact, the process of perception and representation of our own image is updated every day. That is, each one of us sees always the same person in the mirror, when brushing the teeth every morning. However, this does not guarantee that two schoolmates meeting after 30 years would be able to recognize each other. In this case, a symmetrical bifurcation entails a lack of validity in the information stored in these two persons about the pattern recognition and representation of the other. If the two had been required to give an input for an identikit (facial sketch) of both themselves and the other, they would have provided updated information for their own face, but completely wrong input about the other.

In this example, the two nonequivalent observers are not using different detectors to look at each other (they are using exactly the same hardware and software for making their observations), but they simply are adopting a different time differential to update their perceptions and representations of changes. The difference in the time differential implies a completely different selection of relevant qualities to be included in the perception and representation of changes. For the daily observation, coarse features (remaining stable over a time duration of months) are ignored in favor of finer-grain resolution of changes over details. This implies that coarse feature changes are ignored in daily observation. The updating of the image of a friend after 30 years (which has lost a lot of details in the storage) on the contrary has to do with updating first the correspondence of coarse characteristics. From this example we can guess the existence of mosaic effects within the information gathered within the holarchy. The interaction of nonequivalent holons within a stable holarchy requires the integration and ability to make effective use of different flows of information coming from nonequivalent observers operating on different hierarchical levels and space–time scales.

Holarchies are characterized by jumps or discontinuities in the rates of activity of self-organization (patterns of energy dissipation) across the levels. Hierarchical levels are, in fact, the result of differences in process rates related to energy conversions stabilized on controlled autocatalytic loops (Holling, 1995; Odum, 1971, 1983). The mechanism of lock-in associated with the generation of an autocatalytic loop is what generates the discontinuities in scales, which are the real root of near decomposability.

The principle of near decomposability explains why scientists are able to study simplified models of natural systems over a wide range of orders of magnitude, from the dynamics of subatomic particles to the dynamics of galaxies in astrophysics. When dealing with hierarchical systems, we can study the dynamics of a particular process on a particular level by adopting a description that seals off higher and lower levels of behavior. In this way, we can obtain a description that is able to provide an operational identity (a finite set of relevant qualities) for the system under investigation. This has been proposed as an operation of triadic reading or triading filtering by Salthe (1985). This means that we can describe, for example, in economics consumer behavior while ignoring the fact that consumers are organisms composed of cells, atoms and electrons. The concept of triading reading refers to the choice made by the scientist of three contiguous levels of interest within the cascade of hierarchical levels through which holarchies are organized. That is, when describing a particular phenomenon occurring within a holarchy, we have to define a group of three contiguous levels, starting with:

1. **Focal level** — This implies the choice of a space–time window of observation by which the qualities of interest of the particular holon (expressed in the formal identity) can be defined and studied by using a set of observables (encoding variables assumed to be proxies of changes.

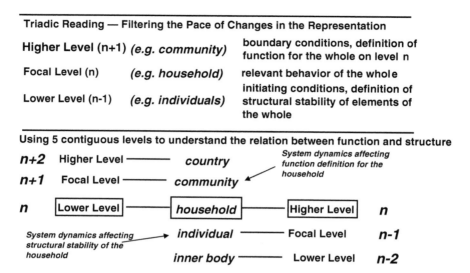

FIGURE 2.5 Triadic reading and the need for five contiguous levels.

in the qualities considered relevant). For example, if we are dealing with consumer behavior, we will not select a space–time scale for detecting qualities referring to electrons. Therefore, the choice of variables able to catch changes in the relevant qualities of our system reflects (1) the goal of our analysis (why we want to represent its behavior) and (2) the characteristics of the measurement scheme (the type of detectors available to generate a data stream and the experimental setting used to extract data from the reality).

2. **Higher level** — The choice of a formal identity for the investigated system at the focal level is based on the assumption that changes of the characteristics of the higher level are so slow when described on the space–time window of the focal level that they can be assumed to be negligible. In this case, the higher level can be accounted for — in the scientific description — as a set of external constraints imposed on the dynamics of the focal level (the given set of boundary conditions).

3. **Lower level** — The difference of time differentials across levels implies that the mechanisms determining the dynamics of lower-level components are not always relevant in relation to the mechanisms determining the behavior on the focal level description. In fact, when considering the aggregate behavior of lower-level elements, lower-level activity can be accounted for in terms of a statistical description of events occurring there. In this way, the individuality of lower-level elements is averaged out by considering such variability as noise. That is, when adopting a triadic reading, the identity of lower-level elements is accounted for in the focal description just in terms of a set of initiating conditions determining the outcome of the studied dynamics.

To give an example of triadic reading, economic analyses describe the economic process by adopting a focal level with a time window (1) small enough to assume changes in ecological processes such as climatic changes or changes in institutional settings (the higher level) negligible and (2) large enough to average out noise from processes occurring at the lower level — e.g., the nonrational consumer behavior of artists, terrorists or Amish is averaged out by a statistical description of population preferences (Giampietro, 1994b).

It should be noted, however, that the trick of the triadic reading works well only when applied to those parts of holarchies that have quite a robust set of identities, that is, in those cases in which the simultaneous interaction of processes occurring on lower, higher and focal levels manages to generate a lock-in (a mechanism of self-entailment across dynamics operating in parallel over different hierarchical levels), which guarantees stability or resilience toward external (from the higher level) and internal (from the lower level) perturbations. This requires holarchies able to generate robust integrated patterns on multiple scales (that can guarantee the validity of a coordinated set of identities over different levels — as in the examples of Figure 1.2). The various patterns expected or recognized on different levels should be stable enough to justify the expression "quasi-steady state" to the perception and representation of the system over the focal level.

Clearly the process of triadic reading can be repeated across contiguous levels through the holarchy (see Figure 2.5). That is, a household can be at the same time (1) the higher level (the fixed boundary context) for an individual belonging to it (for those scientists interested in studying the behavior of individuals — e.g., a psychologist); (2) the focal level (for those scientists interested in describing possible changes in household identity in relation to changes in the social context, or the characteristics of individuals — e.g., anthroplogist); and (3) the lower level (organized structures determining emergent properties on the focal level) for social systems (for those scientists studying the behavior of social systems made up of households). Due to this chain of relations across levels of holons, the issue of sustainability requires the consideration of at least five contiguous hierarchical levels at the same time (Flood and Carson, 1988), as shown in Figure 2.5. When considering five contiguous levels, we can describe those processes that determine the various relevant aspects of the stability of the holon under investigation:

1. The set of identities of lower-level organized structures (parts) that determines with its variability of typologies of components and distribution over possible typologies of the population of components — initiating conditions
2. The pattern referring to the focal level (the whole), for which we can simulate behavior with an appropriate model after receiving the required information about the actual state of boundary conditions (referring to the higher level) and initiating conditions (referring to the lower level)
3. The identity of the environment (the context) that is influencing the admissible behaviors of the system on its focal level

The sustainability of the process represented in the original triadic reading requires verification of the compatibility of changes occurring at different speeds on these five contiguous levels. This means that if we later want to scale up or down the effects of changes induced at any of these five levels in the holarchy, or if we want to establish links among nonequivalent descriptions referring to the nonreducible identities defined on different levels, it is crucial to have adequate information about the various identities interacting among these five levels. This is at the basis of the concepts developed in Part 2 and applied in Part 3.

It is important to recall here the warning about the epistemological implications of the trick of triadic reading across levels, as shown in Figure 2.5. This is where the epistemological predicament of holons enters into play. As observed by O'Neill et al. (1989), biological systems have the peculiar ability of both being in quasi-steady state and becoming at the same time. Their hierarchical nature makes possible this remarkable achievement. In fact, biological systems are easily described, as in the quasi-steady state on small space–time windows (when dealing with the identity of cells, individual organisms, species), that is, on the bottom of the holarchy. However, the more we move up in the holarchic ladder, the more we find entities that are becoming. When using a much larger space–time window (moving to the perception and representation of higher holarchic levels), we are forced to deal with the process of evolution (e.g., ecosystem types co-evolving within Gaia). At this level, new essences (roles or types) are continuously added to the open information space of the holarchy. This implies that the near decomposability of hierarchical systems works well only (1) when the observer is focusing on a well-defined part of the holarchy at the time and (2) when the set of qualities of interest to be isolated (the identity we decide to adopt to describe the holon) can be assumed to be stable on the time window considered relevant for the analysis (the lower we are in the ladder of the holarchy, the better). When dealing with the sustainability of human societies, this rarely occurs. The reader can recall here the example of the difference in difficulty faced when trying to reach an agreement on a formal identity to be used for particles or for mothers. The higher one goes in the holarchies, the richer the set of categories included in the semantic identity becomes. This implies that it becomes tougher to reach an agreement on how to compress this semantic identity into a finite, limited, closed formal identity.

2.3.4 Types Are out of Scale and out of Time, Realizations Are Scaled and Getting Old

A type is a given set of relations of qualities of a system associated with the ability to express some emergent property in a given associative context. Koestler (1968) in his *The Ghost in the Machine* uses the term *associative context* in a description of the relative cognitive process. The term *associative context* indicates that the characteristics of a given type are always associated with the actual possibility of performing a given, expected function. The type is assumed to operate in its expected environment (e.g., niche for a population of a given species). Living fish have water as their associative context; birds have air; humans cannot operate in melted iron.

In the same way, epistemic categories used to organize our perceptions and communicate meaning require, imply, or entail the existence of the right associative context for the type to which they refer. Change the expected associative context to a word and you get a joke (Koestler, 1968). For example, the old joke "I met a guy with a wooden leg called Joe Smith. What was the name of the other leg?" is based on the violation of the basic association of the name Joe Smith with a person. In the joke, the label Joe Smith is instead associated with the word *leg*.

The concept of a required associative context for a given realization of an essence applies to both the epistemological (e.g., words) and ontological (e.g., members of an equivalence class) sides. In fact, in the real operation of a dissipative system, the existence and survival of an organism also invoke the unavoidable association with an appropriate environment. An *admissible environment* is the concept used by Rosen (1958a) for biological systems.The environment must be a source of admissible input and a sink for admissible output. Prigogine (1978), when introducing the rationale of dissipative systems, uses the same concept. There is no realized organized structure of dissipative systems that can perform a given function (or keep its own individuality) without favorable boundary conditions (without operating within its required associative context).

As noted by Allen and Hoekstra (1992), the definition of a type per se does not carry a scale tag. A given ratio between the relative size of the head, the body and the legs of a given shape of organism can be realized at different scales (this is the basis of modeling). It is only when a particular typology is realized that the issue of scale enters into play. At that point, scale matters in relation to (1) the definition of the identity of lower-level elements (level $n - 1$) responsible for the structural stability of the system (at the focal level n) — what the realization is made of and (2) the definition of the identity of the context (level $n + 1$) in which the system has to be able to express its function — what the realization is interacting with.

The special status of types that are out of scale means that they are also out of time. Models are made with types, and therefore the validity of models requires the validity and usefulness of the relation between type, associative context and goal of the analysts. As discussed later, this is why a quality control on the validity of a given model always has to be based first on a semantic check about its usefulness at a given point in space and time. Models have to make sense; they must convey meaning and make possible the organization of perceptions of a group of observers about what is known about a given problem.

The discussion of the dual nature of holons is reminiscent of the principles of quantum mechanics articulated in the 1920s, those of indeterminacy and complementarity. Complementarity refers to the fact that holons, due to their peculiar functioning on parallel scales, always require a dual description. The relational functional nature of the holon (focal-higher-level interface) provides the context for the structural part of the holon (focal-lower-level interface), which generates the behavior of interest on the focal level. Therefore, a holarchy can be seen as a chain of contexts and relevant behaviors in cascade. The niche occupied by the dog is the context for the actions of individual organisms, but at the same time, any particular organism is the context for the activity of its lower-level components (organs and cells dealing within organisms with viruses and enzymes).

Established scientific disciplines rarely acknowledge that the unavoidable and prior choice of perspective determines what should be considered the relevant action, and its context, which is indicated by the adoption of a single model (no matter how complicated), implies a bias in the consequent description of complex systems behavior (Giampietro, 1994b). For example, analyzing complex systems in terms of organized structures or incumbents (e.g., a given doctor in a hospital) implicitly requires assuming for the validity of the model (1) a given set of initiating conditions (a history of the system that affects its present behavior) and (2) a stable higher level on which functions or roles are defined for these structures to make them meaningful, useful and thus stable in time (Simon, 1962). That is, the very use of the category doctors implies, at the societal level, the existence of a job position for a doctor in that hospital, together with enough funding to run the hospital.

Similarly, to have functions at a certain level, one needs to assume the stability at the lower levels where the structural support is provided for the function. That is, the use of the category hospital implies that something (or someone) must be there to perform the required function (Simon, 1962). In our example the existence of a modern hospital (at the societal level) implies the existence of a supply of trained doctors (potential incumbents) able to fill the required roles (an educational system working properly). All these considerations become quite practical when systems run imperfectly, as when doctors are in short supply, have bogus qualifications, are inadequately supported, etc.

Hence, no description of the dynamics of a focus level, such as society as a whole, can escape the issues of structural constraints (*what/how*, explanations of structure and operation going on at lower levels) and functional constraints (*why/how*, explanations of finalized functions and purposes, going on at or in relation to the higher level). The key for dealing with holarchic systems is to deal with the difference in the space–time domain that has to be adopted for getting the right pattern recognition. Questions related to the why/how questions (to study the niche occupied by the *Canis familiaris* species or the characteristics of the U.S. presidency) are different from those required for the what/how questions (to study the particular conditions of our neighbor's dog related to her age and past, or the personal conditions of President Bush this week). They cannot be discussed and analyzed by adopting the same descriptive domains. Again, even if the two natures of the holon act as a whole, when attempting to represent and explain both the why/how and what/how questions, we must rely on complementary nonequivalent descriptions, using a set of nonreducible and noncomparable representations.

2.4 Conclusion: The Ambiguous Identity of Holarchies

Holons and holarchies require the use of several nonequivalent identities to be described, even though they can be seen and perceived as a single individuality. The couplets of types (or roles) and individual realizations (or incumbents) overlap in natural systems when coming to specific actions (e.g.. President Bush and the president of the U.S.). However, the two parts of the holon have different histories, different mechanisms of control and diverging local goals. For example, the case of Monica Lewinsky, which led to impeachment proceedings for President Clinton, was about legitimate contrasting interests expressed by the dual nature of that specific holon: the wants of President Clinton as a human being in a particular moment of his life diverged from the goals of the institutional role associated with the U.S. presidency. Unfortunately, scientific analyses trying to model holons operating within holarchies have no other option but to consider a single formal identity for each acting holon at the time. At this point models referring to just one of the two relevant identities associated with the label can only be developed within the particular descriptive domain associated with the selected identity (referring to either the role or the incumbent).

The existence of a multiplicity of roles for the same natural system operating within holarchies shows the inadequacy of the traditional reductionist scientific paradigm for modeling them. For the assumption of a single goal and identity for the acting holon (which is necessary for mapping its behavior within a given inferential system) restricts it to a particular model (descriptive domain), to the exclusion of all others.

2.4.1 Models of Adaptive Holons and Holarchies, No Matter How Validated in the Past, Will Become Obsolete and Wrong

To get a quantitative characterization of a particular identity of a holon, one has to assume the holarchy is in steady state (or at least in quasi-steady state). That is, one has to choose a space–time window at which it is possible to define a clear identity for the system of interest (the triadic reading is often expressed using the more familiar term of *ceteris paribus assumption*). However, as soon as one obtains the possibility of quantifying characteristics of the system after "freezing it" on a given space–time window, one loses, as a consequence of this choice, any ability to see and detect existing evolutionary trends (recall here the Jevons' paradox discussed in Chapter 1). Evolutionary trajectories are detectable only by using a much larger space–time scale than that of the dynamic of interest (Salthe, 1993). This involves admitting that sooner or later the usefulness of the current descriptive domain and the validity of the selected modeling relation will expire. By choosing an appropriate window of observation, we can isolate and describe, in simplified terms, a domain of the reality — the behavior of a system within a descriptive domain — the one we are interested in. In this way, it is possible to define boundaries for a specified system that can then be considered an independent entity from the rest of the holarchy to which it belongs. The side effect of this obliged procedure, however, is the neglect, either consciously or unconsciously, of (1) dynamics and other relevant features that are occurring outside the space–time differential selected in the focal descriptive domain and (2) changes in other systems' qualities that were not included in the original set of observable qualities and encoding variables used in the model.

When dealing with becoming systems, we should expect that it will be necessary to continuously update the identities used in evolving descriptive domains (we should expect that useful definitions of the state space will change in time). Georgescu-Roegen (1971) used to say that modeling means a "heroic simplification of reality." Each model reflects a given application of a triadic filtering to the reality based on a previous definition of a time duration for the system and the dynamic of interest included in the analytical representation. When we do that, we are choosing just one of the possible nonequivalent descriptive domains for perceiving and representing our system. This explains why there can be no complete, neutral, objective study of a holarchic system, and why these systems are complex in the sense of having multiple legitimate meaningful relations between perception and representation.

This is particularly important when dealing with the issue of sustainability. The existence of wide differences in timescales in problems of sustainability is well known. For example:

1. The process of biological evolution (e.g., the becoming of ecological holons) requires the use of relevant time differentials of thousands of years.
2. The process of evolution of institutional settings of human societies requires the use of relevant time differentials of centuries.
3. The process of evolution of human technology requires the use of relevant time differentials of decades.
4. When dealing with price formation, we are dealing with a time differential of 1 year or less.
5. Preferences and feelings of individuals can change in a second.

Obviously the epistemic categories, formal identities and relative models required for representing changes over these different time windows cannot be mixed.

To make things more complicated, complex adaptive systems tend to pulse and operate in cyclic attractors. This indicates an additional problem. Scientific analyses should be able to avoid confusing movements of the system over predictable trajectories in a given state space (e.g., the trajectory of a perfect pendulum) with changes due to the genuine emergence of new evolutionary patterns. As will be discussed in Chapter 8, we can detect genuine emergence by the fact that we have to update identity of the state space used in the analysis. Emergence implies the use of new epistemic categories and new modeling relations in the observer–observed complex.

References

Ahl, V. and Allen T.F.H., (1996), *Hierarchy Theory*, Columbia University Press, New York.

Allen, T. and Starr, T., (1982), *Hierarchy: Perspectives for Ecological Complexity*, University of Chicago Press, Chicago.

Allen, T.F.H. and Hoekstra, T.W., (1992), *Toward a Unified Ecology*, Columbia University Press, New York.

Bailey, K.D., (1990), *Social Entropy Theory*, State University of New York Press, Albany.

Barthes, R., (1964), *Elements de semiologie*, Editions du Seuil, Paris.

Flood, R.L. and Carson, E.R., (1988), *Dealing with Complexity: An Introduction to the Theory and Application of Systems Science*, Plenum Press, New York.

Georgescu-Roegen, N., (1971), *The Entropy Law and the Economic Process*, Harvard University Press, Cambridge, MA.

Giampietro, M. 1994a. Using hierarchy theory to explore the concept of sustainable development. Futures 26 (6): 616–625.

Giampietro, M. 1994b. Sustainability and technological development in agriculture: A critical appraisal of genetic engineering. BioScience 44 (10): 677–689.

Giampietro, M., Bukkens, S.G.F., and Pimentel, D., (1997), Linking technology, natural resources, and the socioeconomic structure of human society: examples and applications, in *Advances in Human Ecology*, Vol. 6, Freese, L., Ed., JAI Press, Greenwich, CT, pp. 131–200.

Giampietro, M. and Pastore, G., (2001), Operationalizing the concept of sustainability in agriculture: characterizing agroecosystems on a multi-criteria, multiple-scale performance space, in *Agroecosystem Sustainability: Developing Practical Strategies*, Gliessman, S.R., Ed., CRC Press, Boca Raton, FL, pp. 177–202.

Gibson, C., Ostrom E., and Toh-Kyeong, A., (1998), *Scaling Issues in the Social Sciences*, working paper published on the Internet by the International Human Dimensions Programme on Global Environmental Change (IHDP), available at www.uni-bonn.de/ihdp.

Glansdorff, P. and Prigogine, I., (1971), *Thermodynamics Theory of Structure, Stability and Fluctuations*, John Wiley & Sons, New York.

Gomiero, T. 2003 Multi-Objective Integrated Representation (Moir) As a Tool To Study and Monitor Farming System Development and Management. Ph.D. Thesis to be submitted in Environmental Science, Universitat Autonoma de Barcelona, Spain.

Grene, M., (1969), Hierarchy: one word, how many concepts? in *Hierarchical Structures*, Whyte, L.L., Wilson, A.G., and Wilson, D., Eds., American Elsevier Publishing Company, New York, pp. 56–58.

Holling, C.S., (1995), Biodiversity in the functioning of ecosystems: an ecological synthesis, in *Biodiversity Loss: Economic and Ecological Issues*, Perring, C., Maler, K.G., Folke, C., Holling, C.S., and Jansson, B.O., Eds., Cambridge University Press, Cambridge, U.K., pp. 44–83.

Home, D. and Robinson, A., (1995), Einstein and Tagore: man, nature and mysticism, *J. Conscious.*, 2, 167–179.

Hospers, J., (1997), *Introduction to Philosophical Analysis*, 4th ed., Taylor & Francis, London, 288 pp.

Koestler, A., (1968), *The Ghost in the Machine*, MacMillan Co., New York, 365 pp.

Koestler, A., (1969), Beyond atomism and holism: the concept of the Holon, in *Beyond Reductionism*, Koestler, A. and Smythies, J.R., Eds., Hutchinson, London, pp. 192–232.

Koestler, A., (1978), *Janus: A Summing Up*, Hutchinson, London.

Mandelbrot, B.B., (1967), How long is the coast of Britain? Statistical self-similarity and fractal dimensions, *Science*, 155, 636–638.

Maturana, H.R. and Varela, F.J., (1980), *Autopoiesis and Cognition: The Realization of the Living*, D. Reidel Publishing, Dordrecht.

Maturana, H.R. and Varela, F.J., (1998), *The Tree of Knowledge: The Biological Roots of Human Understanding*, Shambhala Publications, Boston.

Mayr, E., (1982), *The Growth of Biological Thought: Diversity, Evolution and Inheritance*, Belknap Press, Cambridge, MA.

Nicolis, G. and Prigogine, I., (1977), *Self-Organization in Nonequilibrium Systems*, Wiley-Interscience, New York.

Odum, H.T., (1971), *Environment, Power, and Society*, Wiley-Interscience, New York.

Odum, H.T., (1983), *Systems Ecology*, John Wiley, New York.

O'Neill, R.V., (1989), Perspectives in hierarchy and scale, in *Perspectives in Ecological Theory*, Roughgarden, J., May, R.M., and Levin, S., Eds., Princeton University Press, Princeton, NJ, pp. 140–156.

O'Neill, R.V., DeAngelis, D.L., Waide, J.B., and Allen, T.F.H., (1986), *A Hierarchical Concept of Ecosystems*, Princeton University Press, Princeton, NJ.

Pattee, H.H., Ed., (1973), *Hierarchy Theory*, George Braziller, New York.

Polanyi, M., (1958), *Personal Knowledge*, Routledge & Kegan Paul, London, U.K.

Polanyi, M., (1977), *Meaning*, University of Chicago Press, Chicago.

Popper, K.R., (1993), *Knowledge and the Mind-Body Problem*, M.A. Notturno, Ed., Routledge, London, U.K.

Prigogine, I., (1978), *From Being to Becoming*, W.H. Freeman and Co., San Francisco.

Prigogine, I. and Stengers, I., (1981), *Order out of Chaos*, Bantam Books, New York.

Rosen, R., (1958a), A relational theory of biological systems, *Bull. Math. Biophys.*, 20, 245–260.

Rosen, R., (1958b), The representation of biological systems from the standpoint of the theory of categories, *Bull. Math. Biophys.*, 20, 317–341.

Rosen, R., (1977), Complexity as a system property, *Int. J. Gen. Syst.*, 3, 227–232.

Rosen, R., (1985), *Anticipatory Systems: Philosophical, Mathematical and Methodological Foundations*, Pergamon Press, New York.

Rosen, R., (1991), *Life Itself: A Comprehensive Inquiry into the Nature, Origin and Fabrication of Life*, Columbia University Press, New York, 285 pp.

Rosen, R., (2000), *Essays on Life Itself*, Columbia University Press, New York, 361 pp.

Salthe, S.N., (1985), *Evolving Hierarchical Systems: Their Structure and Representation*, Columbia University Press, New York.

Salthe, S., (1993), *Development and Evolution: Complexity and Change in Biology*, MIT Press, Cambridge, MA.

Simon, H.A., (1962), The architecture of complexity, *Proc. Am. Philos. Soc.*, 106, 467–482.

Whyte, L.L., Wilson, A.G., and Wilson, D., Eds., (1969), *Hierarchical Structures*, American Elsevier Publishing Company, New York.

3

Complex Systems Thinking: New Concepts and Narratives

This chapter provides practical examples that illustrate the relevance of the concepts introduced in Chapter 2 to the challenges faced by scientists working in the field of sustainable agriculture. In fact, it is important to have a feeling of practical implications of complexity, in terms of operation of scientific protocols of analysis, before getting into an analysis of the challenges faced by those willing to do things in a different way (Chapters 4 and 5), and before exploring innovative concepts that can be used to develop new analytical approaches (Part 2).

3.1 Nonequivalent Descriptive Domains and Nonreducible Models Are Entailed by the Unavoidable Existence of Multiple Identities

3.1.1 Defining a Descriptive Domain

Using the rationale proposed by Kampis (1991, p. 70), we can define a particular representation of a system as "the domain of reality delimited by interactions of interest." In this way, one can introduce the concept of descriptive domains in relation to the particular choices associated with a formal identity used to perceive and represent a system organized on nested hierarchical levels. A descriptive domain is the representation of a domain of reality that has been individuated on the basis of a preanalytical decision on how to describe the identity of the investigated system in relation to the goals of the analysis. Such a preliminary and arbitrary choice is needed to be able to detect patterns (when looking at the reality) and to model the behavior of interest (when representing it).

To discuss the need of using in parallel nonequivalent descriptive domains, we can again use the four views given in Figure 1.2, this time applying to them the metaphor of sustainability. Imagine that the four nonequivalent descriptions presented in Figure 1.2 were referring to a country (e.g., the Netherlands) rather than a person. In this case, we can easily see how any analysis of its sustainability requires an integrated use of these different descriptive domains. For example, by looking at socioeconomic indicators of development (Figure 1.2b), we "see" this country as a beautiful woman (i.e., good levels of gross national product (GNP), good indicators of equity and social progress). These are good system qualities, required to keep low the stress on social processes. However, if we look at the same system (same boundary) using different encoding variables (e.g., a different formal identity based on a selection of biophysical variables) — Figure 1.2d in the metaphor — we can see the existence of a few problems not detected by the previous selection of variables (i.e., sinusitis and a few dental troubles). In the metaphor this picture can be interpreted, for the Netherlands, as an assessment of accumulation of excess nitrogen in the water table, growing pollution in the environment, excessive dependency on fossil energy and dependence on imported resources for the agricultural sector. Put another way, when considering the biophysical dimension of sustainability, we can "see" some bad system qualities that were ignored by the previous selection of economic encoding variables (a different definition of formal identity for perception and representation). Analyses based on the descriptive domain of Figure 1.2a are related to lower-level components of the system. In the Dutch metaphor, this could be an analysis of technical coefficients (e.g., input/output) of individual economic activities (e.g., the CO_2 emissions for producing electricity in a power plant). Clearly, the knowledge obtained when adopting this descriptive domain is crucial to determine the viability and sustainability of the whole system (the possibility of improving or adjusting the overall performance of the Dutch economic process if and when changes are required). In the same way, an analysis of the relations of the system with its larger context can indicate the need to consider a descriptive domain based on pattern recognition referring to a larger space–time domain (Figure 1.2c). In the Dutch metaphor this could be an analysis of institutional settings, historical entailments or cultural constraints over possible evolutionary trajectories.

3.1.2 Nonequivalent Descriptive Domains Imply Nonreducible Assessments

The following example refers to four legitimate nonreducible assessments and can again be related to the four views presented in Figure 1.2. This is to show how general and useful is the pattern of multiple identities across levels. The metaphor this time is applied to the process required to obtain a specific assessment, such as kilograms of cereal consumed per capita by U.S. citizen in 1997. The application of such a metaphor to the assessment of cereal consumption per capita is shown in Figure 3.1. Let us imagine that to get such a number a very expensive and sophisticated survey is performed at the household level. By recording events in this way, we can learn that, in 1997, each U.S. citizen consumed 116 kg of cereal/person/year.

On the other hand, by looking at the Food and Agriculture Organization (FAO) Food Balance Sheet (FAO Agricultural Statistics), which provides for each FAO member country a picture of the flow of food consumed in the food system, we can derive other possible assessments for the kilograms of cereal consumed per capita by U.S. citizens in 1997.

A list of nonequivalent assessments could include:

1. **Cereal consumed as food, at the household level.** This is the figure of 116 kg/year/capita for U.S. citizens in 1997, discussed above. This assessment can also be obtained by dividing the total amount of cereal directly consumed as food by the population of U.S. in that year.

2. **Consumption of cereal per capita in 1997 as food, at the food system level.** This value is obtained by dividing the total consumption of cereal in the U.S. food system by the size of the U.S. population. This assessment results in more than 1015 kg (116 kg directly consumed, 615 kg fed to animals, almost 100 kg of barley for making beer, plus other items related to industrial processing and postharvest losses).

3. **Amount of cereal produced in U.S. per capita in 1997, at the national level, to obtain an economic viability of the agricultural sector.** This amount is obtained by dividing the total internal production of cereal by population size. Such a calculation provides yet another assessment: 1330 kg/year/capita. This is the amount of cereal used per capita by the U.S. economy.

4. **Total amount of cereal produced in the world per capita in 1997, applied to the humans living within the geographic border of the U.S. in that year.** This amount is obtained by dividing the total internal consumption of cereal at the world level in 1997 (which was 2×10^{12} kg) by the world population size that year (5.8 billion). Clearly, such a calculation provides yet another assessment: 345 kg/year/capita (160 kg/year direct, 185 kg/year indirect). This is the amount of cereal used per capita by each human being in 1997 on this planet. Therefore, this would represent the share assigned to U.S. people when ignoring the heterogeneity of pattern of consumption among countries.

The four views in Figure 1.2 can be used again, as done in Figure 3.1, to discuss the mechanism generating these numerical differences. In the first two cases, we are considering only the direct consumption of cereal as food. On a small scale (assessment 1 reflecting Figure 1.2a in the metaphor) and on a larger scale (assessment 2 referring to Figure 1.2b in the metaphor), the logic of these two mappings is the same. We are mapping flows of matter, with a clear identification in relation to their roles: food as a carrier of energy and nutrients, which is used to guarantee the physiological metabolism of U.S. citizens. This very definition of consumption of kilograms of cereal implies a clear definition of compatibility with the physiological processes of conversion of food into metabolic energy (both within fed animals and human bodies). This implies that, since the mechanism of mapping is the same

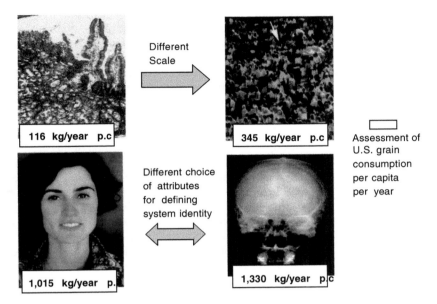

FIGURE 3.1 Nonequivalent descriptive domains equals nonreducible models.

(in the metaphor of Figure 1.2a and b, we are looking for pattern recognition using the same visible wavelength of the light), we can bridge the two assessments by an appropriate process of scaling (e.g., life cycle assessment). This will require, in any case, different sources of information related to processes occurring at different scales (e.g., household survey plus statistical data on consumption and technical coefficients in the food systems). When considering assessment 3, we are including kilograms of cereal that are not consumed either directly or indirectly by U.S. households in relation to their diet. The additional 315 kg of cereal produced by U.S. agriculture per U.S. citizen for export (assessment 3 minus assessment 2) is brought into existence only for economic reasons. But exactly because of that, they should be considered as "used" by the agricultural sector and the farmers of the U.S. to stabilize the country's economic viability. The U.S. food system would not have worked the way it did in 1997 without the extra income provided to farmers by export. Put another way, U.S. households indirectly used this export (took advantage of the production of these kilograms of cereal) for getting the food supply they got, in the way they did. This could metaphorically be compared to the pattern presented in Figure 1.2d. We are looking at the same head (the U.S. food system in the analogy) but using a different mechanism of pattern recognition (x-rays rather than visible light). The difference in numerical value between assessments 1 and 2 is generated by a difference in the hierarchical level of analysis, whereas the difference between assessments 2 and 3 is generated by a bifurcation in the definition of indirect consumption of cereal per capita (a biophysical definition vs. an economic definition). Finally, Figure 1.2c would represent the numerical assessment obtained in assessment 4, when both the scale and the logic adopted for defining the system are different from the previous one (U.S. citizens as members of humankind).

The fact that these differences are not reducible to each other does not imply that any of the assessments are useless. Also, in this case, depending on the goal of the analysis, each one of these numerical assessments can carry useful information.

3.2 The Unavoidable Insurgence of Errors in a Modeling Relation

3.2.1 Bifurcation in a Modeling Relation and Emergence

To introduce this issue, consider one of the most successful stories of hard science in this century: the claimed full understanding achieved in molecular biology of the mechanism through which genetic information is stored, replicated and used to guarantee a predictable behavior in living systems. This example is relevant not only for supporting the statement made in the title of this section, but also for pointing to the potential risks that a modeling success can induce on our ability to understand complex behaviors of real systems.

To cut a long and successful story very short, we can say that, in terms of modeling, the major discoveries made in this field were (1) the identification of carriers of information as DNA bases organized into a double helix and (2) the individuation and understanding of the mechanisms of encoding based on the use of these DNA bases to (i) store and replicate information in the double helix, and (ii) transfer this information to the rest of the cell. This transfer of information is obtained through an encoding and decoding process that leads to the making of proteins. Due to the modulation of this making of proteins in time, the whole system is able to guide the cascade of biochemical reactions and physiological processes. In particular, four basic DNA bases were identified (their exotic names are not relevant here, so we will use only the first letter of their names: C, G, T and A), which were found to be the only components used to encode information within the DNA double helix.

Two pairs of these bases map onto each other across helixes. That is, whenever there is a C on one of the two helices, there is a G on the other (and vice versa); the same occurs with A and T. This means that if we find a sequence CCAATGCG on one of the two helices of the DNA, we can expect to find the complementing sequence GGTTACGC on the other. This self-entailment (loop of resonating mappings in time) across linked sequences of bases is the mechanism that explains the preservation of a given identity of DNA in spite of the large number of replications and reading processes. By applying a system of syntactic rules to this mechanism of reciprocal mapping, it is also possible to explain, in general terms, the process of regulation of the biochemical behavior of cells (some parts of DNA strings made up of these four bases have regulative functions, whereas other parts codify the actual making of proteins).

At this point this process of handling information from what is written in the DNA to what is done by the cells can be represented in a simplified form (modeled) by using types. That is, there is a closed set of types (triplets of bases) that are mapping onto a closed set of types (the amino acids used to make proteins). Obviously, there are a lot of additional specifications required, but details are not relevant here. What is relevant is the magnificent success of this modeling relation. The model was so good in explaining the behavior of interest that nowadays humans can not only manipulate genetic information within living systems to interfere with their original systems of storage of information and regulation, but also make machines that can generate sequences of DNA following an input given by the computer.

The big success of this model is also reason for concern. In fact, according to this modeling relation, we are told at school — when learning about DNA behavior — that a mutation represents an error in the mechanism handling genetic information. The expression "error" refers to the fact that a given type on one side of the mapping (e.g., a given base A) is not generating the expected type on the other side of the mapping (e.g., the complementing base T). This can imply that a given triplet can be changed and therefore generate an incorrect insertion of an amino acid in the sequence making up a protein.

According to Rosen (1977, p. 231) what the expression error means:

> is something like the following: DNA is capable of many interactions besides those involved in its coding functions. Some of these interactions can affect the coding functions. When such an interaction occurs, there will be a deviation between what our simple model tells us ought to be coded, and what actually is coded. This deviation we call a mutation, and we say that the DNA has behaved erroneously.

Even with a cursory reflection we can immediately see that any system handling genetic information within a becoming system must have, to keep its ability to evolve, an open information space that has to be used to expand the set of possible behaviors in time (to be able to become something else). Such a system therefore must admit the possibility of inducing some changes in the closed set of syntactic entailments among types that represents its closed information space. The closed information space is represented by what has been expressed up to now by the class of individual organized structures that have been produced in the past history of the biological system with which the studied DNA is associated. To evolve, biological systems need mutations to expand this closed set; therefore, they must be able to have mutations. The existence and characteristics of this function (the ability to have mutations), however, can only be detected over a space–time window much larger than the one used to describe mechanical events in molecular biology. Being a crucial function, the activity of inducing changes on the DNA to expand the information space requires careful regulation. That is, the rate of mutations must not be too high (to avoid the collapse of the regulative mechanisms on the smaller space–time window of operations within cells). On the other hand, it has to be large enough to be useful on an evolutionary space–time window to generate new alternatives when the existing structures and functions become obsolete. The admissible range of this rate of mutation obviously depends on the type of biological system considered; for example, within biological systems high in the evolutionary rank, it is sexual reproduction of organisms that takes care of doing a substantial part of this job with fewer risks.

In any case, the point relevant in this discussion is that mutations are not just errors, but rather the expression of a useful function needed by the system. The only problem is that such a function was not included in the original model used to represent the behavior of elements within a cell type, adopted by molecular biology in the 1960s and early 1970s. These models were based on a preliminary definition of a closed set of functions linked to the class of organized structures (DNA bases, triplets, amino acids) considered over a given descriptive domain. Within the descriptive domain of molecular biology (useful to describe the mechanics of the encoding of amino acids onto triplets in the DNA), the functions related to evolution or co-evolution of biological systems cannot be seen. This is what justifies the use of the term *error* within that term of reference.

The existence of machines able to generate sequences of DNA is very useful, in this case, to focus on the crucial difference between biological systems and human artifacts (Rosen, 2000). When a machine making sequences of DNA bases includes in the sequence a base different from that written on the string used as input to the computer, then we can say that the machine is making an error. In fact, being a human artifact, the machine is not supposed to self-organize or become. A mechanical system is not supposed to expand its own information space. Machines have to behave according to the instructions written before they are made for (and used by) someone else. On the space–time scale of its life expectancy, the organized structure of a machine has no other role but that of fulfilling the set of functions assigned by the humans who made it. Living systems are different.

Going back to the example of DNA, the more humans study the mechanism storing and processing genetic information, the more it becomes clear in both molecular and theoretical biology that the handling of information in DNA-based systems is much more complex than the simple cascade of a couple of mappings: (C↔G, T↔A) and (closed set of triplets types → closed set of amino acid types).

The lesson from this story is clear: whenever a model is very useful, those who use it tend to sooner or later confuse the type of success (the representation of a relevant mechanism made using types, that is, by adopting a set of formal identities) with the real natural system (whose potential semantic perceptions are associated with an open and expanding information space) that was replaced in the model by the types.

Due to this unavoidable generation of errors, every time we make models of complex systems, Rosen (1985, chapter on theory of errors) suggests using the term *bifurcations* whenever we face the existence of two different representations of the same natural system that are logically independent of each other.

The concept of bifurcation in a modeling relation entails the possibility of having two (or more) distinct formal systems of inference, which are used on the basis of different selections of encoding variables (selection of formal identities) or focal level of analysis (selection of scale) to establish different modeling relations for the same natural system. As noted earlier, bifurcations are therefore also entailed by the existence of different goals for the mapping (by the diverging interests of the observer), and not only by intrinsic characteristics of the observed system.

The concept of bifurcation implies the possibility of a total loss of usefulness of a given mapping. For example, imagine that we have to select an encoding variable to compare the size of London with that of Reykjavik, Iceland. London would be larger than Reykjavik if the selected encoding for the quality size is the variable population. However, by changing the choice of encoding variable, London would be smaller than Reykjavik if the perception of its size is encoded as the number of letters making up the name (a new definition of the relevant quality to be considered when defining the sizes of London and Reykjavik). Such a choice of encoding could be performed by a company that makes road signs.

In this trivial example we can use the definition of identity discussed in Chapter 2 to study the mechanism generating the bifurcation, in this case, two nonequivalent observers: (1) someone characterizing London as a proxy for a city will adopt a formal identity in which the label is an epistemic category associated with population size and (2) someone working in a company making road signs, perceiving this label as just a string of letters to be written on its product, will adopt a formal identity for the name in which the size is associated with the demand for space on a road sign. The proxy for the latter system quality will be the number of letters making up the name. Clearly, the existence of a different logic in selecting the category and proxy used to encode what is relevant in the quality size is related to a different meaning given to the perception of the label "London." This is the mechanism generating the parallel use of two nonequivalent identities for the same label. Recall here the example of the multiple bifurcations about the meaning of the label "segment of *coastal line*" in Figure 2.1.

This bifurcation in the meaning assigned to the label is then reflected in numerical assessments that are no longer necessarily supposed to be reducible into each other or directly comparable by the application of an algorithm. A bifurcation in the system of mapping can be seen as, as stated by Rosen (1985, p. 302), "the appearance of a logical independence between two descriptions." As discussed in Chapter 2, such a bifurcation depends on the intrinsic initial ambiguity in the definition of a natural system by using symbols or codes: the meaning given to the label "London" (a name of a city made up of people or a six-letter word). As observed by Schumpeter (1954, p. 42), "Analytical work begins with material provided by our vision of things, and this vision is ideological almost by definition."

3.3   The Necessary Semantic Check Always Required by Mathematics

Obviously, bifurcations in systems of mappings (reflecting differences in logic) will entail bifurcations also in the use of mathematical systems of inference. For example, a statistical office of a city recording the effect of the marriage of two singles already living in that city and expecting a child would map the consequent changes implied by these events in different ways, according to the encoding used to assess changes in the quality population.

The event can be described as $1 + 1 \rightarrow 1$ (both before and after the birth of the child) if the mapping of population is done using the variable number of households. Alternatively, it can be described as $1 + 1 \rightarrow 3$ (after the birth of the child) if the mapping is done in terms of number of people living in the city. In this simple example, it is the definition of the mechanism of encoding (implied by the choice of the identity of the system to be described, i.e., households vs. people) that entails different mathematical descriptions of the same phenomenon.

The debate about the possibility of replacing external referents (semantic) with internal rules (syntax) is a very old one in mathematics. The Czech-born mathematician Kurt Gödel demonstrated that, in mathematics, it is impossible to define a complete set of propositions that can proven true or false on the basis of a preexisting internal set of rules and axioms. Depending on the meaning attributed to the statements about numbers within a given mathematical system, one has to go outside that system looking for external referents. This is the only way to individuate the appropriate set of rules and axioms. However, after such an enlargment of the system, we will face a new set of unprovable statements (that would require an other enlargement in terms of additional external referents). This is a process that leads to an infinite regress.

Any formalization always requires a semantic check, even when dealing with familiar objects such as numbers:

> The formalist program was wrecked by the Gödel Incompleteness Theorem which showed that Number Theory is already nonformalizable in this sense. In fact, Gödel (1931) showed that any attempt to formalize Number Theory, to replace its semantic by syntax, must lose almost every truth of Number Theory. (Rosen, 2000, p. 267)

3.3.1   The True Age of Dinosaurs and the Weak Sustainability Indicator

To elaborate on the need of a continuous semantic check when using mathematics, it can be useful to recall the joke proposed by Funtowicz and Ravetz (1990) exactly for this purpose. The subject of the

SKELETON of FUNTRAVESAURUS
AGE: 250,000,00⃞8 years

FIGURE 3.2 The author's drawing of the "true" age of the dinosaur. (From Funtowicz and Ravetz, 1990.)

joke is illustrated in Figure 3.2. The skeleton of a dinosaur (an exemplar of Funtravesaurus) is in a museum with a sign saying "age 250,000,000 years" on the original label. However, the janitor of the museum corrected the age to reflect 250,000,008. When asked about the correction, he gave the following explanation: "When I got this job 8 years ago, the age of the dinosaur, written on the sign, was 250,000,000 years. I am just keeping it accurate."

The majority of the people listening to this story do interpret it as a joke. To explain the mechanism generating such a perception, we can use the same explanation about jokes discussed before about the leg named Joe. When dealing with the age of a dinosaur, nobody is used to associating such a concept with a numerical measure that includes individual years. However, as noted by Funtowicz and Ravetz (1990), commenting on this joke, when considering the formal arithmetic relation $a + b = c$, there are no written rules in mathematics that prevent the summing of a expressed in hundreds of millions to b expressed in units. Still, common sense (semantic check) tells us that such an unwritten rule should be applied. The explanation given by the janitor simply does not make sense to anybody familiar with measurements.

In this case, it is the incompatibility in the two processes of measurement (that generating a and that generating b) that makes their summing impossible. A more detailed discussion of this point is provided at the end of this chapter (Section 3.7). What is bizarre, however, is that very few scientists operating in the field of ecological economics seem to object to a similar summation proposed by economists when dealing with the issue of sustainability. For example, the weak sustainability indicator proposed by Pearce and Atkinson (1993) is supposed to indicate whether an economy is sustainable. According to this formal representation of changes in an economy, weak sustainability implies that an economy saves more than the combined depreciation on human-made and natural capital. The formal representation of this rule is given in Equation 3.1:

$$S \geq dHMC/dt + dNC/dt \qquad (3.1)$$

where S is savings, HMC is human-made capital and NC is natural capital.

There are very good reasons to criticize this indicator (for a nice overview of such a criticism, see Cabeza-Gutés (1996)), mainly related to the doubtful validity of the assumptions that it implies, that is, a full substitutability of the two different forms of capital mapped by the two terms on the right (e.g., technology cannot replace biodiversity). But this is not the argument relevant here. The epistemological capital sin of this equation is related to its attempt to collapse into a single encoding variable (a monetary variable) two nonequivalent assessments of changes referring to system qualities that can only be recognized using different scales, and therefore can only be defined using nonequivalent descriptive domains. An assessment referring to dNC/dt uses a formal identity expressing changes using 1996-U.S.$ as the relevant variable. Such an assessment can only be obtained by using a measurement scheme operating with a time differential of less than 1 year (assuming the validity of the *ceteris paribus* assumption for no more than 10 years). On the contrary, changes in natural capital refer to qualities of ecosystems or biogeochemical cycles that have a time differential of centuries. The semantic identity of natural capital implies qualities and epistemic categories that, no matter how creative the analyst is, cannot be expressed in 1996-U.S.$ (a measurement scheme operating on a dt of years). In the same way, changes measured in 1996-U.S.$ cannot be represented when using variables able to catch changes in qualities with a time differential of centuries. Each of the two terms dHMC/dt and dNC/dt cannot be detected when using the descriptive domain useful to define the other.

In conclusion, the sum indicated in relation 1, first, does not carry any metaphorical meaning since the two forms of capital are not substitutable and, second, in any case could not be used to generate a normative tool, since it would be impossible to put meaningful numbers into that equation.

3.4   Bifurcations and Emergence

The concept of bifurcation also has a positive connotation. It indicates the possibility of increasing the repertoire of models and metaphors available to our knowledge. In fact, a direct link can be established between the concepts of bifurcation and emergence. Using again the wording of Koestler (1968) we have a *discovery* — Rosen (1985) suggests using for this concept the term *emergence* — when two previously unrelated frames of reference are linked together. Using the concept of equivalence classes for both organized structures and relational functions, we can say that emergence or discovery is obtained (1) when assigning a new class of relational functions (which indicates a better performance of the holon on the focal–higher level interface) to an old class of organized structures, or (2) when using a new class of organized structures (which indicates a better performance of the holon on the focal–lower level interface) to an existing class of relational functions. We can recall again the example of the joke, in which a new possibility of associating words is introduced, opening new horizons to the possibility of assigning meaning to a given situation.

An emergence can be easily detected by the fact that it requires changing the identity of the state space used to describe the new holon. A simple and well-known example of emergence in dissipative systems is the formation of Bénard cells (a special pattern appearing in a heated fluid when switching from a linear molecular movement to a turbulent regime). For a detailed analysis of this phenomenon from this perspective, see Schneider and Kay (1994). The emergence (the formation of a vortex) requires the use in parallel of two nonequivalent descriptive domains to properly represent such a phenomenon. In fact, the process of self-organization of a vortex is generating in parallel both an individual organized structure and the establishment of a type. We can use models of fluid dynamics to study, simulate and even predict this transition. But no matter how sophisticated these models are, they can only guess the insurgence of a type (under which conditions you will get the vortex). From a description based on the molecular level, it is not possible to guess the direction of rotation that will be taken by a particular vortex (clockwise or counterclockwise). However, when observed at a larger scale, any particular Bénard cell, because of its personal history, will have a specific identity that will be kept as long as it remains alive (so to speak). This symmetry breaking associated with the special story of this individual vortex will require an additional source of information (external referent) to determine whether the vortex is rotating clockwise or counterclockwise. Thus, we have to adopt a new scale for perceiving and representing the operation of a vortex (above the molecular one) to detect the direction of rotation. This implies also the use of a new epistemological category (i.e., clockwise or counterclockwise) not originally included in the equations. To properly represent such a phenomenon, we have to use a descriptive domain that is not equivalent to that used to study lower-level mechanisms. Put another way, the information required to describe the transition on two levels (characterizing both the individual and the type) cannot all be retrieved by describing events at the lower level. More about this point is discussed in Section 3.7 on the root of incommensurability between squares and circles.

Another simple example can be used to illustrate the potential pitfalls associated with the generation of policy indications based on extrapolation to a large scale of findings related to mechanisms investigated and validated at the local level. Imagine that an owner of a sex shop is looking for advice about how to expand business by opening a second shop. Obviously, when analyzing the problem at a local level (e.g., when operating in a given urban area), the opening of two similar shops close to each other has to be considered as bad policy. The two shops will compete for the same flow of potential customers, and therefore the simultaneous presence of two similar shops in the same street is expected to reduce the profit margin of each of the two shops. However, imagine now the existence of hundreds of sex shops in a given area. This implies the emergence of a new system property, which is generally called a red-light district. Such an emergent property expresses functions that can only be detected at a scale larger than the one used to study the identity of an individual sex shop. In fact, red-light districts can also attract potential buyers from outside the local urban area or from outside the city. In some cases, they can even draw customers from abroad. In technical jargon we can say that the domain of attraction for potential customers of a red-light district is much larger than the one typical of an individual sex shop. This can imply that — getting back to the advice required by the owner of an individual sex shop — there is a trade-off to be considered when deciding whether to open a new shop in a red-light district. The reduction of profit due to the intense competition has to be weighed against the increase of customer flow due to the larger basin of attraction. Such a trade-off analysis is totally different if the shop will be opened in a normal area of the city.

In conclusion, whereas it is debatable whether the concept of emergence implies something special in ontological terms, it is clear that it implies something special in functional and epistemological terms. Every time we deal with something that is *more than* and *different from* the sum of its parts, we have to use in parallel nonequivalent descriptive domains to represent and model different relevant aspects of its behavior. The parts have to be studied in their role of parts, and the whole has to be studied in its role as a whole. Put another way, emergence implies for sure a change (a richer information space) in the observer–observed complex. The implications of this fact are huge. When dealing with the evolution of complex adaptive systems (real emergence), the information space that has to be used for describing how these systems change in time is not closed and knowable *a priori*. This implies that models, even if validated in previous occasions, will not necessarily be good in predicting future scenarios. This is especially true when dealing with human systems (adaptive reflexive systems).

3.5 The Crucial Difference between Risk, Uncertainty and Ignorance

The distinction proposed below is based on the work of Knight (1964) and Rosen (1985). Knight (1964) distinguishes between cases in which it is possible to use previous experience (e.g., record of frequencies) to infer future events (e.g., guess probability distributions) and cases in which such an inference is not possible. Rosen (1985), in more general terms, alerts us to the need to always be aware of the clear distinction between a natural system, which is operating in the complex reality, and the representation of a natural system, which is scientist-made. Any scientific representation requires a previous mapping, within a structured information space, of some of the relevant qualities of the natural system with encoding variables (the adoption of a formal identity for the system in a given descriptive domain). Since scientists can handle only a finite information space, such a mapping results in the unavoidable missing of some of the other qualities of the natural system (those not included in the selected set of relevant qualities).

Using these concepts, it is possible to make the following distinction between risk and uncertainty:

Risk — Situation in which it is possible to assign a distribution of probabilities to a given set of possible outcomes (e.g., the risk of losing when playing roulette). The assessment of risk can come either from the knowledge of probability distribution over a known set of possible outcomes obtained using validated inferential systems or in terms of agreed-upon subjective probabilities.

ECLIPSE PATH
4 DEC 6-8 PM
OBSERVE
SPEED LIMITS

The quality to be guessed
(relative position in time) can
be described by a valid model

Photo by Manfred Bruenjes

Information space known,
associative context given,
nothing changes in time

FIGURE 3.3 (a) Guessing Eclipse's predictive power is very high. (b) Conventional risk assessment prediction using frequencies to estimate probabilities.

In any case, risk implies an information space used to represent the behavior of the investigated system, which is (1) closed, (2) known and (3) useful. The formal identity adopted includes all the relevant qualities to be considered for a sound problem structuring. In this situation, there are cases in which we can even calculate with accuracy the probabilities of states included in the accessible state space (e.g., classic mechanics). That is, we can make reliable predictions of the movement in time of the system in a determined state space (Figure 3.3a).

The concept of risk is useful when dealing with problems that are (1) easily classifiable (about which we have a valid and exhaustive set of epistemological categories for the problem structuring) and (2) easily measurable (the encoding variables used to describe the system are observable and measurable, adopting a measurement scheme compatible in terms of space–time domain with the dynamics simulated in the modeling relation). Under these assumptions, when we have available a set of valid models, we can forecast and usefully represent what will happen (at a particular point in space and time). When all these hypotheses are applicable, the expected errors in predicting the future outcomes are negligible. Alternatively, we can decide to predict outcomes by using probabilities derived from our previous knowledge of frequencies (Figure 3.3b).

Uncertainty — Situation in which it is not possible to generate a reliable prediction of what will happen. That is, uncertainty implies that we are using to make our prediction an information space that is (1) closed, (2) finite and (3) partially useful, according to previous experience, but at the same time, there is awareness that this is just an assumption that can fail.

Therefore, within the concept of uncertainty we can distinguish between:

*   **Uncertainty due to indeterminacy** — There is a reliable knowledge about possible outcomes and their relevance, but it is not possible to predict, with the required accuracy, the movement of the system in its accessible state space (e.g., the impossibility to predict the weather 60 days from now in London) (Figure 3.4a). Indeterminacy is also unavoidable when dealing with the reflexivity of humans. The simultaneous relevance of characteristics of elements operating on different scales (the need to consider more than one relevant dynamic in parallel on different space–time scales) and nonlinearity in the mechanisms of

(a)

Image by EUMETSAT

information space known
but depending on the time
horizon considered changes
cannot be predicted with
the required accuracy

"Butterfly effect" - Nobody can predict the weather in London in 60 days . . .

**It is not about being unable to guesstimate probabilities.
Rather, it is about ignoring the relevant attributes that
will matter for us in the future.**

Alice wondering about the "DRINK-ME" bottle

(b)

FIGURE 3.4 (a) Prediction facing uncertainty. (b) Prediction facing ignorance.

controls (the existence of cross-scale feedback) entails that expected errors in predicting future outcomes can become high (butterfly effect, sudden changes in the structure of entailments in human societies — laws, rules, opinions). Uncertainty due to indeterminacy implies that we are dealing with problems that are classifiable (we have valid categories for the problem structuring), but that they are not fully measurable and predictable.

Whenever we are in the presence of events in which emergence should be expected, we are dealing with a new dimension of the concept of uncertainty. In this case, we can expect that the structure of causal entailments in the natural system simulated by the given model can change or that our selection of the set of relevant qualities (formal identity) to be used to describe the problem can become no longer valid. This is a different type of uncertainty.

- **Uncertainty due to ignorance** — Situations in which it is not even possible to predict what will be the set of attributes that will be relevant for a sound problem structuring (an example of this type of uncertainty is given in Figure 3.4b). Ignorance implies that (1) awareness that the information space used for representing the problem is finite and bounded, whereas the information space that would be required to catch the relevant behavior of the observed system, is open and expanding and (2) our models based on previous experience are missing relevant system qualities. The worst aspect of scientific ignorance is that it is possible to know about it only through experience, that is, when the importance of events (attributes) neglected in a first analysis becomes painfully evident. For example, Madame Curie, who won two Nobel Prizes (in physics in 1903 and in chemistry in 1911) for her outstanding knowledge of radioactive materials, died of leukemia in 1934. She died "exhausted and almost blinded, her fingers burnt and stigmatised by 'her' dear radium" (Raynal, 1995). The same happened to her husband and her daughter. Some of the characteristics of the object of her investigations, known nowadays by everybody, were not fully understood at the beginning of this new scientific field, not even by the best experts available.

There are typologies of situations with which we can expect to be confronted in the future with problems that we cannot either guess or classify at the moment, for example, when facing fast changes in existing boundary conditions. In a situation of rapid transition we can expect that we will soon have to learn new relevant qualities to consider, new criteria of performance to be included in our analyses and new useful epistemological categories to be used in our models. That is, to be able to understand the nature of our future problems and how to deal with them, we will have to use an information space different from the one we are using right now. Obviously, in this situation, we cannot even think of valid measurement schemes (how to check the quality of the data), since there is no chance of knowing what encoding variables (new formal identities expressed in terms of a set of observable relevant qualities) will have to be measured.

However, admitting that ignorance means that it is not possible to guess the nature of future problems and possible consequences of our ignorance does not mean that it is not possible to predict at least when such an ignorance can become more dangerous. For example, when studying complex adaptive systems it is possible to gain enough knowledge to identify basic features in their evolutionary trajectories (e.g., we can usefully rely on valid metaphors). In this case, in a rapid transitional period, we can easily guess that our knowledge will be affected by larger doses of scientific ignorance.

The main point to be driven home from this discussion over risk, uncertainty and ignorance is the following: in all cases where there is a clear awareness of living in a fast transitional period in which the consequences of scientific ignorance can become very important, it is wise not to rely only on reductionist scientific knowledge (Stirling, 1998). The information coming from scientific models should be mixed with that coming from metaphors and additional inputs coming from various systems of knowledge found among stakeholders. A new paradigm for science — postnormal science — should aim at establishing a dialogue between science and society moving out from the idea of a one-way flow of information. The use of mathematical models as the ultimate source of truth should be regarded just as a sign of ignorance of the unavoidable existence of scientific ignorance.

3.6   Multiple Causality and the Impossible Formalization of Sustainability Trade-Offs across Hierarchical Levels

3.6.1   Multiple Causality for the Same Event

The next example deals with multiple causality: a set of four nonequivalent scientific explanations for the same event are listed in Figure 3.5 (the event to be explained is the possible death of a particular individual). This example is particularly relevant since each of the possible explanations can be used as input for the process of decision making.

Explanation 1 refers to a very small space–time scale by which the event is described. This is the type of explanation generally looked for when dealing with a very specific problem (when we have to do something according to a given set of possibilities, perceived here and now — a given and fixed associative context for the event). Such an explanation tends to generate a search for maximum efficiency. According to this explanation, we can do as well as we can, assuming that we are adopting a valid, closed and reliable information space. In political terms, these types of scientific explanations tend to reinforce the current selection of goals and strategies of the system. For example, policies aimed at maximizing efficiency imply not questioning (in the first place) basic assumptions and the established information space used for problem structuring.

Explanation 2 refers again to a small space–time scale by which the event is described. This is the type of explanation generally looked for when dealing with a class of problems that have been framed in terms of the what/how question. We have an idea of the how (of the mechanisms generating the problem), and we want to both fix the problem and understand better (fine-tuning) the mechanism according to our scientific understanding. Again, we assume that the basic structuring of the available information space is a valid one, even though we would like to add a few improvements to it.

Explanation 3 refers to a medium to large scale. The individual event here is seen through the screen of statistical descriptions. This type of explanation is no longer dealing only with the

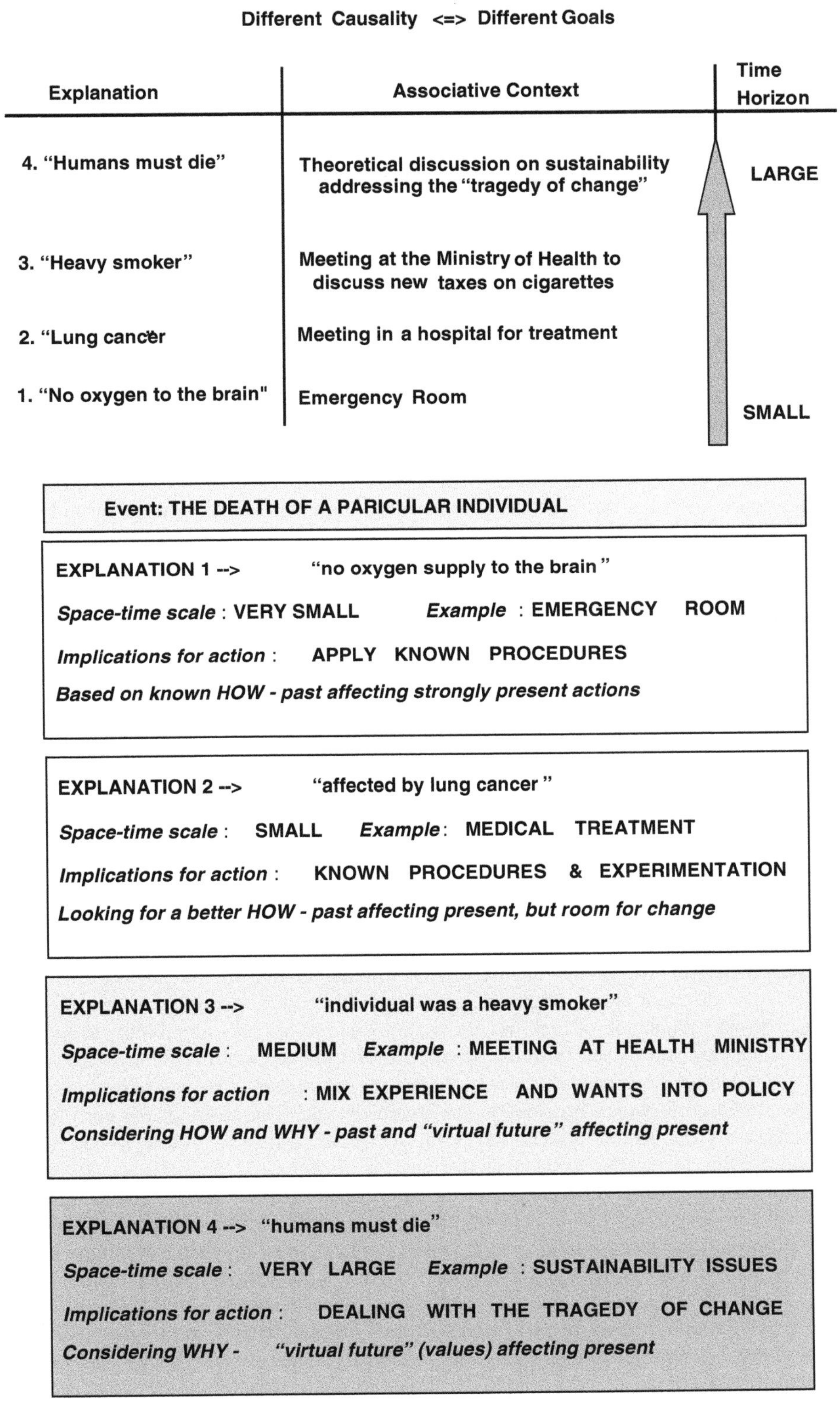

Different Causality <=> Different Goals

Explanation	Associative Context	Time Horizon
4. "Humans must die"	Theoretical discussion on sustainability addressing the "tragedy of change"	LARGE
3. "Heavy smoker"	Meeting at the Ministry of Health to discuss new taxes on cigarettes	
2. "Lung cancer	Meeting in a hospital for treatment	
1. "No oxygen to the brain"	Emergency Room	SMALL

Event: THE DEATH OF A PARICULAR INDIVIDUAL

EXPLANATION 1 --> "no oxygen supply to the brain "

Space-time scale : VERY SMALL   *Example* : EMERGENCY   ROOM

Implications for action :   APPLY   KNOWN   PROCEDURES

Based on known HOW - past affecting strongly present actions

EXPLANATION 2 --> "affected by lung cancer "

Space-time scale :   SMALL   *Example* : MEDICAL   TREATMENT

Implications for action :   KNOWN   PROCEDURES   &   EXPERIMENTATION

Looking for a better HOW - past affecting present, but room for change

---

EXPLANATION 3 --> "individual was a heavy smoker"

Space-time scale :   MEDIUM   *Example* : MEETING   AT HEALTH   MINISTRY

Implications for action   : MIX EXPERIENCE   AND WANTS   INTO POLICY

Considering HOW and WHY - past and "virtual future" affecting present

---

EXPLANATION 4 --> "humans must die"

Space-time scale :   VERY LARGE   *Example* : SUSTAINABILITY ISSUES

Implications for action :   DEALING   WITH THE TRAGEDY   OF CHANGE

Considering WHY -   "virtual future" (values) affecting present

FIGURE 3.5 Multiple explanations for an event — in this case, the death of a particular person.

what/how question, but also, in an indirect way, with the why/what question. We want to solve the problem, but, to do that, we have to mediate between contrasting views found in the population of individuals to which we want to apply policies. In this particular example, dealing with the trade-offs between individual freedom of smoking and the burden of health costs for society generated by heavy smoking. We no longer have a closed information space and a simple mechanism to determine optimal solutions. Such a structuring of the problem requires an input from the stakeholders in terms of value judgments (for politicians this could be the fear of losing the next election).

Explanation 4 refers to a very large scale. This explanation is often perceived as a joke within the scientific context. My personal experience is that whenever this slide is presented at conferences or lessons, usually the audience starts laughing when it sees the explanation "humans must die" listed among the possible scientific explanations for the death of an individual. Probably this reflects a deep conditioning to which scientists and students have been exposed for many decades. Obviously, such an explanation is perfectly legitimate in scientific terms when framing such an event within an evolutionary context. The question then becomes why is it that such an explanation tends to be systematically neglected when discussing sustainability? The answer is already present in the comments given in Figure 3.5. Such an explanation would force scientists and other users of it to deal explicitly and mainly with value judgments (dealing with the why or what for question rather than with the how question). This is probably why this type of question seems to be perceived as not scientifically correct according to Western academic rules.

Also in this example we find the standard predicament implied by complexity: the validity of using a given scientific input depends on the compatibility of the simplification introduced by the problem structuring with the context within which such information will be used. A discussion of pros and cons of various policies restricting smoking would be considered unacceptable by the relatives of a patient in critical condition in an emergency room. In the same way, a physiological explanation on how to boost the supply of oxygen to the brain would be completely useless in a meeting discussing the opportunity of introducing a new tax on cigarettes.

3.6.2   The Impossible Trade-Off Analysis over Perceptions: Weighing Short-Term vs. Long-Term Goals

The example given in Figure 3.6 addresses explicitly the importance of considering the hierarchical nature of the system under investigation. This example was suggested to me by David Waltner-Toews (personal communication; details on the study are available at Internet address of International Development Research Centre). The goal of this example is to illustrate that when reading the same event on different levels (on different space–time horizons), we will see different solutions for the very same problem. A compared evaluation of these potential alternative solutions is impossible to formalize.

Very briefly, the case study deals with the occurrence of a plague in a rural village of Tanzania. The plague was generated by the presence of rats in the houses of villagers. The rats moved into the houses following the stored corn, which previously was stored outside. The move of the corn to inside the houses was necessary due to the local failure of the social fabric (it was no longer safe to store corn outside). Such a collapse was due to the very fast process of change of this rural society (triggered by the construction of a big road). Other details of the story are not relevant here, since this example does not deal with the implications of this case study, but just points to a methodological impasse.

A simple procedure that can be used to explore the implication of the fact that human societies are organized in holarchic ways is indicated in Figure 3.6. After stating the original problem as perceived and defined at a given level, it is possible to explore the causal relations in the holarchy by climbing the various levels through a series of "why and because" (upper part of Figure 3.6). Upon arriving at an explanation that has no implications for action we can stop. Then we can descend the various levels by answering new types of questions related to the "how and when" dimension (lower part of Figure 3.6).

Looking at possible ways of structuring the problem experienced by the villagers following this approach, we are left with a set of questions and decisions typical of the science for governance domain:

Phase 1: climbing the Holarchy through resonating WHY and BECAUSE

WHY there is the plague in the village?

BECAUSE rats and humans interact too much — **WHY?**

BECAUSE the structure of local society and ecosystem were disturbed — **WHY?**

BECAUSE an exogenous model of development was imposed on them — **WHY?**

BECAUSE of a lack of empowerment of local communities (changes too much influenced by Western models) — **WHY?**

BECAUSE two socioeconomic systems with a big difference in their level of development are interacting too much (North/South). This is generating a strong friction on lower-level holons (changes are so fast, that marginal social groups operating on a very small scale have no power of negotiation) -**WHY?**

BECAUSE of historical accidents and different boundary conditions

→ answer indicating that there is nothing we can do about it

Phase 2: redescending the holarchy through resonating HOW and WHEN

HOW/WHEN can we eliminate the problem?

HOW/WHEN — existing difference between North and South can be eliminated? Which one of the two models of development we like most? Which model is feasible for the entire world (compatible with biophysical constraints, compatible with different cultures)? What is the lag time needed for expected changes?

HOW/WHEN — can we reestablish a fair negotiation among the elements of the holarchy in spite of the large differences between North and South? (It is possible to empower local communities? How? What is the expected lag time to get relevant changes?)

HOW/WHEN — can we generate room for expressing local aspirations within the constraints given by the evolutionary trajectory of the larger system to which the community belongs?

HOW/WHEN — the disturbance to the local ecosystem and the local social system can be reduced to acceptable levels? what are the possible options? What is the expected lag time to get results? What is the level of uncertainty on the options?

HOW/WHEN — Can we eliminate the rats? what is the expected lag time? What are the costs? What are possible negative side effects? If rats are symptoms rather than the real cause, then is there any negative feedback in curing symptoms?

FIGURE 3.6 Multiple causality for a given problem: plague in the village.

- *What is the best level that should be considered when making a decision about eliminating the plague? Who is entitled to decide this?* The higher we move in the holarchy, the better is the overview of parallel causal relations and the richer (more complex) is the explanation. On the other hand, this implies a stronger uncertainty about predicting the outcome of possible policies, as well as a longer lag time in getting a fix (prolongation of sufferance of lower-level holons — those affected by the plague in the village, in this case, mainly women). The smaller the scale, the easier is the identification of direct causal relations (the easier the handling of specific projects looking for quick fixes). However, faster and more reliable causal relations (leading to rapid solutions) carry the risk of curing symptoms rather than causes. That is, the adoption of a very small scale of analysis risks the locking-in of the system in the same dynamic that generated the problem in the first place, since this main dynamic, operating on a larger scale, has not been addressed in the location-specific analysis.

- *How should the trade-offs linked to the choice of one level rather than another be assessed?* Using a very short time horizon to fix the problem (e.g., kill the rats while keeping the society and ecosystem totally unbalanced) is likely to sooner or later generate another problem. If rats were just a symptom of some bigger problem, the cause is still there. On the other hand, using too large of a time horizon (e.g., trying to fix the injustice in the world) implies a different risk — that of attempting to solve the perceived problem very far in the future or distant in space, basing our policies on present knowledge and boundary conditions (perceptions referring to a very small space–time scale). The very same problems we want to solve today with major structural changes in social institutions could have different and easy solutions in 20 to 50 years from now (e.g., climatic changes that make impossible life in that area). If this is the case, policies aiming only at quick relief for the suffering poor (e.g., killing the rats) can be implemented without negative side effects (e.g., generation of a lock-in of a larger-scale problem). Unfortunately, we can never know this type of information ahead of time.

- *Is our integrated assessment of changes reflecting existing multiple goals found in the system?*
  Any integrated assessment of the performance of a system depends on:
  1. Expectations and related priorities (relevant criteria to be considered and weighing factors among them)
  2. Perception of effects of changes (the level of satisfaction given by a certain profile of values taken by indicators of performance)
- In turn, both these expectations and perceptions heavily depend on:
  1. The level of the holarchy at which the system is described (e.g., if we ask the opinion of the president of Tanzania or of a farmer living in that village).
  2. The identity of the various social groups operating within the socioeconomic system at any given level in the holarchy. For example, farmers of a different village in Tanzania can have different perspectives on the effects of the same new road. In the same way, women or men of the same village can judge in different ways the very same change.
- *What is the risk that cultural lock-in, which is clearly space- and time-specific, is preventing the feasibility of alternative solutions?* It is well known that the past — in the form of cultural identity in social systems — is always constraining the possibility of finding new models of development. This is why changes always imply tragedy (Funtowicz and Ravetz, 1994). As noted earlier, when solving a sustainability problem, socioeconomic systems have to be prepared to lose something to get something else. This introduces one of the most clear dimensions of incommensurability in the analysis of sustainability trade-offs. Decisions about sustainability have to be based on a continuous negotiation. The various stakeholders should be able to reach, in an adequate period of time, a consensus on the nature of the problem and an agreement on how to deal with it. In particular, this implies deciding:
  1. What do they want to maintain about the present situation (e.g., how important is to keep what they are getting now)?
  2. What do they want to change about the present situation (e.g., how important is it to get away as fast as possible from the current state)?
  3. How reliable is the information that is to be used to translate the agreement reached about points 1 and 2 into practical action?
- These questions can be reformulated as: When forced to redefine their identity as a social system, what do they want to become, and at what cost?

Clearly, a total agreement over a common satisficing trade-offs profile is certainly not easy to reach (if not impossible) in any social system. The unavoidable existence of different perceptions about how to answer these questions can only be worked out through negotiations and conflicts. Negotiations and conflicts are crucial for keeping diversity in the social entity. The standard solution of imposing a particular viewpoint (a given best satisficing trade-offs profile) with force (hegemonization) — besides the very high cost in terms of human suffering — carries the risk of an excessive reduction in the cultural diversity, and therefore a dramatic reduction of adaptability, in the resulting social systems. The expression "ancient regime syndrome," proposed by Funtowicz and Ravetz (personal communication), indicates that boosting short-term efficiency through hegemonization in a society is often paid for in terms of lack of adaptability in the long term. Such a typical pattern, leading to the collapse of complex social organization, has been discussed in detail by Tainter (1988). More information on the nature of this dilemma is provided in the next section.

3.6.3 The Impossible Trade-Off Analysis over Representations: The Dilemma of Efficiency vs. Adaptability

Adaptability and flexibility are crucial qualities for the sustainability of adaptive systems (Conrad, 1983; Ulanowicz, 1986; Holling, 1995). They both depend on the ability of preserving diversity (actually, this is also the theoretical foundation of democracy). However, the goal of preserving diversity per se collides with that of augmenting efficiency at a particular point in space and time (this is the problem with total anarchy). Efficiency requires elimination of those activities that are the worst performing according to a given set of goals, functions and boundary conditions, and amplification of those activities that are perceived as the best performing at a given point in space and time. Clearly this general rule applies also to technological progress. For example, in agricultural production, improving world agriculture according to a given set of goals expressed by a given social group in power and according to the present perception of boundary conditions (e.g., plenty of oil) implies a dramatic reduction of the diversity of systems of production (e.g., the disappearance of traditional farming systems). Driven by technological innovations such as the green revolution, more and more agricultural production all over our planet is converging on a very small set of standard solutions (e.g., monocultures of high-yielding varieties supported by energy-intensive technical inputs, such as synthetic fertilizers, pesticides and irrigation (Pimentel and Pimentel, 1996)). On the other hand, the obsolete agricultural systems of production, being abandoned all over the planet, can show very high performance when assessed under a different set of goals and criteria (Altieri, 1987).

When reading the process of evolution in terms of complex systems theory (Giampietro, 1997; Giampietro et al., 1997), we can observe that, in the last analysis, the drive toward instability is generated by the reciprocal influence between efficiency and adaptability. The continuous transformation of efficiency into adaptability, and that of adaptability into efficiency, is responsible for the continuous push of the system toward nonsustainable evolutionary trajectories. This is a different view of Jevons' paradox or the agricultural treadmill, both discussed in Chapter 1. The steps of this cycle (with an arbitrary choice of a starting step) are:

1. Accumulation of experience in the system leads to more efficiency (by amplification of the best-performing activities and elimination of the worst-performing ones).
2. More efficiency makes available more surplus to fuel societal activities.
3. The consequent increase in the intensity and scale of interaction of the socioeconomic system with its environment implies an increased stress on the stability of boundary conditions (more stress on the environment and a higher pressure on resources). This calls for increased investments in adaptability.
4. To be able to invest more in adaptability (expand the diversity of activities, which implies developing new activities that at the moment may not perform well), the system needs to be more efficient — i.e., it has to better use experience to produce more. This can only be achieved by amplifying the best-performing activities and eliminating the worst-performing ones. At this point the system goes back to step 1.

An overview of the coexistence of different causal paths between efficiency and adaptability, described on different timescales, is illustrated in Figure 3.7. When considered on a short timescale, efficiency would imply a negative effect on adaptability and vice versa. When a long-term perspective is adopted, they both thrive on each other. However, the only way to obtain this result (based on a sound Yin–Yang tension) is by continuously expanding the size of the domain of activity of human societies. That is, increases in efficiency are obtained by amplifying the best-performing activities, without eliminating completely the obsolete ones. These activities will be preserved in the repertoire of possible activities of the societal system (as a memory of different meanings of efficiency when adopting a different set of boundary conditions and a different set of goals). When the insurgence of new boundary conditions or new goals requires a different definition of efficiency, the activities amplified until that moment will become obsolete, and the system will scan for new (or old) ones in the available repertoire. In this way, at each cycle, societal systems will enlarge their repertoire of knowledge of possible activities and accessible states (boost their adaptability). This expansion in the computational capability of society, however, requires an expansion of its domain of activity, that is, an increase in its size (no matter how we decide to measure its size – total amount of energy controlled, information processed, kilograms of human mass, GNP) (Giampietro et al., 1997).

Sustainability of societal systems can therefore only be imagined as a dynamic balance between the rate of development of their efficiency and adaptability. This can be obtained by a continuous change of structures to maintain existing functions and a continuous change of functions to maintain existing

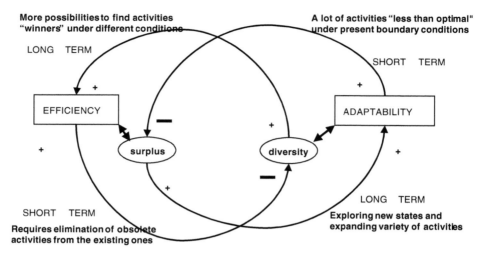

FIGURE 3.7 Self-entailment of efficiency and adaptability across scales.

structures. Put another way, neither a particular societal structure nor a particular societal function can be expected to be sustainable indefinitely.

As noted before, practical solutions to this challenge mean deciding how to deal with the tragedy of change. That is, any social system in its process of evolutions has to decide how to become a different system while maintaining in this process its own individuality (more on this in Chapter 8). The feasibility of this process (which is like changing the structure of an airplane while flying it) depends on the nature of internal and external constraints faced by society. The advisability of the final changes (what the plane will look like at the end of the process, if still flying) will depend on the legitimate contrasting perceptions of those flying it, their social relation and the ability expressed by such a society to make wise changes to the plane at the required speed.

In this way, we found an additional complication to the practical operationalization of the process of decision making for sustainability. In fact, not only is it difficult to find an agreement on what are the most important features to preserve or enhance when attempting to build a different flying airplane, but this decision also has to be made without having solid information about the feasibility of the various possible projects to be followed. As noted earlier, the definition and forecasting of viability constraints are unavoidably affected by large doses of uncertainty and ignorance about possible unexpected future situations. Put another way, when facing the sustainability predicament, humans must continuously gamble trying to find a balance between their efficiency and adaptability. In cultural terms, this means finding an equilibrium between the considerations that have to be given to the importance of the past and the future in shaping a civilization's identity (Giampietro, 1994a).

3.7   Perception and Representation of Holarchic Systems (Technical Section)

3.7.1   The Fuzzy Relation between Ontology and Epistemology in Holarchic Systems

The goal of this section is to wrap up the discussion on the epistemological and ontological meltdown implied by the mechanism of autopoiesis of languages and holarchies. This will be done by using the concepts introduced in this chapter and in Chapter 2.

First of all, let us get back to the peculiar implications of the concept of holon in relation to ontology and epistemology. Such a concept has been introduced by Koestler to focus and deal with the meltdown found when dealing with holons. Recalling one of his famous examples, Koeslter (1968, p. 87) says that it is impossible to individuate what a given opera of Puccini (e.g., *La Bohème*) is in reality. In fact, we can provide various representations of it (individual realizations), all of which would be different from each other. At the same time, the very same representation is always perceived as *La Bohème,* even if in different ways by different spectators (nonequivalent observers). Such an opera was conceived as an individual essence by Puccini, but then it was formalized (encoded) into a set of formal identities (e.g., manuscripts with lyrics, musical scores, description of costumes and set decorations). After that, various directors, musicians, singers and costume designers willing to represent *La Bohème* have adopted different semantic interpretations of such a family of formalizations. To make things more intriguing, it is exactly this process of semantic interpretation of formal identities and consequent actions that results in the generation of formalizations, which manage to keep alive such individuality. The individuality of *La Bohème* will remain alive only in the presence of a continuous agreement among (1) those providing representations (producing realizations of it), that is, musicians, singers, administrators of opera theaters, etc. and (2) those making the production possible (those supporting the process of realization), that is, the spectators paying to attend these representations and the decision makers sponsoring the opera. The survival of the identity of *La Bohème* depends on the ability to preserve the meaning assigned to the label "La Bohème" by interacting nonequivalent observers. This keeps alive the process of resonance between semantic interpretations of previous formalizations of the relative set of identities to generate new formalizations to be semantically interpreted.

In this example, we recognize the full set of concepts discussed in Chapter 2. A given opera by Puccini refers to an equivalence class of realizations all mapping onto the same (1) label "La Boheme" and (2) essence — the universe of images of that opera in the mind of those sharing the meaning assigned to that label. This process of resonance between labels, realizations and shared perception about meaning was started by an individual event of emergence when Puccini wrote the opera (adding a new essence to the set of musical operas). The validity of this essence requires a continuous semantic check based on the valid use of formal identities required to generate equivalence classes of realization.

In this process, the individuality of *La Bohème* is preserved through a process of convergence of meaning in a population of nonequivalent observers and agents who must be able to recognize and perceive the various realizations they experience in their life as legitimate members of that equivalence class. It should be noted that such an identification can be obtained using nonequivalent mappings (an integrated set of identities that make possible the successful interaction of agents in preserving the meaning of such a process). "Some can recognize the words of a famous aria — 'your tiny hand is frozen' after having lost the melody, whereas others can recognize the melody after having lost the world" (Koeslter, 1968, p. 87). Others can use a key code for recognizing operas' special costumes or situations (e.g., a person totally deaf sitting in an opera hall can associate the presence of elephants on the stage with the representation of Verdi's *Aida*).

Obviously, the same mechanism generating or preserving a semantic identity of pieces of art can be applied to other types of artistic objects such as a play of Shakespeare or a famous painting of Picasso (a photographic reproduction would then be a formal identity of it, which is missing relevant aspects found in the original).

The rest of this section makes the point that this fuzzy relation between ontology and epistemology (the existence of valid epistemic categories requires the existence of an equivalence class made up of realizations, and an equivalence class of realizations requires the existence of a valid essence) is not only found when dealing with the perceptions and representations of artistic objects. Rather, this is a very generic mechanism in the formation of human knowledge. To make this point, I propose we have another look at a simple, "innocent" geometric object defined within classic Euclidean geometry (e.g., a triangle) using the various concepts introduced so far.

A triangle is clearly an essence associated with a class of geometric entities. The definition of such an essence is based on a specified, given relation among lower-level elements (the three segments representing the sides of the triangle). Put another way, a triangle is by definition a whole made up of parts. That is, it expresses emergent properties at the focal level (those of being a triangle) due to the organization of lower-level elements (the segments making up its sides). The triangle is an emergent property in the sense that the epistemic category "triangle" refers to a descriptive domain (a two-dimensional plane) nonequivalent to the descriptive domain used to perceive and represent segments (which is one-dimensional). It would be impossible to make a distinction between a triangle and a square for an observer living in a one-dimensional world unless they were to walk around the observer rotating at a given pace (recall the famous book *Flatland* by Edwin Abbott (1952)). Still, this implies using a two-dimensional plane to be able to get around the triangle.

When considering a triangle, we deal with its identity in terms of (1) a label (*triangle* in English and *triangulo* in Spanish), (2) a mental image common to people sharing the meaning assigned to this label and (3) a class of objects that are perceived as being realizations of this essence. Obviously, a class of triangular objects (realizations of such an essence) must exist; in fact, this is what made it possible for humans to share the meaning given to the label in the form of the mental image of it.

In terms of hierarchy theory, we can say that a triangle is a hierarchically organized entity. That is, to have a realization of such an identity, you must have first the realization of three segments (lower-level components defined in one dimension), which must be organized into a two-dimensional figure on a plane. In turn, a segment is perceived and represented as made up of points. Because of this hierarchical nature, there is a double set of constraints associated with the existence of such an object: (1) on the relative length of each of the three segments making up the sides and (2) on the shape of the three angles determined on and determining the corners. The existence of these constraints on the relative size of lower-level elements and their relative position is related to the required closure of the geometric object. Put another way, the very identity of this complex object implies a self-entailment among the various identities of its component elements: (1) lower-level identities of segments (relative length) and (2) focal-level identity of the triangle (relative position of the sides within the whole). The existence of these constraints makes it possible to compress the requirement of information to represent such an object. That is, knowing about the identity of two of the angles entails knowing about the identity of the third one. Knowing the length of the various segments makes it possible to know about the angles. Actually, the existence of this self-entailment among the characteristics of the focal level (level n) and the characteristics of lower-level elements (level $n - 1$) is the subject of elementary trigonometry. When expressed in these terms, a triangle is an essence referring to various possible types (relational definition of the whole based on the definition of a relation among lower-level elements), and therefore is without scale. However, to make it possible for humans to share the meaning given to the label "triangle" — to abstract from their interaction with the reality a mental image of triangles — humans must have seen in their daily life several practical realizations of this type. Moreover, when discussing and studying triangles, humans must make realizations (representations) of the types related to this essence (e.g., different types of triangles, such as an isosceles right triangle) to be able to check with measurements the validity of their theorems.

Therefore, even when dealing with a very abstract discipline such as geometry, humans cannot get rid of the duality between essences and realizations. As soon as an observer either represents a geometric object or perceives it as a natural system belonging to the class of triangles (by associating the shape of a real entity to the mental image of triangles), we are dealing with a case in which a particular type (belonging to the essence) has been realized at a certain scale. The very possibility of doing geometry therefore requires the ability to make and observe realizations of essences at different scales. That is, the operation of pattern recognition based on mental images (related to a preanalytical definition of types or essences) per se would not imply a given scale. However, both processes — observation and measurement (using a detector of signals, which implies a scale because of the interaction associated with the exchange of signals) and fabrication (making and assembling of lower-level structural elements) — are necessarily scaled.

The perception and representation of a triangle refer in parallel to two definitions of scales: (1) the scale required to perceive, represent, measure and fabricate lower-level structural elements (e.g., segments) and (2) the scale required to perceive, represent, measure and fabricate focal-level elements (e.g., triangles) when putting together these lower elements into a whole. A series of triangles of the same shape (type) and different sizes is shown in the top of Figure 3.8. The given typology can be defined by using the relation between the relative measures of angles and sides. When dealing with a type, we are talking of relative measures — a concept that assumes (requires or entails) compatibility in the accuracy of the process of:

1. Actual measurement of both angles and side lengths
2. Fabrication of the geometric object

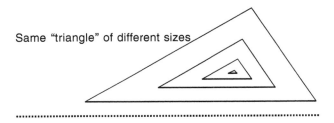

Same "triangle" of different sizes

....

Different realizations of the same triangle at different scales

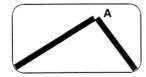
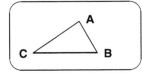

FIGURE 3.8 Types and realizations of a triangle.

Coming to possible realizations of this set of triangles with different sizes (but all belonging to the same type), we can imagine (Figure 3.8):

1. A triangle having a size in the order of centimeters (made by drawing with a pencil three segments on a paper on our desk and observed while sitting at the desk)
2. A triangle with a size in the order of meters (made by putting together wooden sticks in the backyard and observed from the window of the attic of our house)
3. A triangle with a size in the order of kilometers (made by the connection of three highways and observed from an airplane)

Imagine now that we want to assess the different sizes of these three realizations, which are all mapping onto the same type of triangle. This assessment must be associated with a process of measurement. Measuring implies the definition of a differential, which is related to the accuracy of the measurement scheme, that is, the smallest difference in length that can be detected by the measuring device for segments (e.g., dx) and the small difference that can be detected by the measuring device for angles (e.g., $d\alpha$). In the same way, we can define another relevant space differential, related to the process of representation of such an object (fabrication of individual realization of this essence). This would be the smallest gradient in length and angle definition that can be handled in the process of representation. An idea of the potential problems faced when constructing a triangle in relation to the relative scale of segments and the whole is given in the bottom of Figure 3.8. Those familiar with PowerPoint or other graphic software will immediately recognize the nature of this problem faced when zooming too much into the drawing.

We can talk about the measures of a triangle (what makes useful the identity of a triangle throughout trigonometry) only after assuming that the two nonequivalent differentials are compatible. The two differentials in fact are not necessarily always compatible: (1) one refers to the operation of measurement of structural elements at the level $n-1$ and (2) one refers to the representation of the structural elements into a whole at the level n. Using the vocabulary introduced in the previous sections, the compatibility of the two differentials means that lower-level elements and the whole must belong to the same descriptive domain.

Robert Rosen (2000, chapter 4) discusses the root of incommensurability. His example is that it is possible to express without problems the identity of a particular type of triangle (e.g., isosceles right triangle) in terms of a given relation between its parts. A consequence of this definition is that according to the Pythagorean theorem, the size of the square built on the hypotenuse is equal to the sum of the two squares built on the other two sides. That is, when dealing with a two-dimensional object (a triangle), it is possible to express in general terms the relations among its elements using a mapping referring to a two-dimensional representation (squares with squares).

The problem begins when we try to use a one-dimensional descriptive domain for mapping the relation between elements of a two-dimensional object. This led to the introduction of irrational numbers: the length of the hypotenuse is equal to the length of one of the other two sides times the square root of 2 and real numbers that are — as proved by the Zeno paradox — uncountable (Rosen, 2000, p. 74). In this case, the three nonequivalent concepts — measuring, constructing and counting — can no longer be considered compatible or reducible to each other (Rosen, 2000, p. 71). This example raises the following question: Why, when we adopt as a descriptive domain a two-dimensional plane, can we express the relation among the sides of this triangle in terms of squares without problems, whereas when we use a one-dimensional descriptive domain — using the relations among segments — things get more difficult? An explanation can be given by using the concepts discussed before. A two-dimensional descriptive domain assumes the parallel validity of two nonequivalent external referents: (1) the possibility of constructing and measuring lower-level elements (with a dx related to the lengths of segments) and, at the same time, (2) the possibility of constructing and measuring another characteristic quality of such a system, namely, the angles formed by the sides. The forced closure of the triangle on the corners implies assuming the compatibility between the differential dx related to the measuring and representation of the lengths of lower-level elements (segments) and the differential dα related to the measuring and representation of their relative positions (angles) within the whole object. When we talk of the sum of the squares constructed on the various sides, we are assuming the compatibility of dx and dα in the various operations required for making such a sum. In fact, in the Pythagorean theorem squares are assumed to be a valid measuring tool, supposing that all the squares have in common a smaller square U that can be used as a unit to count off an integral a number of times their area. This is "a kind of quantization hypothesis, with the U as a quantum" (Rosen, 2000, p. 71). That is, the hypothesis of an associative descriptive domain for two-dimensional geometric objects is based on the assumption of an agreed-upon representation of triangles and squares in which the two segments in a corner are seen as perfectly touching. This means that there is a total agreement among various observers about the exact relative position of lower-level elements within the geometric figure in relation to both the operation of measurement and representation (in terms of lengths of sides and measurement of angles). This requires a compatibility between the accuracy in the measurement of angles and the accuracy in the determination of the length (again, see the example provided at the bottom of Figure 3.8) in relation to the process of fabrication. As soon as we want to compress the perception and representation of a hierarchically organized object (referring to a two-dimensional epistemic category) into a numerical relation among segments (referring to a one-dimensional epistemic category), we lose the assumed parallel validation of two external referents, and therefore the possibility of generalizing this mapping.

The same mechanism (nonequivalence of the associative descriptive domain) can be used to explain the incommensurability between squares and circles. The very definition of a circle entails a type based on a given relation among lower-level elements (points) that by definition do not have a scale and therefore should be considered other types (they cannot have scaled realizations that are measurable). This is where the systemic incompatibility between the two descriptive domains enters into play. The descriptive domain of a square (as well as triangles and polygons) must provide compatibility between the differentials referring to the construction and measure (perception and representation) of both segments (at the level $n - 1$) and angles (at the level n). On the other hand, the descriptive domain of the circle refers only to a measure of the distance of the various points making up the circumference from the center (at the level n). Put another way, the definition of the identity of a square entails more information than a circle (e.g., in relation to the making and measurement of segments and in relation to the making and measurement of the square). Elementary geometry uses this entailment to infer a lot of relations between parts and wholes in geometric objects, whereas the essence of a circle is based on a definition of lower-level elements (points) that must be dimensionless. Because of this, it leaves open the issue of how to judge the compatibility between the space differential referring to the perception and representation of circles. A square is a real holarchic system; a circle is an image, which can always be realized, but with a weaker identity.

As will be discussed in Part 2, the definition of an identity for a triangle or a square is the result of an impredicative loop. The processes of realization (representation) and measurement (perception) of the various holons in parallel on different scales must converge on a compatible definition of them on relative descriptive domains. Segments are whole, made up of points, and, at the same time, are parts making up the square. This very mechanism of definition of identities in cascade entails that the characteristics of elements expressed or detectable at one level do affect (and are affected) by the characteristics of other elements expressed or detectable at a different level.

A more detailed discussion on how to perform the analysis of impredicative loops, the basis of integrated assessment on multiple levels for complex adaptive holarchic systems, is given in Chapters 6 and 7.

This discussion about abstract geometric objects is very important for two reasons. First, it proves that the approach based on complex systems theory discussed so far is very general and makes possible the gaining of new insights, even when getting into very old and familiar subjects. Second, mathematics and geometry are the most important repertoires of metaphors used by humans to build their epistemic tools. So, from the existence of various types of geometry, we can learn a crucial distinction about different classes of epistemic tools as follows:

- There are epistemic tools — like those provided by classic Eucledian geometry — that consist of a set of definitions of essences out of scale and not becoming in time (e.g., the ideas of Plato). An example would be classic geometric objects (like triangles, squares and circles), which are scale independent. This set of essences and relative types is out of time not only in relation to its relative self-entailment, but also in relation to its usefulness as epistemic tools. The fact that these essences can become irrelevant for the observer is not even considered possible.

- There are epistemic tools that — like the objects defined in fractal geometry — consist of a set of definitions of essences not becoming in time. However, the identity of these fractal objects (e.g., the Julia set) implies the existence of multiple identities, depending on how the observer looks at them. When perceiving and representing a fractal object on different scales, we have to expect the coexistence of different views of it. A set of different views of the Mandelbrot set referring to different levels of resolution (by zooming in and out of the same geometric object) is given in Figure 3.9. Also in this case, the definition of this set of multiple identities is not related to the relevance (the usefulness) that the knowledge of this set of identities can have for the observer.

- There are epistemic tools — referring to the perception and representation of dissipative learning holarchic systems — that consist of a set of integrated identities that are scaled, since

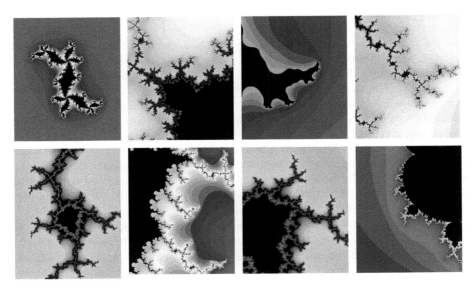

FIGURE 3.9 Different identities of the Mandelbrot set. (Images from Julia and Mandelbrot Explorer by D. Joyce. http://aleph0.clarku.edu/~djoyce/julia/explorer.html)

they require an implicit step of realization to be preserved as types. This explains the existence of the natural set of multiple identities integrated across scales found in biological and human systems (e.g., the different views of a person shown in Figure 1.2). In these systems, the stability of higher-level holons — individuals — is based on the validity of the identity of the class of realizations of lower-level holons — organs — (consistency of the characteristics of the members of equivalence classes of dissipative systems). In turn, lower-level holons — organs — depend for their structural stability on the stability of the identity of higher-level holons — individuals. As soon as one puts in relation this dependency across levels, one is forced to admit that, depending on where one selects the focal level, the identity of a given level is affected by the set of identities determining boundary conditions (on the top) and structural stability (on the bottom). To make things more challenging:

1. This integrated set of identities across scales (the different views shown in Figure 1.2 and Figure 3.1) evolves in time.
2. The ability to have an updated knowledge of this integrated set of identities across scales that are becoming in time is crucial for the survival and success of the observer.

This second observation leads to the concept of complex time — discussed in Chapter 8 — which entails (1) the need to use various differentials for an integrated representation of the system on different levels (the simultaneous use of nonequivalent models) at a given point in time; (2) the continuous updating of these models over a certain time horizon; and (3) the continuous updating of the selection of relevant characteristics used to characterize the identity of the individuality of the holarchy, in relation to the changing goals of the observer, on a larger time horizon.

3.7.2   Dealing with the Special Status of Holons in Science

> Wholes and parts in this absolute sense just do not exist anywhere either in the domain of living organisms or of social organization. (Koestler, 1968, p. 48)

After this long discussion we are left with a huge question: Why, in the first place, do we get equivalence classes of organized structures in the ontological realm? That is, why are a lot of natural systems organized in classes of organized structures that share common sets of characteristics?

Two quick answers to this question are:

1. Systems organized in hierarchies and equivalence classes are easier to perceive, represent and model (to be represented with compression and anticipation). Therefore, systems that base their own stability on the validity of anticipatory models for guiding their action (what Rosen (1985, 1991) calls self-organizing systems generating life) will be advantageous if operating in a universe made up of facts (events, behaviors) organized over typologies. Actually, there is more. Even if the reality is/were made up of facts, events and behaviors generated by entities that are both organized and nonorganized in typologies (just special individual events), the ability to perceive and represent essences by interacting observers developing anticipatory models can only be developed in relation to facts, events and behaviors organized in typologies. In fact, all the rest (special individual entities and behaviors) can only be perceived as noise, since the data stream could not be interpreted or compressed through mapping (for more, see the work of Herbert Simon (1976)).
2. Systems organized in hierarchies are more robust against perturbations. This is true both in relation to the process of their fabrication (recall the metaphor of the two clock makers given by Simon (1962) — the one assembling clocks using a hierarchical approach (using subunits in the process of assembly) was much more resilient to perturbations than the other) and in relation to their operation. Indeed, hierarchical structures operating in parallel on different scales can modulate their level of redundancy of organized structures and functions (buffer against perturbations), which can be diversified in relation to critical functions on different

scales, different tasks in different situations (e.g., critically organized across levels and scales). This dramatically helps the building of an effective filter against perturbations across scales.

Therefore, the concepts of holons and holarchies, even if still quite esoteric to the general reader, seem to be very useful for dealing with the handling of the epistemological challenge implied by complexity. These concepts entail the existence of natural multiple identities in complex adaptive systems. These multiple identities are generated not only by epistemological plurality (by the unavoidable existence of nonequivalent observers deciding how to perceive and describe reality using different criteria of categorization and different detectors), but also by ontological/functional characteristics of the observed systems, which are organized on different levels of structural organization.

From this perspective, there is a lot of free information carried out by a set of natural identities of a holon belonging to a biological or human holarchy. This is at the basis of the concept of the multi-scale mosaic effect.

A holarchy can be seen as a set of natural identities assigned to its own elements by the peculiar process of self-organization over nested hierarchical elements. According to the metaphor proposed by Prigogine (1978), dissipative autopoietic holarchies remain alive, using recipes (information stored in DNA) to stabilize physical processes (the metabolism of organisms carrying that DNA) and physical processes to stabilize recipes — this chicken–egg loop has to be verified and validated at the global and local scales. The same concept has been called a process of self-entailment of natural identities by Rosen (1991), which implies a continuous process of validation of the set of natural identities assigned to the various holons by this process of autopoiesis. This translates into a continuous validity check on:

1. The information referring to the essence of various elements — at the large scale. This is the mutual information that the various elements carry about each other, resulting in the ability to keeping coherence and harmony in the interaction of the various elements.

2. The ability of a given process of fabrication informed by a blueprint to effectively express specimens of the same class of organized structure with a good degree of accuracy — at the local level.

Therefore, a set of multiple identities indicates the past ability to keep correspondence between:

1. The definition of essence for the class, that is, the large-scale validity of the information referring to the characteristics of the equivalence class — the function of the holon in the larger context

2. The viability of structural characteristics of the class, that is, the ability to keep coherence in the characteristics of members of an equivalence class (a set of organized structures sharing the same blueprint and process of fabrication) within their admissible associative context.

3.8 Conclusions

In this chapter I tried to convince the reader that there is nothing transcendent about complexity, something which implies the impossibility of using sound scientific analyses (including reductionist ones). For sure, when dealing with the processes of decision making about sustainability, we need more and more rigorous scientific input to deal with the predicament of sustainability faced by humankind in this new millennium.

On the other hand, complexity theory can be used to show clearly the impossibility of dealing with decision making related to sustainability in terms of optimal solutions determined by applying algorithmic protocols to a closed information space. When dealing with complex behaviors, we are forced to look for different causal relationships among events and keep the information space open and expanding. The various causal relations found by scientific analyses will depend not only on the intrinsic characteristics of the investigated system, but also on decisions made in the preanalytical steps of problem structuring. We can only deal with the scientific representation of a nested hierarchical system by using a strategy of stratification (by using a triadic reading based on the arbitrary selection of a focal space–time differential able to catch one dynamic of interest at the time).

To be able to use science fruitfully, when discussing sustainability, humans should stop pretending that their processes of decision making are based on the ability to detect the best possible course of action after applying standard protocols based on reductionist analyses. This has never been done in the past, it is not done at the present and it will never be done in the future. Any decision always implies a political dimension, since it is based on imperfect information in relation to a given set of goals. Otherwise, it should be called computation (R. Fesce, personal communication).

The confusion on this point is often generated by the fact that, in the last decades, in Western countries the elite in power, for various reasons, decided to pretend that they were making decisions based on substantive rationality. Clearly, this was simply not true, and the clash of reductionist analyses against the issue of sustainability in these decades is clearly exposing such a faulty claim. Complex systems theory can help in explaining the reasons of such a clash. Any definition of priorities among contrasting indicators of performance (reflecting legitimate nonequivalent criteria) is affected by a bias determined by the previous choice of how to describe events (the ideological choices in the preanalytical step). That is, such a choice reflects the priorities and system of values of some agent in the holarchy.

When dealing with the problem of how to do a sound problem structuring, we are in a chicken–egg situation. The results of scientific analyses will affect the selection of what is considered relevant (how to do the next preanalytical step), and what was considered relevant (what was done in the past preanalytical step) affected the results of scientific analyses. This chicken–egg pattern simply explains the coexistence of alternative, nonequivalent and legitimate structuring of sustainability problems found in different human groups separated by geographic and social distances. Different groups went through different paths related to how to organize their perceptions and representations of their interactions with the world. After acknowledging this fact, we cannot expect scientists operating within a given set of assumptions (e.g., those given by an established disciplinary field) to be able to boost the quality of any process of societal problem structuring by forcing their own view on others. The only viable way out of this epistemological predicament is the establishment of procedures of integrated assessment based on transdisciplinary analyses and participatory techniques, that is, by establishing an iterative interaction between scientists and stakeholders as implied by the concept of procedural rationality.

The unavoidable existence of reciprocally irreducible models and the goal of increasing the richness of scientific representation, however, should not be misunderstood as invitations to avoid decisions on how to compress in a useful way the set of analytical tools used to represent and structure our problems. On the contrary, the innate complexity of sustainability issues requires a rigorous filter on sloppy scientific analyses, poor data and inadequate discussion of basic assumptions.

Reciprocally irreducible models can have significant overlap in their descriptive domains. In this case, the parallel use of nonequivalent models dealing with the same system can be used not only to increase the richness of scientific representation, but also to help uncover inconsistencies in the basic hypotheses of the different models, numerical assessments and predicted scenarios. An application of this rationale in terms of biophysical analyses of sustainability is discussed in Parts 2 and 3. In my view, this is a crucial application of complexity in relation to integrated assessment on multiple scales.

Unfortunately, the problem of how to improve the quality of a decision process has not been considered relevant by hard scientists in the past. They have been focused only on finding the best solutions. However, the new nature of the problems faced by humankind in this third millennium implies a new challenge for science. This new term of reference is especially important for those working in the field of integrated assessment of human development.

References

Abbott, E.A, (1952), *Flatland: A Romance of Many Dimensions*, Dover Publications, New York.
Altieri, M., (1987), *Agroecology: The Scientific Basis for Alternative Agriculture*, Westview Press, Boulder, CO.
Cabeza-Gutés, M., (1996), The concept of weak sustainability, *Ecol. Econ.*, 17, 147–156.

Conrad, M., (1983), *Adaptability: The Significance of Variability from Molecule to Ecosystem*, Plenum Press, New York.

FAO Agricultural Statistics, available at http://apps.fao.org/cgi-bin/nph-db.pl?subset=agriculture.

Funtowicz, S.O. and Ravetz, J.R., (1990), *Uncertainty and Quality in Science for Policy*, Kluwer, Dordrecht.

Funtowicz, S.O. and Ravetz, J.R., (1994), Emergent complex systems, *Futures*, 26, 568–582.

Giampietro, M., (1997a), Linking technology, natural resources, and the socioeconomic structure of human society: a theoretical model, in *Advances in Human Ecology*, Vol. 6, Freese, L., Ed., JAI Press, Greenwich, CT, pp. 75–130.

Giampietro, M., Bukkens S.G.F., and Pimentel D., (1997), Linking technology, natural resources, and the socioeconomic structure of human society: examples and applications, in *Advances in Human Ecology*, Vol. 6, Freese, L., Ed., JAI Press, Greenwich, CT, pp. 131–200.

Holling, C.S., (1995), Biodiversity in the functioning of ecosystems: an ecological synthesis, in *Biodiversity Loss: Economic and Ecological Issues*, Perring, C., Maler, K.G., Folke, C., Holling, C.S., and Jansson, B.O., Eds., Cambridge University Press, Cambridge, U.K., pp. 44–83.

International Development Research Centre (IDRC). Website, information on plague in Tanzania, at http://www.idrc.ca/books/reports/V214/plague.html and http://www.idrc.ca/ecohealth/casestudies_e.html.

Kampis, G., (1991), *Self-Modifying Systems in Biology and Cognitive Science: A New Framework for Dynamics, Information, and Complexity*, Pergamon Press, Oxford, 543 pp.

Knight, F.H., (1964), *Risk, Uncertainty and Profit*, A.M. Kelley, New York.

Koestler, A. (1968), *The Ghost in the Machine*. MacMillan. New York.

Pearce, D.W. and Atkinson, G.D. (1993), Capital theory and the measurement of sustainable development: An indicator of "weak" sustainability. *Ecol. Econ.* 8(2): 103–108.

Pimentel, D. and Pimentel, M., (1996), *Food, Energy and Society*, rev. ed., University Press of Colorado, Niwot.

Prigogine, I., (1978), *From Being to Becoming*, W.H. Freeman and Co., San Francisco.

Raynal, F., (1995), A Nobel Prize pioneer at the Pantheon, Label France Mazine n. 21 8/95, http://www.france.diplomatie.fr/label_france/ENGLISH/SCIENCES/CURIE/marie.html.

Rosen, R., (1977), Complexity as a system property, *Int. J. Gen. Syst.*, 3, 227–232.

Rosen, R., (1985), *Anticipatory Systems: Philosophical, Mathematical and Methodological Foundations*, Pergamon Press, New York.

Rosen, R. (1991), *Life Itself: A Comprehensive Inquiry Into the Nature, Origin and Fabrication of Life* Columbia University Press, New York.

Rosen, R., (2000), *Essays on Life Itself*, Columbia University Press, New York, 361 pp.

Schneider, E.D. and Kay, J.J., (1994), Life as a manifestation of the second law of thermodynamics, *Math. Comput. Model.*, 19, 25–48.

Schumpeter, J.A., (1954), *History of Economic Analysis*, George Allen & Unwin, Ltd., London.

Simon, H.A., (1962), The architecture of complexity, *Proc. Am. Philos. Soc.*, 106, 467–482.

Simon, H.A., (1976), From substantive to procedural rationality, in *Methods and Appraisal in Economics*, Latsis, J.S., Ed., Cambridge University Press, Cambridge, U.K.

Stirling, A., (1998), Risk at a Turning Point? *J. Risk Res.* 1(2): 97–110.

Tainter, J., (1988), *The Collapse of Complex Societies*, Cambridge University Press, Cambridge, U.K.

Ulanowicz, R.E., (1986), *Growth and Development: Ecosystem Phenomenology*, Springer-Verlag, New York.

4

The New Terms of Reference for Science for Governance: Postnormal Science

This chapter addresses the epistemological implications of complexity. In fact, according to what has been discussed so far, hard science, when operating within the reductionist paradigm, is not able to handle in a useful way the set of relevant perceptions and representations of the reality used by interacting agents, which are operating on different scales. No matter how complicated, individual mathematical models cannot be used to represent changes on a multi-scale, multi-objective performance space. To make things worse, it must be acknowledged that there are two relevant dimensions in the discussion about science for governance: one related to the *descriptive* side (the ability to represent the effect of changes in different descriptive domains by using an appropriate set of indicators) and one related to the *normative* side (the ability to reach an agreement on the individuation of an advisable policy to be implemented in the face of contrasting values and perspectives). As noted in Chapters 2 and 3, these two dimensions are only apparently separated, since, due to the epistemological implications discussed so far, even when operating within the descriptive domain, there are a lot of decisions that are heavily affected by power asymmetry. Who decides how to simplify the complexity of the reality? Who decides whose perceptions are the ones to be included in the analysis? Who chooses the appropriate language, relevant issues and significant proofs? Put another way, the very definition of a problem structuring (how to describe the problem) entails a clear bias for the normative step. The reverse is also obviously true (policies are determined by the agreed-upon perceptions of costs, benefits and risks of potential options). In conclusion, the issue of science for governance requires addressing the issue of how to generate procedures that can be used to perform multi-agent negotiations aimed at getting compromise solutions on a multi-criteria performance space. The general implications of this fact are discussed in this chapter, whereas technical aspects related to the role of scientists in this process are discussed in Chapter 5.

4.1 Introduction

There is a very popular family of questions that very often are used when discussing sustainability. For example, Richard Bawden often makes the point that both the scientists in charge of developing scenarios, models, indicators and assessments and the stakeholders in charge of the process of decision making should first of all address the following three questions: (1) What constitutes an improvement? (2) Who decides? (3) How do we decide? Joe Tainter's list of questions includes: (1) Sustainability for whom? (2) For how long? (3) At what cost? The group of ecological economics in Barcelona has another variant: (1) What do we want to sustain? (2) Who decided that? (3) How fair was the process of decision? Remaining in the field of ecological economics, Dick Noorgard has been using for more than a decade his own list of a similar combination of questions.

These are just a few samples taken from a large and expanding family. In fact, the same semantic message can be found over and over when looking at the work of different groups of sustainability analysts. The meaning of this family of questions is that, to produce relevant and useful scientific input (before getting into the steps of formalization with models, based on a selection of variables and thresholds and benchmarks on indicators), scientists have to first answer a set of semantic questions that are difficult to be formalized.

By "semantic" I mean the ability to share the meaning assigned to the same set of terms by the population of users of those terms. Very often the task of checking on the semantic of the problem structuring (validity of assumptions and relevance of the selection of encoding variables) is not included among the activities of competence of reductionist scientists. However, when dealing with legitimate contrasting views, uncertainty and ignorance, multiple identities of systems operating in parallel on different scales, such as a quality check, become an additional requirement for the scientists willing to deal with sustainability.

This statement is so obvious to appear trivial. However, looking at the huge amount of literature dealing with the optimization of the performance of farming systems or the optimization of techniques of production, one can only wonder. If scientists are operating in a situation in which they cannot specify with absolute certainty what is the output of agriculture (commodities? quality food? clean water? preservation of desirable landscapes? preservation of biodiversity? other outputs for other people?), then it is not possible to calculate any indicator of absolute efficiency (leading to the individuation of the best strategy of maximization) using classical reductionistic approaches.

The message given in the previous chapters is that the concept of multifunctionality in agriculture translates into the impossibility of (1) representing in a coherent way different typologies of performance (on the descriptive side) and (2) optimizing simultaneously different types of performance (on the normative side). The analyst has to deal with different assessments, which requires the use of nonreducible models (the modeling of different causal mechanisms operating at different scales). The simultaneous use of nonreducible models (referring to logically independent choices of meaningful representations of shared perceptions) implies incommensurability and incomparability of the information used in the integrated assessment.

Talking of a quality check, there is another practical impasse found when considering the reliability of scientific inputs to the process of decision making, which is related to the timing imposed on the scientific process by external circumstances. If scientists are forced by stakeholders to tackle specific problems at a given point in space and time (according to a given problem structuring), and the pace and the identity of the scientific output are imposed on them by the context, then scientists could face a mission impossible in delivering high-quality output in this situation. Depending on the speed at which the mechanisms generating the problem to be studied are changing in time or the speed at which the relevance of issues changes in time, it can become impossible even for smart and dedicated scientists to develop a sound scientific understanding.

The question of how to improve the quality of a decision process that requires a scientific input that is affected by uncertainty has to be quickly addressed by both scientists and decision makers. In 2002 the Royal Swedish Academy of Sciences gave the Nobel Prize in economics to Professor Kahneman for his pioneering work on integrating insights from psychology into economics, "especially concerning human judgment and decision making under uncertainty, where he has demonstrated how human decisions can systematically depart from those predicted by standard economic theory," as said in the official citation. As noted earlier, traditional reductionist theory posits human beings as rational decision makers. But in reality, according to Kahneman, people cannot make rational decisions because "we see only part of every picture."

When science is used in policy, laypersons (e.g., judges, journalists, scientists from another field or just citizens) can often master enough of the methodology to become effective participants in the dialogue. This necessary step will be easier to take if scientists make an effort to package in a more user-friendly way their scientific input. This effort from the scientists is unavoidable since this extension of the peer community is essential for maintaining the quality of the process of decision making when dealing with reflexive complex systems.

It is in relation to this goal that Funtowicz and Ravetz (1992) developed the new epistemological framework called postnormal science. The message is clear: science in the policy domain has to deal with two crucial aspects — *uncertainty* and *value conflict*. The name "postnormal" indicates a difference from the puzzle-solving exercises of normal science, in the Kuhnian sense (Kuhn, 1962). Normal science, which was so successfully extended from the laboratory of core science to the conquest of nature through applied science, is no longer appropriate for the solution of sustainability problems. Sending a few humans for a few hours on the moon is a completely different problem than keeping in harmony and decent conditions in the long run 8 billion humans on this planet. In sustainability problems social, technical and ecological dimensions are so deeply mixed that it is simply impossible to consider them as separate, one at the time, as done within conventional disciplinary fields.

4.2 The Postnormal Science Rationale

4.2.1 The Basic Idea

To introduce the basic concepts related to postnormal science, we use a presentation given by Funtowicz and Ravetz in the book *Chaos for Beginners* (Sardar and Abrams, 1998, pp. 157–159):

> In pre-chaos days, it was assumed that values were irrelevant to scientific inference, and that all uncertainties could be tamed. That was the "normal science" in which almost all research, engineering and monitoring was done. Of course, there was always a special class of "professional consultants" who used science, but who confronted special uncertainties and value-choices in their work. Such would be senior surgeons and engineers, for whom every case was unique, and whose skill was crucial for the welfare (or even lives) of their clients.
>
> But in a world dominated by chaos, we are far removed from the securities of traditional practice. In many important cases, we do not know, and we cannot know, what will happen, or whether our system is safe. We confront issues where facts are uncertain, values in dispute, stakes high and decisions urgent. The only way forward is to recognize that this is where we are at. In the relevant sciences, the style of discourse can no longer be demonstration, as for empirical data to true conclusions. Rather, it must be dialogue, recognizing uncertainty, value-commitments, and a plurality of legitimate perspectives. These are the basis for post-normal science.
>
> Post-normal science can be illustrated with a simple diagram [Figure 4.1].
>
> Close to the zero-point is the old-fashioned "applied science." In the intermediate band is the "professional consultancy" of the surgeon and engineer. But further out, where the issues of safety and science are chaotic and complex, we are in the realm of "post-normal science." That is where the leading scientific challenges of the future will be met.
>
> Post-normal science (PNS) has the following main characteristics: Quality replaces Truth as the organizing principle.

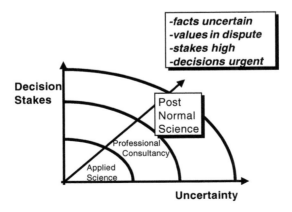

FIGURE 4.1 Postnormal science.

In the heuristic phase space of PNS, no particular partial view can encompass the whole. The task now is no longer one of accredited experts discovering "true facts" for the determination of "good policies." PNS accepts the legitimacy of different perspectives and value-commitment from all the stake-holders around the table on a policy issue. Among those in the dialogue, there will be people with formal accreditation as scientists or experts. They are essential to the process, for their special experience is used in the quality control process of the input. The housewife, the patient, and the investigative journalist, can assess the quality of the scientific results in the context of real-life situation. We call these people an "extended peer community." And they bring "extended facts," including their own personal experience, surveys, and scientific information that otherwise might not have been in the public domain.

PNS does not replace good quality traditional science and technology. It reiterates, or feedbacks, their products in an integrating social process. In this way, the scientific system will become a useful input to novel forms of policy-making and governance.

4.2.2   PNS Requires Moving from a Substantial to a Procedural Definition of Sustainability

It is often stated that sustainable development is something that can only be grasped as a fuzzy concept rather than expressed in an exact definition. This is because sustainable development is often imagined as a static concept that could be formalized in a definition out of time applicable to any specific situation that does not need external semantic referents to get an operational meaning. To avoid this trap, we should move to a definition of sustainability that requires or implies the ability of a society to perform external semantic quality checks on the correct use of all adjectives and terms in the definition. When this ability exists, sustainable development can be defined as the ability of a given society to move, in an adequate time, between satisficing, adaptable, and viable states. Such a definition explicitly refers to the fact that sustainable development has to do with a process of social learning (procedural sustainability) rather than a set of once-and-for-all definable qualities (substantial sustainability). This distinction recalls that made by another Nobel Prize winner in economics, Herbert Simon (1976, 1983), about the different types of rationality used by humans when deciding in the economic process.

Put another way, it is not possible to provide a syntactic representation and formulation of sustainability — both in descriptive and normative terms — that can be applied to any practical case. On the contrary, the idea of post-normal science entails the need to always use a semantic check to arrive at a shared meaning among stakeholders about how to apply general principles to a specific situation (when deciding in a given point in space and time).

A procedural sustainability implies the following points:

1. Governance and adequate understanding of present predicaments, as indicated by the expression "the ability to move in an adequate time."
2. Recognition of legitimate contrasting perspectives related to the existence of different identities for stakeholders. This implies:
    a. The need for an adequate integrated representation reflecting different views (quality check on the descriptive side)
    b. An institutional room for negotiation (quality check on the normative side),
   as indicated by the expression "satisficing"
3. Recognition of the need to adopt an evolutionary view of the events we are describing (strategic assessment over possible scenarios). This implies the unavoidable existence of uncertainty and indeterminacy in the resulting representation and forecasting of future events. When discussing adaptability (the usefulness of a larger option space in the future), reductionistic analyses based on the *ceteris paribus* hypothesis have little to say, as indicated by the expression "adaptable."
4. Recognition of the need to rely on sound reductionistic analyses to verify within different scientific disciplines the viability of possible solutions in terms of technical, economic, ecological and social constraints, as indicated by the expression "viable states."

This definition of sustainable development implies a paradigm shift in the process that is used to generate and organize the scientific information for decision making and that can be related to the very concept of postnormal science.

4.2.3   Introducing the Peircean Semiotic Triad

The validity of models, indicators, criteria and data used in a process of decision making can be checked only against their usefulness for a particular social group — at a given point in space and time — in guiding action. This implies viewing the process of generation of knowledge as an iterative process occurring across several space–time windows at which:

1. It is possible to define a validity for the modeling relation
2. It is possible to generate experimental data sets, through measurement schemes
3. The knowledge system within which the scientist is operating is able to define itself in relation to:
    a. Goals
    b. Perceived results of current interaction with the context
    c. Experience accumulated in the past

An overview of such an iterative process across scales is given in Figure 4.2 using the Peircean semiotic triad as a reference framework (Peirce, 1935). The cyclic process of resonance among the three steps — pragmatics, semantic, syntax — is seen as a process of iteration that goes in parallel in two opposite directions (double asymmetry). The two loops operating in opposite directions on different space–time windows are shown in Figure 4.2. Recall the need to use two nonequivalent external referents in the iterative process of convergence of shared meaning about identities in holarchies (or words in the formation of languages) in Chapter 2 (Figure 2.4).

Starting with the smaller one (the clockwise one in the inside of the scheme), out of the existing reservoir of known models that have been validated in the past, the box labeled "syntax" provides the tools needed to generate numerical assessments (reflecting the identities assigned to relevant systems to be modeled) — *represent*. This makes it possible to recognize patterns and organized structures as types and members of an equivalence class. This is what provides a set of descriptive tools that makes it

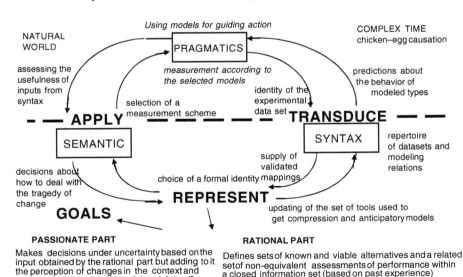

FIGURE 4.2 Self-entailing process of generating knowledge.

possible to run models to generate useful predictions. To get into the *apply* step, however, we have to first go through a semantic check, which implies defining the validity of the selected models (from syntax) in relation to the given goal and context. Gathering data is an operation belonging to the pragmatics domain and implies a direct interaction with the natural world. In this step the system of knowledge is gathering information about the world, organizing the perceptions through the existing set of known epistemic tools. The result of this interaction is the experimental data set. *Transduce* here means that the system of knowledge is internalizing the information obtained when interacting with the natural world according to the two steps represent and apply.

The larger, counterclockwise loop is related to events occurring on a larger scale. Starting from the same box, "syntax," this time the operation *transduce* implies generating predictions about expected behaviors on the basis of the scientific knowledge available to guide action. The interaction with the natural world (belonging to pragmatics) is based on the *apply* of these scientific predictions for guiding actions in relation to the existing set of goals. At this point a semantic check is needed to assess whether the scientific input was useful for guiding such interaction.

If the perceived results of the interaction with the natural world are consistent with the existing set of goals, the scientific input is judged adequate. In this case, the system of knowledge (which is the result of a converging process over the diagram) confirms such a system of models as one of the tools in the repertoire of validated models (to be applied in the same situation) and will rely on it again for future decisions — *represent*. If, on the contrary, important gaps are found between the qualities that are perceived to be relevant for achieving the existing set of goals and the set of qualities mapped by the chosen set of models (the scientific input failed in helping to achieve the goals), then the semantic check declares a particular system of models obsolete, implying an updating in the step *represent*.

It is obvious that the diagram described in Figure 4.2 is no longer describing only the process of making models. Rather, it addresses also the effects that the use of models induces on those using the validated knowledge in the interaction with their context. This is why scientists have to be told whether the scientific input they are generating is relevant. In the diagram, in fact, there are several scales and actors supposed to generate the emergent property of the whole. There are individual scientists developing competing models within individual scientific fields. There are groups of scientists expanding and adjusting the identity of competing scientific fields. Then, the various stakeholders and social actors of the society interact in different ways to legitimize the use of science within the processes of decision making. According to this frame, we should view any system of modeling just as a component of a larger system of knowledge that is in charge of operating an endless process of convergence and harmonization of heterogeneous flows of information referring to (1) a common experience (given past) and (2) a set of different and legitimate goals (possible virtual futures), which must always be linked to an evaluation of (3) present performance in relation to the existing goals and the context. Such a continuous filtering of information across scales and in relation to the need to continuously update the identity of the various components of the society implies again a fuzzy chicken–egg type of process (impredicative loop) rather than a clear-cut, once-and-for-all describable process. Scientists are operating within an existing system of knowledge, and because of that, they are affected in their activity by its identity and are affecting its identity with their activity.

4.2.4   A Semiotic Reading of the PNS Diagram

The problem of governance of human systems can be related to the necessity of selecting components of the holarchy that have to (or should be) sacrificed for the common good. Thanks to the duality of the nature of holons, components to be sacrificed do not necessarily have to be real individual organized structures. Holons to be sacrificed can be jobs, firms, traditions, values, cities. In other cases, however, the sacrifice is tougher, and it can entail destroying resources and, in some cases, even individual humans (e.g., in the case of war). On the other hand, this process of elimination and turnover is related to life. Within adaptive holarchies components have to be continuously eliminated (turnover on the lower-level holons within the larger holon) to guarantee the stability of the whole. The term *governance* refers to the human system, which can be characterized as reflexive systems. This means that human will does affect the pattern of selective elimination of holons within a human holarchy, which therefore is no longer determined only by external selection and pure chance. Evolution, progress and, more in general, the unavoidable process of becoming imply for human systems the necessity of continuously facing the tragedy of change (a term coined for postnormal science by Funtowicz and Ravetz). Even the most innocent and laudable intention framed within a given problem structuring — e.g., the elimination of poverty — will end up by eliminating from our universe of discourse identities relevant within a nonequivalent problem structuring (e.g., eliminating poverty entails the elimination of the various identities taken by the poor). Holons and holarchies can survive only because of their innate tension (a real Yin–Yang tension) between the need of preserving identities and the need of eliminating identities. This means that conflicting interests and conflicting goals are unavoidable within holarchic systems. The search for a win–win solution valid on different timescales and in relation to the universe of the agents is just a myth. The problem is therefore how to handle these tensions within systems that express awareness and reflexivity in parallel at different hierarchical levels (e.g., individual human beings, households, communities, regions, countries, international bodies).

The holarchic nature of human societies implies two major problems related to their capability of representing themselves and individuating rational choices. Robert Rosen, an important pioneer in the applications of complex systems thinking to the issue of sustainability and governance, can be quoted here:

> (1) Life is associated with the interaction of non-equivalent observers. Legitimate and contrasting perceptions and representations of the sustainability predicament are not only unavoidable but also essential to the survival of living systems.

> The most unassailable principle of theoretical physics asserts that the laws of nature must be the same for all the observers. But the principle requires that the observers in question should be otherwise identical. If the observers themselves are not identical; i.e., if they are inequivalent or equipped with different sets of meters, there is no reason to expect that their descriptions of the universe will be the same, and hence that we can transform from any such description to any other. In such a case, the observers' descriptions of the universe will bifurcate from each other (which is only another way of saying that their descriptions will be logically independent; i.e., not related by any transformation rule of linkage). In an important sense, biology depends in an essential way on the proliferation of inequivalent observers; it can indeed be regarded as nothing other than the study of the populations of inequivalent observers and their interactions. (Rosen, 1985, p. 319)

This passage makes a point related to biology, which obviously would be much stronger when related to the status of sciences dealing with the behavior of social systems.

> (2) The sustainability of a holarchy is an emergent property of the whole that cannot be perceived or represented from within. The sustainability predicament cannot be fully perceived by any of the components of social systems.

> The external world acts both to impose stresses upon a culture and to judge the appropriateness of the response of the culture as a whole. The external world thus sits in the position of an outside observer. Since selection acts on the culture as a whole, there is only an indirect effect of selection on the members of the culture and hence on their internal models of the culture. This is indeed, a characteristic property of aggregates like multi-cellular organisms or societies; namely, that selection acts not directly on the individual members of the aggregate, but on the aggregate as a whole. We have seen that the behaviors of the aggregate as a whole are not clearly recognizable by any of the members of the aggregate and therefore none of the internal models of the aggregate can comprehend the manner in which selection is operating. Stated another way, the members of a culture respond primarily to each other, and to each other's models, rather than to the stresses imposed on the culture by the external world. They cannot judge the behaviour of the culture in terms of appropriateness at all, but only in terms of deviation from their internal models. (Rosen, 1975, p. 145)

These two passages beautifully summarize what was said before about the impossibility to define in absolute terms the optimal way to sustainability. It is impossible to define in an objective way what is the right mix between efficiency and adaptability or — expressed in a nonequivalent way — between the respecting and the breaking of the rules (recall here the example of mutations on DNA, which are errors when considered on one scale and useful functions when considered on another). Within the same holarchy, the very fuzzy nature of holons, which are vertically coupled to form an emergent whole, implies that there is a hierarchical level at which humans express awareness (individual humans being) that does not coincide with the hierarchical level at which they express systems of knowledge (culture that is a property of societal groups). In turn, none of these levels coincides with the hierarchical level at which the mechanisms generating biophysical constraints — the mechanisms relevant in relation to sustainable development — are operating (e.g., global stability for ecological, economic and social processes).

Put another way, the growing integration of various human activities over the planet requires a growing ability to represent, link, assess and govern, which in turn requires an increased harmonization of behaviors expressed by different actors/holons (national governments, international bodies, individual human beings, communities, households). This translates into the need to develop nonequivalent meaningful perceptions and representations of processes occurring in parallel on different space–time scales.

To make things more difficult, these integrated representations must be useful in relation to the existing diversity of systems of knowledge. This is where, in these decades, the drive given by reductionist science to technical progress got into trouble. As remarked by Sarewitz (1996, p. 10):

> The laws of nature do not ordain public good (or its opposite), which can only be created when knowledge from the laboratory interacts with the cultural, economic, and political institutions of society. Modern science and technology is therefore founded upon a leap of faith: that the transition from the controlled, idealized, context-independent world of the laboratory to the intricate, context-saturated world of society will create social benefit.

The global crisis of governance can be associated with the fact that science and technology are no longer able to provide all the useful inputs required to handle in a coordinated way (1) the process of economic expansion (which is represented and regulated with a defined set of tools — economic analyses — that worked well only for a part of humankind in the past); (2) the discussion of how to deal with the tragedy of change occurring within fast-becoming cultural identities in both developed and developing countries; and (3) the challenge of handling the growing impact of human activity on ecological processes (which, at the moment, is not understood and represented well enough, especially for large-scale processes such as those determining the stability of entire ecosystems and of the entire biosphere).

At this point it can be useful revisit the diagram of postnormal science given in Figure 4.1, trying this time to frame the basic message using the semiotic triad of Peirce. The original diagram proposed by Funtowicz and Ravetz is a very elegant and powerful descriptive tool able to catch and communicate to a general audience, in an extremely compressed way, the most relevant features of the challenges implied by PNS. Any attempt to present a different version implies certainly the risk of losing much of its original power of compression. However, exploring more in detail the insights given by this diagram can represent a useful complementing input. The complementing diagram (certainly more crowded with information and much less self-explanatory) is presented in Figure 4.3.

4.2.4.1   *The Horizontal Axis* — The horizontal axis, called uncertainty in Figure 4.1, is the axis

that refers to the dimension *represent* of the triad. This has to do with the descriptive role of scientific input (e.g., multi-scale integrated analysis). Moving from the origin rightward means changing the size and nature of the descriptive domain used to represent the event. The label "simple" on the left side of the axis indicates that in this area we are dealing with only one relevant space–time differential when representing the main dynamics of interest. This also implies that we can describe the behavior of interest without being forced to use simultaneously nonreducible, nonequivalent descriptions (the model adopted is not affected by significant bifurcations). In this situation, we can ignore the problems generated by (1) the unavoidable indeterminacy in the representation of initiating conditions of the natural system represented (the triadic filtering is working properly) and (2) the unavoidable uncertainty in any predicted behavior of the natural system modeled (the assumptions of a quasi-steady-state description under the *ceteris paribus* hypothesis are holding). Simple models work well for handling simple situations (e.g.,

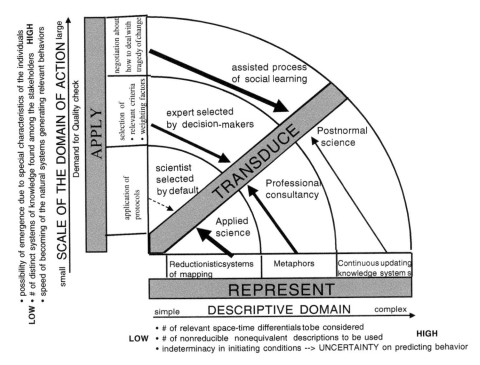

FIGURE 4.3 Evolutionary processes out of human control.

the building of an elevator). Moving to the right means a progressive increase of epistemological problems: the relevant qualities to be considered in the problem structuring require the consideration of nonequivalent perceptions of the reality, and therefore the relative models can be represented only by adopting different space–time windows and using nonequivalent descriptive domains (e.g., maximization of economic profit and minimization of impact on ecological integrity, or in a medical situation, deciding between contrasting indications about costs, risks and expected benefits, both in the short and long term). The more we move to the right, the more we need to use a complex representation of the reality. This implies considering a richer mosaic of observers–observed complexes. A system's behavior must be based on the integrated use of various relevant identities of the system of interest, which in turn translate into the use of several space–time differentials, nonequivalent descriptions and nonreducible models. An unavoidable consequence of this is that the levels of indeterminacy and uncertainty in the prediction of causality (e.g., between the implementation of a policy and the expected effect) become so high that the system requires the parallel use of different typologies of external semantic checks. Recalling the discussion in Chapter 3, uncertainty can be due to two different mechanisms: (1) lack of inferential systems that are able to simulate causal relations among observable qualities on the given descriptive domain (uncertainty due to indeterminacy) and (2) lack of knowledge about relevant qualities of the system (already present but ignored or that will appear as emergent properties in the future) that should be included in one of the multiple identities used to represent the system in the integrated analysis (uncertainty due to ignorance).

4.2.4.2   The Vertical Axis — The vertical axis, which is called decision stakes in Figure 4.1, is the axis that refers to the dimension *apply* of the triad. This has to do with the normative aspect (e.g., multi-criteria evaluation) of the process of decision making. Moving from the origin toward upper values means changing the scale of the domain of action. The label "demand of quality check," which is changing between low (close to the origin) and high (up in the axis), indicates the obvious fact that a change in the scale of the domain of action requires a different quality in the input coming from the step *represent*. The scientific input has to be adequate both in (1) extent (covering the larger space–time window of

relevant patterns to be considered); that is, large-scale scenarios must forget about the *ceteris paribus* hypothesis and look at key characteristics of evolutionary trajectories and (2) resolution (being able to consider all lower-level details that are relevant for the stability of lower-level holons). When operating at a low demand for quality check — close to the origin of the axis — we are dealing with very well established relational functions performed by very robust types within a very robust associative context. When dealing with the description of the behavior of reflexive systems we (humans) face additional problems, due to the unavoidable presence of (1) various systems of knowledge found among social actors that entail the existence of different and logically independent definitions of the set of relevant qualities to be represented, reflecting past experiences and different goals and (2) the high speed of becoming of the social system under analysis, which is generating the relevant behavior of interest (human systems tend to co-evolve fast within their context). This implies the need to establish an institutional activity of quality control and patching and restructuring of the models and indicators used in the process of decision making to perform the step *represent*.

As noted earlier, the fast process of becoming is an unavoidable feature of human societies. Every time we consider representing their behavior on a large space–time domain and an equally expansive domain of action, we have to expect that on the upper part of the holarchy, larger holons cannot be assumed to be in steady state. That is, the *ceteris paribus* assumption becomes no longer reliable. Rather, the holons should be expected to be in a transitional situation in continuous movement over their evolutionary trajectory (and therefore impossible to predict with simple inferential systems).

4.2.4.3   Area within the Two Axes — In the graph of the PNS presented in Figure 4.2 a third diagonal axis is required to complete the semiotic triad of Peirce — an axis related to *transduce* — that wants to indicate the peculiar and circular (egg–chicken) relation between the activities related to *represent* (descriptive side) and *apply* (normative side). Various arrows starting from the two axes and clashing on the diagonal axis indicate the different directions of influence that the various activities of the semiotic triad have on each other over different areas of the diagram.

4.2.4.3.1   Applied Science

When simple descriptive domains are an acceptable input for guiding action (e.g., specific technical problems studying elementary properties of human artifacts — the design and the safety of a bridge), we are in the area of applied science. In this case, (1) the qualities to be considered relevant for the step *represent* are given (that is, reflected in a selection by default of criteria and variables to be used to represent the problem — a standard-type bridge — operating in its expected associative context) and (2) the weight to be given to the various indicators of performance is also assumed to be given to the scientist by society (e.g., design and action must optimize efficiency or minimize costs).

All other significant dimensions of the problem have been taken care by the scientific framing of the problem (problem structuring) given to the engineer (in the case of the bridge). Reductionist models are the basis for the step representation in this area. They imply the generation of a clear input for guiding action within well-specified and known associative contexts (e.g., the application of protocols for building and maintaining bridges). Under these conditions, the specific identity of scientists providing such an input to the process is really not relevant. Their personal values cannot affect the identity of the representation input in a relevant manner. Therefore, any information about the cultural or political identity of the scientists in charge of delivering the descriptive input to the process of decision making is not considered relevant.

4.2.4.3.2   Professional Consultancy

When simple descriptive domains are no longer fully satisficing for guiding actions (e.g., when dealing with problems requiring the consideration of several noncommensurable criteria), we are in the area of professional consultancy. In this case, the step *represent* is based on the use of metaphors (applications of models that were verified and applied before but, in the case of analysis, cannot be backed up by an experimental scheme). This is always the case when dealing with the specific performance of a specific natural system at a particular point in space and time, and that implies important stakes for the decision maker (e.g., advice asked of a surgeon about a delicate medical situation). This situation implies the mixing of (1) a generic input for guiding actions (expressed in terms of metaphors or principles to be used in a defined class of situations) derived from existing and codified knowledge and (2) a tailoring of such useful information, which is asked of the expert for dealing with the specific case. When asked to put her or his head in the mouth of a tiger (facing very high stakes), the tamer needs (1) a basic knowledge about tigers' behavior (retrievable from books), (2) experience and direct knowledge about past interactions with that particular specimen of tiger and (3) a guesstimate based on intuition (reading or feeling) about what this particular specimen of tiger will do at this particular point in space and time. Obviously, in these cases, the particular identity of the scientist providing the input about a meaningful perception or representation is no longer irrelevant for the stakeholder getting advice from the scientist. If you put a crucial decision about your life into the hands of a surgeon, you want to know about that doctor's character and special aspects as a person.

In the field of professional consultancy, both the individual natural system to be modeled (belonging to an equivalence class associated with the type) and the individual modeler (belonging to an equivalence class of scientists — medical doctors) are considered special, since the particular combination of the two can make an important difference. In this case, value judgments are essential on both sides; when perceiving and representing on the descriptive side (since the scientist will apply the available metaphors according to their particular perception of the specific situation) and when applying on the normative side (since the decision maker will select the scientist to hire, according to various criteria that are related not only to the nature of the specific problem to be solved). Moreover, the decision maker can decide not to follow the advice received by the consultant if such advice is not convincing enough. In this case, the selection of the particular scientists to be used in the step *represent* will be based on their perceived ability to (1) use a set of incommensurable criteria (the set of qualities to be considered relevant for the step *represent*) that reflects those relevant for the decision maker and (2) tailor their profiles of weighing factors over incommensurable criteria (the final advice about what to do) according to the ideas expressed by the decision maker who hired them as consultants.

4.2.4.3.3 Postnormal Science

When several descriptive domains are required to consider various nonequivalent causal relations and the domain of action includes the unavoidable interaction of agents adopting heterogeneous systems of knowledge, we no longer can expect to have a unique objective input about how to perceive and represent the problem that can be used as a basis for structuring the process of decision making. As noted earlier, the unformalizable tension between compression and relevance (on the descriptive side), as well as between adaptability and efficiency (on the normative side), makes it virtually impossible to handle the search for the best course of action in human systems in a rational (syntactic) way. Actually, if we move too much to the upper-right corner of the graph, we can arrive at an area in which the whole system is evolving by continuously generating emergent properties. This frontier area escapes any possible definition of improvement or worsening. A useful problem structuring is simply not possible because:

1. On the descriptive side (horizontal axis), we cannot have a prediction and therefore a useful understanding of possible future scenarios. This prevents the development of adequate (relevant and therefore useful) descriptive domains. Ignorance means that we do not and cannot know about future emergent properties. We cannot know what new indicators referring to new relevant qualities have to be used now to decide what to do (Figure 3.4b). In this last analysis, we lack the ability to assign useful identities to the future organized structures and relational functions (holons) of becoming systems. Obviously, this makes unthinkable any attempt to represent our future perceptions of them.

2. On the normative side, we have no possibility of reaching an agreement about what should be a shared meaning of perceptions of future events in relation to the future cultural identities that will be associated with distinct systems of knowledge. Not only do we lack any input on the descriptive side, but also we lack crucial input (goals, wants, fears of future generations) on the normative side. In conclusion, within this area — when considering very large-scale and

evolutionary processes — things just happen out of human control. It is important to note, as remarked by Rosen above, that any qualification of scenarios in the very long term is unavoidably affected by our limited ability to represent now the future perceptions that humans will have of their reality then. The present generation must necessarily base its reasoning on an image biased by perspectives belonging to lower-hierarchical-level perceptions (the various social groups now expressing their systems of knowledge within the existing holarchy).

The area called postnormal science is living dangerously on the border, with professional consultancy on one side and the impossible handling of an agreed-upon representation of future perceptions on the other. On this frontier, new meanings are being generated within human systems of knowledge, but these new meanings still have not been fully internalized by existing cultures. In this part of the plan the complexity of description to be used in the process of decision making is so high that it implies a continuous process of learning (patching and updating of the systems of knowledge used) in the *represent* step through the iterative process shown in Figure 4.2. The scale with which humans try to keep coherence between perceptions and representations is so large that social holarchies at that level are becoming something else at a speed that prevents the consolidation of agreed-upon validated and established common knowledge of the whole. The continuous creation of meaning can be interpreted as the ability to find new combinations of identities for the definition of a shared problem structuring that make sense (is useful) in relation to the successful interaction of nonequivalent agents.

Clearly, when dealing with the semiotic triad at this level, the only way to verify the efficacy of any knowledge system is by checking its usefulness in guiding actions. The semiotic triad can be seen as a process of social learning aimed at tuning the patching and updating of the various systems of knowledge to the process of co-evolution of different social systems within their contexts. What is important here is to be aware that the context for a knowledge system is never a biophysical environment (i.e., ecological process), but rather the activities controlled by other systems of knowledge within a given biophysical environment. This is why no one can see the whole. The demand for adaptability implies that, in this process of becoming, social systems should be able to increase their cultural diversity (to avoid hegemonization of one system of knowledge over the others). This can be obtained only by sharing the stress generated by the tragedy of change. That is, the preservation of cultural identity requires an institutionalized process of negotiation among different social holons to guarantee the diversity of systems of knowledge from massive extinction (Matutinovic, 2000). Put another way, this view implies that conflicts are important, since they are the sign of the existence of diversity. This is a must for guaranteeing the sustainability of human development in the long run.

In this view, a full reliance on rational choices (e.g., maximizing performance according to a representation of benefits supposed to be substantive since it reflects existing perceptions of a given social group in power) cannot be a wise decision. Humans can and should decide to go for suboptimal solutions (even when facing an easy definition of local optimum) just to preserve and avoid the collapse of cultural diversity. This is what is done, for example, when allocating resources on marginalized social groups.

On the other hand, this rule does not always require an express enforcement, since human systems already go for suboptimal solutions. They do so because they do not have the time or means to look for rational choices (maximization of utility under a given set of assumptions and institutions) or, rather simply, because they like suboptimal solutions — when standing for values, going for romantic escapades, drinking with friends, fighting for justice, smoking, doing the "right" thing right now, gambling etc. From this perspective, it is possible to see that passionate choices (as opposed to rational choices) are perfectly defendable in scientific terms. Ethics and rebellion, vices and deep values affecting human choices in fact play a crucial role in keeping alive the process of evolution of social systems, and this tension between mechanisms keeping order and generating disorder can be easily related to the unavoidable uncertainty related to the process of evolution of becoming systems. Errors and mutations so dangerous at one level are needed and useful at a higher one. Long-term sustainability of reflexive systems requires that agents decide to use a mix of rational and passionate choices. In fact, in the long term, and in relation to the whole complex of interacting nonequivalent observers, at any given point in space and time (within a given descriptive domain) not only it is impossible to select optimal solutions on a purely rational basis, but also it is impossible to select suboptimal solutions.

In conclusion, when dealing with postnormal science:

1. The right set of relevant criteria to be used to represent a problem is not known or knowable *a priori*. A satisficing set of relevant criteria can be obtained only as the result of negotiation among various stakeholders who are collectively dealing with the stress implied by the tragedy of change.

2. The weight to be given to incommensurable and contrasting criteria of performance cannot be defined once and for all after considering existing knowledge system(s) and applicable over the planet to different location-specific situations.

3. It is impossible to have an objective definition of the best thing to do, even at a particular point in space and time. After all, the very choice of assigning different weights to nonreducible indicators of performance in a specific situation for a specific social element is equivalent to a choice of the best suboptimal solution, which by definition is a non-sense.

4.3 Quality Replacing Truth: Science, Sustainability and Decision Making

In the last two centuries, hard sciences have focused only on those situations that were easy to represent and model with success. Problems requiring too many relevant variables and nonequivalent descriptions were considered uninteresting (or even nonscientifically relevant) since they were not easy to compress through "heroic" simplifications. On the other hand, in recent decades, the speed of industrialization and globalization is posing new types of challenges to the process of governance of human development. Under these circumstances, a curiosity-driven science (picking up only those topics that happen to be tractable) is no longer useful for the stabilization of current systems of control. The sustainability predicament is asking scientists to fill, as fast as possible, an increasing knowledge deficit generated by the lack of useful representation tools and useful predictive models to be used in the process of decision making. However, when framing the problem in this way, we must notice that the epistemic tools required to perceive and represent problems and predictive models to assess potential solutions should include in their descriptive domains the very same human holons who are making the representation and those who are making the decisions. This implies a shortcut in the semiotic triad (something that can be called epistemological hypercomplexity).

As observed in the introduction, it is interesting to note that, until now, the official academic world has tried to ignore this problem as much as possible, trying to stick to the business-as-usual paradigm. Paradoxically, it is on the side of the decision makers (and even more so where the other stakeholders are operating) that uneasiness about this situation is becoming more and more evident. Governments and NGOs are the more active actors in putting on the agenda the discussion of new paradigms to be used in the perception and representation of the sustainability predicament. This growing concern of decision makers about the loss of confidence in public opinion in the conventional academic establishment has been generated by repeated situations in which the general public refused to accept as reliable and verified the various scientific truths proposed by the academic world. For example, in many European countries, the public openly mistrusted the proclaimed safety of nuclear energy after the Chernobyl accident, as well as the proclaimed safety of eating meat after the epidemic of mad cow disease. This led to a disbelief in the safety of eating food produced with genetically engineered organisms, even in the absence of a major crisis so far.

This is a crucial point for our discussion, since a loss of confidence in the "truth" as individuated and certified by scientists can have very dangerous consequences in terms of loss of legitimization of the system of control operating within a social contract. This is another basic idea put forward by Funtowicz and Ravetz in relation to the concept of postnormal science. In these days of sweeping changes and transitions, traditional science is no longer able to play the same role in the legitimization (linked to social stability) of systems of control in Western societies as it did in the last two centuries. This basic idea is illustrated in Figure 4.4.

The organization of the state in Europe before the scientific revolution of the 17th century was based on a system of control (a hierarchy of power) that was getting its legitimization directly from God

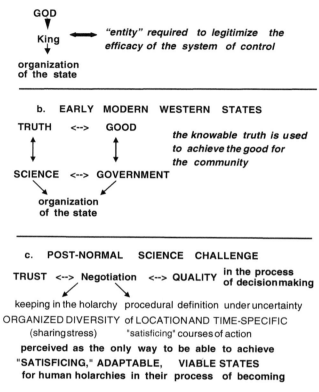

FIGURE 4.4 Different ways of legitimizing systems of control (hierarchies of power) within human societies.

(Figure 4.4a). This "absolute" source of legitimization was then reflected in the figure of the king, who was the entity in charge of performing semantic checks (quality control) over the validity of the specific standardized societal system of knowledge adopted in the given society. Clearly, in his job, the king was using a certain number of advisors who were selecting, storing and refining better representative tools whenever convenient for the interests of the king. When the modern states began, the process of democratization and the reduction of the influence of religions in determining the legitimization of systems of control (hierarchy of power) implied that, in many Western states, a different mechanism of legitimization was needed (Figure 4.4b).

Truth (linked to the assumed possibility of relying on a unique, verified and reliable standardized system of knowledge — substantive rationality) was assumed to lead directly to the possibility of acting for the common good of the community. Governance therefore was assumed to be linked to the activity of making the right choices according to such a truth. In this way, it was still possible to obtain the legitimization of the system of control (linked to the organization of the state) according to the set of relations indicated in Figure 4.4b. It should be noted that the birth of these Western modern states coincided with an evolutionary phase of fast expansion (colonization, industrial revolution) and therefore with the need of massive investments in efficiency that were continuously paying back. The fact that Western countries were in an evolutionary phase of rapid expansion implied a demand for a large degree of hegemonization of the winning patterns. Therefore, the underlying idea that sound governance could be obtained relying on just one unique standardized system of knowledge was not particularly disturbing.

In this way it was assumed that a quality control (semantic check on the efficacy in relation to the existing set of goals — of the existing repertoire of representation and normative tools) on the validity of such a unique and generalized system of knowledge was actually possible. This quite bold assumption can be held only for a short time, that is, only (1) during an evolutionary phase of rapid expansion,

(2) for societies characterized by a quite homogeneous distribution of cultural identities and (3) for societies operating on a relatively small space scale (i.e., Western countries during the industrial revolution). When dealing with issues such as sustainability on a global scale, this assumption simply no longer works.

This is where the need for a paradigm shift enters into play. The call of postnormal science to replace truth with quality means moving from substantial rationality (forgetting about the validity of two assumptions: (1) the existence of solutions and accessible states for human holarchies that are optimal in absolute terms and (2) the possibility of finding and moving to them in a finite amount of time) to a procedural rationality (assuming that it is not possible to represent or define optimal solutions; due to the epistemological predicament of real life, we can only look for satisficing solutions — perceived suboptimal ones).

The consequences of this paradigm shift are very important for determining alternative ways for organizing scientific information in the process of decision making (Figure 4.4c). Trust, which entails reciprocity, loyalty and shared ethical values among the various stakeholders involved in the process of negotiation, becomes the crucial input in the new challenge of governance.

4.4 Example: What Has All This to Do with the Sustainability of Agriculture? The Challenge of Operationalizing the Precautionary Principle

The example discussed in this section refers to the application of the precautionary principle to the regulation of genetically modified organisms (GMOs). The unavoidable arbitrariness in the application of the precautionary principle can be related to the deeper epistemological problems affecting scientific analyses of sustainability discussed in the previous chapters. Hence, traditional risk analysis (probability distributions and exact numerical models) becomes powerless. The precautionary principle entails that scientists move away from the concept of substantive rationality (trying to indicate to society optimal solutions) to that of procedural rationality (trying to help society in finding satisficing solutions).

4.4.1 The Precautionary Principle

The precautionary principle was explicitly recognized during the United Nations Conference on Environment and Development (UNCED) in Rio de Janeiro in 1992, and was included in the Protocol on Biosafety signed in the Convention on Biological Diversity, January 28, 2000 (CEC, 2000). It justifies early action in the case of uncertainty and ignorance to prevent potential harm to the environment and human health: "the principle states that potential environmental risks should be dealt with even in the absence of scientific certainty" (Macilwain, 2000). Obviously, its very definition introduces a certain ambiguity in its possible enforcement. *How* do we decide if the potential environmental risk is sufficient to warrant action? In spite of the difficulty in its application, the precautionary principle has recently been restated as a key guiding concept for policy in a communication from the European Commission (CEC, 2000). This move has increased tension among stakeholders because there is "considerable confusion and differing perspectives, particularly on different sides of the Atlantic, amongst scientists, policymakers, business people and politicians, as to what the precautionary principle does, or should, mean" (EEA, 2001). Given the difficulty in obtaining reliable cost–benefit quantifications for uncertain future scenarios of environmental hazards, the precautionary principle is often regarded as a disguised form of protectionism (ACCB, 2000; Foster et al., 2000) or even as a Trojan horse used by activists moved by ideological biases against technological progress (Miller, 1999).

Indeed, the message of the precautionary principle is clear in its substance, but extremely vague when it comes to practical applications. Its implicit demand for a more effective way to manage hazards than traditional scientific risk assessment (Macilwain, 2000) is generating heated discussions in the scientific community of traditional risk analysis. There are those who still demand numbers and hard proofs as requisite for action, while others call for the adoption of a new paradigm in science for governance (Funtowicz and Ravetz, 1992).

Against this background, in the rest of this section I elaborate on the following points:

1. To understand the practical problems faced when trying to operationalize the precautionary principle, one should be aware of the clear distinctions between the scientific concepts of risk, uncertainty and ignorance. I discuss these concepts and question the current practice of using only traditional risk analysis when discussing the large-scale release of GMOs into the environment, and sustainability in general.

2. Alternative analyses can be used to deal with the ecological hazards of the large-scale release of GMOs. I illustrate the possible use of metaphors derived from systems analysis and network analysis, and of general ecological principles.

3. A paradigm shift is needed when dealing with integrated assessment of sustainability. I will argue that the scientific community should move from the paradigm of substantive rationality (trying to indicate to society optimal solutions) to that of procedural rationality (trying to help society in finding satisficing solutions).

4.4.2 Ecological Principles and Hazards of Large-Scale Adoption of Genetically Modified Organisms

Is there someone that can calculate the risks (e.g., probability distributions) for a world largely populated by genetically modified organisms? Given the definitions of risk, uncertainty and ignorance introduced in Chapter 3 (in particular Figure 3.4), the answer must be no. Nobody can know or predict the consequences of a large-scale alteration of genetic information in plants and animals. The consequences of this perturbation should be considered on various different hierarchical levels and nonequivalent dimensions of interest (human health, health of local ecosystems, health of economies, health of communities and health of the planet as a whole) (Giampietro, 2003).

The mad cow disease epidemic nicely illustrates this issue. In the discussion of the use of animal protein to feed herbivores in the 1980s, with the aim of augmenting the efficiency of beef production, nobody could have calculated the risk of the insurgence of bovine spongiform encephalopathy (BSE). To do that, one should have known that a hitherto unidentified brain protein, known nowadays as *prion*, could lead to an animal disease that also affects human beings (for an overview of this issue, see www.mad-cow.org/). When dealing with a complex problem such as the forecasting of possible side effects of a change imposed on an adaptive self-organizing system, *metaphors* (even if developed within other scientific disciplines) can be more useful than validated models developed in the field of interest. In the specific case of animal-feed regulation, for example, indications from the field of network analysis could have been found. Network analysis shows that a hypercycle in a network is a source of trouble — e.g., microphone feedback to the amplifier to which it is connected — (Ulanowicz, 1986). Indeed, also in dynamic system analysis, it is known that a required level of accuracy in predictions cannot be maintained in the presence of autocatalytic loops. That is, when an output feeds back as input, even small levels of indeterminacy can generate unpredictable large effects — the so-called butterfly effect.

For example, the idea of cows eating cows implies a clear violation of basic principles describing the stability of ecological food webs (probable troubles). Therefore, the need for extreme precaution when implementing such a technique of production could have been guessed before knowing the technicalities regarding the mad cow disease (the specific set of troubles). The lesson to be learned is clear: when dealing with a new situation, it is not wise to rely only on the assessment of probabilities provided by experts who claim that there is negligible risk. Numerical assessments of risks must necessarily assume that the old problem structuring will remain valid in the future. This is usually an incorrect assumption for complex adaptive systems (Rosen, 1985; Kampis, 1991; Ulanowicz, 1986; Gell-Mann, 1994). Thus, in these cases, systems thinking can be more useful because it shows that large-scale infringing of systemic principles will lead, sooner or later, to some yet-unknowable, and possibly unpleasant, events. Below, I further elaborate on an example of systems thinking to characterize potential problems related to large-scale adoption of genetically modified organisms in agricultural production.

4.4.3 Reduction of Evolutionary Adaptability and Increased Fragility

As discussed in Section 3.6.6, efficiency, in fact, requires (1) elimination of those activities that are worse performing according to a given set of goals, functions and boundary conditions and (2) amplification of those activities perceived as best performing at a given point in space and time. This general rule applies also to technological progress in agricultural production. Improving world agriculture, according to a given set of goals expressed by the social group in power and to the present perception of boundary conditions, has led to a reduction of the diversity of systems of production (e.g., abandoning traditional systems of agricultural production). On the other hand, these obsolete systems of production often show high performance when different goals or criteria of performance are adopted (Altieri, 1987).

Several ecologists, following the pioneering work of E.P. Odum (1989), H.T. Odum (1983), and Margalef (1968), have pointed to the existence of systemic properties of ecosystems that are useful for studying and formalizing the effects of changes induced in these systems. Recent developments of these ideas within the emerging field of complex systems theory led to the generation of concepts such as ecosystem integrity (Ulanowicz, 1995) and ecosystem health (Kay et al., 1999). Methodological tools to evaluate the effect of human-induced changes on the stability of ecological processes focus on structural and functional changes of ecosystems (Ulanowicz, 1986; Fath and Patten, 1999) using the relative size of functional compartments, the value taken by parameters describing expected patterns of energy and matter flows, the relative values of turnover times of components, and the structure of linkages in the network. In particular, network analysis can be usefully applied in the analysis of ecological systems (Fath and Patten, 1999; Ulanowicz, 1997) to:

1. Explore the difference between development (harmony between complementing functions, including efficiency and adaptability, reflected into the relative size of the various elements) and growth (increase in the throughput obtained by a temporary takeover of efficiency over diversity)
2. Estimate the relative magnitudes of investments in efficiency and adaptability among the system processes

When looking from this perspective at possible large-scale effects from massive use of GMOs in agricultural production, current research in genetic engineering goes against sound evolutionary strategies for the long-term stability of terrestrial ecosystems (Giampietro, 1994). That is, the current direction of technological development in agriculture implies a major takeover of efficiency over adaptability (based on the representation of benefits on a short-term horizon and using a limited set of attributes of performance). For example, the number of species operating on our planet is on the order of millions, within which the edible species used by humans are on the order of thousands (Wilson, 1988). However, due to the continuous demand for more efficient methods of production, 90% of the world's food is produced today using only 15 vegetal and eight animal species (FAO Statistics). Within these already few species used in agriculture, the continuous search for better yields (higher efficiency) is reducing the wealth of diversity of varieties accumulated through millennia of evolution (Simmonds, 1979). FAO estimates that the massive invasion of commercial seeds resulted in a dramatic threat to the diversity of domesticated species. In fact, available data on genetic erosion within cultivated crops and domesticated animals are simply scary (FAO, 1996; Scherf, 1995). This is a good example of an important and unexpected negative side effect generated by the large-scale application of the green revolution (Giampietro, 1997).

C.S. Holling, one of the fathers of modern ecology, uses a famous line to indicate the negative consequences of the lack of diversity of ecological processes in terms of increased fragility: "a homogeneous ecological system is a disaster waiting to happen" (Holling, 1986, 1996). Technological progress in agriculture can easily generate the effect of covering our planet with a few best-performing high-tech biologically organized structures (e.g., a specific agent of pest resistance coded in a piece of DNA). In this case, it will almost ensure that, due to the large scale of operation, something that *can* go wrong (even if having a negligible probability in a laboratory setting) *will* go wrong. The resulting perturbation (e.g., some unexpected and unpleasant feedback) could easily spread through the sea of homogeneity (i.e., genetically modified monocultures), leaving little or no time for scientists to develop mechanisms of control.

The threat of reduction of biodiversity applies also to the diversity of habitats. Moving agricultural production into marginal areas (in agronomic terms) hitherto inaccessible to traditional crops is often listed among the main positive features of GMOs. In this way, humans will destroy the few terrestrial ecosystems left untouched (escaping excessive exploitation) that provide the diversity of habitats essential for biodiversity preservation. In this regard, note that humans already appropriate a significant fraction of the total biomass produced on earth each year (Vitousek et al., 1986).

But even when looking at potential positive effects, one is forced to question the credibility of the claim of GMO developers that they will be able to increase the ecocompatibility of food production for 10 billion. Given the basic principles of agroecology (Gliessman, 2000; Pimentel and Pimentel, 1996), one is forced to question the idea that simply putting a few high-tech seeds of genetically modified crop plants in the soil could stabilize nutrient cycles within terrestrial ecosystems at a pace dramatically different from the actual ones. This is like trying to convince a physician that, by manipulating a few human genes, it will be possible to feed humans 30,000 kcal of food per day (10 times the physiological rate) without incurring any negative side effects. An ecological metaphor can also be used to check this idea (Giampietro, 1994). Even if we engineer a super spider potentially able to catch 10 times more flies than the ordinary species, the super spider will be limited in its population growth by the availability of flies to eat. Flies, in turn, will be limited, in a circular way, by other elements of the terrestrial ecosystem in which they live. Unless we provide an extra supply of food for these super spiders, their enhanced characteristics will not help them expand in a given ecological context. This concept can be translated to agriculture: if one uses super harvests to take away tons of biomass per hectare from an ecosystem, then one has to put enough nutrients and water back into the soil to sustain the process in time (to support the relative photosynthesis).

This is why high-tech agriculture is based on the systematic breaking of natural cycles (independently from the presence of GMOs). That is, high-tech agriculture necessarily has to be a high-input agriculture (Altieri, 1998). Talking of the green revolution, E.P. Odum notes: "cultivation of the 'miracle' varieties requires expensive energy subsidies many underdeveloped countries cannot afford" (Odum, 1989, p. 83). Because of the high demand for technical capital and know-how of high-input agriculture, many agroecologists share the view of the difficulty of implementing high-tech, GMO-dependent production in developing countries (Altieri, 2000).

As soon as one looks at the ecological effects of innovations in agriculture, one finds that important side effects often tend to be ignored. For example, 128 species of the crops that have been intentionally introduced in the U.S. have become serious weed pests and are causing more than $30 billion in damage (plus control costs) each year (Pimentel et al., 2000). When dealing with ecological systems, and in particular with the growing awareness of the possible impact of GMOs on nontarget species and additional ecological side effects (Cummins, 2000), one should always keep the following (old) aphorism in mind: "You can never do just one thing."

4.4.4 Precautionary Principle and the Regulation of Genetically Modified Organisms

The economic implications of national regulations for the protection of the environment and human, animal and plant health can be huge. In relation to international trade of genetically modified food, the following quotation illustrates this quite clearly: "US soybean exports to EU have fallen from 2.6 billion annually to 1 billion. … Meanwhile, Brazilian exporters are doing a brisk business selling 'GE-free' soybeans to European buyers. … James Echle, who directs the Tokyo office of the American Soybean Association, commented, 'I don't think anybody will label containers genetically modified, it's like putting a skull and crossbones on your product'" (UNEP/IISD, 2000).

This is why the trade dispute between the European Union (EU) and the U.S. over genetically modified food is bringing the precautionary principle to the top of the political agenda. In this particular example, in spite of the increasing attention given to the relationship between environment and trade (UNEP/IISD, 2000), the interpretation of the various key agreements on international trade is still a source of bitter controversy (Greenpeace, 2000).

Also within the EU, the precautionary principle is generating arguments between the commission and individual member states in relation to the moratorium on field trials on GM crops (Meldolesi, 2000), as well as among different ministers within national governments in relation to the funding of research on GMOs (Meldolesi, 2000). Again, it is easy to explain such a controversy. The simple fact that there is a hazard associated with large-scale adoption of GMOs in agriculture does not imply per se that research and experimentation in this field should be stopped all together. Current demographic trends clearly show that we are facing a serious hazard (social, economic, ecological) related to future food production, although the successful and safe translation of high-tech methods to an appropriate agricultural practice is widely recognized as being problematic.

Such a hazard applies to all forms of agricultural development, even when excluding GMOs. However, deciding whether there is sufficient scientific evidence to justify action requires a broader perspective on the hazards, a perspective that goes beyond reductionist science. In particular, the weighing of evidence must be explicit, as well as the inclusion of issues of actual practice, technology, environment and culture.

Life is intrinsically linked to the concept of evolution, which implies that hazards are structural and crucial features of life. Current debates on the application of the precautionary principle to the regulation of GMOs are simply pointing to a deep and much more general dilemma faced by all evolving systems. Any society must evolve in time, and as a consequence, it must take chances when deciding how and when to innovate. This predicament cannot be escaped, regardless of whether society decides to take action — because of what is done or because of what is not done (Ravetz, 2001). Technical innovations have an unavoidable component of gambling. Possible gains have to be weighed against possible losses in a situation in which it is not possible to predict exactly what can happen (Giampietro, 1994). This implies that, when dealing with processes expressing genuine novelties and emergence, we are moving into a field in which traditional risk analysis is basically helpless.

The challenge for science within this new framework becomes that of remaining useful and relevant even when facing an unavoidable degree of uncertainty and ignorance. The new nature of the problems faced in this third millennium (due to the dramatic speed of technical changes and globalization) implies that more and more decision makers face "PNS situations" (facts are uncertain, values in dispute, stakes high and decisions urgent) (Funtowicz and Ravetz, 1992). That is, when the presence of uncertainty/ignorance and value conflict is crystal clear from the beginning, it is not possible to individuate an objective and scientifically determined best course of action. Put another way, when dealing with this growing class of problems, the era of closed expert committees seems to be over. Crucial to this change of paradigm is the rediscovery of the old concept of scientific ignorance, which goes back to the very definition of scientists given by Socrates: "Scientists are those who know about their ignorance."

These are relevant points since, on the basis of the Agreement on the Application of Sanitary and Phytosanitary Measures (valid since January 1995), the World Trade Organization authorizes (or prevents) (Article 2.2) all member countries to enforce the precautionary principle if there is (not) enough scientific evidence (see http://www.wto.org/english/tratop_e/sps_e/spsagr_e.htm). In particular, Article 2.2 has been used to oppose compulsory labeling of genetically modified food. The reasoning behind this opposition appears to be based on the concept of substantive rationality, and it is well illustrated by Miller's (1999) paper published on the policy forum of science. Labeling requirements should be prevented since they "may not be in the best interest of consumers" (Miller, 1999). The same paper identifies the best interest of consumers with lower production costs, the possibility of achieving economies of scale and keeping at maximum speed research and development of GMOs (i.e., maximization of efficiency). Referring to a decision of the U.S. Court of Appeals (against labeling requirements), Miller (1999) comments: "Labeling cannot be compelled just because some consumers wish to have the information."

Two questions can be used to put in perspective the difference between the paradigm of substantive rationality and that of procedural rationality when dealing, as in this case, with scientific ignorance and legitimate contrasting perspectives:

1. What if the perception of the best interest of consumers adopted by the committee of experts does not coincide with the set of criteria considered relevant by the consumers themselves? For example, assume that there is a general agreement among scientists that the production of pork is more efficient and safer than the production of other meats. Should then the government deny Muslims or Jews the right to know — through a label — whether the meat products they buy include pork? If we agree that Muslims and Jews have a right to know, then why should consumers

who are concerned with the protection of the environment have less right to know — through a label — if the food products they buy contain components that are derived from GMOs?

2. What if the assessment of better efficiency and negligible risk provided by the committee of experts turns out to be wrong? Actually, this is exactly what happened in the European Union with the decisions regulating the use of animal protein feeds for beef production (the move that led to the rise of mad cow disease).

4.5   Conclusions

Globalization implies a period of rapid transition in which the global society as a whole has to learn how to make tough calls finding the right compromise between too much and too little innovation. Since nobody can know *a priori* the best possible way of doing that, satisficing solutions (Simon, 1976), rather than optimal solutions to this challenge can be found only through a process of social learning on how to better perceive, describe and evaluate the various trade-offs of sustainability. Scientists have a crucial role to play in this process. But to do that, they have to learn how to help rather than hamper this process. In this respect, the concept of scientific ignorance is very useful for putting scientists back into society (procedural rationality requires a two-way dialogue) rather than above society (substantive rationality implies a one-way flow of information).

References

ACCB (American Chamber of Commerce in Belgium, EU, Committee of the), (2000), Position Paper on the Precautionary Principle, available at http://www.eucommittee.be/pop/pop2000/Env/env.

Altieri, M.A., (1987), *Agroecology: The Scientific Basis for Alternative Agriculture*, Westview Press, Boulder, CO.

Altieri, M., (1998), *The Environmental Risks of Transgenic Crops: An Agro-Ecological Assessment*, Department of Environmental Science, Policy and Management, University of California, Berkeley.

Altieri, M., (2000), Executive Summary of the International Workshop on the Ecological Impacts of Transgenic Crops, available at http://www.biotech-info.net/summary1.pdf.

CEC (Commission of the European Communities), (2000), Communication from the Commission on the Precautionary Principle, COM(2000)1, Brussels, 02.02.2000, available at http://europa.eu.int/comm/off/com/health_consumer/precaution.htm.

Cummins, R., (2000), GMOs Around the World, paper presented at IFOAM: Ecology and Farming, May–August 9.

EEA (European Environment Agency), (2001), *Late Lessons from Early Warnings: The Precautionary Principle 1898–1998*, European Environment Agency, Copenhagen.

FAO Statistics, available at http://apps.fao.org/page/collections?subset = nutrition.

FAO (Food and Agriculture Organization), (1996), The State of the World's Plant Genetic Resources for Food and Agriculture, background documentation prepared for the International Technical Conference on Plant Genetic Resources, Leipzig, Germany, June 17–23, 1996.

Fath, B.D. and Patten B.C., (1999), Review of the foundations of network environ analysis, *Ecosystems*, 2, 167–179.

Foster, K.R., Vecchia, P., and Repacholi, M.H., (2000), Science and the precautionary principle, *Science*, 288, 979–981.

Funtowicz, S.O. and Ravetz, J.R., (1992), Three types of risk assessment and the emergence of post-normal science, in *Social Theories of Risk*, Krimsky, S. and Golding, D., Eds., Praeger, Westport, CT, pp. 251–273.

Gell-Man, M., (1994), *The Quark and the Jaguar*, Freeman, New York.

Giampietro, M. (1994), Sustainability and technological development in agriculture: a critical appraisal of genetic engineering. *BioScience* 44(10): 677–689.

Giampietro, M., (1997), Socioeconomic constraints to farming with biodiversity, *Agric. Ecosyst. Environ.*, 62, 145–167.

Giampietro, M. (2003), Complexity and scales: The challenge for integrated assessment. In: J. Rotmans and D.S. Rothman (Eds.), *Scaling Issues in Integrated Assessment*. Swets & Zeitlinger B.V., Lisse, The Netherlands. pp. 293–327.

Gliessman, S.R., (2000), *Agroecology: Ecological Processes in Sustainable Agriculture*, Lewis Publishers, Boca Raton, FL.

Greenpeace, (2000), WTO Must Apply the Precautionary Principle, available at http://www.green-peace.org/majordomo/index-press-release/1999/msg00121.html.

Holling, C.S., (1986), The resilience of terrestrial ecosystems: local surprise and global change, in *Sustainable Development of the Biosphere*, Clark, W.C. and Munn, R.E., Eds., Cambridge University Press, Cambridge, U.K., pp. 292–317.

Holling, C.S., (1996), Engineering resilience vs. ecological resilience, in *Engineering within Ecological Constraints*, Schulze, P.C., Ed., National Academy Press, Washington D.C., pp. 31–43.

Kampis, G. (1991), *Self-Modifying Systems in Biology and Cognitive Science: A New Framework for Dynamics, Information, and Complexity*. Pergamon Press, Oxford, U.K.

Kay, J.J., Regier, H., Boyle, M., and Francis, G., (1999), An ecosystem approach for sustainability: addressing the challenge of complexity, *Futures*, 31, 721–742.

Kuhn, T.S., (1962), *The Structure of Scientific Revolutions*, University of Chicago Press, Chicago.

Macilwain, C., (2000), Experts question precautionary approach, *Nature*, 407, 551.

Margalef, R., (1968), *Perspectives in Ecological Theory*, University of Chicago Press, Chicago.

Matutinovic, I., (2002), Organizational patterns of economies: an ecological perspective, *Ecol. Econ.*, 40, 421–440.

Meldolesi, A., (2000), Green ag minister wreaks havoc on Italy's ag biotech, *Nat. Biotechnol.*, 18, 919 – 920.

Miller, H., (1999), A rational approach to labeling biotech-derived foods, *Science*, 284, 1471–1472.

O'Connor, M. and Spash, C., Eds., (1998), *Valuation and the Environment: Theory, Methods and Practice*, Edward Elgar, Cheltenham, U.K.

Odum, E.P., (1989), *Ecology and our Endangered Life-Support Systems*, Sinauer Associated, Sunderland, MA.

Odum, H.T., (1983), *Systems Ecology*, John Wiley, New York.

Peirce, C.S., (1935), *Collected Papers 1931–35*, Harvard University Press, Cambridge, MA.

Pimentel, D. and Pimentel, M., (1996), *Food, Energy and Society*, rev. ed., University Press of Colorado, Niwot.

Pimentel, D., Lach, L., Zuniga, R., and Morrison, D., (2000), Environmental and economic costs of non-indigenous species in the United States, *Bioscience*, 50, 53–65.

Ravetz, J.R., (2001), Safety in the globalising knowledge economy: an analysis by paradoxes, *J. Hazard. Mater.*, (special issue on risk and governance), De Marchi, B. (Ed.), forthcoming.

Rosen, R. (1975), Complexity and error in social dynamics. *Int. J. Gen. Syst.* Vol, 2: 145–148.

Rosen, R., (1985), *Anticipatory Systems: Philosophical, Mathematical and Methodological Foundations*, Pergamon Press, New York.

Sardar, Z. and Abrams, I., (1998), *Chaos for Beginner*, Icon Books Ltd., Cambridge, U.K.

Sarewitz, D., (1996), *Frontiers of Illusion: Science and Technology, and the Politics of Progress*, Temple University Press, Philadelphia.

Scherf, B.D., (1995), *World Watch List for Domestic Animal Diversity*, FAO, Rome.

Simmonds, N.W., (1979), *Principles of Crop Improvement*, Longman, U.K.

Simon, H.A., (1976), From substantive to procedural rationality, in *Methods and Appraisal in Economics*, Latsis, J.S., Ed., Cambridge University Press, Cambridge, U.K.

Simon, H.A., (1983), *Reason in Human Affairs*, Stanford University Press, Stanford, CA.

Ulanowicz, R.E. (1986), *Growth and Development: Ecosystem Phenomenology*. Springer-Verlag, New York.

Ulanowicz, R.E., (1995), Ecosystem integrity: a causal necessity, in *Perspectives on Ecological Integrity*, Westra, L. and Lemons, J., Eds., Kluwer Academic Publishers, Dordrecht, pp. 77–87.

Ulanowicz, R.E., (1997), *Ecology, The Ascendent Perspective*, Columbia University Press, New York, 242 pp.

UNEP/IISD, (2000), Environment and Trade: A Handbook, available at http://www.unep.ch/etu/ or ht-pp://iisd.ca/trade/handbook.

Vitousek, P.M., Ehrlich, P.R., Ehrlich, A.H., and Matson, P.A., (1986), Human appropriation of the products of photosynthesis, *Bioscience*, 36, 368–373.

Wilson E.O. (Ed.). (1988), *Biodiversity*. National Academy Press, Washington D.C.

5

Integrated Assessment of Agroecosystems and Multi-Criteria Analysis: Basic Definitions and Challenges

This chapter addresses the specific challenges faced by scientists willing to contribute to a process of integrated assessment. Integrated assessment, when applied to the issue of sustainability, has to be associated with a multi-criteria analysis (MCA) of performance, which, by definition, is controversial. This in turn requires (1) a preliminary institutional and conflict analysis (to define what are the relevant social actors and agents whose perceptions and values should be considered in the analysis, and what are the power relations among them); (2) the development of appropriate procedures able to be involved in the discussion about indicators, options and scenarios on the largest number of relevant social actors; and (3) the development of fair and effective mechanisms of decision making. The continuous switching of causes and effects among the activities related to both the descriptive and normative dimensions makes this discussion extremely delicate. Scientists describe what is considered relevant by social actors, and social actors consider relevant what is described by scientists. The two decisions — (1) who are the social actors included in this process and (2) what should be considered relevant when facing legitimate but contrasting views among the social actors — are key issues that have to be seriously considered by the scientists in charge of generating the descriptions used for the integrated assessment. This is why, in this chapter, I decided to provide an overview of terms and problems related to this relatively new field.

5.1 Sustainability of Agriculture and the Inherent Ambiguity of the Term *Agroecology*

The two terms included in the title of this chapter — *integrated assessment* and *agroecosystems* — are terms about which it is almost impossible to find definitions that will generate consensus. In fact, integrated assessment is a neologism that is becoming more and more popular in the scientific literature dealing with sustainability. An international journal (http://www.szp.swets.nl/szp/frameset.htm?url=%2Fszp%2Fjournals%2Fia.htm) and a scientific society bear this name, to which one should add a fast-growing pile of papers and books dedicated to the subject. This term, however, is mainly gaining popularity outside the field of scientific analysis of agricultural production. Very little use of the term can be found in journals dealing with the sustainability in agriculture. The other term, *agroecosystems*, is derived from the concept of agroecology, which is another neologism that was introduced in the 1980s. Unlike the first term, this one is very popular in the literature of sustainable agriculture. At this point in the book, it is possible to make an attempt to justify the abundant use of neologisms so far. Nobody likes using a lot of neologisms or, even worse, "buzzwords" in scientific work. A simple look at the two definitions of neologism found in the *Merriam-Webster Dictionary* explains why:

> *Neologism* — (1) a new word, usage, or expression; (2) a meaningless word coined by a psychotic.

Introducing a lot of neologisms without being able to share their meaning with the reader tends to classify the user or proponent of these neologisms in the category of psychotic. On the other hand, when an old scientific paradigm is no longer able to handle the challenge (and I hope that at this point the reader is convinced that this is the case with integrated analyses of sustainability), it is necessary to introduce new concepts and words to explore and build new epistemological tools. Moreover, a lot of new words and concepts are already used in the fields of integrated assessment and multi-criteria analysis (and this author has nothing to do with this impressive flow of neologisms), so I find it important to share with the reader the meaning of these new terms. In particular, what is relevant here is the application of the concept of integrated assessment to the concept of agroecosystems. Before getting into this discussion, let us start with the definition of the term *agroecosystem*, which implies dealing with the concept of agroecology.

The term *agroecology* was proposed in a seminal book by Altieri (1987). This was an attempt to put forward a new catchword pointing to the need to introduce a paradigm shift in the world of agricultural research when taking seriously the issue of sustainability. In that book, Altieri focuses on the unavoidable existence of conflicts linked to the concept of sustainability in the field of agriculture. His main point is that if we define the performance of agricultural production only in economic terms, then other dimensions such as the ecological, health and social dimensions will be the big losers of any technical development in this field. When mentioning conflicts here, we do not refer only to conflicts between social actors, but also to conflicts between optimizing principles derived by the adoption of different scientific analyses of agriculture (when getting into the normative side by using different definitions of costs and benefits). For example, an anthropologist, a neoclassical economist and an ecologist tend to provide very different views of the performance of the very same system of shifting cultivation in Papua New Guinea.

Two main lines of action were suggested by Altieri:

1. The concept of agroecology has to be associated with a total rethinking of the terms of reference of agriculture. (What should be considered an improvement in the techniques of production? Improvement for whom? In relation to which criterion? Which time horizon should be adopted to assess improvements?)
2. The concept of agroecology requires expanding the universe of possible options (technical solutions, technical coefficients, socioeconomic regulations) for agricultural development. This can be obtained in two ways:
   a. By exploring new alternative techniques of production (changing the existing set of available technical coefficients)
   b. Studying and preserving the cultural diversity of agricultural knowledge already existent in the world (preserving techniques guaranteeing technical coefficients, which could be useful when adopting different optimizing functions)

It should be noted that the majority of groups using the term *agroecology*, especially in the developed world, endorse basically the second line, without fully addressing the implications of the first. The basic idea of this position can be characterized as follows: The sustainability predicament and the existing difficulties experienced by agriculture in both developed and developing countries are just because humans are not using the most appropriate technologies and not relying on a given set of sound principles. Put another way, this second historical interpretation of agroecology assumes a substantive definition of it. The vast majority of the people using this interpretation tend to associate agroecology with concepts like organic farming, low-external-input agriculture, "small is beautiful," and empowerment of family farms. They are assuming that the way out of the current lack of sustainability in agriculture can be found by relying on sound principles and by studying how to produce more profit with (1) less environmental impact and (2) happier farmers.

The problem with this position is that it does not address (1) the unavoidable existence of conflicts implicit in the concept of sustainable development and (2) the unavoidable existence of uncertainty and ignorance about our knowledge of future scenarios. Put another way, the very concept of sustainability entails an unavoidable dialectic between actors and strategies. When discussing the development of agricultural systems, there is no single set of most appropriate technologies. At each point in space and time, the objectives (goals, targets), constraints (resources, laws, taboos), the available sets of options and of acceptable compromises among which to choose must first be explicitly defined for the scientists. Only at this point does it become possible for them to identify a set of appropriate technologies based on either politically defined priorities among the different objectives or a negotiated consensus on a compromise solution that realizes all the various goals (as expressed by relevant social actors) to some extent.

This is why, in the last two decades, the first direction of research suggested by Altieri, "totally rethinking the terms of reference of agriculture," has also been gaining attention. This radical position seems to be supported by those working on scenarios about the future of agriculture (e.g., within the U.S. to avoid the Blank hypothesis (Blank, 1998)). It is also shared by those working on *ex post* evaluation of agricultural policies (e.g., the massive failure of development programs of UN agencies in developing countries and that of agricultural policies in the EU). In fact, a complete recasting is at the moment the official position of the European Commission for the future of European agriculture (e.g., http://www.newscientist.com/news/news.jsp?id=ns99991854).

In the face of this mounting pressure, the forces for business as usual (economic and political lobbies, academic institutions) are trying to develop a strategy of damage control. Many within the agricultural establishment say that a total rethinking is not really needed. They suggest that a few technical adjustments and a little more talking with the farmers will suffice. They also recommend a few new regulations to internalize some of the externalities that have until now escaped market mechanisms. This position has important ideological implications. It accepts the notion that technical development of agriculture should be driven, by default, by the maximization of productivity and profit (bounded by a set of constraints to take care of the environment and the social dimension).

I have no intention of getting into an ideological discussion of this type. This chapter and book are written assuming that the emerging paradigm that perceives the development of rural areas in terms of integrated resource management carried out by multifunctional land use systems is valid. In this paradigm, flexibility in the management strategy and participatory techniques for defining what should be the desirable characteristics of the system are assumed to be necessary steps to achieve such a goal. Therefore, in the rest of this chapter, I will not deal with the question, "Why should we do things in a different way when perceiving and representing the performance of agriculture?" but rather with the question, "How can we do things in a different way?"

In fact, acknowledging the need for a total rethinking of agriculture is just the first step. To act, we must first reach an agreement as to how things should be done differently. This can be achieved only by answering some tough questions such as: Who is supposed to rethink the terms of reference of agriculture? How might we change the shape of the plane on which we are flying? What do we do if different social actors have different views on how to make changes? An acute problem in this regard is that both colleges of agriculture and reputable scholars, in general, are less than fully willing to engage in this debate, perhaps because they view totally rethinking the terms of reference of agriculture as a threat to their present agenda. This is, however, not reasonable: If we acknowledge that changes on the societal side resulted in a shift in the priorities among objectives and, in some cases, led to the formulation of completely new objectives in agriculture, then we are forced to accept the following conclusions: (1) We have to do things differently in agriculture, and to do that (2) we have to perceive and represent things differently in the scientific disciplines dealing with the description of agricultural performance.

As soon as one tries to draw this logical consequence, however, one crashes against one of the mechanisms generating the lock-in on business as usual. Much funding of colleges of agriculture is channeled through private companies with a clear agenda (maximizing profit through maximization of productivity). Even public funding is heavily affected by lobbies that are operating within the conventional paradigm. These lobbies perceive agriculture as just an economic sector producing commodities and added value.

To the best of my knowledge, the only big agricultural university that is working hard on a radical and dramatic restructuring of its courses (to reflect a total rethinking of the terms of reference for agriculture) is Wageningen University in the Netherlands. Actually, the restructuring started with its very name. It used to be the glorious WAU (Wageningen Agricultural University) until 2 years ago, and then they dropped the A.

A very quick summary of relevant events leading to this restructuring is that, in the early 1990s, the big departments resisted any friendly attempts at change from the inside. Actually, they reacted to signals of crisis by continuing to do more of the same thing. The concept of "ancient regime syndrome," proposed by Funtowicz and Ravetz (when facing a crisis, do more of the same, even though it is not working) , discussed in Chapter 4 should be recalled here. The fatal response of agricultural departments was better and more complicated, optimizing models to get additional economies of scale and increases in efficiency. At the very moment when the basic assumptions of agriculture as an economic sector just producing commodities were under revision, the credibility of these assumptions was stretched even further. The catastrophe came when the rest of society (e.g., consumers, farmers, politicians) imposed a new research agenda in a quite radical way. They were told, "No more money for models that optimize the ratio of milk produced per unit of nitrogen and phosphorus in the water table." And the edict was given almost overnight.

Central to any discussion about a different way to perceive and represent the performance of agricultural systems is the idea that agricultural production is not the full universe of discourse for any of the relevant agents operating at different levels (households, local communities, counties, states, countries, international bodies). Then it becomes obvious that analytical approaches aimed at optimizing production techniques do not represent the right way to go. When we analyze the livelihood of households, local communities, counties, states, countries and international bodies, a sound representation of the performance of agricultural activities (how to invest a mix of production factors to alter ecosystems to produce food and fibers) is just a part of the story. That is, (1) the mix of relevant activities considered in the analysis has to include more than just the production of crops and animal products and (2) the list of consequences considered in the analysis has to include more than the economic and biophysical productivity of agricultural techniques (e.g., additional relevant indicators should address social, health and ecological impacts and quality of life). Performing this integrated analysis does not require the introduction of new revolutionary analytical tools, but rather the ability to provide new packages for existing tools.

In engineering, for example, it is possible to have a rigorous treatment of decision support analysis for design. The terms used there are multi-objective decision making and multi-attribute decision making (e.g., http://design.me.uic.edu/~mjscott/papers/95f.pdf). The great advantage of industrial design is that all the relevant information for defining the performance of the designed system is supposed to be available to the designer. The same approach is explored in other fields dealing with the issue of sustainability (e.g., ecological economics, science for governance (participatory integrated assessment), evaluation of sustainability, natural resources management). The application of these concepts is generally indicated under a family of names like integrated assessment, sustainability impact assessment, strategic environmental assessment and extended cost–benefit analysis (CBA).

However, when applying these tools to self-organizing systems, especially when dealing with reflexive systems (humans), a multi-criteria evaluation has to deal with three very large systemic problems:

- It is not possible to formalize a procedure to define in a substantive way (outside of a specific and local context of reference) what is the right set of relevant criteria of performance that should be considered for a sound analysis.

- It is unavoidable to find legitimate contrasting views on what should be considered an improvement or what should be the best alternative to select. Social agents will always have divergent opinions. For example, it is unavoidable to find different opinions on whether it is good or bad to have nuclear weapons or use genetically modified organisms.

- It is not possible to get rid of uncertainty and ignorance in the various scientific analyses that are required. This implies that not all the data, indicators and models required to consider different dimensions of analysis (the views of different agents at different levels) have the same degree of reliability and accuracy.

Because of these three major problems, there is a general convergence in the field of integrated assessment and multiple-criteria analysis that it is not possible to achieve the right problem structuring of a sustainability problem without the integrated and iterative use of two types of tool kits:

1. Discussion support systems (term introduced by H. van Keulen)

In this activity scientists are the main actors and social actors are the consultants; the goal is the development of integrated packages of analytical tools required to do a good job on the descriptive side. The resulting information space used in the decision-making process has to represent the system of interest, in scientific terms, on different scales and dimensions of analysis. This information space has to be constructed according to the external input received from the social actors of what is relevant and what is good and bad. The social actors, as consultants, have to provide a package of questions to be answered. But the scientists are those in charge of processing such an input according to the best available knowledge of the issue.

This is a new academic activity, which implies a strong scientific challenge: keeping coherence in an information space made up of nonequivalent descriptive domains (different scales and different models). This requires an ability to make a team of scientists coming from different disciplines interact on a given problem structuring provided by society. This is what we will introduce later on under the label of multiple-scale integrated analysis (MSIA).

2. Decision support systems

In this activity, social actors are the main actors and scientists the consultants; the goal is the development of an integrated package of procedures required to do a good job on the normative side. The resulting process should make it possible to decide, through negotiations:

a. What is relevant and what should be considered good and bad in the decision process

b. What is an acceptable quality in the process generating the information produced by the scientists (e.g., definition of quality criteria — relevance, fairness in respecting legitimate contrasting views, no cheating with the collection of data or choice of models)

c. Deciding on an alternative (or a policy to be implemented)

This process requires an external input (given by scientists) consisting of a qualitative and quantitative evaluation of the situation on different scales and dimensions. In their input, scientists also have to include information about expected effects of changes induced by the decision under analysis (discussion of scenarios and reliability of them), but the social actors are those in charge of processing such an input. This is what we will introduce later as social multi-criteria evaluation (SMCE), following the name proposed by Munda (2003).

Since the scientific process associated with the operation of tool kit 1 affects the social process associated with the operation of the tool kit 2 and vice versa, the only reasonable option for handling this situation is to establish some form of iteration between the two. In doing this, however, it must be clear that process 1 is a scientific activity (which requires an input from social actors) and process 2 is a social activity (which requires input from scientists). Each, however, depends on the other. This is where the need of a new type of expertise enters into play. To have such an iterative process, it is necessary to implement an adequate procedure.

The rest of this chapter is divided into three sections. Section 5.2 discusses the systemic problems faced when considering agriculture in terms of multifunctional land use. Any analysis based on indicators reflecting legitimate but contrasting views and referring to events described at different scales implies facing serious procedural problems. This section makes the point that, when dealing with the sustainability of agriculture, we do face a postnormal science situation. Section 5.3 provides an overview of concepts and tools available for dealing with such a challenge (e.g., integrated assessment, multi-criteria evaluation, and a first view at multi-objective multi-scale integrated analysis), as well as practical examples of problems associated with their use. Section 5.4 briefly describes existing attempts to establish procedures able to generate the parallel development of discussion support systems and decision support systems, and then an iteration between the two (e.g., the soft systems methodology proposed by Checkland, 1981, Checkland and Scholes, 1990)) Section 5.5 provides a practical example (the current making of farm bills) in which we can appreciate the need of developing these procedures as soon as possible.

FIGURE 5.1 Nonequivalent observers of agroecosystems. (Photo by M.G Paoletti.)

5.2 Dealing with Multiple Perspectives and Nonequivalent Observers

In this section I elaborate on the two points discussed in the introduction:

1. It is unavoidable to find legitimate contrasting views on what should be considered an improvement or what should be considered the best alternative to select (Section 5.2.1).
2. It is not possible to formalize a procedure to define in a substantive way what is the right set of relevant criteria that should be considered to perform a sound analysis (Section 5.2.2).

5.2.1 The Unavoidable Occurrence of Nonequivalent Observers

The lady shown in Figure 5.1 is performing a very old traditional technique of Chinese farming. She is applying "night soil" (human excrement) to her garden, making sure that as little as possible of this valuable resource gets lost in the recycling. This is why she carefully pours only small amounts of the organic fluid on each plant. There are plenty of such pictures of this woman, since the colleagues (i.e., ecologists and experts of organic agriculture) who were working with me on a project there were delighted by this image. They took about 50 pictures of her in different moments of her daily routine. For Westerners, this picture is a vivid metaphor of the ultimate ecological wisdom of ancient agriculture — the closure of the cycle of nutrients between humans and nature. The unexplained mystery associated with such a vivid metaphor, though, is that this image is disappearing from this planet pretty quickly.

Later on, when talking to that woman, I asked about the explanations for the abandonment of this and other ecologically friendly activities (such as digging silt out of channels) so valuable for the preservation of Chinese agroecological landscapes. She replied abruptly, "Have you been in Paris?" "Of course I have been in Paris" was my immediate (and careless) answer. At that point she could go for it: "I have never been in Paris. None of those living in this village have ever been in Paris. None of my daughters will ever go to Paris. You want to know why? Because we have been digging channels and carefully pouring night soil to preserve this agroecosystem instead. Personally, I don't want to do that anymore. If things will not change during my lifetime, I want that at least my great-grandchildren will

FIGURE 5.2 Models presented at Beijing's fashion week 2002. (Photo by Wilson Chu, Reuters. With permission.)

have the option to go to Paris. If this agroecosystem is going to hell, I am happy about that, the sooner the better."

The three relevant points about this story are:

1. A clear disagreement about basic goals and strategies among different actors. Our team of scientists was in China with the goal of preserving that agroecosystem, whereas the lady had the goal of getting rid of it (she was forced to keep recycling night soil, but for her this was only a temporary solution needed for feeding her family).

2. The parallel use of different and logically independent indicators of performance for a given agroecosystem. The agroecologists in our project were happy about her recycling according to the indications given by bioindicators (earthworms) assessing changes in the health of the soil. The lady was unhappy about night soil in relation to her impossibility to go to Paris, used as indicator of the performance of agronomic activities.

3. The tremendous speed at which human systems can redefine what is desirable and acceptable. Our local students told us, to explain her reaction, that a TV set had just arrived in the village, and this generated a communal daily watching. The soap opera in fashion at that moment featured two Chinese yuppies living in Paris and drinking champagne from cold flutes. This was enough for the villagers watching the show to update their representation of what should be considered a desirable and acceptable socioeconomic performance of agricultural activities. The picture that the woman pouring night soil had in mind for the future of her great-granddaughter was more related to what is shown in Figure 5.2.

5.2.2 Nonreducible Indicators and Nonequivalent Perspectives in Agriculture

When dealing with sustainable agriculture, we have to expect a representation of performance that is based on different criteria (reflecting the different values and goals) and different hierarchical levels (requiring a mix of nonequivalent descriptive domains). Without using a multi-level analysis, it is very easy to get models that simply suggest shifting a particular problem between different descriptive domains. Put another way, optimizing models based on a simplification of real systems within a single descriptive domain just tends to externalize the analyzed problem out of their own boundaries. For

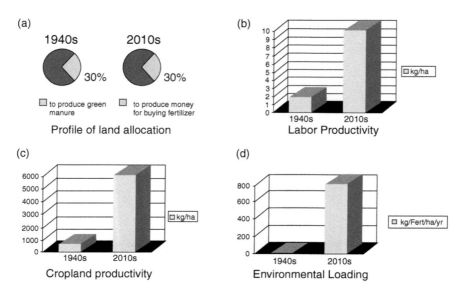

FIGURE 5.3 Different indicators that can be used to characterize historical trends in rice farming in China.

example, economic profit can be boosted by increasing ecological or social stress. In the same way, ecological impact can be reduced by reducing economic profit, and so on. That is, conventional scientific analyses in general provide policy suggestions that are based on the detection of some benefits by a given model referring to a certain descriptive domain and by the neglecting of other costs ignored by the model, since they are detectable only on different descriptive domains (when adopting a different selection of variables). This epistemological cheating can be avoided only by adopting a set of different descriptive domains able to see those costs externalized (put under the carpet) by a given mechanism of accounting. By using an integrated set of indicators, we can observe that problems externalized by the conclusions suggested by one model (based on an optimizing variable defined on a given scale — e.g., when describing things in economic terms over a 10-year horizon) reappear amplified in one of the parallel models (based on a different optimizing variable defined on a different scale — e.g., when describing the same change in biophysical terms on a 1000-year horizon). As discussed in Chapters 2 and 3, the ability of any model to see and encode some qualities of the natural world implies that the same model cannot see other qualities detectable only on different descriptive domains.

To provide an example of nonequivalent indicators that can be used to characterize historical changes in a farming system, Figure 5.3 provides examples of four numerical assessments that characterize the dramatic developments of farming systems in rural China.

5.2.2.1 *Land Requirements for Inputs* — The first indicator used in Figure 5.3a is related to the profile of land use. In particular, the numerical assessment indicates the percentage of cropland invested by farmers with the aim of guaranteeing nutrient supply to crop production. In the 1940s, about 30% of cropland was allocated to green manure cultivation, and hence, this land was unavailable for subsistence or cash crop production. The intensification of crop production, driven by population growth and socioeconomic pressure, led to a progressive abandonment of the use of green manure (too expensive in terms of land and labor demand) and general switching to synthetic fertilizer use. This resulted in a sensible increase in multiple-cropping practices and, consequently, in a dramatic improvement of agronomic indices of crop production (e.g., yields per hectare), that is, a dramatic increase in crop production for self-sufficiency and freeing land for cultivation of cash crops (Li et al., 1999). However, according to current trends, a further increase in demographic and economic pressure can lead to further intensification of agricultural throughputs (Giampietro, 1997a, b). In this case, depending on the ratio of sales price of crops and cost of fertilizer, as well as technical coefficients, we could easily return — in the first decade of the third millennium — to the 30% mark, the same as it was in the 1940s. That is, about

30% of the land invested in cash crops will be used just to pay for technical inputs. Put another way, when considering the criterion "land requirement for stabilizing agricultural production" (resource eaten by an internal loop within the system of production), the two solutions requiring a 30% investment of the total budget of available land to make available the required production inputs are equal for the farmer. According to farmers' perception, the same fraction of land is lost whether it is to green manure production or to crop production to purchase chemical fertilizer. The characterization (mapping of system qualities) given in Figure 5.3a is not able to catch the difference implied by these two solutions. Other criteria (and therefore indicators) are needed if we want to obtain a richer characterization (a better explanation) of such a trend.

5.2.2.2 Household's Perspective

*5.2.2.2 Household's Perspective — *When considering the parameter "productivity of labor" as an indicator of performance (Figure 5.3b), we see that the solution of chemical fertilizer implies a much higher labor productivity than the green manure solution. Higher labor productivity in this case translates into a higher economic return of labor. Depending on the budget of working time available to the household, it is possible to reduce, in this way, the fraction of working time allocated to self-sufficiency and, as a consequence, to increase the fraction of working time allocated to cash flow generation (either on or off the farm) and leisure. Thus, even if 30% of the available budget of land is lost to fertilization, according to the new criterion "labor productivity," farmers will prefer the solution of chemical fertilizer because it enables a better allocation of their time budget.

*5.2.2.3 Country's Perspective — *When considering the parameter "productivity of food of cropped land" as the indicator of performance (Figure 5.3c), we see that the solution of chemical fertilizer implies a much higher land productivity than the green manure solution. In fact, the land used to produce crops for the market to pay for chemical fertilizer (perceived as lost by farmers), when considered at the national level, is seen as land that produces food for the urban population. On the contrary, green manure production is seen by the national government as a use of cropping area that does not generate food. Indeed, the goal of the central government of China to boost food surplus in rural areas, making it possible to feed the growing urban population, can actually lead to a promotion of policies of intensification of agricultural production by boosting the use of technical inputs. Given this goal, an excessive fraction of farmers' land budget eaten by the cost of purchasing chemical fertilizer would discourage farmers from intensive use of technical inputs. Therefore, the central government can decide to subsidize the use of these inputs. As seen from the farmers' perspective, a lower cost of fertilizer reduces the fraction of their land that has to be invested in procuring fertilizer and therefore induces an intensification of agricultural production. Note, however, that the reduction of land lost to buy chemical fertilizer (as detected by the farmers' perception) and an increase in cropland productivity (as detected by the central government) obtained by subsidization of fertilizer, in turn increase another relevant indicator — the economic cost of internal food production (yet another relevant criterion for the Chinese government when deciding about policies of agricultural development). That is, the advantage given by the use of subsidies to fertilizer — characterized by the indicator "cropland productivity" — induces a side effect that can be detected only by using an additional criterion (and relative indicator) referring to the country level: the economic burden of subsidizing technical inputs (note that this is a relevant indicator that is not given in Figure 5.3).

*5.2.2.4 Ecological Perspective — *When considering the ecological perspective, we find a totally different picture of the consequences of the two "30% of land budget allocation to fertilizer" solutions. The use of green manure in the 1940s was certainly benign to the environment because the flow of nutrients in the cropping system was kept within a range of values of intensity close to those typical of natural flows. Put another way, the acceleration of nutrient throughputs induced by the use of synthetic fertilizers dramatically increased the environmental stress on the agroecosystems. When bio-physical indicators of environmental stress are considered to characterize the trend, we obtain an assessment of performance that is totally unrelated (logically independent) to assessments based on the use of economic variables. For example (Figure 5.3d), 800 kg of synthetic fertilizer applied per hectare per year (due to the high multiple-cropping index) is too much fertilizer for healthy soil, no matter how the economic cost of fertilizer compares with its economic return.

A couple of points can be driven home from this example: (1) The same criterion (land demand per output) can require different indicators, when reflecting the perspective of performance related to different hierarchical levels. The indicators in Figure 5.3a and c are giving contrasting indications about the solution of green manure vs. that of synthetic fertilizer in relation to use of land. Farmers see no difference between the two solutions; the government of the country sees the two solutions as dramatically different. (2) Criteria and indicators referring to different descriptive domains (Figure 5.3b and d) (environmental loading assessed in kilograms of fertilizer per hectare vs. labor productivity expressed either in added value per hour or kilograms of crop per hour) reflect not only incommensurable qualities, but also the existence of unrelated (logically independent) systems of control. As a consequence, when dealing with trade-offs defined on different descriptive domains, we cannot expect to work out simple protocols of optimization able to compare and maximize relative costs and benefits. Recalling the examples provided in Chapter 3, we can say that the existence of multiple relevant hierarchical levels, nonequivalent descriptive domains, can imply a nonreducibility of models on the descriptive side. This leads to a problem that Munda (2003) calls *technical incommensurability* (the impossibility of establishing a clear link between nonequivalent definitions of costs and benefits obtainable only on nonreducible descriptive domains). A difference in the perception about priorities (the two different views about the future of agriculture shown in Figures 5.1 and 5.2) found in social actors carrying conflicting goals and values should be associated with *social incommensurability* (Munda, 2003). There will be more on this in the following section.

5.3 Basic Concepts Referring to Integrated Analysis and Multi-Criteria Evaluation

In this section I provide an overview of concepts and definitions that is an attempt to frame the big picture within which the various pieces of the puzzle belong. A more detailed discussion about how to build an analytical tool kit for integrated analysis of agroecosystems is provided in Part 3.

5.3.1 Definition of Terms and Basic Concepts

5.3.1.1 *Problem Structuring Required for Multi-Criteria Evaluation* — This refers to the identification of relevant qualities of the system under investigation that have to be characterized, modeled and assessed in relation to the specified set of goals expressed by relevant social actors. This integrated appraisal leads to the individuation of a set of relevant issues to be considered in the formal problem structuring in terms of a list of options, criteria, and indicators and measurement scheme that will be used to decide about the action.

5.3.1.2 *Multi-Scale Integrated Analysis (Multiple Set of Meaningful Perceptions/ Representations)* — This is the simultaneous consideration of a set of system qualities (judged relevant for the goals of the study in the first step of problem structuring) that must be observable and can be encoded into variables used in the set of selected models. Depending on the set of relevant criteria, MSIA might require the parallel use of indicators referring to different scales and dimensions of analysis, e.g., gross national product (GNP) in U.S. dollars, life expectancy, megajoules of fossil energy, level of food intake, fractal dimension of cropfields, Gini index for equity, efficiency indices and nitrogen concentration in the water table.

5.3.1.3 *Challenge Associated with the Descriptive Side (How to Do a MSIA)* — This is the study of nonequivalent typologies of (1) performance indicators and (2) mechanisms generating relevant constraints, in relation to a given problem structuring.

The standard objective of MSIA is the simultaneous consideration of economic viability, ecological compatibility, social acceptability and technical feasibility. This requires the ability to simultaneously:

1. Describe different effects in relation to the selected set of relevant constraints using different indicators
2. Understand the various mechanisms generating relevant features and patterns using in parallel nonreducible models
3. Gather the adequate information required to operate the selected sets of indicators and models
4. Assess the quality of the results obtained in the steps 1, 2 and 3.

5.3.1.4 Challenge Associated with the Normative Side (How to Compare Different Indicators, How to Weight Different Values, How to Aggregate Different Perspectives — Social Multi-Criteria Evaluation)

— From a philosophical perspective, it is possible to distinguish between two key concepts (Martinez-Alier et al., 1998; O'Neill, 1993): *strong comparability* and *weak comparability.*

With strong comparability it is possible to find a single comparative term by which all different policy options can be ranked. Strong comparability can be divided into (1) strong commensurability (it is possible to obtain a common measure of the different consequences of a policy option based on a quantitative scale of measurement) and (2) weak commensurability (it is possible to obtain a common measure of the different consequences of a policy option but based only on a qualitative scale of measurement). The concept of strong comparability implies the assumption that the value of everything (including your mother) can be compared with the value of everything else (including someone else's mother) by using a single numerical variable (e.g., monetary or energy assessments).

Weak comparability implies incommensurability; i.e., there is an irreducible value conflict when deciding what common comparative term should be used to rank alternative actions.

As noted in previous chapters, complex systems exhibit multiple identities because of epistemological plurality (nonequivalent observers see different aspects of the same reality) and ontological characteristics (nested hierarchical systems can only be observed on different levels using different types of detectors and different typologies of pattern recognition). This is what leads to the distinction proposed by Munda about:

1. **Social incommensurability** — referring to the existence of a multiplicity of legitimate values and points of views in society. It is not possible to decide in a substantive way that a set of values of a social group is more valuable than a set of values of another social group.

2. **Scientific or technical incommensurability** — referring to the nonreducibility of nonequivalent models. This is justified by hierarchy theory and can be related to the impossible task of representing multiple identities (as resulting from analysis on different scales) in a single descriptive model. It is not possible to reduce in a substantive way a given system description related to either a particular level of analysis or the use of a certain disciplinary view to another.

5.3.1.5 The Rationale for Societal Multi-criteria Evaluation

— It is important to note that weak comparability does not imply at all that it is impossible to use rationality when deciding. Rather, it implies that we have to move from a concept of substantive rationality (based on strong comparability) to that of procedural rationality (based on weak comparability and SMCE). Procedural rationality is based on the acknowledgment of ignorance, uncertainty and the existence of legitimate nonequivalent views of different social actors (Simon, 1976, 1983). "A body of theory for procedural rationality is consistent with a world in which human beings continue to think and continue to invent: a theory of substantive rationality is not" (Simon, 1976, p. 146).

Concepts like welfare and sustainability are multidimensional in nature. Therefore, the evaluation of technological progress, policies, public plans or projects has to be based on procedures that explicitly require the integration of a broad set of various and conflicting points of view and the parallel use of nonequivalent representations. Consequently, multi-criteria methods are, in principle, an appropriate modeling tool for policy issues, including conflicting socioeconomic and nature conservation objectives.

5.3.2   Tools Available to Face the Challenge

In recent years the use of multi-criteria methods has been gaining popularity at an increasing pace. Their major strength is their ability to address problems marked by various conflicting evaluations (Bana e

Alternatives

Criteria	Units	FORD Mondeo	HONDA Civic	VW Golf	NISSAN Micra
Fuel Consumption	US$	a_{11}	a_{12}	...	a_{14}
Maintenance cost	US$	...	...	...	...
Price	US$			a_{33}	
Road Handling	Index				
Reliability	Index	...	a_{52}	...	...
Safety devices	Index				
Power	HP	a_{71}			
Comfort	Index	...	...	...	...
Noise	db			a_{93}	
Design	Index		$a_{10\,3}$		
Status Symbol	Index	...	...	. ...	..
Colour	Index	a_{121}			a_{124}

FIGURE 5.4 Integrated assessment — the formalist perspective. Closing the information space into a formal problem structuring (how to choose a car among to many options, computers can handle it ...)

Costa, 1990; Beinat and Nijkamp, 1998; Janssen and Munda, 1999; Munda, 1995; Nijkamp et al., 1990; Vincke, 1992; Voogd, 1983; Zeleny, 1982).

  To clarify the idea of multi-criteria analysis in relation to the concepts presented before, let us discuss a very simple illustrative example. Imagine that one wishes to buy a new car and wants to decide among the existing alternatives on the market. Also imagine that the choice would depend on four main criteria: economy, safety, aesthetics and driveability. To describe the mechanism of decision, it is necessary to first specify the criteria (dimensions of performance) taken into account by a given buyer, since it is not possible to know all the potential criteria that are used by the universe of nonequivalent buyers operating in this world. Whatever criteria are considered, however, it is sure that some (measured by their relative indicators) will be (1) *technically incommensurable* (price in dollars, speed in kilometers per hour, fuel consumption in liters of gasoline used for 100 km and so on) and (2) *conflicting in nature* (e.g., the higher the safety characteristics required, the higher the economic cost). The performance of any given alternative, according to the set of relevant criteria, can be characterized through a multi-criteria impact profile, which can be represented either in matrix form, as shown in Figure 5.4, or in graphic form, as shown in Figure 5.5. These multi-criteria impact profiles can be based on quantitative, qualitative or both types of criterion scores.

  Another crucial feature related to the available information for decision making concerns the uncertainty contained in this information (How reliable are the criterion scores contained in the impact matrix?). Whenever it is impossible to exactly establish the future state of the problem faced, one can decide to deal with such a problem in terms of either *stochastic uncertainty* (thoroughly studied in probability theory and statistics) or *fuzzy uncertainty* (focusing on the ambiguity of the description of

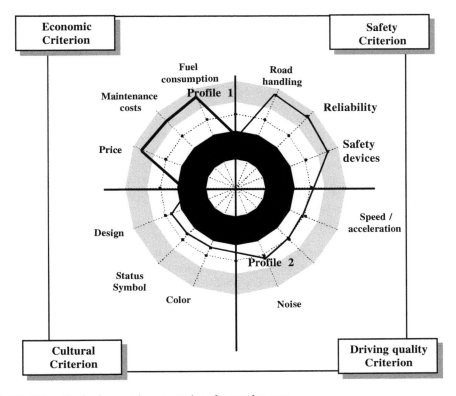

FIGURE 5.5 Multi-objective integrated representation of car performance.

the event itself) (Munda, 1995). However, one should be aware that genuine ignorance is always present too. This predicament is particularly relevant when facing sustainability issues because of large differences in scales of relevant descriptive domains (e.g., between ecological and economic processes) and the peculiar characteristics of adaptive systems (adaptive systems are self-modifying and becoming systems — see the relative discussions in Chapters 2 and 3). In this case, it is unavoidable that the information used to characterize the problem is affected by subjectivity, incompleteness and imprecision (e.g., ecological processes are quite uncertain and little is known about their sensitivity to stress factors such as various types of pollution). A great advantage of multi-criteria evaluation is the possibility of taking these different factors into account.

5.3.2.1 *Formalization of a Problem Structuring through a Multi-Criteria Impact Matrix* — A very familiar example of an impact matrix related to the structuring of a decision process is provided in Figure 5.4. This is a typical multi-criteria problem (with a discrete number of alternatives) that can be described in the following way: A is a finite set of n feasible policy options (or alternatives); m is the number of different evaluation criteria g_i ($i = 1, 2, ..., m$, considered relevant in a decision problem), where action a is evaluated to be better than action b (both belonging to the set A) according to the i-th criterion if $g_i(a) > g_i(b)$. In this way, a decision problem can be represented in a tabular or matrix form. Given the two sets of A (of alternatives — in this case, models of car to buy) and G (of evaluation criteria — in this case, four criteria), and assuming the existence of n alternatives and m criteria, it is possible to build an $n \times m$ matrix P called an evaluation or impact matrix (see Figure 5.4), whose typical element p_{ij} ($i = 1, 2, ..., m$; $j = 1, 2, ..., n$) represents the evaluation of the j-th alternative by means of the i-th criterion. Obviously, to have a process of decision in a finite time, n and m in such an impact matrix have to be finite and data should be available to characterize the various options.

5.3.2.2   *A Graphical View of The Impact Matrix: Multi-Objective Integrated*
Representation — The graph shown in Figure 5.5 (a different representation of the information presented in the impact matrix given in Figure 5.4) is an example of a multi-objective integrated representation (MOIR) (a set of different indicators reflecting different criteria of performance selected in relation to different objectives associated with the analysis). In this way, we can visualize in graphical form the information given in Figure 5.4. This form of graphic representation is becoming quite popular in the literature of integrated analysis.

The popularity of this graphic form is because some additional features are possible on the resulting problem structuring. In the graph in Figure 5.5 (starting with the same problem structuring given in Figure 5.4), there are 12 *indicators*, shown by the 12 axes on the radar diagram (e.g., price, maintenance costs, fuel consumption). These indicators can be grouped into four main dimensions of performance or *criteria* (economy, safety, aesthetics and driveability). *Goals* (for each indicator) can be represented as target values over the set of selected indicators. In Figure 5.5, they are indicated by the bullets on the various indicators in the radar diagram.

In this way, it is possible to bridge three different hierarchical levels of analysis:

1. The definition of performance in general terms obtained by selecting the set of different relevant dimensions. This is associated with the answers given to a set of semantic questions about sustainability: Sustainability of what? For whom? On which time horizon?

2. The formulation of general objectives in relation to the selection of indicators: What should be considered an improvement or a worsening in relation to the different criteria and indicators? What are the goals? What should be considered acceptable? This makes it possible to reflect on the perspectives found among the stakeholders.

3. Translation of these general principles into a numerical mapping of performance over a set of indicators and measurement schemes required for data collection that are necessarily context specific (location-specific description). At this point a multi-scale integrated analysis based on the simultaneous scientific analysis of different attributes (using nonequivalent descriptive domains) requires a tailoring of the semantic of the problem structuring into a context-specific formalization (required to perform scientific analyses).

When dealing with a graphic representation of this type, it becomes possible to discuss the definitions of:

1. Special threshold values (e.g., a limited budget for buying the car) implied by the existence of *constraints* on the value that can be taken by the criteria or attributes. In this case, a set of constraints defines a *feasibility region* (i.e., a set of constraints defines what can be done or carried out). In the example given in Figure 5.5, the feasibility region would be the area on the radar diagram.

2. Areas in the admissible range of values associated with qualitative differences in performance. This requires a previous process of normalization on benchmark values within the viability domains. For example, the flag model developed in the school of Nijkamp (e.g., http://www.tinbergen.nl/discussionpaper/9707.pdf) proposes three sections within the viability domain: (1) good (in green) — data in this area indicate a good state of the investigated system in relation to a given indicator; (2) acceptable (in yellow); and (3) unsatisfactory (in red).

Also in this case, things look good on paper, but as soon as one tries to get into the process of definition of the various viability domains, one is forced to admit the limitations implied by the epistemological predicaments already discussed in Chapters 2 and 3: (1) Any procedure of normalization and definition of performance score over areas in the admissible range is unavoidably affected by value judgment and (2) any assessment of viability, compatibility and acceptability into the future is affected by an unavoidable dose of uncertainty and ignorance.

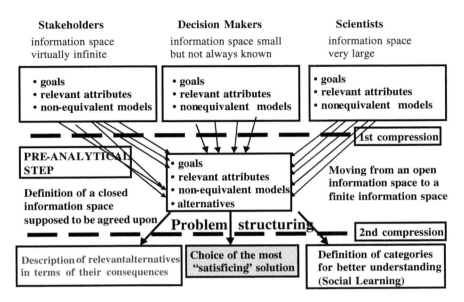

FIGURE 5.6 Problem structuring as a heroic compression of the information space.

5.4 The Deep Epistemological Problems Faced When Using These Tools

5.4.1 The Impossible Compression of Infinite into Finite Required to Generate the Right Problem Structuring

As noted earlier, to be able to provide the right set of data and models, scientists working on MSIA require an agreed-upon and closed problem structuring from the social actors, that is, a formal definition of what the problem is in the form of a specification of what type of scientific information is required for characterizing it. A closed problem structuring in turn requires a previous clear definition of the goals of the analysis. This implies that every social process used to select a policy or rank options requires, in the first place, an operational definition of an agreed-upon set of common values for the community of social actors. In the example of Figure 5.5, this would be a preliminary definition of what would be a valuable car for the household buying it. On the other hand, the very concept of the unavoidable existence of nonequivalent observers or agents entails the existence of different interests, differences in cultural identities, different fears and goals. Even individuals within the same household can have different definitions of what a valuable car is for them. As a consequence of this, when considered one at a time, social actors would provide different definitions of what is the right set of criteria and indicators that should be used to reflect their own definition of value in the decision. This set of values is then difficult to aggregate to reflect the set of values adopted by the household as a whole when deciding what car to buy.

When assessing policies or ranking technical options, we are first of all making a decision about what is important for the community of the social actors (as a whole), as well as what are the relevant characteristics of the problem described in the models. This requires addressing three different problems: (1) exploring the variety of legitimate nonequivalent perspectives found among the social actors (this is especially relevant for normative purposes), (2) generating the best possible representation of the state-of-the-art knowledge relevant to the decision to be made (this is especially relevant for descriptive purposes) and (3) trying to find a fair process of aggregation of contrasting preferences and values (this is crucial to have a fair process of governance).

An overview of the challenge faced when attempting to generate a fair and effective problem structuring, within a process of decision making is given in Figure 5.6. Very little explanation is needed to illustrate this overview. Three relevant points are:

1. Any problem structuring implies an impossible mission of compressing a virtually infinite and unstructured universe of discourse and values (goals, organized perceptions, meanings, epistemological categories, alternative models) that could be used in the problem structuring into a finite and structured information space. It is therefore sure that each problem structuring is missing relevant aspects of the problem and is reflecting a power struggle among social actors

2. The preanalytical step of compression of the virtually infinite and unstructured universe of discourse into a finite and structured information space is the most crucial step of the whole process of decision making. In this step, basically whatever is relevant for determining the decision is already decided. That is:

    a. Whose perspectives count?

    b. Whose alternatives should be considered among the possible choices?

    c. What are the criteria and indicators to use in the characterization of the possible alternatives?

    d. What are the models to use to represent causality and construct scenarios?

    e. What data should be considered reliable?

    It is remarkable that this step is not the subject of any discussion by reductionist scientists. Reductionist science, to operate, must have a closed problem structuring as a starting input. The discussion of how to generate this finite and structured information space, however, is not included in the realm of scientific activities. It is important to keep in mind that when one is working on formal models, everything that is relevant for a discussion about how to help social actors with different perspectives to negotiate a compromising solution is already gone.

3. It is impossible to do this compression in a satisficing way (suggested by H. Simon, 1976, 1983 instead of an optimal way) in a single step. Therefore, we should expect that any sound process of decision making related to sustainability cannot be the result of a single process of individuation of the optimal alternative. Rather, we should expect an iterative process of problem structuring and discussion (exploring different possible ways of compressing and structuring the universe of discourse into a finite information space). This can imply going over and over the compression performed in step 2. This would be the process of negotiation among different stakeholders with legitimate nonequivalent perspectives to arrive at an agreed-upon problem structuring. The usefulness of scientific analyses based on the finite information space i — obtained in step i — is mainly related to the possibility of generating a better compression of the universe of discourse into a different finite information space $i + 1$ — obtained in the step $i + 1$. This goal should be considered more important than that of individuating the best course of action in the final step n — within the final finite information space n. In becoming systems, it is impossible to reach the final step determining the most suitable information space to be used in decision making. Therefore, we should rely on the metaphor of the Peircean triad (Figure 4.2) visualizing a continuous process of learning how to make better decisions.

5.4.2 The Bad Turn Taken by Algorithmic Approaches to Multi-Criteria Analysis

The implications of the first compression shown in Figure 5.6 have always been clear to smart economists. For example, Georgescu-Roegen (1971) talks of heroic compression implied by the choices made by scientists when representing the complexity of reality into a given model. Schumpeter (1954, p. 42) observes that "analytical work begins with material provided by our vision of things, and this vision is ideological almost by definition." Myrdal (1966), who was awarded the Nobel Prize in economics, states in his book *Objectivity in Social Science* "that ignorance as the knowledge is intentionally oriented."

But even when ignoring the implications of this heroic compression, as done by many neoclassical economists nowadays, a lot of problems remain. In fact, things are still quite messy when dealing with the second compression indicated in Figure 5.6. How do we decide the best alternative in the face of uncertainty, legitimate contrasting views and incommensurable indicators, which still are affecting the information space considered in the given problem structuring? Put another way, even if one can start from a multi-criteria information space that is finite, discretized and assumed to be valid — as shown in Figure 5.4 or Figure 5.5 — things are still not easy when arriving at a final decision. Such a second compression still requires the ability to deal with information coming from a heterogeneous information space made up of a set of indicators referring to nonequivalent descriptive domains (dealing with technical incommensurability). This requires handling and comparing several dimensions of performance that can be analyzed only by using nonreducible models (models assessing profits are not reducible to models assessing ecological integrity).

The trouble associated with formalizing the universe of potential perceptions and existing values into a closed and finite problem structuring points to an additional problem. Not only is the double compression indicated in Figure 5.6 and obtained at a given point in time impossible, but also one should be aware that the universe of potential perceptions and existing values is open and expanding. As already observed in Chapter 4 and as discussed again in the section on complex time in Chapter 8, when dealing with sustainability we must acknowledge that both the observed and the observer are becoming in time.

The reaction of reductionism when facing this challenge followed (and is still following in different contexts) the standard strategy. First, there is total denial: there is nothing that cannot be reduced to cost–benefit analysis. That is, try to ignore the problem until it disappears. In fact, the majority of neoclassical economists working on cost–benefit analyses to deal with problems that would require MSIA and SMCE operate under such an assumption. They seem to believe that it is possible (1) to reduce all types of costs and benefits into a single mapping expressed monetarily (e.g., U.S. dollars of 1987) and (2) to aggregate in a neutral (objective) way all the different perspectives found among the stakeholders about what should be considered a cost and what should be considered a benefit. In spite of their clear untenability, these assumptions are needed to escape such an impasse. This has led to a situation in which even experts in cost–benefit analysis such as E.J. Mishan complain about the misuse of such a tool. CBA is a useful tool, but it should not be applied well outside its original domain of competence (e.g., see Mishan, 1993). These two assumptions, however, are held because of their huge ideological implications. They are required to defend the claim that it is possible to handle in a scientific way (neutral, value-free assessment) the weighing of different typologies of performance (equity vs. profit, social stress vs. ecological integrity, values of a social group vs. values of another social group). A huge amount of literature is available providing technical arguments attacking these assumptions (e.g., an overview in O'Connor and Spash, 1998; Mayumi, 2001). Personally, I do not believe that a lot of disciplinary discussions are required to assess their credibility. A simple practical reflection can do it. This means assuming that, when facing a tough decision related to an important conflict in social systems (e.g., a dispute about world trade of GMOs), the happiness and the health of your children, the value of your mother, and the memory of your cultural heritage can be (1) first measured and expressed in U.S. dollars of 1987, and then (2) compared with the value of someone else's children, mother and cultural heritage. Very few people really believe that this is possible.

This is why smarter reductionists realized that the reduction or collapse of different typologies of performance using a single variable like U.S. dollars of 1987 (or megajoules of fossil energy) is impossible. They realized that those assumptions, in spite of their ideological relevance, cannot be held any longer. This is why the second attempt to keep the claim of the neutral, value-free input of science in the process of decision making was aimed at operationalizing multi-criteria analysisin a technocratic way. The gospel always remained the same: If the human mind cannot handle the simultaneous analysis of non-equivalent indicators characterizing multiple options, computers will. The impact matrix represented in Figure 5.4 is an example of a formalization of the problem structuring associated with a multi-criteria evaluation of cars at the moment of purchasing one. I have neither competence nor space enough to get into a detailed analysis of the formalist or algorithmic approach to MCA. Such a field is well established, with a huge amount of literature available. Even manuals sponsored by governments are available nowadays (e.g., Dodgson et al., 2000). I am dealing in this section only with an analysis of the impossibility of using the information provided by this impact matrix in an algorithmic way to calculate the best possible car to buy. The main point I want to drive home is that, in spite of its reassuringly formal look, this impact matrix hides a lot of problems.

To shortcut long discussions, let us use a couple of trivial examples:

Incommensurability among the different indicators of performance and coexistence of legitimate contrasting perspectives. In my life, I have adopted the two profiles of satisficing performance (the two illustrated in Figure 5.5) when buying a car. A profile of satisficing performance reflects the particular selection of factors chosen to weight the priorities among different attributes of performance. Actually, the two profiles shown in Figure 5.5 can be used to study the differences in my mechanisms of evaluation associated with different historical moments in my life. When I was a student with a low income (line 1), none of the criteria represented in Figure 5.5 except that of being cheap was relevant for the purchasing of a car. However, as soon as I became a father of two with a tenured position, the profile of my multi-criteria satisfaction for the buying of a car changed (the second profile — line 2 — is self-explanatory). In this example, we can see that even for the same person it is not possible to define a default weighting profile among different indicators of performance referring to different criteria. As a consequence, a committee made up of the best experts in the world cannot decide what should be considered an optimal car (Used where? For whom? For doing what? When?).

Unavoidable handling of nonreducible assessments referring to nonequivalent descriptive domain. Any numerical assessment refers to our representation of our perception of the reality and therefore is affected by choices made by the observer. As discussed in the example of Figure 3.1, even a simple "hard" measurable number, such as "kilograms of cereal consumed in 1997 per capita in the U.S.," can exhibit multiple identities depending on the reason we want such an assessment.

As already noted at the end of Section 3.1, the differences in the four nonreducible assessments of U.S. cereal consumption per capita in 1997 given in Figure 3.1 do not imply that any of these assessments is wrong or useless. Each of those numerical assessments could be the right one (useful information) depending on the interests of the social actors. This is why it is important to develop procedures that enable the integrated handling of a heterogeneous information space. Otherwise, we could find two different scientists fighting over a multi-criteria impact matrix about the numerical value to be assigned to a cell, without understanding that the number to be used in the process of benchmarking on a given indicator depends on a lot of assumptions that have to be explicitly discussed in the process of generation of the integrated assessment.

Unavoidable existence of uncertainty and ignorance. Going back to the example of the problem structuring related to the choice of a car to buy: When going for a short sabbatical in Madison with my family in January 2002, I had to actually get into a process involving an MSIA and SMCE for buying a car. In the process that led to the closure of our information space (the formal problem structuring adopted by our household), we — as a family — did not include a lot of indicators that other people might have included. For example, we did not consider the environmental impact of our choice (we selected a big old car for safety and economic reasons). Probably this was because we were buying a car for only 5 months to be operated abroad. Another explanation could be that we did not have available a valid model to associate our personal choice of a car to possible consequences on the health of the environment in the short, medium and long term. That is, for the environment, is it better to buy an big old car that is recycled or a new car that is more energy efficient?

Another important criterion or indicator was missing in our selected problem structuring: the relationship between the size of the garage of our rented house and the size of the car to be purchased. Our big car did not fit into our garage, and we had to leave it out through the freezing winter months of Madison. Every morning from January to late March, when de-icing the windows and removing the snow before using the car, I regretted our ignorance about the relevance of that indicator.

Unavoidable existence of conflict among social actors. What is illustrated in Figure 5.7 (also called social impact matrix) is a complementing analysis of the matrix shown in Figure 5.4. This has to do with the study of conflict analysis. This time, the matrix is constructed by judging the

OPTIONS → STAKEHOLDERS ↓	Ford Mondeo	VW Golf	Honda Civic	Nissan Micra	
Sandra (my wife)	Yes +	No	Yes/No	Yes ++	**Veto Power !!!**
Mario (me)	Yes +	Yes/No	No	Yes/No	**Relevant**
Olga (older daughter)	No	Yes/No	Yes ++	Yes ++	**Relevant only in fine tuning**
Sofia (younger daughter)	**It must be red**	**It must be red**	**It must be red**	**It must be red**	**Ignored, but what if...**

FIGURE 5.7 Multi-criteria evaluation requires more information.

different options in relation to the interests and opinions of the various social actors. Obviously, to do that, it is necessary to ask them. Also in this case (I am again applying this matrix to our family decision to buy a car in Madison), the figure is self-explanatory. The different stakeholders in my family had different power in the process of decision making. My wife, as designated taxi driver, had veto power. Then the other members of the family had decreasing influence on the basis of their accumulated experience about cars. Obviously, our process of decision making was strongly influenced by such a ranking. But what if the concern of my younger daughter, Sofia ("it must be red"), would have suddenly become relevant? Imagine that Sofia were suddenly given (by a political decision) veto power on the selection process. Obviously, the whole problem structuring, starting with the selection of the set of alternatives considered in the matrix and the data collection in the field, would have to be completely different. Actually, none of the used cars we evaluated in the process of selection was red. Deciding the validity of the scientific information included in the problem structuring (often considered to be scientifically substantive input) has a lot to do with power relations among the social actors. In fact, this is what determines the identity of the option space about which the scientific input is required.

Also in this case, I am providing trivial examples of very sophisticated procedures and tools. Several methods have been developed for introducing conflict analysis in a frame of multi-criteria decision support. One example is NAIADE (Novel Approach to Imprecise Assessment and Decision Environments), which is structured in software — for applications see http://alba.jrc.it/ulysses/voyage-home/naiade/naiade.htm. For an overview of similar methods, see http://www.dodccrp.org/Proceedings/DOCS/wcd00000/wcd00091.htm.

5.4.3 Conclusion

Several protocols for decision making can be based on the application of predetermined algorithms to an impact matrix like the one shown in Figure 5.4. These protocols often require input from the social actors (e.g., how to weigh the differences in priorities in relation to different attributes and nonequivalent criteria). However, the adoption of algorithmic protocols must assume (1) the validity of a given problem structuring (as if this were a substantive definition of the right problem structuring) and (2) the possibility of selecting an optimal solution in a single process. On the contrary, from the examples discussed so far, I claim that the two processes of multi-scale integrated analysis and societal multi-criteria evaluation are different activities depending on each other. They cannot be either collapsed into a single one or held separated in time. They have to be performed in an iterative process. This does not mean, obviously, that it is impossible to find cases in which algorithms and software can be very useful in a process of decision making based on the adoption of multi-criteria methodologies. Rather, this is a warning against the application of these formal protocols without an adequate quality control on the relative semantic.

Coming to the representation of different matrices in Figures 5.4 and 5.7 and the two heroic compressions illustrated in Figure 5.6, three tasks are crucial in relation to this process:

1. Definition of the identity of the two matrices in terms of legends of the matrices. For the matrix shown in Figure 5.4, what are the relevant criteria? What are the indicators and the target values used to assess the performance on each indicator? What are the alternatives to be considered? For the matrix shown in Figure 5.7, who are the relevant actors? What is their relative power? Who has veto power? How acceptable is, in ethical terms, the present situation?

2. What has to be included as a valid data set inside the cells of the matrix? How should the values taken by the various indicators in the various alternatives considered be measured? How reliable is the assessment included in the various cells?

3. How should it be decided what is the wisest course of action on the basis of a given problem structuring (the representation of options, criteria and indicators obtained in tasks 1 and 2).

The term *wisest solution* has been suggested by Bill Bland (personal communication) as opposed to *optimal solution*. The term *wisest*, in fact, refers to the need to reach an agreement on the definition of something that is perceived by the various social actors (after a process of negotiation) as feasible, desirable, satisficing and reasonable according to previous knowledge, and prudent in relation to the unavoidable existence of uncertainty and ignorance.

5.5   Soft Systems Methodology: Developing Procedures for an Iterative Process of Generation of Discussion Support Systems (Multi-Scale Integrated Analysis) and Decision Support Systems (Societal Multi-Criteria Evaluation)

5.5.1   Soft Systems Methodology

In this section, we provide a quick summary of crucial concepts introduced by Checkland and others (Checkland, 1981; Checkland and Scholes, 1990; Röling and Wagemakers, 1998) dealing exactly with the impasse typical of science for governance discussed in this and the previous chapters. Since Checkland has done outstanding work in this area for more than 30 years, it is wise to use his own words to present his approach.

A paragraph taken from the introduction of Checkland and Scholes's (1990, p. xiii) book explains beautifully the basic rationale of SSM:

> Soft Systems Methodology (SSM) was developed in the 1970's. It grew out of the failure of established methods of "systems engineering" (SE) when faced with messy complex problem situations. SE is concerned with creating systems to meet defined objectives, and it works well in those situations in which there is such general agreement on the objectives to be achieved and the problem can be thought of as simply the selection of efficacious and efficient means to achieve them. A good example would be the USA's programme in the 1960s with its unequivocal objective of "landing a man on the Moon and returning him safely to Earth" (President Kennedy's words). Not many human situations are as straightforward as this, however, and SSM was developed expressly to cope with the more normal situation in which the people in a problem situation perceive and interpret the world in their own ways and make judgments about it using standards and values which may not be shared by others.

Taking advantage of the extraordinary clarity of the introductory chapters of that book, I will again use groups of statements that are related to the points discussed so far in this chapter and the previous ones. The special approach of SSM is crucial when dealing with science for governance:

Thus theory must be tested out in practice; and practice is the best source of theory. In the best possible situation the two create each other in a cyclic process in which neither is dominant but each is the source of the other. (p. xiv)

To "manage" anything in everyday life is to try to cope with a flux of interacting events and ideas which unrolls through time. The "manager" tries to "improve" situations which are seen as problematical — or at least as less than perfect — and the job is never done (ask the single parent!) because as the situation evolves new aspects calling for attention emerge, and yesterday's "solutions" might now be seen as today's "problems." (p. 1)

Mankind finds an absence of meaning unendurable. We are meaning-endowing animals, on both the global long-term and the local short-term level. Members of organizations, for example, tend to see the world in a particular way, to attribute at least partially shared meaning to their world. (p. 2)

But what an observer sees as wisdom can to another be blinkered prejudice (p. 3).

His definition of system "a set of elements mutually related such that the set constitutes a whole having properties as an entity" (p. 4), or "a whole with emergent properties" (p. 21)

Pruzan (1988) lists a number of the shifts entailed in a move from "classic" to "soft" Operational Research (though he himself does not use that phrase): from optimization to learning; from prescription to insight; from "the plan" to the "planning process"; from reductionism to holism … from an approach aimed at optimizing a system to an approach based on articulating and enacting a systemic process of learning (p. 15).

The lessons that led to the peculiar characteristics of SSM:

The Lancaster researchers started their action by taking hard systems engineering as a declared framework and trying to use it in unsuitable situations, unsuitable, that is, in the sense that they were very messy problem situations in which no clear problem definition existed (about the emergence of SSM, p. 16).

If the system and its objectives are defined, then the process is to develop and test models of alternative systems and to select between them using carefully defined criteria which can be related to the objectives. … Systems engineering looks at "how to do it" when "what to do" is already defined. … This was found to be the Achilles' heel of systems engineering, however, when it was applied in the Lancaster research programme, to ill-defined problem situations. Problem situations, for managers, often consist of no more than a feeling of unease, a feeling that something should be looked at. … This means that naming a system to meet a need and defining its objective precisely — the starting point of systems engineering — is the occasional special case. (p. 17)

What was found to be needed was a broad approach to examining problem situations in a way which would lead to decisions on action at the level of both "what" and "how." The solution was a system of enquiry. In it a number of notional systems of purposeful activity which might be "relevant" to the problem situation are defined, modelled, and compared with the perceived problem situation to articulate a debate about change, a debate which takes in both "whats" and "hows." (p. 18)

The basic features of SSM in relation to postnormal science:

The description of any purposeful holon must be from some declared perspective or worldview. This stems from the special ability of human beings to interpret what they perceive. Moreover, the interpretation can, in principle, be unique to a particular observer. This means that multiple perspectives are always available. (p. 25)

We have a situation in everyday life which is regarded by at least one person as problematical. There is a feeling that this situation should be managed ... to bring about "improvement." The whats and the hows of the improvement will all need attention, as will consideration of through whose eyes "improvement" is to be judged. The situation itself, being a part of human affairs, will be a product of a particular history, a history of which there will always be more than one account. ... We have to learn from the relative failure of classical management science, since that is surely due to its attempt to be ahistorical [based on the characterization of situations based on the use of typologies out of time]. ... We are not indifferent to that logic, but are concerned to go beyond it to enable action to be taken in the full idiosyncratic context of the situation, which will always reveal some unique features [all real situations are special]. (p. 28)

A number of purposeful holons in the form of models of human activity are represented in the form of systems which are named, modelled, and used to illuminate the problem situation. This is done by comparing the models with perceptions of the part of the real world being examined. What is looked for in the debate is the emergence of some changes which could be implements in the real world and which would represent an accommodation between different interests. It is wrong to see SSM simply as consensus-seeking. That is the general case within the general case of seeking accommodation in which the conflicts endemic in human affairs are still there, but are subsumed in an accommodation which different parties are prepared to "go along with." (p. 30)

Which selected "relevant" human activity systems are actually found to be relevant to people in the problem situation will tell us something about the culture we are immersed in. And knowledge of that culture will help both in selection of potentially relevant systems and in delineation of changes which are culturally feasible. (p. 30)

No human activity system is intrinsically relevant to any problem situation, the choice is always subjective. ... In the early years of SSM development, much energy was wasted in trying at the start to make "the" best possible choice. (This at least was better than the very earliest attempts to name the relevant system, in the singular!) (p. 31)

About the proposed procedure (CATWOE) to apply SSM:

Pay close attention to the formulation of the names of relevant systems. These had to be written in such a way that they made possible to build a model of the system named. The names themselves became known as "root definitions" since they express the core or essence of the perception to be modelled. (p. 33) [In the previous chapters I proposed the expressions "identity" and "multiple identities" to indicate the set of names given to our nonreducible perceptions of a given system.]

The positive aspect of the use of *more complex models* is that it might enrich the debate when models are compared with the real world. The negative aspect is that the increased *complexity of the models* might lead to our slipping into thinking in terms of models of part of the real world, rather than models relevant to debate about change in the real world. (p. 41) [In Chapter 2 I suggested the expression "complicated models" to indicate models with a large number of variables, more parameters and nonlinear dynamics. Complexity in the view proposed in this book has to do with addressing the semantic aspect related to the use of inferential systems. Adopting the vocabulary used in this book, the authors are referring in this passage to complicated models.]

Once a model of a purposeful holon exists ... then it can be used to structure enquiry into the problem situation. However, before using the model as a tool ... most modellers will probably be asking themselves if their intellectual construct is adequate, or valid. Since the model does not purport to be a description of part of the real world — but rather — merely a holon [the author means with this term the representation of a given shared perception] relevant to debating perceptions of the real world, adequacy and validity cannot be checked against the world. Such models are not, in fact, "valid" or "invalid," only technically defensible or indefensible. (p. 41)

Models are only a means to an end which is to have a well structured and shared representation of the perception of a problem situation to be used in the debate about how to improve it. That debate is structured by using the models based on a range of worldviews to question perceptions of the situation. (p. 43)

5.5.2 The Procedural Approach Proposed by Checkland with His Soft System Methodology

A quick presentation of the procedural approach proposed by Checkland is given below. This presentation is taken from the book by Allen and Hoekstra (1992), *Toward a Unified Ecology*. We decided to use this narrative because of two points: (1) Allen and Hoekstra propose in their book an epistemology framed within complex systems theory and (2) the reference to SSM as a problem-solving engine is directly related, in their book, to the issue of sustainability with a specific reference to multiple land use and ecological compatibility.

The steps identified by Allen and Hoekstra (1992, p. 308–316) follow in the ensuing sections.

5.5.2.1 Step 1: Feeling the Disequilibrium, Recognizing That There Is a Problem Even if It Is Not Clearly Expressed — If we accept that a problem is the existence of a gradient between our perception of the reality and our expectation about the reality, it becomes immediately clear that even reaching an agreement on the existence of a problem is anything but trivial. The denial of the existence of problems that would require a discussion of the identity of those in power is a well-known phenomenon in human affairs (e.g., the ego denying the process of aging, academic institutions denying the need to change the way a disciplinary field is taught, the government denying the existence of economic problems).

5.5.2.2 Step 2: Generate Actively as Many Points of View for the System as Possible — After intuiting that there is a mess (mess is regarded here as a technical word "that couches the situation in terms that recognize conflicting interests" — p. 309), the second step is to actively generate as many points of view for the system as possible. Checkland calls this stage painting the rich picture, or the problem situation expressed. The distinctive feature here is not the building of a model that has a particular point of view, but rather taking into account as many explicitly conflicting perspectives as possible. It is the richness of the picture that is important at this stage, not the restricted mental categories one might create to deal with it.

We can use an analogy with the quality of digital images. We know that, depending on the number of pixels per cell, we can have a better quality in the image, no matter what type of image will be shown on the screen. In the same way, the ability to perceive the same process or facts using a wider set of nonequivalent detectors, mechanisms of mapping and epistemic categories will provide more robustness to the final image. This is a characteristic of the process of representation that will hold, whatever we decide should be the subject on which the camera focuses.

5.5.2.3 Step 3: Explicit Development of Abstractions, Finding the Root Definitions — The third stage is the most critical and involves the explicit development of abstractions. It puts restrictions on the rich picture in the hope of finding a workable solution. Checkland calls this stage finding the root definitions.

This is the stage at which we have to decide how to identify and represent our problem situation. A formal problem structuring requires (as noted in Chapters 3 and 4) first a semantic problem structuring. That is, the analyst must be able to answer a family of basic questions: What is the system of interest? What is this system doing? Why is this relevant? Relevant to whom? What are the criteria used to decide that? What are the system attributes that produce the conflicts and the unease that generated the willingness to get into the first step of the process?

As noted in Chapters 2 and 3, depending on the various perceptions of the physical structure of the system of interest, we should expect to find different identities for the same system. These identities will change depending on the scale or the points of view adopted.

About the existence of multiple identities required for a useful problem structuring, Allen and Hoekstra (1992, p. 313) observe: "*It is important to realize that the several different sets of root definitions are not only possible, but desirable.*"

The heuristic tool suggested by Checkland for dealing with the delicate step of deciding about a set of useful root definitions (identities) for the problem structuring is based on the use of the acronym CATWOE. The six letters of the acronym stand for (quoted from Allen and Hoekstra, 1992):

C — The client of the system and analysis. For whom does the system work? Sometimes the client is the person for whom the system does not work, namely, the victim.

A — The actors in the system.

T — The transformations or underlying processes. What does the system do? What are the critical changes? These critical transformations are generally performed by the actors.

W — *Weltanschauung* (worldview); identifies the implicit worldview invoked when the system is viewed in a particular manner. This defines the set of phenomena of interest.

O — The owner of the system. Who can pull the plug on the whole thing?

E — The environment, that is, what the system takes as given. By default, the environment defines the scale of the system's extent by being everything that matters that is too large to be differentiated.

To bridge this analysis to what was presented in the previous chapters, we can now translate this vocabulary into what is generally found in the literature of integrated assessment and multi-criteria analysis.

Three of these letters — C, A and O — refer to different categories of relevant social actors (which are not mutually exclusive, but rather overlapping). Before discussing the differences, let us first start with the standard definition of *stakeholder* (a technical term often used to refer to relevant social actors) found in the literature of MCA: Stakeholders are those actors who are directly or indirectly affected by an issue and who could affect the outcome of a decision-making process regarding that issue or are affected by it.

The suggested nonequivalent categories of social actors can be interpreted as follows:

Clients — The stakeholders who are ethically relevant in relation to the *Weltanschauung* in which the process of decision making is taking place. For example, when dealing with sustainability, these clients could be the future generations. They can be nonagents; they cannot have power in the negotiation, but their perspective can still be relevant and should be included for ethical reasons. Obviously, it is essential to start the process with a clear picture of who are the ethically relevant stakeholders, since the integrated assessment has to include indicators able to detect the effects that our decision will have on them.

Actors — The stakeholders who are relevant agents within the mechanisms determining the set of phenomena of interest. Also in this case, it is essential to start with a clear picture of who are the stakeholders who are agents, to be able to consider in our models and future scenarios the possible reactions of these agents to the changes implicit by the selected actions or policies. Agents do not necessarily have strong negotiating power in the process. For example, poor marginal farmers can decide to increase the pressure on free natural resources because of bad policies of central government. In this case, if the result of this overexploitation is just a decrease in their local sustainability and material standard of living (e.g., deforestation or soil erosion), we are dealing with an example of relevant agents without negotiating power in the process of decision making.

Owner – The stakeholders with a clear power asymmetry in the process of negotiation used to define what are the perceptions that count when defining the problem structuring. Also in this case, it is essential to start with a clear picture of the existing power structure among the considered set of (1) relevant stakeholders (clients) and (2) agents (actors). In fact, going back to the scheme presented in Figure 5.6 and the example of social impact matrix presented in Figure 5.7, such a relationship will be essential in determining whose perceptions, goals and vetoes will be more important in the process of selection of a semantic problem structuring that will then be translated into a relative formal problem structuring.

The definitions of these three categories (C, A and O) have to do with relevant issues in relation to the normative side (*apply* in Figure 4.2).

The other two letters — T and E — refer to choices related to our representation of facts. Also in this case, it is possible to relate these letters to concepts presented and discussed in previous chapters:

Transformations — The set of modeled behaviors (e.g., inputs transformed into outputs) resulting from the choice of encoding variables and inferential systems used to describe the reality within the selected representation of relevant perceptions. That is, these are the transformations included in the simplified representation of the reality obtained through modeling. Each one of these transformations represented within individual models refers to a particular dynamic that is simulated using simple time (following a triadic reading of nested holarchies).

Environment — the set of assumptions about the compatibility of initiating conditions (stability of structural elements) and admissibility of boundary conditions (stability of the meaning of a given function in a given associative context). These are the assumptions required for the triadic reading (modeling in a given descriptive domain) of complex systems organized in nested hierarchies. The definitions of what should be considered as environment have to do with choices made by the modelers about the potential obsolescence of the models used to represent transformations and therefore the scale (time differential and time horizon for the validity of the model). When the becoming reality changes in relation to the selected models, the representation of transformations loses validity.

The definition of these two categories (T and E) has to do with the descriptive side (*represent* in Figure 4.2).

Finally, the last letter, W, directly relates to how the descriptive and normative sides are affecting each other:

Weltanschauung — the preanalytical set of choices about (1) what should be considered the universe of relevant facts (the universe of discourse within which analysts look for explanations and models) and (2) how to structure the representation in this universe (after deciding what has to be given priority over the rest in relation to agreed-upon goals). This preanalytical set of choices is related to the evolution of the system of knowledge within which the process is taking place. This is where the history of the social group enters into play, affecting how a particular social group will end up representing its shared perception of the reality (Figure 4.2). According to this history (the past experience of the human group) and the virtual future (aspirations and wants expressed at the level of the whole group), the modelers have to define who belongs to the three categories of C, A and O when organizing the normative side, and choose how to define T and E when deciding about the descriptive side. As observed several times, the definitions of C, A and O will affect the way the modelers select and define T and E. The reverse is also true: the definitions of T and E will affect the way C, A and O are perceived and individuated. The result of this convergence in the past is what determines the starting point of this reciprocal definition now (the current *weltanschauung*). However, when discussing complex time, we already addressed the problem of the potential obsolescence of the validity of the preanalytical choices required for selecting identities and multiple identities (or root definitions) in any problem structuring.

The definition of this category therefore has to do with the challenge of keeping coherence in the process leading to a shared perception of the reality in relation to action and the relative representation. This has to do with *transduce* (Figure 4.2).

Before getting back to the remaining five steps, it is opportune to have a look at the overview of the representation of the iterative process given by Allen and Hoekstra, which is illustrated in Figure 5.8.

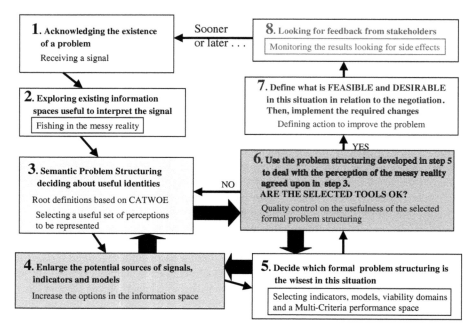

FIGURE 5.8 The iterative process suggested by Checkland. (Adapted from Allen, T.F.H. and Hoekstra, T.W., *Toward a Unified Ecology*, Columbia University Press, New York, 1992.)

The first three steps belong to a large iterative loop (indicated by small arrows), whereas step 3 (reaching an agreement on the definition of a semantic problem structuring), which resonates with step 6 (deciding the validity of the formalization of the semantic problem structuring), belongs to an additional internal loop (indicated by the large arrows) related to the need of a quality control on the effective convergence between the information referring to the normative and descriptive sides.

5.5.2.4   Step 4: Building the Models — The fourth stage is that of building the models. There will need to be a different model for each set of alternative root definitions (nonreducible models for nonequivalent identities).

In this step, the various modelers coming from different scientific disciplinary backgrounds will bring into the process the specific know-how of their expertise. That is, thanks to this expertise, it will be possible to individuate constraints and crucial mechanisms affecting the behavior of the system in relation to general patterns well known in different academic fields (economic, social, technical, ecological). Obviously, the potential contribution of scientists coming from different backgrounds will depend on the selection of relevant aspects to be included in the problem structuring.

5.5.2.5   Step 5: Returning to Observations of the World and Checking the Model against What Happens — The fifth stage returns to observations of the world and the model is checked against what happens. If the actors are people (when dealing with ecological systems, this is not necessarily the standard case), one can ask them their opinion of the model and modify it to be consistent with their special knowledge.

In this step, there is the decision of selecting among all the possible models, indicators, and metaphors and principles that could be applied to this situation only on an information space that can usefully be handled as the formal problem structuring to be adopted in the process of decision making.

5.5.2.6   Step 6: Exploration of Feasible and Desirable Changes — This step is present in the figure given by Allen and Hoekstra in the book (Figure 9.9 on p. 310), but it is not described in the text, which mentions only seven steps.

In this step, feasible and desirable changes are explored using the analytical tools developed so far. If nothing feasible and desirable is found in the process of negotiation, it is obvious that the original problem statement is wrong (useless) and has to be revisited. There are several ways out of this impasse. Check the validity of the perceptions and the expectations, and try to change the terms of reference determining constraints and feasible solutions.

5.5.2.7 Step 7: Identification of Desirable and Feasible Changes for the System —

At this stage, one identifies desirable and feasible changes for the system. Decisions about selecting and implementing a policy are taken at the end of this step.

5.5.2.8 Step 8: Evaluation, Widening the View of the Whole Process — After the

implementation of a given policy, it is necessary to widen the view of the whole process. Evaluation — did it work?

Again also in this case, a more effective monitoring of the results looking for unexpected side effects can be obtained by relying on nonequivalent observers, which can generate new types of signals about the expected and unexpected effects induced by the choice (or policy). Increasing the diversity of nonequivalent observers performing such monitoring can increase the ability to detect the consequences of the unavoidable ignorance load of the semantic and formal problem structuring adopted in the decision-making process. This explains why this step naturally leads into a new step 1 for another cycle.

An important feature of the procedure suggested by Checkland, stressed by Allen and Hoekstra, is the ability to compress and expand the information space considered in the process during the various steps. The implications in terms of compression and expansion of the information space (number of attributes, variables or indicators used in the set of identities used in the various models; number of models; number of relevant issues; and anticipatory systems adopted in the analysis) are shown in Figure 5.9. This has to do with the challenge of handling the heroic compression illustrated in Figure 5.6. Since it is impossible to even dream of performing such a compression in a satisficing way in a single step, it is wise to compress and expand several times in an iterative process. This can make it easier to perform a quality control of the choices made during such a process (normative vs. descriptive, technical feasibility vs. social acceptability, accuracy vs. relevance). This is especially true when dealing with the unavoidable existence of conflicts and power asymmetry. In this regard, the distinction introduced by Checkland about the three relevant attributes that should be used to classify stakeholders in relation to their relevance in this process (C, A and O) is particularly useful.

Coming to the descriptive part, the part that is relevant for the rest of this book, it should be noted that a continuous shift between semantic (the use of metaphors required for the sharing of meaning about a situation — definition of classes of models) and formal models (translating the meaning of the perceptions associated with a class of models in relation to a location-specific situation to generate data related to variables that can be used as indicators) is the only way out of the impasse of reductionism described so far.

5.5.3 Looking at This Procedure in Terms of an Iteration between Discussion Support System and Decision Support System

The expression "discussion support system" was suggested by Herman van Keulen to indicate the activity of handling of nonequivalent descriptive domains, sets of indicators referring to legitimate contrasting perspectives and nonreducible models during the process of selection of a problem structuring. The output of this process has been called multi-scale integrated analysis and will be the subject of the next part of this book. The activity referring to the generation of an MSIA refers to the loop indicated by the large arrows in Figure 5.8.

The *quality* of a given problem structuring provided by scientists to the stakeholders involved in a process of decision making depends on the ability to represent in analytical terms sustainability trade-offs (on different scales) in relation to the set of legitimate views considered relevant (multi-objective) and to the set of nonequivalent dimensions of viability considered relevant (multidimensional). This is

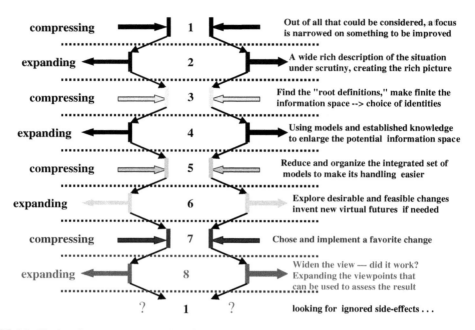

compressing ➡ 1 ⬅ Out of all that could be considered, a focus is narrowed on something to be improved

expanding ⬅ 2 ➡ A wide rich description of the situation under scrutiny, creating the rich picture

compressing ⇨ 3 ⇦ Find the "root definitions," make finite the information space --> choice of identities

expanding ⬅ 4 ➡ Using models and established knowledge to enlarge the potential information space

compressing ⇨ 5 ⇦ Reduce and organize the integrated set of models to make its handling easier

expanding ⇦ 6 ⇨ Explore desirable and feasible changes invent new virtual futures if needed

compressing ➡ 7 ⬅ Chose and implement a favorite change

expanding ⬅ 8 ➡ Widen the view — did it work? Expanding the viewpoints that can be used to assess the result

? 1 ? looking for ignored side-effects . . .

FIGURE 5.9 The iterative process suggested by Checkland. (Adapted from Allen, T.F.H. and Hoekstra, T.W., *Toward a Unified Ecology*, Columbia University Press, New York, 1992.)

why such an input is called multi-scale integrated analysis (multi-objective and multidimensional). Scientists should be able to characterize the performance of socioeconomic systems in relation to envisioned changes, considering parallel short-, medium- and long-term perspectives of the various stakeholders. To do that, they have to tailor their descriptive tools on what is required by multi-criteria methods of evaluation.

To accomplish that, they must perform several quality checks on the validity and usefulness of the representative tools used in the process. If we accept that the organization of the information space needed for SMCE can no longer be based on traditional (reductionist) descriptive tools (as discussed in Chapters 2 and 3), that is, it can no longer be based on the assumption that it is possible to provide substantive definitions of optimal solutions, then we have to look for new descriptive tools (e.g., those presented in Section 5.4) and adequate procedures (those presented in this section).

These new tools have the goal of organizing the scientific representation of the problem, after acknowledging the existence of the serious epistemological challenges found when attempting to generate relevant, reliable and transparent scientific input for the process of decision making. Such scientific input has to be realized in the form of an agreed-upon multi-scale integrated analysis (i.e., a set of assessments of different typologies or costs and benefits in relation to different definitions of costs and benefits), which has to be referred to a preliminary definition of an option space (i.e., a given set of scenarios reflecting possible alternative choices).

Therefore, any procedure for obtaining a useful MSIA should be based on a genuine iterative process between scientists and the rest of the social actors aimed at generating an evolving discussion on how to better represent and structure the problem to be tackled. This procedure has to deal with two major problems: (1) On the normative side, it has to deal with the unavoidable ambiguity in the definition of terms and identification of common goals (social incommensurability). Everyone can agree on the need to look for peace and freedom; on the other hand, there is an unavoidable ambiguity when translating these general concepts into action within a given context. That is, those considered terrorists by one side are the freedom fighters of the other side (Tim Allen, personal communication). The definition of actions required to obtain and preserve peace often bifurcates when coming to the decision of what to do next in front of an existing confrontation. (2) On the descriptive side, it has to deal with the unavoidable nonre-

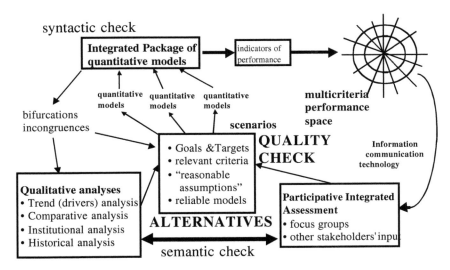

FIGURE 5.10 Iterative process mixing quantitative and qualitative analyses in a participatory discussion of sustainability dialectics.

ducibility of models built by using variables defined in nonequivalent descriptive domains (technical incommensurability).

This section focuses on the implications of this fact from the perspective of the scientist willing to get involved in such a process. An overview of the iterative process that is required to define a multi-scale integrated analysis to be generated by a discussion support system is given in Figure 5.10. This MSIA has to be based on:

1. **Useful representations** (multi-objective, multi-scale, multidimensional) of relevant features of the system able to reflect legitimate perspectives found among relevant social actors and information required for decision making of relevant agents. This translates into the selection of a set of models that use nonequivalent identities and boundaries for the same system, which is able to represent the problem over different descriptive domains in relation to different dimensions of analysis. To start such an iterative process, a first draft of a formal problem structuring (to be criticized and changed later on) can be used to start a discussion about the semantic questions required for the problem structuring (step 3 in Figure 5.8).

2. **Definition of the feasibility space** (range of admissible values) for each of the selected indicators of performance. Feasibility should reflect the reciprocal effect of constraints across hierarchical levels. The most common selection of constraints includes economic, biophysical, institutional and social constraints. This is the enrichment of information obtained during step 4 (Figure 5.7), when individual reductionist scientists can bring into the discussion metaphorical knowledge (validated on class of models) that can be applied to the specific situation.

3. **Integrated representation of system performance** in relation to the selected set of incommensurable criteria. This requires selecting a package of indicators referring to the desirability of feasible changes in relation to different relevant criteria. This requires assessing the value of each indicator included in the package. In this way, it becomes possible to represent (1) what should (or could) be considered an improvement or a worsening in relation to each attribute, (2) how the system compares with appropriate targets and other similar systems and (3) what are potential critical threshold values of certain variables where nonlinear effects can play a crucial role. This is the first part of the activity included in step 5, illustrated in Figure 5.8.

4. **Strategic assessment of possible scenarios** done by addressing the problem of uncertainty and general evolutionary trends that can be expected. In relation to this point, the scientific representation no longer has to be based only on a steady-state view and on a simplification

of the reality represented according to a single dimension at a time (*ceteris paribus*). That is, conventional reductionist analyses providing the picture of the position of the system on a multi-objective performance space (as the radar diagrams shown in Figure 5.5) have to be complemented by analyses of (1) evolutionary trends (related to adaptability implying a virtual future–present causal perspective), (2) the crucial effects of the particular history of a system determining lock-in and behavioral constraints and (3) the parallel consideration of processes and mechanisms operating simultaneously on several levels and scales. An evolutionary analysis should make it possible to classify the investigated system in relation to its position within the domain of possible evolutionary trajectories (by comparing it with other similar systems within the same state space used for the integrated assessment). Only in this way it is possible to enable the stakeholders to provide an input about the validity of the scientific representation of the issue. The quality check of the stakeholders has to deal also with the issue of overall credibility of scenarios and simulation, besides that of relevance of the analyses included in the package. This activity is in step 5 of Figure 5.8.

5. **Mosaic effects providing robustness to the scientific input, obtained through redundancy in the information space.** Mosaic effects can be obtained by bridging of nonequivalent descriptions using the forced congruence of numerical assessment across scales. This mechanism that can be used to boost the reliability of data and models will be introduced and discussed in Parts 2 and 3. The mosaic effect can be used not only to perform a congruence check on the validity of the database used in the description, but also to fill empty spaces in the database when gaps occur. When dealing with practical problems of sustainability of human societies, the bridge across descriptive domains can be obtained by forcing the congruence of flows of money, energy, matter and human time. The condition of congruence implies that nonequivalent descriptions (organs within humans and humans within households) adopted when generating and calculating the set of different indicators included in the package must still exhibit coherence in relation to congruence checks (the total weight of organs and the total weight of humans must be congruent with each other). Recall the example of the multiple assessments of cereal consumption per capita in the U.S. in 1997, according to micro, macro, biophysical and economic analyses. By establishing bridges among nonequivalent readings, we can check whether assessments coming from an analysis performed at a given level (the household level) are consistent — when scaled up or down from a different level — with indications coming from the analysis based on a nonequivalent description performed at a different level. This congruence check can be obtained in terms of both numerical values and known trends. The generation of a mosaic effect patching together nonreducible models can make possible the filtering out of incoherent scenarios or biased policy solutions proposed by interested scientists and stakeholders, as well as forcing transdisciplinarity in the step representation. A more detailed description of this approach is found in Chapter 6 and applications in Chapter 9. Again, this is a type of quality check on the scientific input of representation that can be performed by scientists themselves. As observed by Checkland, this has to do only with the technical defensibility of a given integrated representation. For example, looking at the Jevons' paradox (Chapter 1), by performing an analysis in parallel on two scales, we can immediately check that the inference that an increase in efficiency will lead to a reduction of consumptions (obtained using models based on the *ceteris paribus* hypothesis) clashes against a trend analysis performed at a different scale.

6. **Analysis of the sustainability dialectics,** which is unavoidably implied by multifunctionality and sustainability (a given blanket cannot be pulled in all directions). This analysis has to include an assessment on the uncertainty associated with various scenarios considered in relation to the criteria and alternatives indicated as more relevant in the discussion. This is where the activity of scientists has to be integrated by an interaction with social actors and where the border between descriptive and normative blurs. This is especially important in step 6 of Figure 5.8, when an overall agreement has to be reached on the validity of the existing problem structuring as the agreed-upon scientific input to be adopted in the process of social multi-criteria evaluation.

In conclusion, the scientific aspects of multi-scale integrated analysis, which is the required input to a process of social multi-criteria evaluation, can be related to the activities that are related to the four iterative steps included in the loop indicated by large arrows in Figure 5.8. Even when remaining within the problems associated with the activities required in these four steps (steps 3 to 6), the scientists are forced to address, first of all, the implications of the delicate preanalytical step of compression of the virtual infinite universe of discourse into the finite and formalizable information space represented by the scientific input to the process of decision making (Figure 5.6). In the rest of this book I will deal only with this problem.

Technical issues and examples of how to organize the scientific input in the form of multi-scale integrated analysis of agroecosystems useful for SMCE are discussed in Part 3 of this book Part 2 provides the theoretical rationale of these approaches in relation to complex systems theory.

5.6 What Has All This to Do with the Sustainability of Agriculture? Example: The Making of Farm Bills (Institutionalizing a Discussion of the Social Contract about Agriculture)

5.6.1 What Should Be the Role of Colleges of Agriculture in the New Millennium?

Current basic strategy driving technical progress of humankind, which is aimed at achieving a much higher material standard of life for an increasing number of people, is becoming more and more a sort of gamble. Humankind is risking what already exists to have more (Giampietro, 1994a, 1994b). But even if we accept that gambling cannot be avoided by systems that are forced to evolve in time (this is what life is about), we should at least be able to define the terms of the bet (what can be gained, what can be lost) and the rules of the game (who is calling the bet and who will pay for or gain by it). Unfortunately, at present, these terms of gambling are anything but clear, especially when dealing with the technical progress of agriculture.

These two facts combined are generating a clear paradox. Because of the big challenges and the large dose of uncertainty faced by humankind in this delicate phase of transition, science and technology are needed as never before. However, one should always be aware that technology can be either part of the solution or part of the problem. This is why, in parallel with the development of new technologies, humans have to develop new scientific fields aimed at interfacing the activity of scientists with the activity of the rest of the society as soon as possible. When dealing with science for governance, scientific know-how cannot be taken from the shelf. It must always be applied and tailored to specific situations through an iterative process of interaction of all relevant social actors.

Against this general framework, what is the role that academic institutions and, in particular, colleges of agriculture should play in the field of sustainability of agriculture? We can use a statement taken by the post-Newtonian quantum physicist David Bohm, quoted by Röling (1994, p. 390): "Science consists not in the accumulation of knowledge, but in the creation of fresh modes of perception" (Bohm, 1993). Coming back to the discussion presented in Chapter 4 about the problem of how to update societal perception, representation and regulation of agricultural activities, we can say that this is clearly a task that cannot be performed by scientists alone. Such an updating implies learning how to elaborate better compromises among contrasting values, perceptions of risks and opportunities, and aspirations. Put another way, any assessment of innovations, regulations and policies to be adopted in the food system requires the implementation of procedures of participatory integrated assessment, that is, the development of useful multi-scale integrated assessment used and driven by processes of societal multi-criteria evaluation.

In this perspective, conventional agricultural research should be complemented by a new field of analysis dealing with the issue of science for governance. Basic knowledge about agricultural practices is and always will be required to make possible the continuous readjustment of mixes of activities in relation to different definitions of performance in a multifunctional frame. However, at the same time, a truly innovative field of scientific activity is required to enable a two-way exchange of information between academic institutions and civil society. This is not about generating better programs of extension, but rather looking at a two-way direction of information flows. The ivory tower so useful in the past to preserve the special status of academic research has to become more open. To make this point clear, consider the example of the making of a farm bill.

5.6.2   The Case of the U.S. Farm Bill 2002

The U.S. Farm Bill 2002 represents a clear change in (or a big revision of) the U.S. federal government's attitude toward regulation and intervention in agricultural development. The passing of this farm bill has generated contrasting views and assessments on its overall quality (e.g., http://www.sustainableagriculture.net/summary-5-6-02.htm). We can relate this example of decision making to the discussion of analytical tools presented in Section 5.4. When making this decision, the U.S. federal government must have selected this specific bill out of a set of possible alternative bills. To do that, U.S. decision makers must have used a problem structuring similar to that represented in both Figure 5.4 (in terms of impact matrix) and Figure 5.7 (in terms of social impact matrix). In this case, n is the set of possible alternative ways of spending a certain amount of (billions of) dollars in implementing a package of policies and m is the set of criteria used to represent and assess the expected performance associated with the implementation of each of the alternative policies. Actually, it should be noted that the decision about the amount of money to be spent in a farm bill could be considered itself as a variable, rather than a constraint. In this case, different policies requiring the expenditure of different amounts of billions of dollars should have been considered in the analysis. Obviously, the chosen alternative (the actual "Farm Bill 2000") must have been considered the "wisest" choice in relation to: (1) a set of m multiple goals (e.g., economic viability of the agricultural sectors, food security, quality of the food, environmental impact, social stress in rural community, protection of cultural values, etc.), (2) a set of data and models characterizing n scenarios associated with the implementation of the n alternative policies included in the problem structuring and (3) the legitimate contrasting perspectives of the social actors considered relevant in such an analysis.

Put another way, to make this choice, the U.S. government must have used:

- **A problem structuring,** implying decisions about (1) what is the set of relevant social actors who have to be considered when deciding, (2) a set of criteria that are considered relevant for this choice, (3) what is the set of criteria that are relevant to some stakeholders but that can be neglected to satisfy other conflicting criteria and (4) a mix of policy options that can be combined to generate the set of alternative bills considered.
- **A set of analyses (models and data) and predictions** used to characterize the effects of the possible policies on different descriptive domains through scenarios (e.g., in social terms, economic terms, ecological terms, landscape use terms) characterized at different scales. That is, to make such a decision, it has been necessary to have an idea of what will or could happen when adopting the package of monetary and regulatory policies 1, 2 or 3. The MSIA of different effects of each package must have been characterized using a set of different indicators reflecting the relevant and legitimate contrasting views defined in the problem structuring.
- **A process of political negotiation** among different interests and concerns associated with the various stakeholders in the U.S. food system. The particular choice of one of the possible policy packages (the actual Farm Bill 2002), in fact, implied that some stakeholders got more benefits than others (implying that some of the criteria have been given more priority than others).

Analysts of agriculture as well as social actors might have asked a number of questions about the choices made by the U.S. government in this farm bill:

1. Could U.S. society at large and the various stakeholders in the U.S. food system have had a better chance to form a clearer picture of what was the information space used by the decision makers for organizing such a discussion?
2. Could U.S. society at large and the various stakeholders in the U.S. food system have been involved in a more transparent process of discussion of the basic problem structuring (defining the relevant criteria and defining possible options)?

3. Could U.S. society at large and the various scientists and academic institutions have been involved in a more effective multi-scale integrated analysis of possible effects of the various alternative bills?

4. Could U.S. society at large and the various stakeholders in the U.S. food system have been involved in a more transparent process of negotiation about the weights to be used when dealing with contrasting perspectives about priorities?

If the answer to any of these questions is yes, then U.S. society at large and the various stakeholders in the U.S. food system have lost an important opportunity to learn how to design, discuss, understand and negotiate future farm bills in a better way. In the future, this ability will become extremely important if the existing trends referring to the postindustrialization of a globalized world remain.

The context is changing so fast that the validity of the social contract used to define the various roles that social actors have to play in the food system has to be monitored and negotiated on a regular basis. This is why academic programs willing to deal with the sustainability problems of agriculture should give top priority to the development of new tools and procedures for dealing with MSIA of agriculture. Agriculture should no longer be considered just another economic sector producing commodities. Rather, agriculture should be associated with a multifunctional set of activities associated with land use.

At this point, we can try to answer the question about the role of academic programs dealing with agriculture in the new millennium. The top priorities for academic programs of agricultural colleges within the U.S. should be that of producing scientific information relevant for the discussion of a next farm bill in 2008 and that of becoming able to make a difference in the shaping of the discussion and the societal multi-criteria evaluation of the U.S. farm bill in the year 2014. In fact, future farm bills should be able to better reflect (1) the continuous change in societal perception about what a food system entails and (2) the growing scientific awareness about the ecological and social dimensions of sustainability.

These two lines of research for agriculture can be related to the discussion about the two possible interpretations for the term *agroecology* presented at the beginning of this chapter. The first of the two interpretations (how to totally rethink agriculture) refers to the process of societal learning about how to better design, discuss, understand and negotiate future farm bills in both developed and developing countries. The second interpretation of the term *agroecology* is about the need to expand the option space of the set of possible technical coefficients available to generate mixes of techniques of production in various agroecosystems. This has to do with expanding knowledge about ecological and economic performance profiles of individual techniques of production and integrated systems of production. The usefulness of this second activity is associated with the beneficial effect of increasing the diversity of potential performance profiles to be adopted in a multifunctional framework of land uses.

Obviously, the sustainability predicament of agriculture in both developed and developing countries implies that more research is needed in both directions. For sure, the first direction of research is the one that will provide a higher return in the short term, because of the clear obsolescence and cultural lock-in of current mechanisms of policy interventions in the agricultural endeavor in both developed and developing countries.

References

Allen, T.F.H. and Hoekstra, T.W., (1992), *Toward a Unified Ecology*, Columbia University Press, New York.
Altieri, M., (1987), *Agroecology: The Scientific Basis for Alternative Agriculture*, Westview Press, Boulder, CO.
Bana e Costa, C.A., Ed., (1990),*Readings in Multiple Criteria Decision Aid*, Springer-Verlag, Berlin.
Beinat, E. and Nijkamp, P., Eds., (1998), *Multi-criteria Evaluation in Land-Use Management: Methodologies and Case Studies*, Kluwer, Dordrecht.
Blank, S.C., (1998), *The End of Agriculture in the American Portfolio*, Greenwood Publishing, Westport, CT.
Bohm, D., (1993), "Last words of a quantum heretic" interview with John Morgan, *New Scientist*, 137, 42.
Checkland, P., (1981), *Systems Thinking, Systems Practice*, John Wiley, Chicester, U.K.
Checkland, P. and Scholes, J., (1990), *Soft-Systems Methodology in Action*, John Wiley, Chicester, U.K.

Dodgson, J., Spackman, M., Pearman, A. and Phillips, L. (2000). *Multi-Criteria Analysis: A Manual*. Great Britain Department of the Environment, Transport and the Regions. Office of the Deputy Prime Minister. DETR, London.

Georgescu-Roegen, N. (1971), *The Entropy Law and the Economic Process*. Harvard University Press. Cambridge, MA.

Giampietro, M., (1994a), Using hierarchy theory to explore the concept of sustainable development, *Futures*, 26, 616–625.

Giampietro, M., (1994b), Sustainability and technological development in agriculture: a critical appraisal of genetic engineering, *Bioscience*, 44, 677–689.

Giampietro, M., (1997a), Socioeconomic pressure, demographic pressure, environmental loading and technological changes in agriculture, *Agric. Ecosyst. Environ.*, 65, 201–229.

Giampietro, M., (1997b), Socioeconomic constraints to farming with biodiversity, *Agric. Ecosyst. Environ.*, 62, 145–167.

Janssen, R. and Munda, G., (1999), Multi-criteria methods for quantitative, qualitative and fuzzy evaluation problems, in *Handbook of Environmental and Resource Economics*, van den Bergh, J., Ed., Edward Elgar, Cheltenham, U.K., pp. 837–852.

Li, J., Giampietro, M., Pastore, G., Liewan, C., and Huaer, L., (1999), Factors affecting technical changes in rice-based farming systems in southern China: case study of Qianjiang municipality, *Crit. Rev. Plant Sci.*, 18, 283–298.

Martinez-Alier, J., Munda, G., and O'Neill, J., (1998), Weak comparability of values as a foundation for ecological economics, *Ecol. Econ.*, 26, 277–286.

Mayumi, K. (2001), *The Origins of Ecological Economics: The Bioeconomics of Georgescu-Roegen*. Routledge, London.

Mishan, E.J., (1993), *The Costs of Economic Growth*, rev. ed., Weidenfield and Nicolson, London, 243 pp.

Munda, G., (1995), *Multi-criteria Evaluation in a Fuzzy Environment: Theory and Applications in Ecological Economics*, Physica-Verlag, Heidelberg.

Munda, G., (2003), Social multi-criteria evaluation (SMCE): methodological foundations and operational consequences, *Eur. J. Operation. Res.*, in press.

Myrdal, G., (1966), *Objectivity in Social Research*, Pantheon Books, New York.

Nijkamp, P., Rietveld, P., and Voogd, H., (1990), *Multi-Criteria Evaluation in Physical Planning*, North-Holland, Amsterdam.

O'Connor, M. and Spash, C., Eds., (1998), *Valuation and the Environment: Theory, Methods and Practice*, Edward Elgar, Cheltenham, U.K.

O'Neill, J., (1993), *Ecology, Policy and Politics*, Routledge, London.

Röling, N., (1994), Platforms for decision-making about ecosystems, in *The Future of the Land: Mobilizing and Integrating Knowledge for Land Use Options*, Fresco, L.O., Stroosnijder, L., Bouma, J., and van Keulen, H., Eds., John Wiley & Sons Ltd., New York.

Schumpeter, J.A. (1954), *History of Economic Analysis*. George Allen & Unwin, London.

Röling, N. and Wagemakers, A., Eds., (1998), *Facilitating Sustainable Agriculture Participatory Learning and Adaptive Management in Times of Environmental Uncertainty*, Cambridge University Press, Cambridge, U.K.

Simon, H.A., (1976), From substantive to procedural rationality, in *Methods and Appraisal in Economics*, Latsis, J.S., Ed., Cambridge University Press, Cambridge, U.K.

Simon, H.A., (1983), *Reason in Human Affairs*, Stanford University Press, Stanford, CA.

Vincke, P., (1992), *Multi-criteria Decision Aid*, Wiley, New York.

Voogd, H., (1983), *Multi-criteria Evaluation for Urban and Regional Planning*, Pion, London.

Zeleny, M., (1982), *Multiple-Criteria Decision-Making*, McGraw-Hill, New York.

Part 2

Complex Systems Thinking: Daring to Violate Basic Taboos of Reductionism

6

Forget about the Occam Razor: Looking for Multi-Scale Mosaic Effects*

This chapter first introduces the concept of mosaic effect (Section 6.1) in general terms. It then illustrates the special characteristics of holarchic systems with examples (Section 6.2). This class of systems can generate and preserve an integrated set of nonequivalent identities (defined in parallel on different levels and therefore scales) for their constituent holons. The expected relationship among the characteristics of this integrated set of identities makes it possible to obtain some free information when performing a multi-scale analysis. This is the basic rationale for the multi-scale mosaic effect. A multi-scale analysis requires establishing an integrated set of meaningful relationships between perceptions and representations of typologies (identities) defined on different hierarchical levels and space–time domains. This means that in holarchic systems we can look for useful mosaic effects when considering the relations between parts and the whole.

Multi-scale multidimensional mosaic effects can be used to generate a robust multi-scale integrated analysis of these systems. This is discussed in detail in Section 6.3. In particular, examples are given of a multi-scale integrated analysis of the socioeconomic process. Finally, this chapter closes with a discussion of the evolutionary meaning of this special holarchic organization. Holarchic organization, in fact, provides a major advantage in preserving information and patterns of organization. This is done by establishing a resonating entailment across identities that are defining each other across scales. This concept is discussed — using a very familiar example, the calendar — in Section 6.5. The concept of holarchic complexity has been explored in the field of complex systems theory — under different names — in relation to the possible development of tools useful for the study of the sustainability of complex adaptive systems. An overview of these efforts is provided in Section 6.6. Different labels given to this basic concept are, for example, integrity, health, equipollence, double asymmetry, possible operationalizations of the concept of biodiversity.

6.1 Complexity and Mosaic Effects

Before getting into a definition of this concept, it is useful to discuss two simple examples.

6.1.1 Example 1

Koestler (1968, Chapter 5, p. 85) suggests that the human mind can obtain compression when storing information by applying an abstractive memory (the selective removing of irrelevant details). In Chapter 2 we described this process as the systemic use of epistemic categories (the use of a type — dog — to deal with individual members of an equivalence class — all organisms belonging to the species *Canis familiaris*), based on a continuous switch between semantic identities (an open and expanding set of potentially useful shared perceptions) and formal identities (closed and finite sets of epistemic categories used to represent a member of an equivalence class associated with a type) assigned to a given essence. When dealing with the perception and representation of natural holarchies (such as biological systems

---
* Kozo Mayumi is co-author of this chapter.

of socioeconomic systems), this compression is made easy by the natural organization of these systems in equivalence classes (e.g., the set of organisms of a given species are copies made from the same genetic information, as well as with human artifacts; the set of cars belonging to the same model are copies made from the same blueprint).

Getting back to the ideas of Koestler, the compression obtained with language is not obtained by using a single abstractive hierarchy (in our terms, by using a single formal identity for characterizing a given semantic identity), but rather by relying on a "variety of interlocking hierarchies ... with cross-references between different subjects" (Koestler, 1968, p. 87). This is a first way to look at mosaic effects:

> You can recognize a tune played on a violin although you have previously only heard it played on the piano; on the other hand, you can recognize the sound of a violin, although the last time a quite different tune was played on it. We must therefore assume that melody and timbre have been abstracted and stored independently by separate hierarchies within that same sense of modality, but with different criteria of relevance. One abstracts melody and filters out everything else as irrelevant, the other abstracts the timbre of the instrument and treats the melody as irrelevant. Thus not all the details discarded in the process of stripping the input are irretrievably lost, because details stripped off as irrelevant according to the criteria of one hierarchy may have been retained and stored by another hierarchy with different criteria of relevance. The recall of the experience would then be made possible by the co-operation of several interlocking hierarchies. ... Each by itself would provide only one aspect only of the original experience — a drastic impoverishment. Thus you may remember the words only of the aria "Your Tiny Hand is Frozen," but have lost the melody. Or you may remember the melody only, having forgotten the words. Finally you may recognize Caruso's voice on a gramophone record, without remembering what you last heard him sing. (Koestler, 1968, p. 87)

To relate this quote of Koestler to the epistemological discussions of Chapters 2 and 3, it is necessary to substitute the expression "**abstracting hierarchies**" with the expression "**epistemic categories associated with a formal identity used to indicate a semantic identity**" discussed there. Every time we associate the expected set of characteristics (a set of observable qualities) of members assumed to belong to an equivalence class with a label (a name), we are using types (an abstract set of qualities associated with those individuals assumed to belong to an equivalence class). As noted before, the relative compression in the information space obtained by using the characteristics of types (you say a dog and you include them all) to describe the characteristics of individual members perceived as belonging to the class has the unavoidable effect of inducing errors. Not all dogs are the same. It is not possible to cover the open universe of semantic identities (types of dogs) that can be associated with an essence ("dogginess") with a formal identity (a finite and closed set of relevant observable qualities — a formal definition of a dog). This is why humans are forced to use subcategories (e.g., a fox terrier), sub-subcategories (e.g., a brown fox terrier) and sub-sub-subcategories (e.g., a very young brown fox terrier) in an endless chain of possible categorizations. Adopting this solution, however, implies facing two setbacks: (1) In this way, we reexpand the information space required by individual observers to handle the representation (since more adjectives are required to individuate the new sub-subcategory) and (2) in this way, we lose generality and usefulness of the relative characterization. The class of "a very young brown fox terrier having had a stressful morning because of nasty diarrhea and therefore being now very hungry" is not very useful as an equivalence class. In fact, it is not easy to find a standard associative context that would make its use as a general type convenient. This is why we do not have a word (label) for this class.

What gets us out of this impasse is the observation that within a given situation at a given point in space and time, within a specified context (e.g., children getting out of a given school at 13:30 on Thursday, March 23), a combination of a few adjectives (the tall girl with the red dress) can be enough to individuate a special individual in a crowd. The girl we want to indicate is the only one belonging simultaneously to the three categories: (1) girl (individual belonging to the human species, that is, woman and young at the same time), (2) tall (individual belonging to a percentile on the distribution of height of her age class above average) and (3) with the red dress (individual wearing a red dress). Obviously this mechanism of triangulation, based on the use of a few adjectives (the fewer the better), can be adopted only within the specificity of a given context (only if the triangulation is performed at a given point in space and time). The category "tall girl with the red dress" would represent a totally useless category if used in general to individuate someone within the U.S.

The consequences of this example are very important. We can effectively describe a system using a limited set of categories (indicators) by triangulating them — relying on a mosaic effect — but only when we are sure that we are operating within a valid, finite and closed information space. When describing patterns in general, the type is described in general terms within its standard associative context, or a special system is individuated within a specific local setting (at a given point in space and time). When dealing with a specific description of events, the characteristic and constraints of the given context have to be reflected in the selection and definition of an appropriate descriptive domain.

6.1.2   Example 2

Bohm (1995, p. 187) provides an example of integrated mapping based on the mosaic effect:

> Let us begin with a rectangular tank full of water, with transparent walls. Suppose further that there are two television cameras, A and B, directed at what is going on in the water (e.g. fishes swimming around) as seen through the two walls at right angles to each other. Now let the corresponding television images be made visible on screens A and B in another room.

This is a simple example in which we deal with two nonequivalent descriptions of the same natural system (the movements of the same set of fishes seen in parallel on two TV screens). The nonequivalence between the two descriptive domains is generated by the parallel mapping of events occurring in a tridimensional space into two two-dimensional projections (over the two screens A and B). Again, we have the effect of incommensurability already discussed regarding the Pythagorean theorem (Section 3.7) — in that case, a description in one dimension (a single number) was used to represent the relation of two two-dimensional objects (the ratio of two squares). As a consequence of this incommensurability, any attempt to reconstruct the tridimensional movement using just one of the two-dimensional representations could generate bifurcations. That is, two teams of scientists looking at the two parallel nonequivalent mappings of the same event, but looking at only one of the two-dimensional projections (either A or B), could be led to infer a different mechanism of causal relations between the two different perceived chains of events. In this case, the bifurcation is due to the fact that the step *represent* (what the scientists see over each of the two screens — A and B) is only a part of what is going on in reality in the tridimensional tank. The images moving on the two screens are two different narratives about the same reality. The problem of multiple narratives of the same reality becomes crucial, for example, in quantum physics, when the experimental design used to encode changes of a relevant system's qualities in time can generate a fuzzy definition of simultaneity and temporal succession among the two representations (Bohm, 1987, 1995).

It is important to recall here the generality of the lesson of complexity. The scientific predicament is related to the fact that scientists, no matter how hard they think, can only represent perceptions of the reality. As observed by Allen et al. (2001), "Narratives collapse a chronology so that only certain events are accounted significant. A full account is not only impossible, it is also not a narrative." Put another way, a narrative is generated by a particular choice of representing the reality using a subset of possible perceptions of it. Any set of perceptions is embedded by a large sea of potential perceptions that could also be useful when different goals are considered. This implies that providing sound narratives has to do with the ability to share meaning about the usefulness of a set of choices made by the observer about how to represent events. That is, the very concept of narrative entails the handling of a certain dose of arbitrariness about how to represent reality — a degree of arbitrariness about which the scientist has to take responsibility (Allen et al., 2001). Getting back to our example of fish swimming in a tank in front of two perpendicular cameras, looking at the movements of these fish from camera A (on screen A) implies filtering out as irrelevant all the movements toward or away from that camera. A fish moving in a straight line toward camera A will be seen as moving on screen B but not moving on screen A. However, a sudden deflection from the original trajectory to a side of this fish will be perceived as a dramatic local acceleration from camera A. This will generate a nonlinearity in the dynamic of the fish within the descriptive domain represented on screen A. This dynamic will be difficult to explain in physical terms (and to simulate by a dynamic model) by relaying only the information given by screen A. How did the fish manage to get this huge acceleration in the middle of the water without touching anything — moving suddenly away from total immobility? As soon as we check the information coming from screen B, we can easily explain this perceived nonlinearity. The nonlinear dynamic that is "impossible" to explain on the descriptive domain A is simply an artifact generated by the use of a bad descriptive domain (screen A). That is, the original speed of the fish (perceived when looking at screen B) was simply ignored in the descriptive domain A, since the movement was occurring on the direction considered irrelevant according to the selected set of relevant observable qualities associated with the experimental design.

This is a very plain example of the types of problems related to the difficult interpretation of representation of changes when multiple dimensions have to be considered. In this very simple case, we are dealing only with a relevant observable quality: the position of a given object — a fish — that is moving in time. That is, no other relevant attributes but the vectors associated with speed and acceleration are considered when discussing trajectories. Imagine then, a case in which we were required to deal with a much more complex situation in human affairs that would require a much richer characterization (the simultaneous use of a larger set of relevant attributes), which in turn would require the simultaneous use of nonequivalent descriptive domains.

In conclusion, when dealing with the sustainability of a socioeconomic system, we have to first decide what is relevant and irrelevant for explaining the past history of the system and guessing the future trajectory of development, but above all, we have to decide who (what) are the relevant observers who should be considered clients for the tailoring of the representation provided by the analysis. In fact, any formalization of the representation of complex systems behavior implies (1) a large dose of arbitrariness in deciding which are the nonequivalent descriptive domains to be considered to gather useful information (on different dimensions using different "cameras," as in this example) and (2) the risk of making inferences using one of the possible models (based on what is perceived on just one of the possible screens). It is important not to miss crucial information detectable only when looking at different screens.

6.1.3 Mosaic Effect

The two definitions of mosaic effect given below are taken from the field of analysis of language (Prueitt, 1998, Section 3 of the hypertext):

> *Syntactic mosaic effect* — Occurs when structural parts of a single image or text unit are separated into disjoint parts. Each part is judged not to have a certain piece of information but where the combination of two or more of these units is judged to reveal this information.

> *Semantic mosaic effect* — Occurs when structural parts of a single image or text unit are separated into perhaps overlapping parts. Each part is judged not to imply a certain concept but the combination of two or more of these units is judged to support the inference of this concept.

Both definitions are clearly pointing to a process of emergence (a whole perceived as something different from the simple sum of the parts). The syntactic mosaic effect has more to do with pattern recognition (individuating a similarity within the reservoir of available useful patterns), whereas the semantic mosaic effect has more to do with the establishment of a meaningful contextual relation within the loop *represent–transduce–apply*. In both cases, as done often by famous fiction detectives, we can put together a certain number of clues, none of which can by itself identify the murderer we are looking for (they are not mapping 1:1 to the murderer) with a particular combination that provides enough evidence to clearly identify him or her.

Another important aspect that can be associated with the concept of mosaic effect is that of redundancy in the information space, which can be used to increase its robustness. A good example of the "free ride" that can be obtained by an interlaced or interlocking of different systems of mapping generating internal redundancy (we are using here the expressions suggested by Koestler) is the process of solving crossword puzzles. Due to the given and expected organizational structure of the puzzle, you can guess a lot of missing information about individual words by taking advantage of the internal rules of coherence of the system (by the existing redundancy generated by the organization of the information space in crosswords). Examples of how to apply this principle to integrated analysis of sustainability are discussed in the rest of this chapter.

Before concluding this introductory section, we can briefly recall the discussion (Chapter 2) of the innate redundancy of the information space used when describing dissipative adaptive holarchies. In this case, we are dealing with a Russian dolls' structure (nested hierarchies) of equivalence classes generated by a replicated process of fabrication based on a common set of blueprints (e.g., biological systems made using common information stored in the DNA). This innate redundancy is the reason that we can rely so heavily on type-based descriptions related to expected identities. This means that it is easy to find labels about which the users of a given language can share their organized perceptions of types associated with the expected existence of the relative equivalence class. As discussed in Chapter 2, this mechanism used for organizing human perceptions is very deep. This means that, even when looking for the characterization (representation of a shared perception) of an individual human being, it is necessary to use typologies. For example, consider a famous human — let us say Michael Jordan. We can obtain a lot of free information about him from the knowledge related to equivalence classes to which this individual belongs, even without having a direct experience of interaction with him. For example, since we know that Jordan belongs to the human species, we can guess that he has two arms, two eyes, etc. Actually, we can convey a lot of information about him just by adding after his name the simple information "nothing is missing in the standard package of the higher category — human being — to which this individual belongs."

Within this basic typology of "human being" we can use a more specific subtype characterization linked to his identity as a male of a certain age (a smaller subcategory of human beings). This will provide us with another subset of expected standard characteristics (expected observable qualities) and behaviors (expected patterns) against which it will become easier (and cheaper in terms of information to be gathered and recorded) to track and represent the special characteristics of Mr. Jordan (e.g., he is much taller than the average male of his age; he has excellent physical fitness). It should be noted, however, that every time we get closer and closer to the definition of the special individual Michael Jordan in terms of characteristics of the organized structure generating signals, we remain trapped in the fuzziness of the definition of what should be considered as the relative type, against which to make the identification of the individual. In fact, even when we arrive to the clear characterization of an individual person, we are still dealing with a holon at the moment of representing him. This is due to the unavoidable existence of an infinite regression of potential simplifications linked to the very definition (representation of shared perception) of the same holon Michael Jordan.

The universe of potential meaningful relations between perception and representation can be compressed in different ways to obtain a particular formal representation of Michael. This would remain true, even if we used firsthand experimental information about his anthropometric characteristics and behavioral patterns — e.g., by asking his family or by recording his daily life. Each characterization would still be based on various types related to Michael Jordan determining different sets of expected observable qualities and behaviors. That is, we will still end up using different types, such as sleeping Michael Jordan, full-strength Michael Jordan, angry Michael Jordan, affected-by-a-cold Michael Jordan, etc. Even at this point, we can still split these types into other types, all related to the special subset of qualities and behaviors that the individual Michael Jordan, when in full strength, could take. This splitting can be related to different positions in time during a year (spring vs. winter) or during a day (morning vs. night), or changes referring to a time scale of minutes (surprised vs. pleased), let alone the process of aging.

As noted before, it is impossible to define in absolute terms a formal identity for holons (the right set of qualities and behaviors that can be associated in a substantive way with the given organized structure). Each individual holon will always escape a formal definition due to (1) the fuzzy relation between structure and function, which are depending on each other for their definition within a given identity; (2) the innate process of becoming that is affecting them and (3) the changing interest of the observer. The indeterminacy of such a process translates into an unavoidable openness of the information space required to obtain useful perceptions and representations (holons do operate in complex time). Put another way, holons can only be described (losing part of their integrity or wholeness) in semantic terms using types, after freezing their complex identity using the triadic reading over an infinite cascade of categorizations and in relation to the characteristics of the observer. At this point, a formalization of the semantic description represents an additional simplification, which is unavoidable if one wants to use such an input for communicating and interacting with other observers/agents.

The work of Rosen, Checkland and Allen discussed in Part 1 points to the fact that an observer or a given group of observers can never see the whole picture (the experience about reality is the result of various processes occurring at different scales and levels). At a given point in space and time, observers can see only a few special perspectives and parts of the whole. The metaphor of the group of blind people trying to characterize an elephant by feeling its different parts can be recalled here. Rather than denying this obvious fact, scientists should learn how to better deal with it.

In fact, if it is true that holons are impossible to formalize — a *con* in epistemological terms — it is also true that they are able to establish reliable and useful identities (a valid relation between expected characteristics (types) and experienced characteristics of the members of the relative equivalence class (organized structures sharing the same template)), which is a major *pro* in epistemological terms. This implies that as soon as we are dealing with a known class of holarchic systems (as is always the case when dealing with biological and human systems), we should expect that across levels a few characteristics of the relative types can be predicted. Moreover, the characteristics of nested types are defining each other across levels. This means that, after having selected an opportune set of formal identities for looking at these systems, we can also expect to be able to guesstimate some hierarchical relations between parts and the whole.

6.2 Self-Entailments of Identities across Levels Associated with Holarchic Organization

6.2.1 Looking for Mosaic Effects across Identities of Holarchies

First, we have to look for mechanisms of accounting (assigning a formal identity to the semantic identity of a dissipative system) that will make it possible to establish a link between assessments referring to lower-level components and assessments referring to the whole. The choice of a useful system of accounting is a topic that will be discussed in the next chapter about impredicative loop analysis. The following example has only the goal of illustrating the special characteristics of a nested holarchy. Imagine a holarchic system — e.g., the body of a human being — and imagine that we want to study its metabolism in parallel on two levels: (1) at the level of the whole body and (2) at the level of individual organs belonging to the body. To do that, we have to define a formal identity (a selection of variables) that can be used to characterize the metabolism over these two contiguous levels.

That is, the selected formal identity will be used to characterize two sets of elements defined on different hierarchical levels: (1) the parts of the system (defined at level $n - 1$) and (2) the whole body (defined at level n). This example has as its goal to show that the various identities associated with elements of metabolic systems organized in nested hierarchies entail a constraint of congruence on the relative values taken by intensive and extensive variables across levels.

Let us start with two variables that can be used to describe the sizes of both the whole (level n) and parts (level $n - 1$) in relation to their metabolic activities. The two variables adopted in this example to describe the size of a human body (seen as the black box) in relation to metabolic activity are:

1. Variable 1 — kilograms of human mass (1 kg of body mass is defined at a certain moisture content).
2. Variable 2 — watts of metabolic energy (1 W = 1 J/sec of food metabolized). This assessment refers to energy dissipated for basal metabolism.

These two variables are associated with the size of the dissipative system (whole body) and reflect two nonequivalent mechanisms of mapping. The selection of these two variables reflects the possibility of using two nonequivalent definitions of size. The first definition refers to the perception of the internal structure (body mass), and the second definition refers to the degree of interaction with the environment (flow of food consumed). That is, this second variable refers to the amount of environmental services associated with the definition of size given by variable 1.

The same two variables can be used to characterize the system (human body) perceived and represented over two contiguous hierarchical levels: (1) size of the parts (at the level $n - 1$) and (2) size of the whole (at the level n).

In fact, after having chosen variables 1 and 2 to characterize the size of the metabolism of the human body across levels, we can measure both the size of the whole body (at the level n) and the size of the lower-level organs (at the level $n - 1$) using kilograms of biomass or megajoules of food energy converted into heat. Again, assessment 1 (70 kg of body mass for the whole body) represents a mapping related to the black box in relation to its structural components, whereas assessment 2 (80 W of energy input required over a given time horizon — 1 day — to retain the identity of the whole body) represents a mapping of the dependency of the identity of the system (black box) on benign environmental processes (stability of favorable boundary conditions). The fact that this second assessment is expressed in watts (joules per second) should not mislead the reader. Even if the unit of measurement is a ratio (an amount of energy per unit of time), it should not be considered an intensive variable when dealing with a metabolic system whose identity is associated by default with a flow of energy. In fact, according to the system of accounting adopted here, the size of these systems is associated with an amount of energy, required in a standard period of reference — either a day or a year, depending on the measurement scheme. That is, this is an assessment that is related to a given time window (required to obtain meaningful data) that is big enough to assume such an identity constant in relation to lower-level dynamics. The value is then expressed in joules per second, only because of a mathematical operation applied to the data. The value 80 W (for the whole body) has to be considered an extensive variable, since it maps onto an equivalent amount of environmental services (e.g., a given supply of food, amount of energy carriers and absorption of the relative amount of CO_2 and wastes), which must be associated with the metabolism of the system over a given time horizon.

By combining these two extensive variables (1 and 2), we can obtain an average density of energy dissipation per kilograms of body mass, which is 1.2 W/kg. This should be considered, within this mechanism of accounting, an intensive variable (a variable 3 to be added to the set used to characterize metabolism within a formal identity of it). Variable 3 can be seen as a benchmark value (average value for the black box) that can be associated with the identity of the dissipative system considered as a whole at the level n.

If we look inside the black box at individual components (at the level $n - 1$), we find that the average (watts per kilogram, variable 3) assessed at the level n is the result of an aggregation of a profile of different values of energy dissipation per kilogram of lower-level elements (watts per kilogram, variable 3) assessed at the level $n - 1$. For example, the brain, in spite of being only a small percentage of the body weight (around 2%), is responsible for about 20% of the resting metabolism (Durnin and Passmore, 1967). This means that the density of the metabolic energy flow dissipated in the brain per unit of mass (intensive variable 3) is around 12.0 W/kg. The average metabolic rate of the brain per unit of mass is therefore 10 times higher than the average of the rest of the body. If we write an equation of congruence across these two levels, we can establish a forced relation between the characteristics of the elements (whole and parts) across levels.

Level n (the identity of the black box is known)
Total body mass = 70.0 kg   Endosomatic energy = 80.0 W $EMR_n = 1.2$ W/kg

Level $n - 1$ (the identity of the considered lower-level components is known)
Brain = 1.4 kg   Endosomatic energy = 16.2 W $EMR_{n-1} = 11.6$ W/kg

Level $n - 1$ (after looking for a closure we can define a weak identity for other components)
Rest of the body = 68.6 kg   Endosomatic energy = 63.8 W $EMR_{n-1} = 0.9$ W/kg

When we know the hierarchical structure of parts and the whole (how the whole body mass is distributed over the lower-level parts) and the identities of lower-level parts (the characteristic value of dissipation per unit of mass — intensive variable 3 $(EMRn-1)_i$) — we can even express the characteristics of the whole as a combination of the characteristics of its parts:

$$EMRn = \Sigma x_i \, (EMRn-1)_i = 1.2 \text{ W/kg} = (0.02 \times 11.6)_{brain} + (0.98 \times 0.9)_{rest\ of\ the\ body} \qquad (6.1)$$

That is, the hierarchical structure of the system and the previous knowledge of the expected identity of parts make it possible to obtain missing data when operating an appropriate system of accounting. Put another way, we can guess the EMR of the rest of the body (an element defined at the level $n - 1$) by measuring the characteristics of the whole body (at the level n) and the characteristics of other elements at the level $n - 1$ (brain). Alternatively, we can infer the characteristics of the whole body — at the level n — by our knowledge of the characteristics of the lower-level elements (level $n - 1$), provided that the definition of identities (EMR_i) on the level $n - 1$ guarantees the closure over the total mass. This requires that the mapping of lower-level elements in kilograms has to satisfy the relation:

$$\text{Mass "whole body"} = \text{Mass "brain"} + \text{Mass "rest of the body"} \qquad (6.2)$$

This means that the selected system of accounting of the relevant system quality mass must be clearly defined (e.g., body mass has to be defined at a given content of water or on a dry basis) on both levels to obtain closure. In this example, only two compartments were selected (i = 2), but depending on the availability of additional external sources of information (data or experimental settings available) we could have decided to assign more known identities to characterize what has been labeled here as the "rest of the body." That is, we could have used additional identities for compartments at the level $n - 1$ (e.g., brain, liver, heart, kidneys — see Figure 6.1).

This approach makes it possible to bridge (by establishing congruence constraints) nonequivalent representations of a metabolic system across levels. However, this requires that the formal identities used to characterize lower-level elements have a set of attributes in common with the formal identity used to characterize the whole. That is, it is possible to adopt the same set of variables to characterize a relevant quality (e.g., size) of (1) the black box and (2) its lower-level components. In the example of a multi-scale analysis of the metabolism of the human body — an example is given in Figure 6.1 — the two variables are (1) size in kilograms of mass (extensive variable 1) and (2) size in watts of metabolic energy (extensive variable 2). The combination of these two variables makes it possible to define a benchmark value — the metabolic rate of either the whole or an element expressed in watts per kilograms (intensive variable 3) — that can be used to relate the characteristics of the parts to those of the whole.

Obviously, attributes that are useful to characterize crucial features of the whole body (emergent properties of the whole at level n), such as the ability to remain healthy, cannot be included in the definition of identity applied to individual organs (at level $n - 1$). These characteristics are, in fact, emergent on level n and cannot be detected when using a descriptive domain relative to the parts. This is why variables that are useful for generating a multi-scale mosaic effect are not useful as multi-scale indicators. However, they are very useful to establish a bridge among analyses on different scales providing relevant indicators.

An additional discussion of the possible use of equations of congruence (Equations 6.1 and 6.2) applied to a larger number of lower-level elements (level $n - 1$) is given in the following section (also see Figure 6.1). Obviously, the more we manage to characterize the whole size of the black box (defined at the level n) using information gathered at the lower level (by using data referring to the identity of lower-level elements — parts — at level $n - 1$), the more we will be able to generate a robust description of the system. In fact, in this way we can combine information (data) referring to external referents (measurement schemes measuring the metabolism of organs) operating at level $n - 1$ with nonequivalent information (data) about the black box, which has been generated by a nonequivalent external referent (measurement scheme measuring the metabolism of the person) operating at level n. The parallel use of nonequivalent external referents, in fact, is what makes the information obtained through a cross-scale mosaic effect (avoid the tautology of reciprocal definitions in the egg–chicken process – as discussed in the next chapter) very robust.

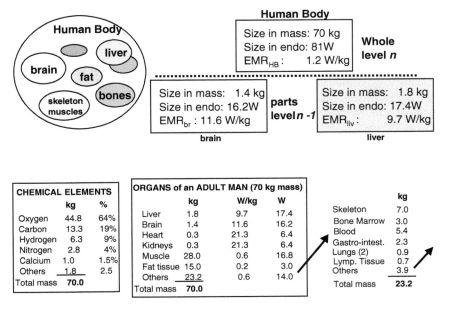

FIGURE 6.1 Constraints on relative values taken by variables within hierarchically organized systems.

6.2.2 Bridging Nonequivalent Representations through Equations of Congruence across Levels

In this section we discuss the mechanism through which it is possible to generate a mosaic effect based on the combined use of intensive and extensive variables describing parts and the whole of a dissipative holarchic system. This operation leads to a process of benchmarking based on the determination of a chain of values for intensive variables 3 across levels. With benchmarking we mean the characterization of the identity of a holon (level n) in relation to the average values referring to the identity of the larger holon representing its context (level $n + 1$) and the lower-level elements that are its components (level $n - 1$).

Let us again use the multi-scale analysis given in Figure 6.1. The two nonequivalent mappings and their ratio (the intensive variable) are defined as follows:

- **Extensive 1** — This is the size of the human body expressed in mass (a mapping linking black-box/lower-level components): 70 kg of body mass.
- **Extensive 2** — This assessment of size measures the degree of dependency of the dissipative system on processes occurring outside the black box, that is, in the context. This can be translated into an number of carriers of endosomatic energy (e.g., kilograms of food) that is required to maintain a given identity (a mapping linking black box/context): 81 W of food energy. This is equivalent to 7MJ/day of food energy to cover resting metabolism.
- **Intensive 3** — The ratio of these two variables is an intensive variable that can be used to characterize the metabolic process associated with the maintenance of the identity of the dissipative system. This ratio can been called the endosomatic metabolic rate of the human body (EMR_{HB}): 1.2 W/kg of food energy/kg of body mass. It is important to note that the values of EMR_i can be directly associated with the identity of the element considered. That is, these are expected values as soon as we know that we are dealing with kilograms of mass of a given element (e.g., brain, liver or heart).

As illustrated in the upper part of Figure 6.1, when considering the human body as the focal level of analysis (level n) — as the black box — we can use this set of three variables (Ext1, Ext. 2 and Int. 3) as a formal identity to characterize its metabolism. The same approach can be used to characterize the identity and metabolism of lower-level elements of the body (level $n - 1$). This implies assuming that it is possible to perceive and represent both the black box and its components as metabolic elements on the same descriptive domain (using a set of reducible variables, even if operating different measurement schemes). This means that the data obtained using nonequivalent measurement schemes can be reduced to each other (e.g., the energy consumed by the brain in form of ATP can be expressed in energy equivalent consumed by the person in the form of kilograms of food). Both assessments refer to resting metabolism. Obviously, this requires having available in parallel two experimental settings — one used to determine the data set for the black box (level n) and one used to determine the same data set but referring to lower-level components (level $n - 1$). The experimental design used to measure the mass and the energy requirement of the whole body, in fact, is different from that adopted to measure the mass and the energy requirement of internal organs.

Let us use this approach to characterize the metabolism of lower-level components of the human body. Obviously, the mass of (extensive variable 1) and the amount of energy dissipated per unit of time (extensive variable 2) in individual components must be smaller than the whole. The same rule does not apply, however, to the value taken by intensive variable 3 describing the level of dissipation per unit of mass. Actually, this is what makes it possible to establish a forced relation between the values taken by the size of the compartments and their levels of dissipation. Looking at Figure 6.1,

Brain: (1) size in mass = 1.4 kg; (2) size in endosomatic energy = 16.2 W; (3) $EMR_{br} = 11.6$ W/kg

Liver: (1) size in mass = 1.8 kg; (2) size in endosomatic energy = 17.4 W; (3) $EMR_{lv} = 9.7$ W/kg

In this example, both elements considered at the level $n - 1$ have an endosomatic metabolic rate much higher than the average found for the human body as a whole. This means that in terms of requirement of input per unit of biomass (e.g., the requirement of input flowing from the environment into the black box), 1 kg of brain is consuming (is equivalent to) almost 10 kg of average human body mass. This ratio reflects the relative value of EMR_i (11.6 W/kg for the brain vs. 1.2 W/kg for the average body mass). This implies the possibility of calculating different levels of embodied ecological activity for ecosystem elements operating at different hierarchical levels (Odum, 1983, 1996). This fact can also be used to calculate biophysical limits of human exploitation of ecological systems (Giampietro and Pimentel, 1991). We can recall here the joke about the national statistics on consumption of chicken per capita (p.c.). If there are parts of the body that consume much more than the average, other parts must consume much less. This implies that the value of the ratio between levels of consumption per unit of mass and the value of the ratio between the sizes of the various parts must be regulated by equations of congruence.

It is important to observe here the crucial role of the peculiar characteristics of nested metabolic elements. They are made up of holons that do have a given identity (they are a realization of a given essence, which implies the association between expected typologies and experienced characteristics in equivalence classes). The brain of a given human being has an expected level of metabolism per kilogram, which we can guesstimate *a priori* from the existing knowledge of the relative type. This level is different from the expected level of metabolism of 1 kg of heart. Both of them, however, can be predicted only to a certain extent. Individuals are just realizations of types (their assessments come with error bars).

The situation would be completely different if we disaggregate the characteristics of the human body — assessed at the level n — by utilizing a selection of mappings based on the adoption of identities referring to much lower levels of organization. For example, imagine using a set of identities referring to the atomic level of organization — as done in the lower part of Figure 6.1, in the white box labeled "chemical elements." In this example, the whole body mass is characterized in terms of a profile of fractions of oxygen, carbon, hydrogen, etc. We can even obtain closure of the size of the whole body (assessed in kilograms) expressed as a combination of lower-level identities (assessed in kilograms). However, with this choice, the distance between the hierarchical levels at which we perceive and represent the characteristics of the metabolism of the human body (at a level we call n) and those at which we can perceive and represent the identity of chemical elements (at a level we call w, with $w \ll n$) is so large that this nonequivalent set of identities — chemical elements — used to describe the components of the human body cannot bring into our descriptive domain (on level n) any free information. In fact, atoms require a descriptive domain for their characterization that is not compatible with the perception and representation of the chemical processes associated with the metabolism of food and the maintenance of the organized structure of the organs through metabolism. That is, with this choice we are not able to reduce the formal definition of these two sets of identities to each other. From our knowledge of the typologies associated with oxygen, carbon and hydrogen (identity of chemical components in relation to lower-level atomic components at the level w and level $w - 1$), we can just estimate the overall mass of the system — nothing about the rate of its metabolism.

On the contrary, when we use a reliable set of identities for organs viewed as metabolic systems — at the level $n - 1$ — for the representation of lower-level elements of the human body, we can use previous knowledge about the given rate of energy dissipation associated with the functions expressed by the relative organized structure. In fact, we expect that both components (organs and the whole) do have a metabolism, so that they can share the same formal identity, even if they require different measurement schemes. This is where we can get some free information from the knowledge of the relative types. To take advantage of this free ride, however, we have to select two sets of identities (e.g., for the whole body and its organs) that can be characterized using the same set of variables. This implies a small distance between hierarchical levels. For example, when using the disaggregating procedure shown in the white box labeled "organs of an adult man" (70 kg of body mass), we can infer, in principle, the dissipation rate of the whole body starting from our knowledge of the dissipation rates of its parts and their relative sizes. Basically, this is the rationale that will be presented later on for generating mosaic effects in the representation of the stability of socioeconomic systems (e.g., multiple-scale integrated analysis of societal metabolism (MSIASM)).

This mechanism, however, requires introducing an additional concept that plays a crucial role in the process: the concept of *closure* over the information space (Dyke, 1988). The bonus obtained when using nonequivalent information derived from nonequivalent observations of a nested hierarchical system in parallel depends in fact on the degree of closure of the relative information space. A mosaic effect is reached when we are able to aggregate the various assessments referring to the parts onto the total size of the whole in a consistent way. In this case, the knowledge of the formal identity of the whole (at the level n) and the knowledge of the formal identities of the various parts (at the level $n - 1$) can be related to each other to gain robustness. This robustness is associated with the congruence of the values taken by the set of different variables used to characterize the two sets of identities across levels (after characterizing the existing relation of parts into the whole, in relation to the selection of formal identity).

To explain this concept, imagine that we want to guesstimate the overall metabolic rate of a human body using only our knowledge of its parts (referring to the level $n - 1$). To do that, we can use the data set included in the white box labeled "organs of an adult man" in Figure 6.1. In this case, we can express the characteristics of the whole body (level n) as a combination of characteristics of seven typologies of lower-level components (level $n - 1$). Of these, six types (liver, brain, heart, kidneys, muscle, fat tissue) have a clear and known (expected) identity. The seventh compartment, which is required for obtaining the closure, is not clearly defined in terms of an established correspondence between an internal mapping (e.g., kilograms of mass) and an external mapping (e.g., energy required for its metabolism), associated with a previous knowledge of this type. Actually, we are all familiar with the label given to this last compartment, which is often found at the bottom of this type of list — "others." Obviously, this solution implies that the identity of the compartment labeled as "others" is not associated with any previous knowledge of an established type at the level $n - 1$. Therefore, the resulting numerical assessment is not obtained by a direct measurement (performed at the level $n - 1$), an external referent, of a sample of members of an equivalence class. Put another way, "others" is not a known type with a given and reliable identity. Rather, the characteristics of this virtual compartment are inferred by considering the difference between (1) the information gathered about the characteristics of the whole human body gathered at level n and (2) the information gathered about the selected set of six identities of lower-level elements, perceived and measured at the level $n - 1$. The characterization of this seventh virtual lower-level element — the identity of "others" (about which we cannot provide any expected value *a priori*) — depends on (1) the values taken by the variables referring to the characteristics of the whole, (2) the selection of identities used to define the various compartments of the whole — the set of lower-level elements used in the disaggregated representation of the whole; and (3) the relative values of the variables describing the selected set of identities of lower-level elements. Getting back to the example of the seven compartments in the white box, we could have used a different selection of six types (e.g., by replacing the 1.8 kg of liver with 7.0 kg of skeleton), and this would have provided a different definition for the virtual identity of the seventh compartment "others." In this case, "others" would have had a mass of 17.8 kg (rather than the 23.3 kg reported in the table) and a different EMR.

This is an important aspect that can be associated with the next concept to be introduced in Chapter 7 — impredicative loop analysis. One of the standard goals of the triangulation of information when dealing with the reciprocal definition of identities across levels in a metabolic holarchy is that of reducing the noise associated with unavoidable presence of informational leftovers as much as possible. The amount of information missed when adopting the category "others" as if it were a real typology can be important. Therefore, the analyst has to choose the most useful way to represent the system (disaggregate it into lower-level elements), trying to reduce as much as possible such a problem. For example, getting back to the example of the disaggregating choices made in the white box labeled "organs of an adult man" (six types plus the seventh compartment labeled "others"), we have a relatively large size of the unaccounted part of the whole in terms of mass (more than 33% of the total mass is included in "others" — that is, 23.2 kg out of 70 kg of the whole body). On the other hand, this mass might not be particularly relevant in terms of metabolic activity, since the resulting level of EMR is quite low (0.6 W/kg). Therefore, by making such a choice, the analyst is ignoring the characteristics of the identity of a big part of the whole in terms of mass. However, if the analyst is concerned only with identifying those organs that are keeping the metabolic rate high, this body part might not be particularly relevant in terms of energy dissipation per unit of mass (e.g., in terms of the qualities associated with extensive variable 2, that is, requirement of services — e.g., sustainable food — from the context). Obviously, any decision about what to include and what to leave out in the virtual category assigned to the "others" compartment will depend on the type of problems faced and the type of questions we want to answer with the study.

When facing a level of closure that is not satisficing for the goal of the analysis, the analyst can decide to get into the remaining parts of the whole labeled as "others" and look for additional typologies (additional valid and useful natural identities). In this way, it becomes possible to reduce the amount of total mass of the whole, which remains unaccounted for in terms of a definition of identities at the lower level. An example of this additional investigation (which implies gathering more information — using additional external referents — at the level $n - 1$) is given in the gray box in the lower part of Figure 6.1. The compartment originally labeled as "others" in the white box (which is covering 23.2 kg of mass of the whole) has been split into seven additional compartments, characterized using an additional six known typologies or identities of lower-level components (skeleton, bone marrow, blood, gastrointestinal tract, lungs, lymph tissue). Also in this new characterization of the black box in terms of an expanded set of lower-level compartments, we still face the presence of a residual compartment labeled "others." However, after this additional injection of information about the identities of elements involved in the metabolism of the human body of an adult man — at the level $n - 1$ — we are able to characterize the metabolism of the whole using previous knowledge related to the characteristics of 12 known typologies or identities of lower-level elements. This reduces the amount of residual unknown body mass not accounted for in terms of expected characteristics of lower-level typologies to only 3.9 kg (over 70 kg). Depending on the questions addressed by the study, the analyst can decide at this point whether this reduction is enough.

Obviously, we cannot expect that it is always possible to keep splitting the residual required information labeled "others" into characteristics associated with known typologies (exploiting in this way preexisting knowledge of additional lower-level identities). The possibility of using this trick has limits.

We can now leave the metaphor of the multi-scale analysis of the metabolism of the human body to get into a more general question. What can be achieved by adopting this approach when studying complex adaptive holarchic systems? What are the advantages of obtaining an adequate closure of the information space, based on a parallel characterization of the identities of metabolic systems organized in holarchies on two contiguous levels (e.g., level n and level $n - 1$)?

We believe that this approach can be used to achieve two important objectives:

1. It provides a general mechanism that can be used for benchmarking (contextualization of an element in relation to the whole to which it belongs). Obviously, any benchmarking will always reflect the previous selection of the formal identity (the set of significant variables) used to check the congruence among flows. For example, a question like, "How well is a farmer doing who makes $1000 per year?" can be answered only after comparing this value (an intensive variable of added value per unit of human time, which can be associated with the identity of a household holon) with the average household income of a given year within a given society in which such a farmer is operating (the identity of the larger holon within which the household holon is operating). In the same way, a yield of 1000 kg of corn/ha can be a remarkable achievement for a farmer operating in a desert area with poor soil when not using any fertilizers, whereas it would be considered a totally unacceptable output if obtained in Iowa in the year 2000. By adopting a multi-scale integrated analysis of agroecosystems, it can be possible to build an integrated mechanism of mappings of flows that can be easily defined and tracked (e.g., flows of food energy, exosomatic energy, added value, water, nitrogen) in relation to (1) the characteristics of the system generating and consuming these flows and (2) the characteristics of the context within which these flows are exchanged. Since the very exchange of these flows is related to the definition and maintenance of an identity for the metabolic elements investigated (across various levels), such an analysis can carry useful free information when addressing the hierarchical structure of the system and when linking identities and indicators referring to wholes and parts.

     As noted earlier, however, to do that we have to be able to express these flows against a matrix (e.g., against human activity or areas) in a way that makes it possible to obtain a closure of the nonequivalent representations of the various identities of compartments across levels. After having done this, we can define whether the values taken by a set of variables used to characterize the performance of a farmer (e.g., level of income, leisure time, life span) or the performance of a particular farming activity (e.g., economic labor productivity, return on the investment, demand of land, associated level of pollution per unit of land) are above or below averages characterizing the equivalence class to which the farmer belongs, and how these values refer to the expected values associated with larger-level holons determining the stability of the context.

2. It provides a general mechanism that can be used for establishing a bridge among nonequivalent descriptive domains, and therefore to boost the coherence and reliability of an integrated package of indicators. The forced relation between parts and the whole (e.g., using the relation between total EMR of the whole body and the various EMR_i of its parts) can be applied to different typologies of flows in parallel (e.g., food produced and required per unit of land, exosomatic energy produced and consumed per unit of land, added value generated and consumed per unit of land) and against different matrices (e.g., human activity and area). This makes it possible to also establish a mosaic effect among nonequivalent readings (definition of different formal identities for the dissipative systems) in relation to the feasibility of the various holons making up the investigated metabolic system. Households, counties, states, macroeconomic reasons, in fact, all do produce and consume (and must produce and consume) flows of money, food, energy. Whenever we map these flows across levels against the same matrix (the same hierarchical frame of unit of lands, or the same profile of allocation of human time), we can establish links among analyses related to different disciplinary fields (e.g., producing the same flow of $10,000/year/ha either by agriculture or by agro-tourism implies different requirements of labor, capital, water, and different environmental impacts). The mosaic effect can also be used to fill knowledge gaps referring to inaccessible information of residuals. Put another way, important facts ignored or heavily underestimated by an economic accounting of farming (e.g., ecological services lost with soil erosion) can be extremely clear when performing a parallel analysis based on a biophysical accounting (e.g., the huge material flow associated with soil erosion). The soil loss, negligible in an economic accounting of profit and revenues per year at the farm level, can become an important factor when adopting a biophysical accounting of matter flows associated with crop production at the watershed level and over a time horizon of 50 years.

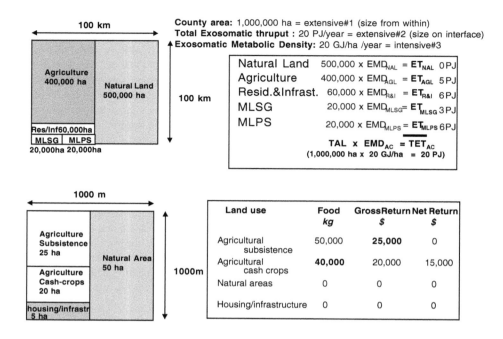

FIGURE 6.2 Constraints on relative values taken by variables within hierarchically organized systems.

6.2.3   Extending the Multi-Scale Integrated Analysis to Land Use Patterns

Linkages among characteristics of typologies belonging to different but contiguous hierarchical levels can also be established by using a spatial matrix, which can provide closure across levels. To explore this option, let us adopt the same approach used in Figure 6.1, but this time mapping densities of flows associated with typologies of land use. An example of this analysis is given in the upper part of Figure 6.2. Imagine a county of a developed country inhabited by 100,000 people and total available land (TAL) of 1 million ha. Assuming a consumption of exosomatic energy per capita (consumption of commercial energy) of 200 GJ/year/person (EMR_{AS} = 2.3 MJ/h), we can calculate a total exosomatic throughput (TET) for that county of 20 PJ (petajoule) of exosomatic energy per year (PJ = 10^{15} J). The possibility of establishing a relation between the values taken by these variables (the characterization of a social system viewed at the focal level *n*) and the values of variables associated with the characteristics of lower-level elements defined at the level *n* – 1 (societal compartments) is discussed in detail in Section 6.3.

In the example of Figure 6.2, the same rationale adopted in Figure 6.1 is applied. The difference in this case is that the common matrix across levels (extensive variable 1) consists of assessments of land area. In practical terms, we have to divide a given amount of TAL (the size of the whole system mapped in terms of units of area), which is the equivalent of the total body mass indicated in Figure 6.1, into a set of typologies of land use (the lower-level compartments to which we assign the characteristics of a typology — the equivalent of organs). In the example of Figure 6.2, the selected set of five typologies is (1) natural land not managed by humans (NAL), (2) residential and infrastructures (R&I), (3) agricultural land (AGL), (4) land used for economic activities belonging to the sector of manufacturing, energy and mining (MLPS) and (5) land used for economic activities belonging to the sector of services and government (MLSG). As noted earlier, we have to obtain closure with this division. That is, TAL (1 million ha) has to be divided according to a given profile of investment of TAL over the five typologies of land use, which provides an arrow of percentages that totals 100. In the example of Figure 6.2, such a profile is (1) NAL, 50%; (2) AGL, 40%; (3) R&I, 6%; (4) MLPS, 2%; and (5) MLSG, 2%.

The breaking down of the whole (TAL) into components, which is done in hectares (or other units of area), provides an internal mapping of the size of the system (TAL defined at the level *n*), which is also used for assessing the size of lower-level components (NAL + AGL + R&I + MLPS + MLSG). That is, the size of component is expressed as a part of the whole. This would be the equivalent of the extensive variable 1 discussed before for the metabolism of the human body.

We now need a nonequivalent assessment of the size of the whole system (the county) in terms of the degree of interaction with the context. We have to define a second extensive variable (variable 2), which makes it possible to adopt the same approach discussed above. The choice adopted in the example provided in the upper part of Figure 6.2 is to use an assessment of exosomatic energy consumption, which is required to guarantee the typical level of metabolism of the county. This is the amount of fossil energy that the county is getting from society in relation to its socioeconomic interaction. As noted earlier, such a value is 20 PJ/year for the whole county. This choice reflects an attempt to keep the analogy with the example provided in Figure 6.1 — this is the equivalent of the extensive variable 2, and with this choice, we also manage to respect the same selection of unit (energy over time). However, as will be discussed later, this approach also works when selecting an economic variable — e.g., assessment of a flow of added value or another biophysical variable, such as water — as extensive variable 2.

Starting from the two values of extensive variables 1 and 2 used to characterize the metabolism of the whole county, we can calculate the intensive variable 3 — the exosomatic metabolic density (average for the county), which is the amount of exosomatic energy consumed per unit of area, referring to the total area occupied by the county. In this example, EMD_{AC} is obtained by dividing TET (20 PJ) by TAL (1 million ha). The result is EMD_{AC} 20 GJ/ha/year.

At this point, we can apply the approach previously illustrated in the multi-scale analysis of the metabolism of the human body. The table in the upper part of Figure 6.2 can be used to get some free ride out of the redundancy existing within this organized information space. Again, this redundancy is generated by our previous knowledge of identities of lower-level typologies, which can be found in our descriptive domain across hierarchical levels. For example, after having structured the information space in this way, we can try to fill the values of the column of EMD_i by using the column of assessments of the amounts of energy consumed by the various sectors used to represent the economic structure of the county. This economic sector would be the equivalent of organs (elements defined at the level $n - 1$). The values taken by extensive variable 2, referring to the identities of the lower elements (reflecting the characteristics of the economic sector of the county), are given in the ET_i column. The values found in the EMD_i column are referring to the typologies of lower-level sectors. That is, the EMD of residential and infrastructure can be calculated by dividing the value of the relative $ET_{R\&I}$ (6 PJ of exosomatic energy per year, which is spent in the residential sector) by the area of 60,000 hectares, which is used by this compartment. In this way, we obtain a value of $EMD_{R\&I}$, which is equal to 100 GJ/ha/year.

On the other hand, we could have found the same value by using a different source of information — a nonequivalent external referent for this assessment, which is related to a nonequivalent perception and representation of events referring to the level $n - 2$. In fact, we can use additional nonequivalent information to define the characteristics of household types at the level $n - 2$. These characteristics will determine the characteristics of the household sector (HH) (at the level $n - 1$). For example, we can start with the average value of consumption per household in the county, related to a given typology of housing (e.g., by looking at the literature, we can find a value of 180 GJ/year/household for the typical houses found in that county). Knowing the average size of the households of that county — e.g., three people — we can estimate an average consumption of 60 GJ/year/person associated with direct energy consumption in the household sector. After using information on the housing typology (e.g., 300 m^2 of house per person and a ratio of 9/1 between the built area of houses and the additional land included in the residential compound), we can assume — for the specific county characterized by such a residential typology — an amount of 3000 m^2 of residential area per person. To this area we have to add an additional area (e.g., 3000 m^2 per person) for infrastructure (roads, parking lots, recreational areas, etc.). Put another way, in this manner we can assess the total request of land per person for the residential or household sector of that particular county.

Using this information in our example of a hypothetical county of the U.S. for which lower-level household typologies are known, we can characterize such a system as composed of 33,000 households (100,000 people divided by 3), generating an aggregate consumption in the residential sector of 6 PJ (180 GJ of exosomatic energy per household spent in the residential compartment), in relation to a requirement of land of 18,000 m² per household (that has to be included in the residential category). Using this set of data (different from what was used before), we can calculate a value of $EMD_{R\&I} = 100$ GJ/ha/year in a nonequivalent way.

In this example we have two nonequivalent ways for calculating $EMD_{R\&I}$:

1. Using information referring to the level $n/(n-1)$. $EMD_{R\&I}$ is the ratio between the total size of the residential sector in terms of exosomatic energy consumption (6 PJ, e.g., as resulting from aggregate record of consumption of that sector in the county) and in terms of land (60,000 ha, e.g., as resulting from remote sensing analysis of land use in the county).

2. Using information referring to the level $(n-1)/(n-2)$. $EMD_{R\&I}$ is calculated from our previous knowledge of the consumption level for a given typology of household, house space requirement per person, house size, ratio between the area of the house and the open space included in the housing compound, ratio between the area occupied by private housing and the area required by common infrastructures.

In a holarchic system made up of nested types, the characteristics of the types making up holons across levels must be compatible with each other. This redundancy is at the root of the existence of free information when dealing with representation across levels of holarchic systems.

This hierarchical structure is very robust; in fact, if more typologies of housing were known in that area — at the level $n-2$ — that is, for example, (1) individual family houses (180 GJ/year/household and 3000 m²/person of area of the residential compound) and (2) condominium apartments (100 GJ/year/household and 500 m²/person of area of the residential compound), the average value (variable 3) for $EMD_{R\&I}$ at the level $n-1$ would have been different. Still, such a characteristic value for the residential sector (at the level $n-1$) can still be expressed in relation to the characteristics of the lower-level typologies (at the level $n-2$). This can be done by considering the profile of distribution of investments of space and energy (using variable 1) within the household sector over the set of possible household types characterized at the level $n-2$ (using information gathered at the level $n-2$) in terms of the intensive variable 3. This mechanism is at the basis of impredicative loop analysis.

Obviously, the example of a redundant definition of a given value is also valid for other typologies of land use. In the same way, starting from characteristics of typologies belonging to the level $n-2$ (e.g., typologies of industrial plants), it is possible to guesstimate the value of EMD_{MLSG} and EMD_{MLPS}, characteristics referring to the level $n-1$. To calculate such a value, it is necessary to study the consumption of different typologies of industrial building and other categories of land use associated with these typologies. The important aspect of this analysis is related to the ability to disaggregate the total area under the various land use categories, in a way that makes it possible to use equations of congruence later on.

This is crucial for another reason. Known typologies (a given typology of housing or a given typology of power plants) make it possible not only to associate a defined density of flows (e.g., the amount of added value per hectare, the amount of food produced per hectare or the amount of exosomatic energy consumed per hectare) per unit of land use in that category, but also to set up a package of different indicators of performance. That is, we can add to the possibility of performing a multi-scale analysis (by simultaneously considering information gathered at different hierarchical levels) the possibility of performing multidimensional analysis (by simultaneously considering the constraints affecting the flows of variables — food, exosomatic energy, added value — referring to different dimensions of sustainability). For example, in the lower part of Figure 6.2 we have an example of a possible characterization of a farm in relation to a given profile of four different typologies of land use: (1) natural area, (2) agriculture for subsistence, (3) agriculture for cash crops and (4) housing and infrastructures. This would be the characterization of the whole and the parts in relation to extensive variable 1.

Imagine now that we are associating the relative mapping of relevant flows with each one of these four typologies defined, using the extensive variable land: (1) a flow of endosomatic energy — food produced by subsistence agriculture and (2) a flow of added value — associated with the production of cash crops. That is, we are using in parallel against the same definition of size (extensive variable 1) two versions of extensive variable 2: an extensive 2 biophysical, which is referring to a biophysical mapping of the interaction with the context in terms of exchange of flows (the flow of food produced, consumed and sold by the farm), and an extensive 2 economic, which is referring to an economic mapping of the interaction with the context in terms of exchange of flows (the flow of added value produced, consumed and spent by the farm).

In this example, we already can extract a set of nonequivalent indicators of performance from this very simple database: (1) the amount of food available for self-consumption (a relevant indicator in those areas in which the market is not reliable), (2) the amount of food supplied by the farm to the rest of society (relevant for determining self-sufficency at the national level) and (3) the amount of added value available to the farmers as net disposable cash (a relevant indicator to determine the potential level of interaction of the household with the rest of society in terms of economic transactions).

Because of the particular structure of this information space, we can establish a link between potential changes in the value taken by these indicators. That is, relative constraints can be studied by using biophysical, agronomic, ecological and socioeconomic variables and models. This example is important to show how the same data set can be used to provide different results according to the adoption of different disciplinary perceptions and representations of changes.

For example, talking of the amount of food available for self-consumption, we have a total of 90,000 kg of grain produced on this farm (50,000 kg of grain for subsistence and 40,000 kg of grain for cash); only 50,000 kg should be counted as an internal supply of food (as a relevant flow for food security for the farmers). The reverse is true when we want to assess the amount of food supply that this farm is providing for the socioeconomic system to which it belongs. That is, the context of this farm is receiving only 40,000 kg of grain of the investment of 100 ha of TAL. Even more complicated is the accounting of economic variables. If we want to assess the income of the farm, we should add the value of the self-consumed grain (the $25,000 indicated in bold) to the value of the net return of cash crop ($15,000). On the other hand, if we want to assess the level of net disposable cash, we have to ignore the $25,000 related to the value of the self-consumed crops. And there still has to be an analysis aimed at assessing the effect of the characteristics of this farming system on the gross national product (GNP) of the country.

The typology natural area (area not managed by humans) is completely irrelevant when dealing with short-term perceptions and representations of economic performance and food security. This typology of area is perceived as not producing anything useful (money or food). This is probably an explanation for the fast disappearance of this typology of land use on this planet. However, as soon as we introduce a new set of relevant criteria and the consequent set of relevant indicators of performance (preservation of biodiversity, support for natural biogeochemical cycles, preservation of soil, quality of the water, etc.), it becomes immediately clear that those typologies of land use that are crucial for determining a high economic performance are at the same time the very same categories that can be associated with the worse performance in ecological terms (see Chapter 10). It is exactly the ability to handle the heterogeneity of information related to different scales and nonreducible criteria of performance that makes the approach of multi-scale integrated analysis of agroecosystems interesting.

Even in this very simple example we can appreciate that a multi-scale integrated analysis is able to handle the information related to indicators that are in a way independent from each other, since they are calculated using disciplinary representations of the reality, which are nonequivalent (e.g., the study of the stability of the loop of food energy spent to generate the labor required for subsistence is independent from the analysis of the economic loop associated with cost and return related to the cultivation of cash crops). However, within this integrated system of accounting across levels, these two representations are indirectly connected. In fact, autocatalytic loops of endosomatic energy (investment of labor to feed the workers), added value (investment of money to pay back the investments), and exosomatic energy (investment of fossil energy to generate the useful energy required for the making of exosomatic devices) all compete for the same budget of limiting resources: human activity and total available land. It is this parallel competition that determines a set of mutual constraint that each one of these autocatalytic loops implies on the others. As noted earlier, the nature of this reciprocal constraint can be explored by considering lower-level characteristics (e.g., technical coefficients) and higher-level characteristics (e.g., economic, social and ecological boundary conditions).

6.3   Using Mosaic Effects in the Integrated Analysis of Socioeconomic Processes

6.3.1   Introduction: The Integrated Analysis of Socioeconomic Processes

In economic terms we can describe the socioeconomic process as one in which humans alter the environment in which they live with their activity (through labor, capital and technology) to increase the efficacy of the process of production and consumption of goods and services. In other words, they attempt to stabilize and improve the structures and functions of their society according to a set of internally generated values and goals (what they perceive and how they represent improvements in the existing situation). In biophysical terms, the process of self-organization of human society can be seen as the ability to stabilize a network of matter and energy flows (defined over a given space–time domain) representing what is produced and what is consumed in the economic process.

To be sustainable, such a process has to be (1) compatible with the aspirations of the humans belonging to the society, (2) compatible with the stability of both natural and human-managed ecosystems, (3) compatible with the stability of social and political institutions and processes, (4) technically feasible and (5) economically viable. The order of these five points does not reflect priorities or relative importance, since each one of these conditions is crucial.

This is to say that when we perform a biophysical analysis of human societies (e.g., using variables such as kilograms of iron or joules of fossil energy), we can see only certain qualities of human societies (e.g., we cannot get any indication about the economic value of commodities), and therefore we can check only a few of the five conditions listed above. The same predicament applies to economic, engineering and political analyses. To be able to see and describe a certain set of system qualities considered relevant in certain disciplines, analysts will have to use a finite set of encoding variables and descriptive domains (they have to assign to the system a formal identity that is useful for applying the relative disciplinary knowledge). However, this choice can imply losing track of other system qualities considered relevant by other disciplines. That is, not everything that is economically viable is, as a consequence, also ecologically compatible. Not every solution that optimizes efficiency is, as consequence, also advisable for keeping social stress low or improving adaptability, and so on.

The integrated use of nonequivalent medical analyses to deal with human health (Figure 6.3) can be used as a good metaphor on how to use scientific analyses in an integrated way when dealing with sustainability. For the same concept, Neurath (1973) proposed the expression "orchestration of sciences." Getting back to the example of Figure 6.3, which is limited only to the challenge of generating a meaningful representation of shared perception at a given point in space and time, if you want to see broken bones you have to use x-rays, but you cannot see in this way soft tissue (for that you need an ultrasound scan). In the same way, if you want to know whether a woman is in the first weeks of pregnancy, you can use a chemical test based on her blood or urine. Endoscopy can be the easiest way in which to look at a local situation, whereas nuclear magnetic resonance (NMR) can also deal with the big picture. In these examples, x-rays, ultrasound scans, NMR, chemical tests and endoscopy are nonequivalent tools of investigation. No matter how powerful or useful any one of them is, when dealing with the behavior of complex systems (i.e., health of humans), we cannot expect that one tool (based on the adoption of a formal identity in the representation of the investigated system) can do all the relevant monitoring. To be able to characterize several relevant nonequivalent aspects of patient behavior, and also for economic criteria (to avoid shooting flies with machine guns), it is wise, depending on the circumstances, to develop and use several nonequivalent analytical tools in different combinations. Sustainability analyses seem to be a classic case in which it is wise to be willing to work with an integrated set of tools. This is the only way to expand the ability of scientists as much as possible to cover the relevant perceptions about sustainability that should be considered, without putting all their eggs in the same basket.

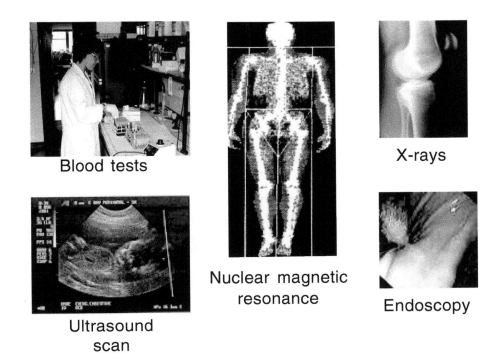

FIGURE 6.3 Nonequivalent complementing views used in medicine. (Photos courtesy of E.B.T. Azzini.)

6.3.2 Redundancy to Bridge Nonequivalent Descriptive Domains: Multi-Scale Integrated Analysis

Redundancy in scientific analysis is often seen as a villain. The axiom used to justify the holy war of science against redundancy is the famous Occam's razor principle: One should not increase, beyond what is necessary, the number of entities required to explain anything. The principle is also called the principle of parsimony. Such a principle requires that scientific analyses follow the goal of obtaining a maximum in the compression of the information space used in their models. That is, sound science must use as few variables and equations as possible. It is worthwhile to observe here that one of the measures of complexity for mathematical objects (computational complexity) is related exactly to the impossibility of compressing the demand of information for their representation (Chaitin, 1987). That is, if you are dealing with a complex object, you cannot expect to compress much in the step of representation (amplify your predictive power) just by developing more sophisticated inferential systems (more complicated models). Without getting into a sophisticated discussion related to this topic, we want to again use a metaphor, that of geographic maps, to question the idea that redundancy should be eliminated as much as possible in scientific analyses.

Using a metaphor based on geographic maps is appropriate since, after all, numerical values taken by variables in an integrated analysis are generated by the application of a selected modeling relation to the representation of a natural system (Rosen, 1985). These assessments are nothing but mappings of selected qualities of the investigated natural system into a given mechanism of representation, which reflects the characteristics of the selected model.

The examples given in Figures 6.4 and 6.5 are related to the discussion in Chapter 3 on nonequivalent descriptive domains. The four different views provided in Figure 6.4 (Catalonia within Europe, a specific county in Catalonia, an area of a national park in the county, and roads within the natural park) reflect the existence of different hierarchical levels at which a geographic mapping can be provided. When the differences in scale are too large, it is almost impossible to relate the nonequivalent information presented

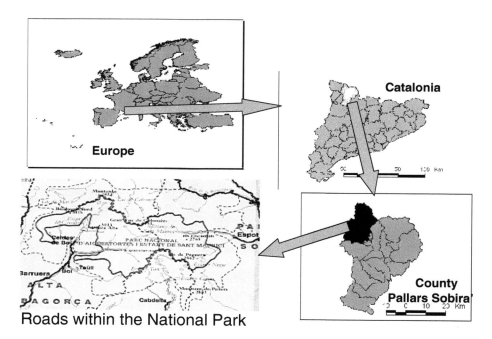

Roads within the National Park

FIGURE 6.4 Nonequivalent descriptive domains due to difference in scale of the map. (Giampietro M. and Mayumi K. (2000a), Multiple-scale integrated assessment of societal metabolism: Introducing the approach – *Popul. Environ.* 22(2): 109–153.)

in distinct descriptive domains (e.g., in Figure 6.4 the upper and lower maps on the left). To link nonequivalent views across different scales, you need a certain level of redundancy among the maps. That is, to be able to appreciate the existing relation between two nonequivalent representations, you must be able to recognize the element (pattern) described within one of the specific descriptive domains in the next one. For example, Catalonia within Europe — in the first map — becomes the whole object within which Pallars Sobira' County is located in the next map. This happens, in Figure 6.4, in all the couplets of maps linked by an arrow.

At this point, if we are able to establish a continuum between the various links across the nonequivalent maps, then it is possible for us to structure the information provided by the set of maps (referring to different hierarchical levels). Distant maps are nonreducible to each other (upper and lower maps on the left); contiguous maps can be bridged. The bridging of nonequivalent information across different maps can be easy, depending on the degree of overlapping of the information contained in them. The higher the level of overlapping, the lower the compression, but the easier it becomes to establish a relation between the information contained in the two maps. On the other hand, very little degree of overlapping (e.g., the two maps on the higher level) implies a more difficult bridging of the meaning conveyed by the maps. By establishing a continuous chain of bridges of meaning across maps, we can relate the information about the layout of the natural park (which is required by someone wanting to drive there) to the information about where such a park is located. Depending on the characteristics of possible users, we have to provide such information in relation to different definitions of such a context. We can say that the park is at the same time in Europe, in Spain, in Catalonia and in a given corner of Pallars Sobira' County.

It should be noted that, in this case, using some redundancy in this integrated system of representations (the partial overlapping of the information in contiguous maps) is the only way to handle such a task. A huge map that would keep the same level of accuracy adopted for the representation of the area within the natural park, applied to the description of all of Europe, cannot be made or operated for theoretical and practical issues (without a hierarchical structuring of the information space, it would not be possible to handle the required amount of bits of information). A map as large as Europe in a scale of 1:1 would simply result in excessive demand of computational capability in both the step of making the represen-

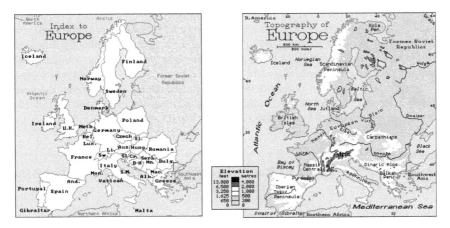

FIGURE 6.5 Same hierarchical level but different categories used in the map (Europe) to characterize system identity. (Giampietro M. and Mayumi K. (2000a), Multiple-scale integrated assessment of societal metabolism: Introducing the approach – *Popul. Environ.* 22(2): 109–153.)

tative tool (encoding the system's qualities) and using it (decoding). Moreover, nobody would find it useful since we already have the original.

In Figure 6.5 we deal with an example of the second source of nonequivalent assessment discussed in Chapter 3 (Figure 3.1): two logically independent systems of mapping of the same geographical entity — Europe. To recall the example given in Figure 3.1, we are in the case of the same system (the head of a given woman) represented by using two different mechanisms of encoding (visible light for the face and x-rays for the skull). In Figure 6.5, a map provides two nonequivalent formal identities: (1) a political mapping, dealing with borders and names of countries and (2) a physical mapping, locating and describing physical elements such as rivers and mountains. In this case, different selections of relevant attributes are used to represent the formal identity of the same system in the map. The two maps are based on two nonequivalent formal identities assigned to the same natural system (Europe) in relation to two possible meaningful relations between shared perception and representation of such a system. Whenever we are in the presence of a bifurcation that generates two useful nonequivalent formal identities, we can no longer compress. It becomes necessary to use and handle these nonequivalent descriptions in parallel.

In this regard, we can recall the main conclusions about the four assessments presented in Figure 3.1, which can be applied to the message given in Figures 6.4 and 6.5. The four examples of assessment given in Figure 3.1 cover the possibility of:

1. Life cycle assessment bridging the assessment consumption per capita at the household level (116 kg/year) and at the food system level (1015 kg/year). This implies the need to use several maps based on the same system of encoding (same set of attributes), but referring to different scales (as in Figure 6.4).

2. The need to use different types of maps — based on different methods of encoding — economic variables (1330 kg/year) and biophysical variables (1015 kg/year) on the same scale (as in Figure 6.5).

But in this case, the two nonequivalent descriptions (two maps based on a different selection of variables) must refer to the same hierarchical element. Otherwise, they could not be used in integrated analysis. Getting back to the various maps shown in Figure 6.4, this implies that we can imagine two versions of the map of Europe (political and physical), as well as two versions of the maps of Spain (political and physical) and Catalonia (political and physical). Put another way, representations that are logically independent in their selection of encoding variables (e.g., political and physical maps — a bifurcation in formal identities) have to be packaged in couplets referring to the same basic definition in terms of space–time domain. Also, in this case, this implies keeping a certain level of redundancy in the representation (e.g., in Figure 6.5 the geographic border of Europe is the same in the two maps).

This observation can appear absolutely trivial when dealing with the example of political and physical representations of geographic entities, as in Figure 6.5. However, when dealing with the parallel reading of socioeconomic systems in biophysical and economic terms, it is very common to face different definitions of border for the same system defined at the same level. For example, the border used to assess the GNP of a country is different from the border used to assess its demand of ecological services — the difference being generated by the effects of import and export.

6.4 Applying the Metaphor of Redundant Maps to the Integrated Assessment of Human Systems

The following two sections present an example of the application of this rationale to a multi-scale integrated analysis of societal metabolism. A detailed presentation of the methodological approach and the database used for generating the material presented here are available in two special issues of *Population and Environment* dedicated to multi-scale integrated assessment of social metabolism (Vol. 22 of the year 2000 and Vol. 22 of the year 2001).

6.4.1 Multi-Scale Analysis of Societal Metabolism: Same Variable (Megajoules), Different Levels

In this section we describe how it is possible to apply the same rationale used for geographic maps in Figure 6.4 to establish a bridge among different numerical values taken by the same variable, when used to represent the multiple identities of a given system resulting from its perception on different hierarchical levels. In this example, we use a system of encoding of the characteristics of the system based on the variable megajoules of energy (this is an example taken from energy analysis).

In the following example we describe a given system (Spain in 1995) in terms of energy flows. To accomplish that, we will build a system of accounting able to establish congruence among different mechanisms of mapping (different perceptions and representations of energy flows) referring to different levels. This requires the use of nonequivalent external referents. Recall again the example of nonequivalent assessments of kilograms of cereal discussed in Figure 3.1.

To represent something in quantitative terms, we must provide root definitions (the identities of the elements that are modeled expressed in terms of encoding variables). More about this step can be found in Chapter 7 (impredicative loop analysis) and in Chapter 9 (applications to agricultural systems). For the moment, it is enough to say that, in our representation, we include a set of energy flows associated with the various activities required to produce and consume goods and services within a socioeconomic system. To define a clear identity for these energy flows, we map them against a reference frame provided by the profile of allocation of total human activity over the set of activities performed within the society (for a more detailed explanation, see the two special issues mentioned above).

A conceptual distinction between exosomatic metabolism (matter and energy flows metabolized by a society outside human body) and endosomatic metabolism (food used to support human physiological processes) was introduced by Lotka (1956), and later on proposed as a working concept for the energetic analyses of bioeconomics and sustainability by Georgescu-Roegen (1975). Such a distinction obviously is based on a previous definition of a given identity for the lower-level converters that are transforming inputs into outputs — i.e., humans, which are considered by this distinction in parallel on two hierarchical levels: (1) endosomatic energy refers to a perception of human metabolism at the level of individuals (physiological conversions) and (2) exosomatic energy refers to a perception of human metabolism at the level of the whole society (technical conversions). The concept of societal metabolism directly addresses the hierarchical structure linking the converters and the whole. In fact, using Lotka's original vision of exosomatic energy, "It has, in a most real way, bound men together into one body: so very real and material is the bond that society might aptly be described as one huge multiple Siamese twin"

(Lotka, 1956, p. 369). The vivid image he proposed explicitly suggests that a hierarchical level of organization higher than the individual converter should be considered when describing the flow of exosomatic energy in modern societies.

- **Endosomatic energy** — Endosomatic means inside the human body. It indicates energy conversions linked to human physiological processes fueled by food energy. Therefore, endosomatic energy implies a clear identity for energy carriers, technical coefficients (power levels are all clustered around the value of 0.1 hp), rates of throughput (energy consumption per capita per day are well known) and output/input ratios.

- **Exosomatic energy** — Exosomatic means outside the human body. It indicates energy conversions that are obtained using sources of power external to human muscles (e.g., machine or animal power) but that are still operated under the control of humans. Depending on the technology available to a given society, exosomatic conversions can imply the existence of huge gradients in power levels. For example, a single farmer driving a 100-hp tractor in the U.S. delivers the same amount of power as 1000 farmers tilling the land by hand in Africa. This is a qualitative difference detected by the existence of huge gradients in power level, which can be totally lost by an assessment of energy flows, since huge tractors are driven only for a few hundred hours per year. In developed societies, exosomatic energy is basically equivalent to commercial energy. In very poor countries, exosomatic energy is less related to the use of commercial energy, but rather related to traditional forms of extra power for humans such as animal power (like mules and buffaloes), wind, waterfalls and fire (used for cooking food, heating dwellings or clearing land).

In both cases, the very idea of *metabolism* requires the mapping of energy flows (megajoules of food per day or gigajoules of tons of oil equivalent/year) against time in relation to the size of a dissipative system. Below, we adopt a mapping of size of energy flows (extensive variable 2) against a mapping of the size of the societal system obtained in terms of human time (extensive variable 1). This variable of size can be divided using different categories at a first sublevel: producing vs. consuming. Then each of these is divided into subcategories of lower-level typologies of activities (e.g., producing in the agricultural sector vs. producing in the industrial sector). Such a mapping can obtain closure across levels of categorization and therefore can be easily used to build a hierarchical matrix against which to frame multi-scale analysis.

There is, however, another important difference that has to be briefly discussed. In the conventional linear representation of the metabolism of a society, energy flows are described as unidirectional flows from left to right (from primary sources to end uses), as illustrated in Figures 6.6 and 6.7. However, it is easy to note that some of the end uses of energy (indicated on the right sides of these two figures) are necessary at the beginning of the chain for obtaining the input of energy from primary energy sources (indicated on the left sides). That is, the problem with a linear representation — as the one adopted in these figures — is generated by the fact that the conversion losses indicated on the left sides of Figure 6.6 and Figure 6.7 for endosomatic and exosomatic energy flows occur before the primary energy sources get into the picture. That is, the stabilization of a given societal metabolism is linked to the ability to establish an egg–chicken pattern within flows of energy (Odum, 1971, 1983, 1996). Activities occurring on the left are not occurring before the one on the right in reality. This is an artifact of the choice made when representing such a metabolism. In reality, all these activities are occurring at the same time within an autocatalytic loop. Unfortunately, this obvious insight is completely lost by a linear representation of energy flows used to generate assessments of outputs and inputs. The possibility of establishing internal links among the values taken by energy variables using the concept of the egg–chicken pattern is discussed in Chapter 7 (in particular, we will return to the discussion of Figures 6.6 and 6.7 in Section 7.3).

6.4.1.1 *Linking Nonequivalent Assessments across Hierarchical Levels* — Let us start by writing a mathematical identity (a redundant definition in formal terms establishing an equivalency statement among names of numbers) of the total exosomatic energy throughput of a society (e.g., tons

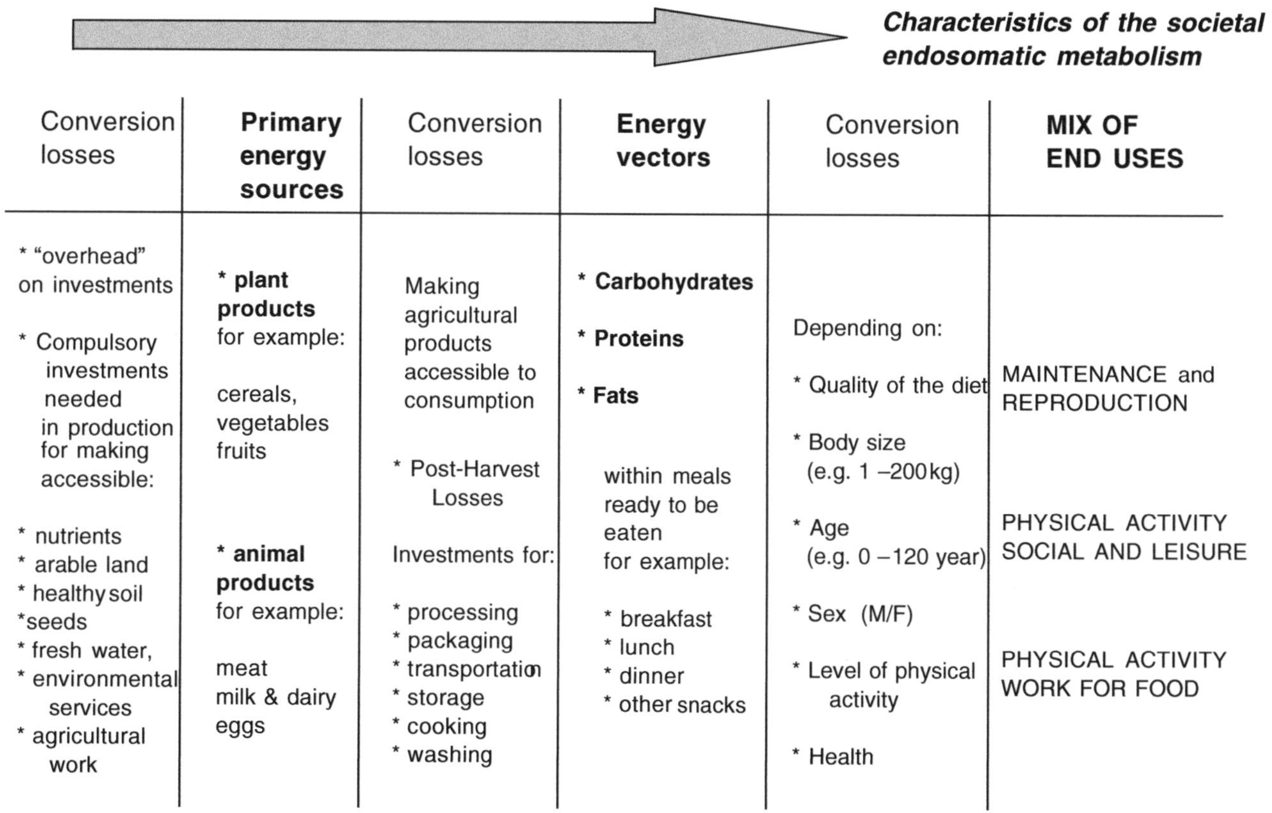

Characteristics of the societal endosomatic metabolism

Conversion losses	Primary energy sources	Conversion losses	Energy vectors	Conversion losses	MIX OF END USES
* "overhead" on investments	* **plant products** for example:	Making agricultural products accessible to consumption	* **Carbohydrates** * **Proteins** * **Fats**	Depending on: * Quality of the diet	MAINTENANCE and REPRODUCTION
* Compulsory investments needed in production for making accessible:	cereals, vegetables fruits	* Post-Harvest Losses	within meals ready to be eaten for example:	* Body size (e.g. 1 –200kg)	PHYSICAL ACTIVITY SOCIAL AND LEISURE
* nutrients * arable land * healthy soil * seeds * fresh water, * environmental services * agricultural work	* **animal products** for example: meat milk & dairy eggs	Investments for: * processing * packaging * transportation * storage * cooking * washing	* breakfast * lunch * dinner * other snacks	* Age (e.g. 0 –120 year) * Sex (M/F) * Level of physical activity * Health	PHYSICAL ACTIVITY WORK FOR FOOD

FIGURE 6.6 Endosomatic energy flow in human societies. (Giampietro M. and Mayumi K. (2000a), Multiple-scale integrated assessment of societal metabolism: Introducing the approach. *Popul. Environ.* 22(2): 109–153.)

Characteristics of the societal exosomatic metabolism

Conversion losses	Primary energy sources	Conversion losses	Energy vectors	Conversion losses	MIX OF END USES
* "overhead" on investments * Compulsory investments needed to enable the flow of energy inputs * extraction from stocks (depending on the quality of reserves) * catching incoming renewable energy (depending on its concentration)	* **fossil energy** for example: oil coal natural gas * **nuclear energy** * **renewable energy** for example: hydroelectric wind power solar heating photovoltaic cells	TWO TYPES of losses: I. Making various forms of energy (derived from primary sources) accessible for final consumption II. Manufacturing exosomatic devices. This is a crucial step for making a massive use of exosomatic energy (losses are = making + repairing discounted on the lifespan of devices) available to people	* **Electricity** * **Fuels** gasoline diesel coal Forms of energy that can be converted into useful power by the consumer: * mechanical power *heating, in relation to various tasks *processing of information	Depending on the characteristics of consumption: *level of economic activity *mix of energy vectors *level of technology *life styles *housing type *climatic conditions *availability of natural resources	RESIDENTIAL (including, heating, cooking leisure and private traveling) PRODUCTIVE SECTORS (industrial, agriculture, energy and mining) SERVICE SECTOR (including retailers) TRANSPORT (roads, railroads, marine and air transport)

FIGURE 6.7 Exosomatic energy flow in human societies. (Giampietro M. and Mayumi K. (2000a), Multiple-scale integrated assessment of societal metabolism: Introducing the approach. *Popul. Environ.* 22(2): 109–153.)

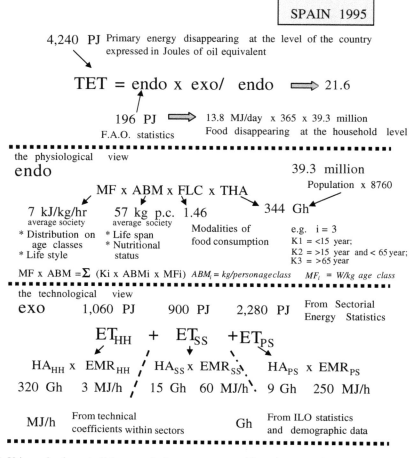

FIGURE 6.8 Using redundancy to link nonequivalent assessments taking advantage of nonequivalent external referents.

of oil equivalent (TET) expressed in gigajoules of tons of oil equivalent consumed by a country per year). For example, we can write that TET —— upper part of Figure 6.8 —— is equal to

$$TET = Endo \times Exo/Endo \qquad (6.3)$$

where:

TET = the total amount of exosomatic energy consumed by the society in a year (expressed in gigajoules of TOE /year). This value can be obtained by checking existing statistics (e.g., *UN Energy Yearbook*); in Spain in 1995 it was 4240 PJ (primary energy expressed in TOE).

Endo = the total amount of food energy consumed by the society in a year (expressed in gigajoules of food/year). This value can be obtained by checking existing data (e.g., FAO statistics); in Spain in 1995 it was 196 PJ (this is food disappearing at the household level).

Exo/Endo = the resulting ratio between the total amount of exosomatic energy metabolized by society and the total amount of endosomatic energy. In Spain in 1995 it was 4240/196 = 21.6.

Obviously, as soon as we calculate the ratio Exo/Endo using only the two numerical values obtained from statistical sources (using the ratio TET/Endo), then we collapse Equation 6.3 to the trivial identity TET = TET. This is why we need to look for external referents bringing in nonequivalent sources of information about the nature of this relation, when perceived and represented across hierarchical levels. This implies adopting different ways of perceiving and representing the qualities indicated by this relation.

6.4.1.2 Looking for Additional External Referents: Endosomatic Flow — the Physiological View

Physiological View — Thanks to the peculiar characteristics of holarchies, we can also express the given system quality — endosomatic energy flow — by using a nonequivalent mechanism of mapping related to the previous knowledge of lower-level typologies. For example, we can write

$$\text{Endo} = \text{THA} \times \text{ABM} \times \text{MF} \qquad (6.4)$$

where:

THA = total human activity (total hours per year of human activity in society). These hours imply a proportional amount of endosomatic energy flow linked to the physiological metabolism of the human body. This assessment, in turn, depends on the variable population. In fact, we can write the variable THA = population × 8760 (where 8760 is the number of hours in a year).

ABM = average body mass (average value of the weight of humans referring to individuals in the society expressed in kilograms).

MF = metabolic flow (average metabolic energy expenditure per kilograms of body mass over the year in the society expressed in watts per kilogram).

These average values (ABM and MF) can be calculated for any given society (or country) starting from the population structure and the record of the mix of physical activities. Put another way, to be able to make these assessments, we must have the knowledge of lower-level characteristics. For example, the average body mass can be estimated from:

1. Age structure of population, which reflects the profile of distribution of individuals over different age classes. An example of this analysis applied to four different types of societies at different degrees of development is given in Figure 6.9.
2. Data on the average body mass within each age class. An example of this analysis applied to four different types of societies at different degrees of development is given in Figure 6.10. Data related to the profiles of values of ABM on different age classes in different countries are available in the field of nutrition (see, for example, James and Schofield, 1990).

This means that we can express the endosomatic flow per capita as

$$\text{Endo} = \text{MJ/h} = \text{ABM} \times \text{MF} = \Sigma(K_i \times \text{ABM}_i \times \text{MF}_i) \qquad (6.5)$$

in which we can use three age classes: $K_1 < 15$ years; 15 years $< K_2 < 65$ years; $K_3 > 65$ years.

It should be noted that the information on the distribution of population over the various age classes implies an additional constraint on the value that can be taken by the dependency ratio of the society (the fraction of population that can be included in the workforce); in fact, it determines the fraction of population that is included in the age brackets of <15 years of age and >65 years of age.

6.4.1.3 Looking for Additional External Referents: Exosomatic Flows — the Technological View

Technological View — Following the same rationale used so far, we can express the value of TET as the sum of the energy consumption of different sectors of the socioeconomic system:

$$\text{TET} = \text{ET}_{\text{HH}} + \text{ET}_{\text{PS}} + \text{ET}_{\text{SG}} \qquad (6.6)$$

where:

ET_{HH} = the exosomatic energy consumption of the household sector

ET_{PS} = the exosomatic energy consumption of the productive sector (including manufacturing, agriculture, energy and mining)

ET_{SG} = the exosomatic energy consumption of the economic sector including (services and government)

Each one of the three assessments included on the right side of Equation 6.6 is expressed in terms of an extensive variable (e.g., gigajoules per year) and can be found on statistical sources directly, but this would imply again using the same external referent used when calculating the Exo/Endo ratio in Equation 6.3. However, assessments of exosomatic energy can be handled in the same way as the assessment of

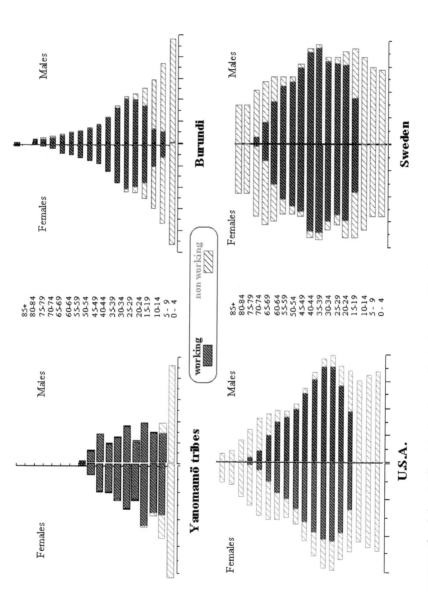

FIGURE 6.9 Population structure of societies at different levels of economic development. (Giampietro M. and Mayumi K. (2000a), Multiple-scale integrated assessment of societal metabolism: Introducing the approach. *Popul. Environ.* 22(2): 109–153.)

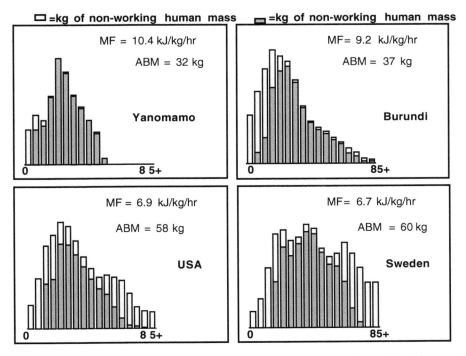

FIGURE 6.10 Assessments of body mass overage classes and average metabolic flow. (Giampietro M. and Mayumi K. (2000a), Multiple-scale integrated assessment of societal metabolism: Introducing the approach. *Popul. Environ.* 22(2): 109–153.)

endosomatic energy in Equation 6.4. That is, we can express each one of these assessments as the product of an intensive and extensive variable, establishing a bridge with a meaningful relation between perceptions and representations of events at the lower level (lower-level assessments).

We have to again use redundancy to link nonequivalent definitions of identities. That is, for each economic sector (productive sector (PS) is used as an example here), we can write

$$ET_{PS} = HA_{PS} \times EMR_{PS} \tag{6.7}$$

where:

HA_{PS} = human activity invested in working in PS (expressed in hours of human activity); being a total amount per year in this sector, this is extensive variable 1 (see Section 6.2).

ET_{PS} = he exosomatic throughput in PS (expressed in gigajoules of exosomatic energy per year); being a total amount per year in this sector, this is extensive variable 2 (see Section 6.2).

EMR_{PS} = the exosomatic metabolic rate of the economic sector (expressed in megajoules of exosomatic energy spent in the sector per hour of human activity invested in the sector); being a ratio — flow of something per unit of something else — this is intensive variable 3 (see Section 6.2).

Note that since we are dealing with dissipative systems (which implies a basic level of energy consumption per year), we are facing a special use of the concept of extensive and intensive variables discussed in the example of Figure 6.1. That is, following our choice of using human activity as extensive variable 1, defining the size of socioeconomic compartments when dealing with the assessment of flows of energy:

1. Extensive variable 2 applies to the class of assessments defining aggregate amounts of either energy or added value per year of a particular element (the size of interaction with the context).

2. Intensive variable 3 applies to the class of assessments referring to ratios of either energy or added value per unit of human activity, as calculated within various elements (the time unit in the case of multi-scale integrated analysis of societal metabolism is 1 h).

Clearly, the same type of relation ($ET_i = HA_i \times EMR_i$) holds for each of the three sectors (ET_{HH}, ET_{PS}, ET_{SG}).

Again, Equation 6.7 can seem useless (a mathematical identity based on redundant information) when considered in isolation within a single descriptive domain (and using an individual data source at the time). That is, you need to know both ET_{PS} and HA_{PS} to calculate EMR_{PS}, and vice versa. However, when this relationship is used in combination with others calculated for the same socioeconomic system but on different descriptive domains (for different elements and at different levels), using different data sources, it provides a powerful mechanism for bridging information of different natures.

In fact, when related to lower-level characteristics, the value of EMR_{PS} depends on (or reflects) the set of technical coefficients and the level of capitalization of the various economic activities performed within PS (characteristics of lower-level elements or subsectors of that sector). When the value of ET_{PS} is related to TET (the hierarchical level of the whole society), this implies a constraint on the possible range of values that can be taken by the other two sectors (ET_{SG} and ET_{HH}) in relation to TET. In fact, the equation of congruence on exosomatic energy flows requires that

$$TET - ET_{PS} = ET_{SG} + ET_{HH} \qquad (6.8)$$

By expanding the original simple identity into lower and higher levels, it is possible to establish a network of relations among various assessments that can be used to characterize the metabolism of human society at different levels. An example of this branching across levels is given in Figure 6.11. We can start with a simple numerical assessment that can be found as a single figure (the numerical value taken by TET in a given year for a given country) on a particular statistical source (e.g., *UN Energy Statistics Yearbook*). Then we can express such a numerical value as related to other numerical values reflecting the characteristics of other elements (and different figures found when using different data sources). When redundancy is used in this way, we can express the same quantity as a function of both intensive and extensive variables, reflecting relations across levels to establish bridges across nonequivalent descriptive domains, as in Figure 6.11.

Equation 6.7 can be applied not only to other economic sectors and subsectors (e.g., service and government (SG) or the manufacturing sector within PS), but also to the characterization of the HH. The exosomatic metabolism of the household sector can be assessed exactly in the same way (using the same approach, $ET_{HH} = HA_{HH} \times EMR_{HH}$) by dividing the energy spent in the (end use residential + private transport) by the hours of human time not allocated in paid work. This assessment can be obtained starting from the population size, dependency ratio and other social parameters determining the amount of work supply in a given society.

When working within the various elements determining the societal metabolism in this way, the more information we add to the network of relations (as when adding more pieces in the solution of a puzzle), the more we will receive free information from the hierarchical structure of the system. This free information is generated by forced congruence of flows across levels. That is, as soon as the loose structure of relations indicated in Figure 6.8 or Figure 6.11 is organized into imposing forcing equations (in terms of energy and human activity), we obtain a clearer network of relations among assessments.

6.4.2 Multi-Scale Integrated Analysis of Societal Metabolism: Two Variables (Megajoules and Dollars) and Different Levels (*Technical Section*)

The approach for a multiple-scale integrated analysis of societal metabolism has been developed to describe socioeconomic systems in parallel using nonequivalent descriptive domains (energy analysis, economics, demography, ecology), while addressing explicitly the issue of multiple scales.

As already shown in the previous section, assessments of human activity (extensive variable 1) can be used as the common matrix against which endosomatic and exosomatic energy flows (two qualities both assessed using megajoules as extensive variables 2) can be assessed at different hierarchical levels over parts and the whole. Total human activity (the total size of the system expressed in terms of the

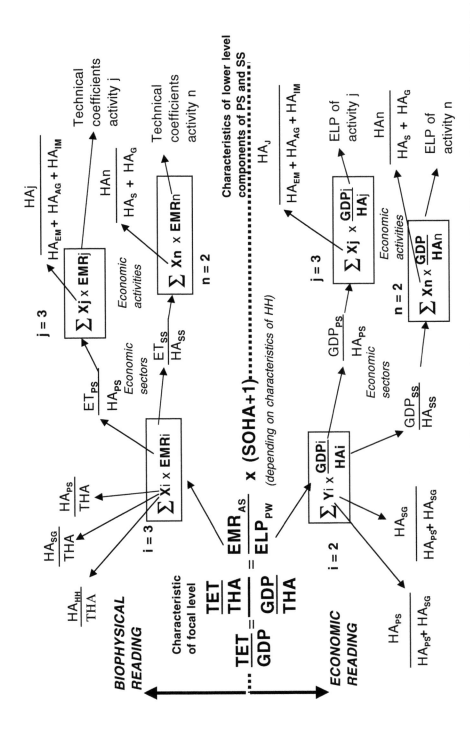

FIGURE 6.11 The nested hierarchical nature of relations in societal metabolism: integrating biophysical and economic variables. (Giampietro, M. and Mayumi K. (2000b), Multiple-Scale Integrated Assessment of Societal Metabolism: integrating biophysical and economic representation across scales. *Popul. Environ.* 22(2), 155–210.)

extensive variable 1) is linearly dependent on population size (when expressed in hours, THA = people × 8760). The profile of distribution of human activity over the various compartments (how THA is invested on lower-level components) depends on or reflects key variables that are socially relevant (age and gender structure of the population, retirement age, compulsory education, access to higher education, participation of women in the economic process, length of workweek).

Chapter 7 shows, with practical examples, that the existence of this mosaic effect across identities of human holarchies makes it possible to establish a set of internal constraints (determined by the need to obtain congruence in the representation of flows across nonequivalent descriptive domains) on the reciprocal value that these variables can take, when considering the dynamic budget of societal metabolism. That is, autocatalytic loops of investments related to food, labor, land and exosomatic energy can be represented in a way that makes it possible to characterize (1) what are feasible costs in relation to existing returns and (2) what are feasible returns in relation to existing costs. This simultaneous nonequivalent characterization is at the basis of what we call impredicative loop analysis. However, before getting into that discussion, it is important to show the possibility of generating a mosaic effect across levels that uses two nonreducible types of extensive variables 2 in parallel: one related to exosomatic energy and the other related to an economic reading of societal metabolism.

That is, we want to show that it is possible to also add the mapping of money flow to the integrated system of mapping across levels illustrated in the previous section. This additional typology of mapping reflects the production and consumption of added value both at the level of the whole society and on lower-level elements (economic sectors of the socioeconomic system and even economic subsectors). To do that, we have to perceive and represent economic entities as being characterized in their functioning by a behavior similar to that of dissipative systems. That is, economic entities must produce and consume a certain amount of added value in time to remain economically viable. Depending on the reference state provided by the context within which the holon is operating (what is the general level of dissipation of the context), a given density of flow of added value occurring within a given holon can be insufficient (e.g., the making of $10/h in 2003 by a farmer in Iowa) or plenty (e.g., the making of $10/h in 2003 by a farmer in China). That is, we can apply the same mechanism used to perceive and represent the relation between the identities of parts and the whole to do benchmarking also applied to economic variables. To do that, we can go through the same mechanism of definitions of extensive and intensive variables illustrated before.

We can start with a characterization of the size of the social system, which will be used to structure the compartments across levels. We can use the same extensive variable 1: total human activity. Then we can assess the size of the economic process (the total amount of added value associated with the production and consumption of goods and services), which represents the extensive variable 2, for example, the total GNP per year, seen as a proxy for the size of the economic system according to extensive variable 2. The relation between these two extensive variables provides an intensive variable 3 (e.g., something similar to GNP p.c. per year — a GNP per hour in our system of accounting), which can be used to characterize the identity of the socioeconomic context in terms of benchmarking. In this way, we can compare the performance of economic sectors within the same level (e.g., level $n - 1$). A much higher level of dollars per hour must be found in an hour of human activity allocated in the compartment paid work, whereas we can expect a much lower level of dollars per hour in an hour of human activity allocated to education. The same benchmarking, based on the values taken by intensive variable 3 — generation of added value per hour — can be used to compare different sectors of the economy (agriculture) in relation to the same value obtained in another sector (manufacturing). This conventional economic analysis, however, can be complemented by (1) a simultaneous comparison based on spatial analysis (e.g., level of generation of added value per hectare of a given category of land use to the average value for a given province — comparing typologies of land use allocated to economic activities) and (2) a simultaneous analysis of biophysical flows.

Obviously, in this way, after calculating the given level of production and the spending of the added value per hour of a socioeconomic system — GNP per THA (dollars per hour) at the level of the country — we can characterize different compartments of the socioeconomic system as handling a higher (those in the productive sectors of the economy) or lower (the household sector) density of money flow per hour of human activity or per hectare of land invested there.

In multi-scale analysis, when taking into account added value flow together with endosomatic and exosomatic flows (using the MSIASM approach), we can use a common skeleton of equations of congruence that maps investments of exosomatic energy and added value (two nonequivalent extensive variables 2) against the same nested hierarchical structure, providing closure, of compartments obtained by dividing THA (the same extensive variable 1). The formation of the skeleton (frame) related to extensive variable 1 is based on four logical steps:

1. The socioeconomic system is divided into a set of relevant compartments, whose sizes are characterized in terms of investments of human activity (the common matrix that provides closure). That is, THA at the level n must be equal to the sum of HA_i (the investments of human activities in the various sectors) defined at the level $n - 1$, following a nested hierarchical structure. That is, for example:

$$(\text{whole country} - \text{level } n) \rightarrow (\text{economic sectors level } n - 1) \rightarrow$$
$$(\text{economic subsectors level } n - 2) \rightarrow \ldots \rightarrow (\text{individual economic activities level } n - x)$$

2. Lower-level compartments must account for the *total* human activity of their upper level (e.g., subsectors making up a sector). That is,

$$\text{THA } n = [\Sigma \, HA_i] \, n - 1 = [\Sigma \, HA_j] \, n - 2$$

With this choice, the definition of economic sectors, subsectors and individual economic activities must also include the fraction of human activity invested in the household sector (nonworking time of the economically active population and all the activities performed by the noneconomically active population, including sleeping).

3. Compartments are then characterized in terms of intensive variables and extensive variables by combining two extensive 2 variables (both exosomatic energy and added value) against the same hierarchical frame provided by the compartmentalization done using extensive variable 1. That is, in this way it is possible to define for different compartments two intensive variables 3:

   Intensive variable 3 biophysical = exosomatic energy per unit of human activity

   Intensive variable 3 economic = added value per unit of human activity

4. After having implemented this mechanism of characterization, it is possible to establish a relation between the size of each compartment, parts (expressed in terms of either GNP or exosomatic energy) and the total of the whole society using (1) the extensive variable 1 (total human activity), which is determined by the population size of the whole society; (2) the fraction of total human activity, which is allocated to the particular compartment; and (3) the two intensive variables 3. For example,

   EMR_i = exosomatic metabolic rate of the compartment i
   (exosomatic energy per unit of human activity in the compartment i)

   ELP_i = economic labor productivity of the compartment i
   (added value per unit of human activity in the compartment i)

The hierarchical frame for the relative relations (assuming α and $\alpha - 1$ as two contiguous hierarchical levels) can be expressed as follows:

   $X_i = HA_i/HA_k$ = the *fraction* of human activity HA_k invested in the i-th sector

   (element i belongs to the level $(\alpha - 1)$; element k belongs to the level α)

   $ET_i = HA_i \times EMR_i$ = the exosomatic energy spent in the i-th sector — at the level $(\alpha - 1)$

   $EMR_i = ET_i/HA_i$ = the exosomatic metabolic rate in the i-th sector — at the level $(\alpha - 1)$

   $GNP_i = HA_i \times ELP_i$ = the added value productivity of the i-th sector — at the level $(\alpha - 1)$

   $ELP_i = GNP_i/HA_i$ = the economic labor productivity of the i-th sector — at the level $(\alpha - 1)$

$$ET_\alpha = \Sigma(ET_i)_{\alpha-1}, \text{ e.g., } TET = (ET_{PS} + ET_{SG} + ET_{HH})$$

$$HA_\alpha = \Sigma(HA_i)_{\alpha-1}, \text{ e.g., } THA = (HA_{PS} + HA_{SG} + HA_{HH})$$

$$GNP_\alpha = \Sigma(GNP_i)_{\alpha-1}, \text{ e.g., } GNP = (GNP_{PS} + GNP_{SG})$$

In these examples, the values referring to the whole country ($TET = ET_{WC}$, $THA = HA_{WC}$, $GNP = GNP_{WC}$) — the α level — are related to the values taken by the same variable in the lower level — ($\alpha-1$) — over the compartments PS (productive sector), SG (service and government), and HH (household sector).

Information (data) about the value taken by both extensive variables (e.g., HA_i extensive variable 1, ET_i and GNP_i extensive variable 2) can be linked using intensive variables (e.g., EMR_i and ELP_i, which are two types of intensive variables 3) across hierarchical levels, about which it is possible to obtained nonequivalent sources of information. The relation among compartments across levels expressed in terms of relation of assessments based on HA_i extensive variable 1 (X_i) provides the congruence requirement. The possibility of using independent external referents to assess the value taken by these variables entails that the redundancy in this set of relations should not be considered a problem (Occam's razor principle). On the contrary, such a redundancy is a plus, since it makes it possible to establish a bridge among nonequivalent forms of perception and representation of societal metabolism (e.g., the social, biophysical and economic) at different scales.

For example, changes in EMR_i and ELP_i in the same compartment (e.g., urban households) are connected because of their common reference frame — the same HA_i. Because of the common matrix represented by the profile of investments of human activity on the different compartments (defined in the same way for the exosomatic energy accounting and the added value accounting), we can generate a holarchic complexity in the structure of relations among data, as illustrated in Figure 6.12. Obviously, the fact that a part of the assessments is in common for the economic and biophysical representations means that changes in the values taken by the two nonequivalent variables (biophysical and economic) are not totally independent.

Applications of the approach of multiple-scale integrated analysis to the analysis of agricultural systems are presented in Part 3.

6.5   Holarchic Complexity and Mosaic Effects: The Example of the Calendar

Finally, it is time to discuss the possible use of the concept of holarchic complexity to provide robustness to a system of mapping. The example discussed below has to do with the encoding of time. A task for its complex nature has been crucial in the shaping of human civilization (e.g., see Duncan, 1999). In the history of their civilization, humans have been forced to develop and continuously update different systems of mappings for keeping time. What is remarkable in this example is the convergence on solutions based on holarchic complexity.

To introduce the example, let us first try to answer the following question: When we say that Cristoforo Colombo (Christopher Columbus in English) arrived for the first time on the American continent on the morning of October 12, 1492, what is the meaning and reliability of such an assessment? Put another way, does it mean that this event took place:

1. In the daily light, when the sun was still rising in its daily trajectory?
2. In a day that was in early autumn?
3. In the same year that the Christian Isabella of Castile and Ferdinando II of Aragon took the city of Granada from the Muslim Emir Al-Zagal?

If the answer to these three questions is yes, then we are confronted with a big question.

According to the assumptions of reductionist science, time encoding is obtained in simple linear terms (e.g., the timekeeping obtained using a clock). That is, a point in time can be individuated by moving

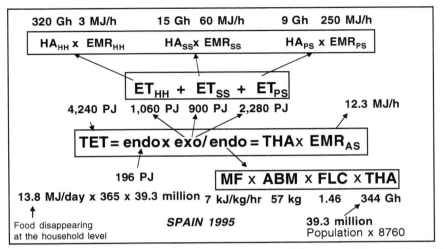

FIGURE 6.12 Equations of congruence establishing a link between nonequivalent assessments.

along a linear trajectory based on a unique variable. In this framework, time durations are defined by differences between the measurement (reading of the clock) at time T2 minus the measurement at time T1.

But if this assumption were true, we would be forced to conclude that the characterization of time — "morning of October 12, 1492" — cannot support all the information required to answer the three questions posed. In fact, when basing our mapping of time on a unidimensional system of encoding (e.g., when using hours or minutes measured against the indications given by an ordinary clock), we have that the difference between

T2 = the morning of the January 1, 2000 and T1 = the morning of October 12, 1492

represents a quantity of time that is 4,454,760 h or 267,285,600 min (the equivalent of almost 509 years after accounting for a few spare months and days, plus the correction for leap years).

Assuming an error of 1% (e.g., generated by the accuracy of the device measuring time and the handling of the information over long periods of time), we would obtain an error bar associated with the assessment "morning of October 12, 1492" of ±5 years. In this case, the information related to concepts morning, early autumn and contemporaneousness of events in Spain in the year 1492 would be completely lost.

Now imagine running the same mental experiment, but trying to fix this problem with a high-tech solution. We can hypothesize that an atomic clock with an incredible degree of accuracy had been available since the year 1450. Would it then be possible to use the linear time assessment T2 (midnight of January 1, 2000) minus T1 (morning of October 12, 1492) to answer the three questions? First, even with an atomic clock, it is difficult to imagine a very reliable mechanism for keeping the record of time in the form of hours and minutes over a time horizon of 500 years. Such a mechanism should be able to go through wars, riots and earthquakes. This is a list of a set of perturbations that can be expected to occur (at least one of them) in a given place over such a large time window. Because of this, such a mechanism would require following the old saying "Do not put all your eggs in the same basket." That is, the only way to obtain reliability of a record over hundreds of years is to make several copies of it and store it in different places using different mechanisms — redundancy and the mosaic effect. Actually, if someone would try to implement this solution by handling the flow of records coming out of a set of synchronized clocks, that person would find that this is not an easy task at all. In fact, the challenge of recording time in terms of hours and minutes in parallel over different places in our planet is anything but simple, and this is part of the fascinating history of the fight that humans have had to wage to master timekeeping.

But there is more direct evidence of the impossibility of tracking the passing of time on a large scale by using a unidimensional mechanism of mapping. Besides the problem of keeping parallel records of aggregate time durations through war, riots and natural disasters, and besides the problem with the accumulated effect of a lack of accuracy of clocks to be kept constant over five centuries, there is another major problem that comes into play in this case. The October 4, 1582, calendar was reformed by Pope Gregory. The original system of accounting of leap years was changed, and 10 days were eliminated from that year.

Because of this perturbation when adopting a reductionist (linear, unidimensional) system of mapping of time, the indication "the morning of October 12, 1492" would no longer make any sense in relation to the first of the three questions listed above. The gradient "morning/night over the day" is much smaller than the error induced by the removing of 10 days from a year. This has nothing to do with the accuracy of the measurement scheme and the reliability of the keeping of records. The 10 days taken away by Pope Gregory would remain missing whether using an old-fashioned solar clock or an atomic one. On the other hand, we all share the feeling that the indication of the morning of October 12, 1492, is conveying useful information about the location in time of this event — that is, the first official European arriving in the new continent landed (1) in the morning, (2) in the fall and (3) in the same year in which Granada was occupied by the Spanish kings. This shared belief is so strong that the date is still a national holiday in the U.S. and Spain.

This paradox can be explained by the mechanism of mapping of time adopted by the calendar, which is not simple or linear. For each of the three statements about the relative location in time of the event (morning, fall, the same year of the occupation of Granada), the calendar is using nonequivalent external referents (the relative position in time is based on the use of distinct markers). That is, with the old-fashioned calendar we are dealing with holarchic complexity in the system of mappings used to encode the passing of time. Such a system is based on an integrated encoding of the same quality (duration of time) obtained by a chain of associative contexts across hierarchical levels. In particular, there is a hierarchical structure based on at least three levels that are linked — for the needed respective semantic checks — to the natural frequencies of three natural processes: (1) the division of a year into subparts (seasons), with an external semantic referent given by cycles of seasons related to the sun; (2) the division of a month into subparts (days), with an external semantic referent given by cycles of the moon; and (3) the division of a day into subparts (e.g., A.M. with daylight, A.M. without daylight, P.M. with daylight, P.M. without daylight), with an external semantic referent given by daily revolutions of the planet Earth on itself.

These subparts can again be divided into smaller parts (hours and minutes) using clocks. Within this structure of Chinese boxes, Russian dolls, or nested interlocking hierarchies (using the expression coined by Koestler), each of these parts can be individuated by using the larger container as the specific associative context (e.g., 30 sec after 45 min after the 5th h of Monday, which is the first day of the 3rd week of the 7th month of …). The simplicity and the validity of each assessment depend on the validity of the assessment referring to the larger associative context of the chain. At this point, the validity of each of these statements can be checked against nonequivalent external referents. The measurement of a time duration of 15 min is quite simple to obtain (with a mechanical clock used as an external referent), as well as the representation of 15 min past the hour. However, to be useful, they both require a higher level (A.M. or P.M.) as a reference — related to the revolution of the Earth on its axis — which in turn can be related to a date (a day within a month, etc.). What is interesting in this system of mapping is its flexibility and internal coherence, which are generated by the use of nonequivalent mechanisms of encoding changes of observable qualities of external referents (sun for years, moon for months and revolving planet Earth for days). This integrated system of assessments is then forced into congruence by a set of semantic checks based on direct perceptions of changes of observable qualities of external referents, which are performed at different space–time windows selected on the basis of the hierarchical structure of the whole.

There are years in which seasons seem to behave in a crazy way. Even in normal years, the very same profile of daylight in both the A.M. and P.M. sections of each day can change dramatically when moving from summer to winter and according to geographic location. There are even cases in which the distinctive quality implied by the very definition of day (the daily shift from night to daylight and vice versa) can totally collapse, disappearing for a while, as in the case of the long nights (or the long days) experienced by those living close to the poles. Still, local failures of individual mappings do not affect the stability and validity of the whole system (the usefulness of using a common calendar for the different countries of this planet). The calendar is working fine for humankind. Whereas individual organisms can be fooled by too simple mechanisms of mapping time (e.g., when they are led to sprout too early by a few warm days, due to the simple mapping "cluster of warm days" = "spring is here"), humans can be aware of its still being winter by looking at the record of days and months on the calendar, even though the outside temperature is extremely pleasant.

The self-entailing structure of the various mappings (a month mapping onto a given number of days, a year mapping onto a given number of months) implies redundancy, and this generates an internal robustness. The number of days (lower-level mapping related to the quality daily revolution of the planet) selected to define a month (in relation to moon cycles — about 28) must be congruent at the year level, with the indications coming from another nonequivalent mapping obtained at that level (the passing of seasons — about 365 days). Bringing into congruence such a system of encoding that is based on nonequivalent mappings of events operating on distinct descriptive domains (into light changes due to daily revolutions, into monthly cycles due to moon revolutions, into season changes due to cycling changes in the inclination of Earth's axis) provides enough redundancy to the system for handling local failure (e.g., days without daylight, warm winters). Actually, such holarchic complexity does much more: it makes possible the patching of individual mappings without losing the functional performance of the whole. That is, with this complex structure, it is possible to have an evolution in time of the integrated system of mapping. In fact, it is well known that the basic problem with the accounting of time for humans has been generated by the fact that the discretization in time units based on moon cycles is not fully compatible with the discretization in time units based on solar cycles (for a detailed account of this issue, see Duncan (1999)). This incommensurability of units related to moon months with units related to solar seasons is what requires the presence of a step of semantic verification when errors or bifurcations occur. As already noted, after the discovery of America, the Gregorian reform of the calendar had the goal of patching the old mechanism of mapping of time, exactly because the semantic check performed on solar cycles was indicating the manifestation of a bifurcation. The position of Eastern — determined according to the system of accounting based on lunar units — was no longer congruent with the position in time expected according to the system of accounting based on solar units. On the occasion, of the reform, not only a more sophisticated mechanism of accounting of the leap year was put in place (the correction factor for dealing with the incommensurability between lunar units and solar units), but, in addition, 10 days were deleted from October 1582 to correct the lack of congruence (the error) generated by the previous operation of the old integrated system of mapping (the old calendar).

The robustness of an integrated system of mapping based on holarchic complexity is so high that the date October 12, 1492, kept its relevance even after the patching done by the Gregorian reform. The perturbation induced by such a correction (a local deletion of 10 days and a systemic change in the accounting of leap years) was absorbed by the complex structure of the traditional calendar without major problems. In fact, the individuation of a given position in time (the morning of October 12, 1492) is obtained through a triangulation of a given position in time related to nonequivalent external referents. This mechanism provides an extreme robustness, since it makes it possible to be anchored to the representation of nonequivalent perceptions relative to different relevant space–time differentials. That is, the decisions taken when reforming the old calendar (adding a lower number of "29th of February" in the next 400 years after having deleted 10 days from the year 1582) did not affect the indication given by the indicator morning over the selected difference $T2 - T1$ at all. This indicator refers to an external referent relevant at a lower-level scale. Therefore, this external referent is operating below the time differential of a day. At that level of the distinction between morning and night, the perturbation associate to the Gregorian reform of the calendar has been simply filtered out. Columbus arrived in the morning no matter which was the individual day of the month.

What about a potential overwhelming error bar on the location in time of October 12, 1492? This could imply, for example, losing the meaningful indication of the season. Again, this preoccupation is just an artifact resulting from the adoption of a reasoning based on the concept of simple time (system of mapping based on only one space–time differential). Within a system of mapping based on the reciprocal entailments between the position of days over months, months over seasons, and cycles of seasons over years, the effect of an aggregation of more or less "29th of February" is simply not taking place. In fact, the reciprocal forced congruence over different levels of nonequivalent encoding in relation to (1) external referents defined over different space–time windows (sun, moon, rotation of the Earth) and (2) semantic checks (Was this in daylight? Is Easter coming in the right expected season?) has exactly the effect of diffusing possible negative effects of aggregation of errors on a single level for a too-long period of time. An error at one level can keep occurring as long as its aggregate effect does not affect the validity of the next external referent in the hierarchy. This built-in default quality check enables a continuous activity of patching within the various range of tolerances specified for each level in the hierarchy. Using hierarchy theory jargon, we can say that external perturbations requiring the patching, and internal perturbations generated by the patching tend to be filtered out by the hierarchical structure of the whole system of mapping. This is why systems hierarchically organized have a much more reliable and robust organization. Actually, both the new mechanism for handling leap years and the deletion of 10 days were introduced exactly to keep congruence between the meaning conveyed by a given date (the location of Eastern within the formal frame of accounting of numbers) and the indications given by the set of external referents (number of moon months and solar seasons). That is, the complex structure of the calendar based on the simultaneous operation of semantic checks and formal representations not only was able to absorb the changes introduced by the reform, but also was the cause of these changes. Its intrinsic robustness was exactly what called for the patching in the first place.

But there is another important point to be made about this example of robustness of the mechanism of mapping time obtained with the calendar. This has to do with the special nature of holarchic objects. When using a holarchic system with a robust set of identities self-entailed across levels, we are forced to admit that its complex structure can also be a reason for trouble. That is, the complex structure cannot be very useful when referring to specific assessments at a given point in space and time. For example, if we try to use the calendar to compare dates and events over a small descriptive domain, we can face clear incongruences and paradoxes. For example, the existence of a geographically determined "day line" imposed on our globe by such a system of mapping implies that people talking on the phone at a distance of a few miles across such a line can be represented as acting simultaneously but on different days. That is, the very structure that makes a coherent overall mapping of time over a set of different space–time windows (the calendar used to compare dates across centuries) reliable is not very useful (should not be used) to map differences in time when dealing with events describable at a given point in space and time. When dealing with small-scale events, simple time (based on only one time differential and only one external referent for semantic checks = a shared indication from a clock) can be more useful.

In conclusion, for a very large descriptive domain (when dealing with the risk of losing relevance in the process of representing the collocation in time of the event discovery of America), we need to use a system of mapping based on holarchic complexity. Over large descriptive domains the application of the reductionist paradigm would not have solved the problem of loss of significance of the expression "morning of October 12, 1492" after 500 years. Rather, a high-tech encoding of time for historic dates would have made the problem of preserving the meaning of this mapping much worse. This is an important point, since technical progress often leads some scientists to believe that the higher accuracy of an electronic clock could dispense with the need to understand the context in which the numerical measurement is used (recall here the true age of the Funtrevesaurus in Figure 3.2). In this way, hard scientists working on integrated assessment can end up neglecting the need for a continuous semantic check about (1) the meaning of each of the encoding variables used in the model and (2) the implications associated with the choice of the relative measurement schemes. In many situations, it is much better to use a primitive clock (a measurement scheme with low accuracy) linked to the use of an old-fashioned calendar (a robust integrated system of mapping) than to rely on an atomic clock (a measurement scheme with high accuracy) linked to a unidimensional accounting of time (an inadequate system of mapping) to record aggregate durations. Obviously, when dealing with small descriptive domains (a single scale in which both the identity of the investigated systems and the interests of the observers do not change in time) the reverse is true. An accurate clock can be much more useful than a calendar for dealing with athletic records.

To resume the main message of this section, we can use a quote from Rosen (1985, p. 317): "Functional reliability (which is different from freedom from errors) can be obtained through an artful employment of redundancy providing robustness to the functions of a network. Functional reliability at some functional level, however, implies the existence of errors at a different level."

6.6   Holarchic Complexity and Robustness of Organization Overview of Literature (*Technical Section*)

In his seminal work, Goldberger (1997) describes human health in terms of the ability to keep harmony between the reciprocal influences of processes occurring in parallel on different scales within the complex web of physiological processes in the human body. The insurgence of a disease is seen as a process decomplexification in which one particular dynamic (a system of control operating over a given relevant space–time differential) — which is therefore related only to a given set of processes going on at one particular hierarchical level — takes over on other dynamics. This makes dynamics expressed on other relevant space–time differentials, which were originally included in the integrated system of controls, irrelevant. In his words, when this occurs, "the system loses its fractal complexity." This leads to pathological manifestations of ordered behavior (e.g., Parkinson's tremors). Again using his words, the human body loses its "organised variability" (balanced mix between functional specialization and functional integration).

The same concept is proposed by O'Neill et al. (1997) when discussing assessment of human interference on ecological processes in relation to spatial configuration of landscapes. When observing natural landscapes, one can find a cascade of scales at which spatial patterns can be found. The negative effects induced by humans can be detected by the disappearance of some of these spatial patterns on certain scales. Those missing patterns indicate the termination processes of self-organization operating at the relative space–time differential. In this perspective, human management can be seen as having the effect of inducing a decomplexification of dynamics within the landscape. In particular, humans are systematic in removing those dynamics (mechanisms and related system of controls) that are not economically relevant within the models assessing the economic performance of various land uses.

The concept of holarchic complexity has been discussed by Grene (1969) in relation to stable hierarchical systems. The term used by Grene is double asymmetry. The concept of double asymmetry indicates the peculiar status of each element of a self-entailing hierarchy that is operating in complex time, which is at the same time ruler and ruled. That is, the existence of a double asymmetry implies that the stability of each level depends on other levels either higher or lower in the hierarchy. The number of predators affects the number of prey and vice versa; governments determine the fate of citizens and vice versa. A dictator or a group of bandits keeping hostages would be an example of social entities lacking double asymmetry, that is, a sick social holarchy. The need of double asymmetry leads, in the long term, to the organization of socioeconomic structures that can be explained in terms of power laws, as noted by Zipf (1948). In modern times, within the complexity literature, we call systems that are organized in this way critically organized (Bak, 1996). The concept of double asymmetry is crucial for the concept of the impredicative loop and will be discussed with practical examples in Chapter 7.

As already discussed in Chapter 2, Koestler (1978) calls the phenomenon of double asymmetry a Janus effect taking place within holarchies. The etymological explanation for the choice of the term *holon* given by Koestler (1968, p. 56) is the following: "a holon is a Janus-faced entity who, looking inward, sees himself as a self-contained unique whole, looking outward as a dependent part." The term is derived (Koestler, 1968, p. 48) from the Greek *holos* (whole), to which the suffix *on* has been added to recall words such as proton or neutron to introduce the idea of particles or parts. Koestler's definition of structural complexity of holarchies is also framed within hierarchy theory, that is, in terms of vertical and horizontal coupling of holons. On their own level of activity, holons have to fight to preserve their identity, both by keeping together their own parts on the lower level and by preventing same-level holons from expanding too much. On the other hand, because of the fact that holons belong to a larger system that provides them with the stability of boundary conditions and lower-level structural organization, they have to support other level processes stabilizing the entire structure of the holarchy. Holons have to contribute to the stability of vertical levels even though this implies negative trade-offs on their own level. A classical example of this tension is represented by ethical dilemmas (personal advantage vs. social advantage). In particular, consider the dilemma of paying taxes. Individual households have a direct short-term return in skipping their tax duty (in this way boosting their horizontal interaction as holons); however, this would weaken the community of which they are a part (this will weaken their vertical integration, the larger holon to which they belong). Another example concerning the role of death in human affairs was discussed in Figure 3.6. For individual human beings, death is a very sad event. On the other hand, from the perspective of the species *Homo sapiens*, a sound turnover of individual realizations within the type is essential for keeping the physiological fitness of this species high. Obviously, different degrees of stress on the horizontal or vertical coupling will dictate the final compromise solution at which the holon will eventually decide to operate. The existence of several distinct relevant space–time differentials makes it impossible to see or represent the relevant attributes reflecting different interests (objectives and goals) of the various relevant wholes and parts within the same representative domain. Even if it is impossible to formally account for it, every time one sees or represents a part of a holarchy, one must imply the whole to which this part belongs and vice versa.

Iberall et al. (1981) introduced a concept very similar to that of double asymmetry by suggesting the term *equipollence*. With this expression they mean that within socioeconomic systems there is a natural tendency toward power balance among components. More powerful components, which are higher in the hierarchy of controls, are fewer in number. This is what generates a balance in power among ruler and ruled on the large scale. The same reasoning applies to the mechanisms generating stability in ecosystems given by H.T. Odum (1971, 1996). The representation of Odum is based on the requirement of a given balance between the mass and rate of energy dissipation of different components of ecosystems. An ecosystem that is in equipollence must have a given ratio between different components (due to the near closure of its matter cycles). Equipollence means that a higher level of energy dissipation, which is found in ecosystem compartments higher in the hierarchy (e.g., 1 kg of biomass of tigers), has to be coupled with a smaller size of such a compartment, when compared with the others stabilizing the whole structure in the lower levels (you must have a lot of kilgrams of biomass of plants per kilograms of biomass of tiger). At this point, we can recall Koestler's idea of a memory for forgetting (our brain when storing concepts tends to remove redundant information — irrelevant details — that are considered, because of their redundancy, as storing less valuable information). Margaleff (1968) suggests the very same idea in relation to the functioning, evolution and organization of ecosystems over hierarchical compartments. In their daily functioning, the different components of an ecosystem, which are organized into trophic chains, literally eat each other at different speeds. In this operation, the more redundant information (the one belonging to the larger compartments made up of a larger number of copies of the same organized pattern) is eaten more to stabilize less redundant information stored on the top of the hierarchy (the one stored in the components making up higher levels). The equipollence of various components of ecosystems, in this frame, can be directly related to the goal of obtaining a balance between functional specialization (increasing the efficacy of metabolic processes at any particular level) and functional integration (increasing the stability of the integrated network in relation to changes on each level and in the larger context).

The NESH network (with D. Waltner-Toews, J. Kay and D.R. Cressman as founding members) working on the concept of ecosystem health, proposes a similar concept to describe sick holarchies. A holarchy is no longer healthy when it loses its ability to share stress among different holons (balancing the focus between horizontal and vertical coupling, maintaining equipollence, being able to act under double asymmetric relations). This happens when policy indications developed according to one perspective (based on the description of what goes on at one level) become dominant over other legitimate and contrasting policy indications coming from descriptions referring to other holons (what has been called decomplexification in human health). This is a sign that the continuous negotiation and interaction among holons within the holarchy (which is needed to avoid the process of decomplexification of the existing dynamics and controls) is no longer effective. Incidentally, we can remember here the point made by Bohm (1995, p. 3) that the word *health* in English is based on an Anglo-Saxon word *hale* meaning "whole." The philosophers of science Jerome Ravetz and Silvio Funtowicz call this same phenomenon hegemonization, which can lead (if not corrected in time) to a state of ancient regime syndrome. When reaching the ancient regime syndrome, adaptive holarchies lose their ability to read and interpret signals coming from the context about the need and urgency of changing their own set of identities. This is due to a systemic error that is generated by the excess of power of higher-level components over the rest of the holarchy. This imbalance of power implies that gradients between expected and experienced patterns occurring in lower levels of the holarchy are ignored (filtered) in the internal communications. Holons on the top manage the whole structure by relying on their perception, which is based on the stored representation of the old identities. In fact, at the higher level, the reality looks different than it does at the lower level. This is due to the active filtering of information that any holarchic structure implies. However, if this systemic error is not corrected in time, this mismatch between what is experienced at the lower level and what is imagined (or denied) at the higher level will make the holarchic structure brittle (less adaptable), and therefore more fragile to even small perturbations. A socioeconomic system in the ancient regime syndrome is much more likely to get into catastrophic events (Tainter, 1988).

References

Allen, T.F.H., Tainter, J.A., Pires J.C., and Hoekstra, T.W., (2001), Dragnet ecology, "just the facts ma'am": the privilege of science in a post-modern world, *Bioscience*, 51, 475–485.

Bak, P., (1996), *How Nature Works: The Science of Self-Organized Criticality*, Copernicus Books/Springer-Verlag, New York, 212 pp.

Bohm, D., (1987), *Causality and Chance in Modern Physics*, University of Pennsylvania Press, Philadelphia, 184 pp.

Bohm, D., (1995), *Wholeness and the Implicate Order*, 3rd ed., Routledge, London, 224 pp.

Chaitin, G.C., (1987), *Algorithmic Information Theory*, Cambridge Tracts in Theoretical Computer Science Vol. 1, Cambridge University Press, Cambridge, U.K.

Duncan, D.E., (1999), *Calendar: Humanity's Epic Struggle to Determine a True and Accurate Year*, Bart Books, London.

Durnin, J.V.G.A., and Passmore, R. (1967), *Energy, Work and Leisure*. Heinemann: London.

Dyke, C., (1988), *The Evolutionary Dynamics of Complex Systems: A Study in Biosocial Complexity*, Oxford University Press, New York.

Georgescu-Roegen, N., (1975), Energy and economic myths, *South. Econ. J.*, 41, 347–381.

Giampietro, M. and Pimentel, D., (1991), Energy efficiency: assessing the interaction between humans and their environment, *Ecol. Econ.*, 4, 117–144.

Giampietro M. and Mayumi K. (2000a), Multiple-scale integrated assessment of societal metabolism: Introducing the approach. *Popul. Environ.* 22(2): 109–153.

Giampietro, M. and Mayumi K. (2000b), Multiple-Scale Integrated Assessment of Societal Metabolism: integrating biophysical and economic representation across scales. *Popul. Environ.* 22(2), 155–210.

Goldberger, A.L., (1997), Fractal variability versus pathologic periodicity: complexity loss and stereotypy in disease, *Perspect. Biol. Med.*, 40, 543–561.

Grene, M., (1969), Hierarchy: one word, how many concepts? in *Hierarchical Structures*, Whyte, L.L., Wilson, A.G., and Wilson, D., Eds., American Elsevier Publishing Company, New York, pp. 56–58.

Iberall, A.H., Soodak H., and Arensburg, C., (1981), Homeokinetic physics of societies: a new discipline: autonomous groups, cultures, politics, in *Perspectives in Biomechanics*, Part 1, Vol. 1, Real, H., Ghista, D., and Rau, G., Eds., Harwood Academic Press, New York.

James, W.P.T. and Schofield, E.C., (1990), *Human Energy Requirement*, Oxford University Press, Oxford.

Koestler, A. (1968), *The Ghost in the Machine*. MacMillan. New York.

Koestler, A. (1978), *Janus: A Summing Up*. Hutchinson, London.

Lotka, A.J., (1956), *Elements of Mathematical Biology*, Dover Publications, New York.

Margalef, R., (1968), *Perspectives in Ecological Theory*, University of Chicago Press, Chicago.

Neurath, O., (1973), *Empiricism and Sociology*, Reidel, Dordrecht.

Odum, H.T., (1971), *Environment, Power, and Society*, Wiley Interscience, New York.

Odum, H.T., (1983), *Systems Ecology*, John Wiley, New York.

Odum, H.T., (1996), *Environmental Accounting: Emergy and Decision Making*, John Wiley, New York.

O'Neill, R.V., Hunsaker, C.T., Jones, K.B., Riitters, K.H., Wickham, J.D., Schwartz, P.M., Goodman, I.A., Jackson, B.L., and Baillargeon, W.S., (1997), Monitoring environmental quality at the landscape scale, *Bioscience*, 47, 513–519.

Prueitt, P.S., (1998), *Manhattan Project*, George Washington University BCN Group, available at http://www.bcngroup.org/area3/manhattan/manhattan.html.

Rosen, R. (1985), *Anticipatory Systems: Philosophical, Mathematical and Methodological Foundations*. Pergamon Press, New York.

Tainter, J., (1988), *The Collapse of Complex Societies*, Cambridge University Press, Cambridge, U.K.

Zipf, G.K., (1941), *National Unity and Disunity: The Nation as a Bio-Social Organism*, The Principia Press, Bloomington, IN.

7

Impredicative Loop Analysis: Dealing with the Representation of Chicken–Egg Processes*

This chapter first introduces the concept of the impredicative loop (Section 7.1) in general terms. Then, to make easier the life of readers not interested in hard theoretical discussions, additional theory has been omitted from the main text. Therefore, Section 7.2 provides examples of applications of impredicative loop analysis (ILA) to three metabolic systems: (1) preindustrial socioeconomic systems, (2) societies basing their metabolism on exosomatic energy and (3) terrestrial ecosystems. Section 7.3 illustrates key features and possible applications of ILA as a heuristic approach to be used to check and improve the quality of multi-scale integrated analyses. That is, this section shows that ILA can be used as a meta-model for the integrated analysis of metabolic systems organized in nested hierarchies. The examples introduced in this section will be integrated and illustrated in detail in Part 3, dealing with multi-scale integrated analysis of agroecosystems. The chapter ends with a two technical sections discussing theoretical aspects of ILA. The first of these two sections (Section 7.4) provides a critical appraisal of conventional energy analysis — an analytical tool often found in scientific analyses of sustainability of agroecosystems. Such a criticism is based on hierarchy theory. The second section (Section 7.5) deals with the perception and representation of autocatalytic loops of energy forms from a thermodynamic point of view (nonequilibrium thermodynamics). In particular, we propose an interpretation of ILA, based on the rationale of negative entropy, that was provided by Schroedinger and Prigogine in relation to the class of dissipative systems. Even though these last two sections do not require any mathematical skills to be followed, they do require some familiarity with basic concepts of energy analysis and nonequilibrium thermodynamics. In spite of this problem, in our view, these two sections are important since they provide a robust theoretical backup to the use of ILA as a meta-model for dealing with sustainability issues.

7.1 Introducing the Concept of Impredicative Loop

Impredicativity has to do with the familiar concept of the chicken–egg problem, or what Bertrand Russel called the vicious circle (quoted in Rosen, 2000, p. 90). According to Rosen (1991), impredicative loops are at the very root of the essence of life, since living systems are the final cause of themselves. Even the latest developments of theoretical physics — e.g., superstring theory — represent a move toward the very same concept. Introducing such a theory, Gell-Mann (1994) makes first reference to the *bootstrap principle* (based on the old saw about the man that could pull himself up by his own bootstraps) and then describes it as follows: "the particles, if assumed to exist, produce forces binding them to one another; the resulting bound states are the same particles, and they are the same as the ones carrying the forces. Such a particle system, if it exists, gives rise to itself" (Gell-Mann, 1994, p. 128). The passage basically means that you have to assume the existence of a chicken to get the egg that will generate the chicken, and vice versa. As soon as the various elements of the self-entailing process — defined in parallel on different levels — are at work, such a process is able to define (assign an identity) to itself. The representation of this process, however, requires considering processes and identities that can only be perceived and represented by adopting different space–time scales.

---

* Kozo Mayumi is co-author of this chapter.

A more technical definition of impredicativity provided by Kleene and related more to the epistemological dimension is reported by Rosen (2000, p. 90):

> When a set *M* and a particular object *m* are so defined that on the one hand *m* is a member of *M*, and on the other hand the definition of *m* depends on *M*, we say that the procedure (or the definition of *m*, or the definition of *M*) is impredicative. Similarly when a property *P* is possessed by an object *m* whose definition depends on *P* (here *M* is the set of objects which possess the property *P*), an impredicative definition is circular, at least on its face, as what is defined participates in its own definition. (Kleene, 1952, p. 42)

It should be noted that impredicative loops are also found in the definition of the identity of crucial concepts in many scientific disciplines. In biology, the example of the definition of the mechanism of natural selection is well known (the survival of the fittest, in which the "fittest" is then defined as "the surviving one"). The same mechanism is found in the basic definition of the first law of dynamics (F = m × a), in which the force is defined as what generates an acceleration over a mass, whereas an acceleration is described, using the same equation, as the result of an application of a force to a given mass. Finally, even in economics we can find the same apparently tautological mechanism in the well-known equation P × Y = M × V (price level times real gross national product (GNP) equal to amount of money times velocity of money circulation), in which the terms define and are defined by each other.

Impredicative loops can be explored by explicitly acknowledging the fact that they are in general occurring across processes operating (perceived and represented) in parallel over different hierarchical levels. That is, definitions based on impredicative loops refer to mechanisms of self-entailment operating across levels and that therefore require a set of representations of events referring to both parts and wholes in parallel over different scales. Exactly because of that, as it is discussed in the technical Section 7.4, they are out of the reach of reductionist analyses. That is, they are out of the reach of analytical tools developed within a paradigm that assumes that all the phenomena of the reality can be described within the same descriptive domain, just by using a set of reducible models referring to the same substantive definition of space and time. However, this does not imply that impredicative loops cannot be explored by adopting an integrated set of nonequivalent and nonreducible models. That is, by using a set of different models based on the adoption of nonequivalent descriptive domains (nonreducible definition of space and time in formal terms — as discussed by Rosen (1985) and in the technical section at the end of this chapter), it is possible to study the existence of an integrated set of constraints. These constraints are generated by the reciprocal effect of agency on different levels (across scales) and are referring to different relevant characteristics of the process (across disciplinary fields). The feasibility of an impredicative loop, with this approach, can be checked on different levels by using nonreducible models taking advantage of the existence of mosaic effects across levels (Giampietro and Mayumi, 2000a, 2000b; Giampietro et al., 2001).

However, this approach requires giving up the idea of using a unique narrative and a unique formal system of inference to catch the complexity of reality and to simulate the effects of this multi-scale self-entailment process (Rosen, 2000). Giving up this reductionist myth does not leave us hopeless. In fact, the awareness of the existence of reciprocal constraints imposed on the set of multiple identities expressed by complex adaptive holarchies (the existence of different dimensions of viability, e.g., chemical constraints, biochemical constraints, biological constraints, economic constraints, sociocultural constraints) can be used to do better analyses.

7.2 Examples of Impredicative Loop Analysis of Self-Organizing Dissipative Systems

7.2.1 Introduction

With the expression "impredicative loop analysis" we want to suggest that the concept of impredicative loop can be used as a heuristic tool to improve the quality of the scientific representation of complex

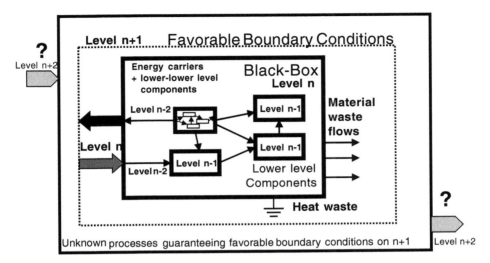

FIGURE 7.1 Hierarchical levels that should be considered for studying autocatalytic loops of energy forms.

systems organized in nested hierarchies. The approach follows a rationale that represents a major bifurcation from the conventional reductionist approach. That is, the main idea is that first of all it is crucial to address the semantic aspect of the analysis. This implies accepting a few points that are consequences of what was presented in Part 1:

1. The definition of a complex dissipative system, within a given problem structuring, entails considering such a system to be a whole made of parts and operating in an associative context (which must be an admissible environment). In the step of representation this implies establishing a set of relations among a set of formal identities referring to at least five different hierarchical levels of analysis: (1) level $n - 2$, subparts; (2) level $n - 1$, parts; (3) level n, the whole black box; (4) level $n + 1$, an admissible context; and (5) level $n + 2$, processes in the environment that guarantee the future stability of favorable boundary conditions associated with the admissible context of the whole. An overview of such a hierarchical vision of an autocatalytic loop of energy forms is given in Figure 7.1. This representation can be directly related to the discussion in Chapter 6 about multi-scale mosaic effects for metabolic systems organized in nested hierarchies.

2. It is always possible to adopt multiple legitimate nonequivalent representations of a given system that are reflecting its ontological characteristics. Therefore, the choice of just one particular representation among the set of potential representations reflects not only characteristics of the observed system, but also characteristics of the observer (goals of the analysis, relevance of system's qualities included in the semantic identity, credibility of assumptions about the models, congruence of nonequivalent perceptions of causal relations in different descriptive domains).

3. A given problem structuring (the system and what it does in its associative context) reflects an agreement about how to perceive and represent a complex adaptive holarchy in relation to the choices of (1) a set of semantic identities (what is relevant for the observer about the observed) and (2) an associated set of formal identities (what can be observed according to available detectors and measurement schemes), which will be reflected into the selection of variables used in the model. It is important to notice that such an agreement about what is the system and what the system is doing in its context is crucial to get into the following step of selection of formal identities (individuation of variables used as proxies for observable qualities). Prior to reaching such an agreement about how to structure in scientific terms the problem of how to represent the system of interest, experimental data do not count as relevant information.

That is, before having a valid (and agreed-upon) problem structuring that will be used to represent the complex system using different models referring to different scales and different descriptive domains, data per se do not exist. The possibility of using data requires a previous validated definition of (1) what should be considered relevant system qualities, (2) which observable system qualities should be used as proxies of these relevant qualities and (3) what is the set of measurement schemes that can be used to assign values to the variables, which then can be used in formal models to represent the system's behavior. The information provided by data therefore always reflects the choices made when defining the set of formal identities adopted in the representation of the reality by the analyst.

Sometimes scientists are aware of the implications of these preanalytical choices, and sometimes they are not. Actually, the most important reason for introducing complex systems thinking is increasing the transparency about hidden implications associated with the step of modeling. The approach of impredicative loop analysis is aimed at addressing this issue. The meat of ILA is about forcing a semantic validity check over the set of formal identities adopted in the phase of representation by those making models.

To obtain this result, it is necessary to develop meta-models that are able to establish typologies of relations among parts and wholes, which can be relevant and useful when dealing with a class of situations. Useful meta-models can be applied, later on, to special (individual) situations belonging to a given typology. These meta-models, to be useful, have to be based on a standard characterization of the mechanism of self-entailment among identities of parts, whole and context, defined on different levels. Actually, this is exactly what is implied by the very concept of impredicative loop. Looking for meta-models, however, implies accepting the consequence that any impredicative loop does have multiple possible formalizations. That is, the same procedure for establishing relations among identities of parts and the whole within a given impredicative loop can be interpreted in different ways by different analysts, even when applied to the same system considered at the same point in space and time. Meta-models, by definition, generate families of models based on the adoption of different sets of congruent formalization of identities. Obviously, at the moment of selecting an experimental design (or a specific system of accounting), we will have to select just one particular model to be adopted (to gather experimental data) and stick with it. Experimental work is based on the selection of just one of the possible formalizations of the meta-model, applied at a specific point in space and time.

This transparent arbitrariness of models that are built in this way should not be considered a weakness of this approach. On the contrary, in our view, this should be considered a major strength. In fact, after acknowledging from the beginning the existence of an open space of legitimate options, analysts coming from different disciplinary backgrounds, cultural contexts or value systems are forced to deal, first of all and mainly, with the preliminary discussion of semantic aspects associated with the selection of models. This certainly facilitates a discussion about the usefulness of models and enhances the awareness of crucial epistemological issues to be considered at the moment of selecting experimental designs.

Below we provide three practical examples of dissipative systems: (1) a preindustrial society of 100 people on a desert island, (2) a comparison of the trajectory of development of two modern societies that base the metabolism of their economic process on exosomatic energy (Spain and Ecuador), and (3) the dynamic budget stabilizing the metabolism of terrestrial ecosystems. For the moment, we just describe how it is possible to establish a relation between characteristics of parts and the whole of these systems in relation to their associative contexts. Common features of the three analyses will be discussed in Section 7.3. More general theoretical aspects are discussed in Section 7.5.

7.2.2   Example 1: Endosomatic Societal Metabolism of an Isolated Society on a Remote Island

7.2.2.1   *Goals of the Example* — As noted earlier, the ability to keep a dynamic equilibrium between requirement and supply of energy carriers (e.g., how much food must be eaten vs. how much food can be produced in a preindustrial society) entails the existence of a biophysical constraint on the relative sizes and characteristics of various sectors making up such a society. The various activities linked

to both production and consumption must be congruent in terms of an analysis based on a combined use of intensive and extensive variables across levels (mosaic effects across levels — Chapter 6). That is, we can look at the reciprocal entailment among the definitions of size and characteristics of a metabolic system organized on nested hierarchical levels (parts and whole). Then we can relate it to the aggregate effect of this interaction on the environment. This is what we call an impredicative loop analysis.

Coming to this first example, we want to make it immediately clear to the reader that the stability of any particular societal metabolism does not depend only on the ability of establishing a dynamic equilibrium between requirement and supply of food. The stability of a given human society can be checked in relation to a lot of other dimensions — i.e., alternative relevant attributes and criteria. For example, is there enough drinking water? Can the population reproduce in the long term according to an adequate number of adult males and females? Are the members of the society able to express coordinated behavior to defend themselves against external attacks? Indeed, using an analysis that focuses only on the dynamic equilibrium between requirement and supply of food is just one of the many possible ways for checking the feasibility of a given societal structure.

However, given the general validity of the laws of thermodynamics, such a check cannot be ignored. As a matter of fact, the same approach (checking the ability of obtaining a dynamic equilibrium between requirement and supply) can be applied in parallel to different mechanisms of mapping that can establish forced relations among flows and sizes of compartments and wholes across levels, in relation to different flows (as already illustrated in Chapter 6), to obtain integrated analysis. The reader can recall here the example of the various medical tests to be used in parallel to check the health of a patient (Figure 6.3). In this first example of impredicative loop analysis we will look at the dynamic budget of food energy for a society. This is like if we were looking at the bones — using x-rays — of our patient. Other types of impredicative loop analysis (next two examples) could represent nonequivalent medical tests looking at different aspects of the patient (e.g., ultrasound scan and blood test). What is important is to have the possibility, later on, to have an overview of the various tests referring to nonequivalent and nonreducible dimensions of performance. This is done, for example, in Figure 7.6, which should be considered an analogous to Figure 6.3.

7.2.2.2   *The Example* — As soon as we undertake an analysis based on energy accounting, we have to recognize that the stabilization of societal metabolism requires the existence of an autocatalytic loop of useful energy (the output of useful energy is used to stabilize the input). In this example, we characterize the autocatalytic loop stabilizing societal metabolism in terms of reciprocal entailment of the two resources: human activity and food (Giampietro, 1997). The term *autocatalytic loop* indicates a positive feedback, a self-reinforcing chain of effects (the establishment of an egg–chicken pattern). Within a socioeconomic process we can define the autocatalytic loop as follows: (1) The resource human activity is needed to provide control over the various flows of useful energy (various economic activities in both producing and consuming), which guarantee the proper operation of the economic process (at the societal level). (2) The resource food is needed to provide favorable conditions for the process of reproduction of the resource human activity (i.e., to stabilize the metabolism of human societies when considering elements at the household level). (3) The two resources, therefore, enhance each other in a chicken–egg pattern. In this example we are studying the possibility of using the impredicative loop analysis related to the self-entailment of identities of parts and the whole, which are responsible for stabilizing the autocatalytic loop of two energy forms: chemical energy in the food and human activity expressed in terms of muscle and brain power.

Within this framework our heuristic approach has the goal of establishing a relation between a particular set of parameters determining the characteristics of this autocatalytic loop as a whole (at level n) and a particular set of parameters that can be used to describe the characteristics of the various elements of the socioeconomic system at a lower level (level $n - 1$). These characteristics can be used to establish a bridge with technological changes (observed on the interface of level $n - 1$/level $n - 2$) and to effect changes on environmental impact at the interface — level n/level $n + 1$ (see Figure 7.1).

In this simplified example, we deal with an endosomatic autocatalytic loop (only human labor and food) referring to a hypothetical society of 100 people on an isolated, remote island. The numbers given in this example per se are not the relevant part of the analysis. As noted earlier, no data set is relevant without a previous agreement of the users of the data set about the relevance of the problem structuring (in relation to a specific analysis performed in a specific context). We are providing numbers — which are familiar for those dealing with this topic — just to help the reader to better grasp the mechanism of accounting. It is the forced relation among numbers (and the analysis of the mechanism generating this relation) that is the main issue here. Different analysts can decide to define the relations among the parts and the whole in different ways, and therefore this could lead to a different definition of the data set. However, when adopting this approach, they will be asked by other analysts about the reasons for their different choices. This then will require discussing the meaning of the analysis.

The following example of ILA presenting a useful metaphor (meta-model) for studying societal metabolism has two major goals:

1. To illustrate an approach that makes it possible to establish a clear link between the characteristics of the societal metabolism as a whole (characteristics referring to the entire loop — level n) and a set of parameters controlling various steps of this loop (characteristics referring to lower-level elements and higher-level elements —defined at either level $n - 1$ or level $n + 1$). Moreover, it should be noted that the parameters considered in this analysis are those generally considered, by default, as relevant in the discussion about sustainability (e.g., population pressure, material standard of living, technology, environmental loading). This example clearly shows that these parameters are actually those crucial in determining the feasibility of the autocatalytic loop, when characterized in terms of impredicative loop analysis.

2. To illustrate the importance of closing the loop when describing societal metabolism in energy terms, instead of using linear representations of energy flows in the economic process (as done with input/output analyses). In fact, the conventional approach usually adopted in energy analysis, based on conventional wisdom, keeps its focus on the consideration of a unidirectional flow of energy from sources to sinks (the gospel says "while matter can be recycled over and over, energy can flow only once and in one direction"). As discussed in Section 7.4, a linear representation of energy flows in terms of input/output assessments cannot catch the reciprocal effect across levels and scales that the process of energy dissipation implies (Giampietro and Pimentel, 1991a; Giampietro et al., 1997). In fact, it is well known that in complex adaptive systems, the dissipation of useful energy must imply a feedback, which tends to enhance the adaptability of the system of control (Odum, 1971, 1983, 1996). Assessing the effect of such feedback, however, is not simple because this feedback can only be detected and represented on a descriptive domain that is different (larger space–time scale) from the one used to assess inputs, outputs and flows (as discussed at length in Sections 7.4 and 7.5). This is what Georgescu-Roegen (1971) describes as the impossibility to perform an analytical representation of an economic process when several distinct time differentials are required in the same analytical domain. Actually, he talks of the existence of incompatible definitions of duration for parallel input/output processes (the replacement of the term *duration* with the term *time differentials* is ours). Our ILA of the 100 people on the remote island provides practical examples of this fact.

The representations given in Figure 6.6 of how endosomatic energy flows in a society is a classic example of the conventional linear view. Energy flows are described as unidirectional flows from left to right (from primary sources to end uses). However, it is easy to note that some of the end uses of energy (indicated on the right side) are necessary for obtaining the input of energy from primary energy sources (indicated on the left side) in the first place. That is, the stabilization of a given societal metabolism is linked to the ability to establish an egg–chicken pattern within flows of energy. In practical terms, when dealing with the endosomatic metabolism of a human society, a certain fraction of end uses (e.g., in Figure 6.6, the physical activity "work for food") must be available and used to produce food. The expression autocatalytic loop actually indicates the obvious fact that some of the end uses must reenter into the system as input to sustain the overall metabolism. This is what implies the existence of internal constraints on possible structures of socioeconomic systems. In practical terms, when dealing with the endosomatic metabolism of a human society, a certain fraction of the end uses must be available and used to produce food before the input enters into the system (as indicated on the lower axis of Figure 7.2).

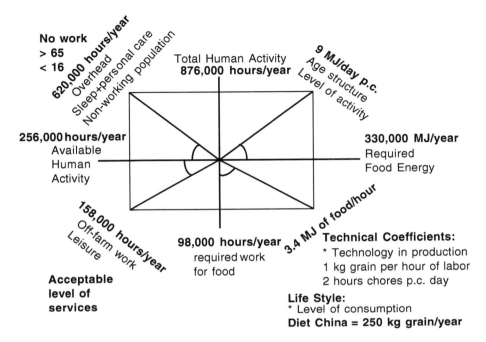

FIGURE 7.2 One hundred people on a remote island.

7.2.2.3 Assumptions and Numerical Data for This Example — We hypothesize that a

society of 100 people uses only flows of endosomatic energy (food and human labor) for stabilizing its own metabolism. To further simplify the analysis, we imagine that the society is operating on a remote island (survivors of a plane crash). We further imagine that its population structure reflects the one typical of a developed country and that the islanders have adopted the same social rules regulating access to the workforce as those enforced in most developed countries (that is, persons under 16 and those over 65 are not supposed to work). This implies a dependency ratio of about 50%; that is, only 50 adults are involved in the production of goods and social services for the whole population. We finally add a few additional parameters needed to characterize societal metabolism. At this point the forced loop in the relation between these numerical values is described in Figure 7.2:

- **Basic requirement of food.** Using standard characteristics of a population typical of developed countries, we obtain an average demand of 9 MJ/day/capita of food, which translates into 330,000 MJ/year of food for the entire population.
- **Indicator of material standard of living.** We assume that the only "good" produced and consumed in this society (without market transactions) is food providing nutrients to the diet. In relation to this assumption we can then define two possible levels of material standard of living, related to two different qualities for the diet. The two possible diets are: (1) Diet A, which covers the total requirement of food energy (3300 MJ/year/capita) using only cereal (supply of only vegetal proteins). With a nutritional value of 14 MJ of energy/kg of cereal, this implies the need to produce 250 kg of cereal/year/capita. (2) Diet B, which covers 80% of the requirement of food energy with cereal (190 kg/year per capita (p.c.)) and 20% with beef (equivalent to 6.9 kg of meat/year p.c.). Due to the very high losses of conversion (to produce 1 kg of beef you have to feed the herd 12 kg of grain), this double conversion implies the additional production of 810 kg of cereal/year. That is, Diet B requires the primary production of 1000 kg of cereal/capita (rather than 250 kg/year of Diet A). Actually, the value of 1000 kg of cereal consumed per capita, in indirect form in the food system, is exactly the value found in the U.S. today (see the relative assessment in Figure 3.1).

- **Indicator of technology.** This reflects technological coefficients, in this case, labor productivity and land productivity of cereal production. Without external inputs to boost the production, these are assumed to be 1000 kg of cereal/hectare and 1 kg of cereal/hour of labor.

- **Indicator of environmental loading.** A very coarse indicator of environmental loading can be assessed by the fraction land in production/total land of the island, since the land used for producing cereal implies the destruction of natural habitat (replaced with the monoculture of cereal). In our example the indicator of environmental loading is heavily affected by the type of diet followed by the population (material standard of living) and the technology used. Assuming a total area for the island of 500 ha, we have an index of EL = 0.05 for Diet A and EL = 0.20 for Diet B (EL = hectares in production/total hectares available on the island).

- **Supply of the resource human activity.** We imagine that the required amount of food energy for a year (330,000 MJ/year) is available for the 100 people for the first year (assume it was in the plane). With this assumption, and having the 100 people to start with, the conversion of this food into endosomatic energy implies (it is equivalent to) the availability of a total supply of human activity of 876,000 h/year (24 h/day × 365 × 100 persons).

- **Profile of investment of human activity of a set of typologies of end uses of human activity** (as in Figure 7.2). These are:

  1. *Maintenance and reproduction* — It should be noted that in any human society the largest part of human activity is not related to the stabilization of the societal metabolism (e.g., in this case producing food), but rather to maintenance and reproduction of humans. This fixed overhead includes:

     a. Sleeping and personal care for everybody (in our example, a flat value of 10 h/day has been applied to all 100 people, leading to a consumption of 365,000 h/year of the total human activity available).

     b. Activity of nonworking population (the remaining 14 h/day of elderly and children, which are important for the future stability of the society, but which are not available — according to the social rule established before — for the production of food). This indicates the consumption of another 255,000 h/year (14 × 50 × 365) in nonproductive activities.

  2. *Available human activity for work* — The difference between total supply of human activity (876,000 h) and the consumption related to the end use maintenance and reproduction (620,000 h) is the amount of available human activity for societal self-organization (in our example, 256,000 h/year). This is the budget of human activity available for stabilizing societal metabolism. However, this budget of human activity, expressed at the societal level, has to be divided between two tasks:

     a. Guaranteeing the production of the required food input (to avoid starvation) — work for food

     b. Guaranteeing the functioning of a good system of control able to provide adaptability in the future and a better quality of life to the people — social and leisure

At this point, the circular structure of the flows in Figure 7.2 enters into play. The requirement of 330,000 MJ/year of endosomatic energy input (food at time t) entails the requirement of producing enough energy carriers (food at time t + 1) in the following years. That is a biophysical constraint on the level of productivity of labor in the activity producing food. Therefore, this characteristic of the whole (the total demand of the society) translates into a nonnegotiable fraction of investment of available human activity in the end use work for food (depending on technology and availability of natural resources). This implies that the disposable fraction of available human activity, which can be allocated to the end use social and leisure, is not a number that can be decided only according to social or political will. The circular nature of the autocatalytic loop implies that numerical values associated with the characterization of various identities defining elements on different hierarchical levels (at the level of individual compartments; extensive — segments on the axis — and intensive variables; wideness of angles) can be changed, but only respecting the constraint of congruence among flows over the whole loop. These constraints are imposed on each other by the characteristics and size — extensive (1 and 2) and intensive (3) variables — of the various compartments.

7.2.2.4   Changing the Value of Variables within Formal Identities within a Given
Impredicative Loop — Imagine to change, for example, some of the values used to characterize this autocatalytic loop of energy forms. For example, let us change the parameter "material standard of living," which in our simplified model is expressed by a formal definition of quality of the diet. The different mix of energy vectors in the two diets (vegetal vs. animal proteins) implies a quantitative difference in the biophysical cost of the diet expressed in terms of both a larger work requirement and a larger environmental loading (higher demand of land). The production of cereal for a population relying 100% on Diet A requires only 25,000 h of labor and the destruction of 25 ha of natural habitat ($EL_A = 0.05$), whereas the production of cereal for a population relying 100% on Diet B requires 100,000 h of labor and the destruction of 100 ha of natural habitat ($EL_B = 0.20$). However, to this work quantity required for producing the agricultural crop, we have to add a requirement of work for fixed chores. Fixed chores are preparation of meals, gathering of wood for cooking, getting water, and washing and maintenance of food system infrastructures in the primitive society. In this example we use the same flat value for the two diets — 73,000 h/year (2 h/day/capita = 2 × 365 × 100). This implies that if all the people of the island decide to follow Diet A, they will face a fixed requirement of "work for food" of 98,000 h/year. If they all decide to adopt Diet B, they will face a fixed requirement of "work for food" of 173,000 h/year. At this point, for the two options we can calculate the amount of disposable available human activity that can be allocated to social and leisure. It is evident that the amount of time that the people living in our island can dedicate to running social institutions and structures (schools, hospitals, courts of justice) and developing their individual potentialities in their leisure time in social interactions is not the result of their free choice. Rather, it is the result of a compromise between competing requirements of the resource "available human activity" in different parts of the economic process.

That is, after assigning numerical values to social parameters such as population structure and a dependency ratio for our hypothetical population, we have a total demand of food energy (330,000 MJ/year) and a fixed overhead on the total supply of human activity, which implies a flat consumption for maintenance and reproduction (620,000 h/year). Assigning numerical values to other parameters, such as material standard of living (Diet A or Diet B) and technical coefficients in production (e.g., labor, land and water requirements for generating the required mix of energy vectors), implies defining additional constraints on the feasibility of such a socioeconomic structure. These constraints take the form of (1) a fixed requirement of the resource "available human activity" that is absorbed by "work for food" (98,000 h for Diet A and 173,000 h for Diet B) and (2) a certain level of environmental loading (the requirement of land and water, as well as the possible generation of wastes linked to the production), which can be linked, using technical coefficients, to such a metabolism (in our simple example we adopted a very coarse formal definition of identity for environmental loading that translates into $EL_A = 0.05$ and $EL_B = 0.20$).

With the term *internal biophysical constraints* we want to indicate the obvious fact that the amount of human activity that can be invested into the end uses "maintenance and reproduction" and "social and leisure" depends only in part on the aspirations of the 100 people for a better quality of life in such a society. The survival of the whole system in the short term (the matching of the requirement of energy carriers' input with an adequate supply of them) can imply forced choices (Figure 7.3). Depending on the characteristics of the autocatalytic loop, large investments of human activity in social and leisure can become a luxury. For example, if the entire society (with the set of characteristics specified above) wants to adopt Diet B, then for them it will not be possible to invest more than 83,000 h of human activity in the end use "social and leisure." On the other hand, if they want, together with a good diet, also a level of services typical of developed countries (requiring around 160,000 h/year/100 people), they will have to "pay" for that. This could imply resorting to some politically important rules reflecting cultural identity and ethical believes (what is determining the fixed overhead for maintenance and reproduction). For example, to reach a new situation of congruence, they could decide to either introduce child labor or increase the workload for the economically active population (e.g., working 10 h a day

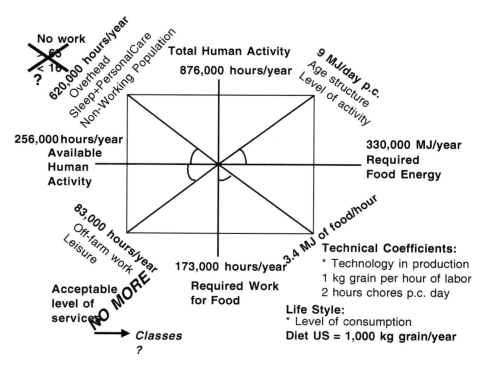

FIGURE 7.3 One hundred people on a remote island.

for 6 days a week) (Figure 7.3). Alternatively, they can accept a certain degree of inequity in the society (a small fraction of people in the ruling social class eating Diet B and a majority of the ruled eating Diet A). We can easily recognize that all these solutions are today operating in many developing countries and were adopted, in the past, all over our planet.

7.2.2.5 Lessons from This Simple Example

The simple assumptions used in this example for bringing into congruence the various assessments related to a dynamic budget of societal metabolism are of course not realistic (e.g., nobody can eat only cereal in one's diet, and expected changes in the requirements of work are never linear). Moreover, by ignoring exosomatic energy, we do not take in account the effect of capital accumulation (e.g., potential use of animals, infrastructures, better technology and know-how affecting technical coefficients), which is relevant for reaching new feasible dynamic points of equilibrium of the endosomatic energy budget. That is, alternative points of equilibrium can be reached by, besides changing population structure and size, changing technology (and the quality of natural resources). Actually, it is easy to make models for preindustrial societies that are much more sophisticated than the one presented in Figure 7.2: models that take into account different landscape uses, detailed profiles of human time use, and reciprocal effects of changes on the various parameters, such as the size and age distribution of society (Giampietro et al., 1993). These models, after entering real data derived from specific case studies, can be used for simulations, exploring viability domains and the reciprocal constraining of the various parameters used to characterize the endosomatic autocatalytic loop of these societies. However, models dealing only with the biophysical representation of endosomatic metabolism and exosomatic conversions of energy are not able to address the economic dimension. Economic variables reflect the expression of human preferences within a given institutional setting (e.g., an operating market in a given context) and therefore are logically independent from assessment reflecting biophysical transformations.

Even within this limitation, the example of the remote island clearly shows the possibility of linking the representation of the conditions determining the feasibility of the dynamic energy budget of societal metabolism to a set of key parameters used in the sustainability discussions. In particular, characterizing societal metabolism in terms of autocatalytic loops makes it possible to establish relations among changes occurring in parallel in various parameters, which are reflecting patterns perceived on different levels and scales. For example, how much would the demand of land change if we change the definition of the diet? How much would the disposable available human activity change if we change the dependency ratio (by changing population structure or retirement age)? In this way, we can explore the viability domain of such a dynamic budget (what combination of values of parameters are not feasible according to the reciprocal constraints imposed by the other parameters).

A technical discussion of the sustainability of the dynamic energy budget represented in Figure 7.2 and Figure 7.3 in terms of potential changes in characteristics (e.g., either the values of numbers on axis or the values of angles) requires using in parallel trend analysis on nonequivalent descriptive domains. In fact, changes that are affecting the value taken by angles (intensive variables) or the length of segments on axes (extensive variables) require considering nonequivalent dynamics of evolutions reflecting different perceptions and representations of the system. These relations are those considered in the discussion about mosaic effect across levels in Chapter 6.

For example, if the population pressure and the geography of the island imply that the requirement of 100 ha of arable land are not available for producing 100,000 kg of cereal (e.g., a large part of the 500 ha of the island is too hilly), the adoption of Diet B by 100% of the population is simply not possible. The geographic characteristics of the island (defined at level $n + 2$) can be, in this way, related to the characteristics of the diet of individual members of the society (at level $n - 2$). This relation between shortage of land and poverty of the diet is well known. This is why, for example, all crowded countries depending heavily on the autocatalytic loop of endosomatic energy for their metabolism (such as India or China) tend to have a vegetarian diet. Still, it is not easy to define such a relation when adopting just one of these nonequivalent descriptive domains.

To make another hypothesis of perturbation within the ILA shown in Figure 7.2, imagine the arrival of another crashing plane with 100 children on board (or a sudden baby boom on the island). This perturbation translates into a dramatic increase of the dependency ratio. That is, a higher food demand, for the new population of 200 people would have to be produced by the same amount of 256,000 h of available human activity (related to the same 50 working adults). In this case, even when adopting Diet A, the larger demand of work in production will force such a society to dramatically reduce the consumption of human activity in the end use related to social and leisure. The 158,000 h/year, which were available to a society of 100 vegetarians (adopting 100% Diet A) for this end use — before the crash of the plane full of children — can no longer be afforded. This could imply that the society would be forced to reduce the investments of human activity in schools and hospitals (to be able to produce more food), at the very moment in which these services should be dramatically increased (to provide more care to the larger fraction of children in the population). This could appear as uncivilized behavior to an external observer (e.g., a volunteer willing to save the world in a poor marginal area of a developing country). This value judgment, however, can only be explained by the ignorance of such an external observer of the existence of biophysical constraints that are affecting the very survival of that society. Survival, in general, gets a higher priority than education.

The information used to characterize the impredicative loop that is determining the societal metabolism of a society translates into an organization of an integrated set of constraints over the value that can be taken by a set of variables (both extensive and intensive). In this way, we can facilitate the discussion and evaluation of possible alternative solutions for a given dynamic budget in terms of trade-off profiles. We earlier defined sustainability as a concept related to social acceptability, ecological compatibility, stability of social institutions, and technical and economic feasibility. Even when remaining within the limits of this simple example, we can see the integrative power of this type of multi-level integrated analysis. In fact, the congruence among the various numerical values taken by parameters characterizing the autocatalytic loop of food can be obtained by using different combinations of numerical values of variables defined at different hierarchical levels and reflecting different dimensions of performance. There are variables or parameters (e.g., technical coefficients) that refer to a very location-specific space–time scale (the yield of cereal at the plot level in a given year) and others (e.g., dependency ratio) that reflect biophysical processes (demographic changes) with a time horizon of changes of 20 years.

Finally, there are other variables or parameters (e.g., regulation imposed for ethical reasons, such as compulsory school for children) that reflect processes related to the specific cultural identity of a society.

For example, data used so far in this example about the budget of the resource human activity (for 100 people) reflect standard conditions found in developed countries (50% of the population economically active, working for 40 h/week × 47 weeks/year). Now imagine that for political reasons we are introducing a working week of 35 h (keeping five or six weeks of vacation per year) — a popular idea nowadays in Europe. Comparing this new value to previous workload levels, this implies moving from about 1800 to about 1600 h/year/active worker (work absences will further affect both). This reduction is possible only if this new value is congruent with the requirement imposed by technical coefficients (the requirement of work for food) and the existing level of investments/consumption in the end use "maintenance and reproduction." If this is not the case, depending on how strong is the political will of reducing the number of hours per week, the society has the option of altering some of the other parameters to obtain a new congruence. One can decide to increase the retirement age (by reducing the consumption of human activity by "maintenance and reproduction," that is, by reducing the amount of nonworking human activity associated with the presence of elderly in the population) or to decrease the minimum age required for entering the workforce (a very popular solution in developing countries, where children below 16 years generally work). Another solution could be that of looking for better technical coefficients (e.g., producing more kilograms of cereal per hour of labor), but this would require both a lag time to get technical innovations and an increase in investments of human work in research and development.

Actually, looking for better technical coefficients is the standard solution to all kinds of dilemmas about sustainability looked for in developed countries (since this makes it possible to avoid facing conflicts internal to the holarchy). This is what we called in Part 1 the search for silver bullets or win–win–win solutions. However, any solution based on the adding of more technology does not come without side effects. It requires adjustments all over the impredicative loop. Moreover, this solution could imply an increase in the environmental impact of societal metabolism (e.g., in our example, increasing the performance of monocultures could increase the environmental impact on the ecosystem of the island). Again, when we frame the discussion of these various options within the framework of integrated analysis of societal metabolism over an impredicative loop, we force the various analysts to consider, at the same time, several distinct effects (nonequivalent models and variables) belonging to different descriptive domains.

To make things more difficult, the consideration in parallel of different levels and scales can imply reversing the direction of causation in our explanations. That is, the direction of causality will depend on what we consider to be the independent definitions of identity (parameters) and the dependent definitions of identities (variables) within the impredicative loop (Figure 7.4). For example, looking at the four quadrants shown in Figure 7.4, we see that physiological characteristics (e.g., average body mass) can be given (e.g., in the example of the plane full of Western people crashing on the island, we are dealing with an average body mass of more than 65 kg for adults). On the other hand, if the average body mass is considered a dependent variable (e.g., in the long term, when adopting the hypothesis of "small and healthy" physiological adaptation to reduce food supply), we can expect that, as occurring in preindustrial societies, in the future we will find on this island adults with a much smaller average body mass. In the same way, the demographic structure can be a variable (when importing only adult immigrants, whenever a larger fraction of workforce is required) or a given constraint (when operating in a social system where emigration or immigration are not an option). The same applies to social rules (e.g., slavery can be abolished and declared immoral when no longer needed or used to boost the performance of the economy and the material standard of living of the masters). In the same way, what should be considered an acceptable level of service is another system quality that can be considered a dependent variable (e.g., if you are in a marginal social group forced to accept whatever is imposed on you) by the system. It becomes an independent variable, though, for groups that have the option to force their governments to do better or that have the option to emigrate. Technical coefficients can be seen as driving changes in other system qualities, when adopting a given timescale (e.g., population grew because better technology made available a larger food supply), or they can be seen as driven by changes in other system qualities when adopting another timescale (e.g., technology changed because population growth required a larger food supply). Every time the analyst decides to adopt a given formalization of

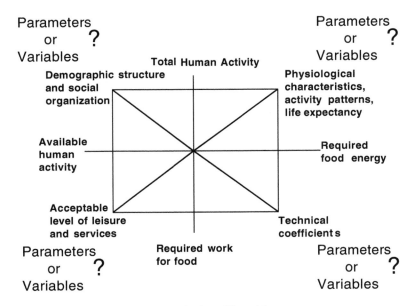

FIGURE 7.4 Arbitrariness associated with a choice of a time differential.

this impredicative loop based on a preanalytical definition of what is a parameter and what is a variable (which in turn implies choosing a given triadic filtering on the perception of the reality), such a decision implies exploring the nature of a certain mechanism (and dynamics) by ignoring the nature of others. Recall the different explanations for the death of a person (Figure 3.5) or the example of the plague in the village in Tanzania (Figure 3.6).

This fact, in our view, is crucial, and this is why we believe that a more heuristic approach to multiscale integrated analysis is required. Reductionist scientists use models and variables that are usually developed in distinct disciplinary fields. These reductionist models can deal only with one causal mechanism and one optimizing function at a time, and to be able to do so, they bring with them a lot of ideological baggage very often not declared to the final users of the models.

We believe that by adopting impredicative loop analysis we can enlarge the set of analytical tools that can be used to check nonequivalent constraints (lack of compatibility with economic, ecological, technical and social processes), which can affect the viability of considered scenarios. This approach can be used to generate a flexible tool bag for making checks based on different disciplinary knowledge, while keeping at the same time an approach that guarantees congruence among the various assessments referring to nonequivalent descriptive domains (some formal check on congruence among scenarios).

7.2.3 Example 2: Modern Societies Based on Exosomatic Energy

Impredicative loop analysis applied to self-entailment among the set of identities — energy carriers (level $n - 2$); converters used by components (on the interface level $n - 2$/level $n - 1$); the whole seen as a network of parts (on the interface level $n - 1$/level n); and the whole seen as a black box interacting with its context (on the interface level n/level $n + 1$) — is required to represent the metabolism of exosomatic energy in modern societies, as illustrated in Figure 7.1. The way to deal with such a task is illustrated in Figure 7.5 (more details in theoretical Section 7.4). The four angles refer to the forced congruence among two different forms of energy flowing in the socioeconomic process: (1) fossil energy used to power exosomatic devices, which is determining/determined by (2) human activity used to control the operation of exosomatic devices. For more on this rationale, see Giampietro (1997).

There two sets of four-angle figures that are shown in Figure 7.5. Two of these four-angle figures (small around the origin of axes) represent two formalizations of the impredicative loop generating the energy budget of Ecuador at two points in time (1976 and 1996). The other two four-angle figures (dotted

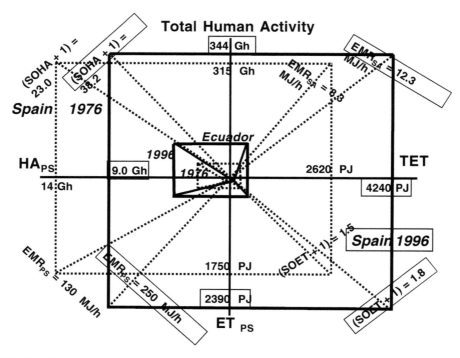

FIGURE 7.5 Total human activity.

and solid squares) represent two formalizations of the impredicative loop generating the energy budget of Spain at the same two points in time: 1976 and 1996. This figure clearly shows that by adopting this approach, it is possible to address the issue of the relation between qualitative changes (related to the readjustment of reciprocal values of intensive variables within a given whole) and quantitative changes (related to the values taken by extensive variables — that is, the change in the size of internal compartments and the change of the system as a whole). The approach used to draw Figure 7.5 is basically the same as that used in Figure 7.2 in terms of the basic rationale. That is, the set of activities required for food production within the autocatalytic loop of endosomatic energy has been translated into the set of activities producing the required input of useful energy for machines (energy and mining + manufacturing).

For a more detailed explanation of the formalization used in the four-angle figures shown in Figure 7.5, see Giampietro (1997), Giampietro et al. (2001) and the two special issues of *Population and Environment* (Vol. 22, pp. 97–254, 2000; and Vol. 22, pp. 257–352, 2001). Moreover, a detailed explanation of this type of analysis will be discussed in Chapter 9 when discussing the concepts of demographic and socioeconomic pressure on agricultural production.

Economic growth is often associated with an increase in the total throughput of societal metabolism, and therefore with an increase in the size of the whole system (when seen as a black box). However, when studying the impredicative loop that is determining an integrated set of changes in the relative identities of different elements (e.g., economic sectors seen as the parts) and the whole, we can better understand the nature and effects of these changes. That is, the mechanism of self-entailment of the possible values taken by the angles (intensive variables) reflects the existence of constraints on the possible profiles of distribution of the total throughput over lower-level compartments. In the example given in Figure 7.5, Spain changed, over the considered period, the characteristics of its metabolism both in (1) qualitative terms (*development* — different profile of distribution of the throughput over the internal compartments; changes in the value taken by intensive 3 variables, i.e., angles) and (2) quantitative terms (*growth* — increase in the total throughput; changes in the value taken by extensive variables, i.e., segments).

On the other hand, Ecuador, in the same period, basically expanded only the size of its metabolism (the throughput increased as a result of an increase in redundancy — more of the same; increase in extensive variable 2, i.e., segments), but maintained the original relation among intensive variables (the same profile of distribution of values of intensive variables 3, i.e., angles reflecting the characteristics of lower-level components; growth without development). In our view, an analytical approach based on an impredicative loop analysis can provide a powerful diagnostic tool when dealing with issues related to sustainability, environmental impact associated with growth and development (e.g., when dealing with issues such as the mythological environmental Kuznet curves). In fact, in these situations it is very easy to extrapolate wrong conclusions (e.g., dematerialization of developed economies) after being misguided by the reciprocal effect of changes among intensive and extensive variables — Jevons' paradox (see Figure 1.3) — leading to the generation of treadmills, as discussed in Table 1.1.

7.2.4 Example 3: The Net Primary Productivity of Terrestrial Ecosystems

7.2.4.1 *The Crucial Role of Water Flow in Shaping the Identity of Terrestrial Ecosystems*

Before getting into the discussion of the next example of ILA applied to the mechanism of self-entailment of energy forms associated with the identity of different types of terrestrial ecosystems, it is useful to quote an important passage of Tim Allen about the crucial role of water in determining the life of terrestrial ecosystems:

> Living systems are all colloidal and, for the narrative we wish to tell, water is the constraining matrix wherein all life functions. Unfortunately, most biological discussion turns on issues of carbon chemistry, such as photosynthesis, and the water is taken for granted. Think then of the amount of water that is in your head as you think about these ideas. Your thoughts are held in a brain that is over 80% water. Might it not be foolish to take that water for granted. Water is the medium in which life is constrained. Mars is a dead planet because it has insufficient liquid water. The controls of Gaia (Lovelock, 1986) on this planet work through water as a medium of operation. There is no life on Mars because there is no water to get organized. When we take water seriously, as a matrix of life, living systems are an emergent property not of carbon and its chemistry, but an emergent property of planetary water. Thus I misspeak when I tell my students that terrestrial animals are zooplankton that have brought their water with them. Rather they are pieces of ocean water that has brought its zooplankton with it. In the time between the next to last breath of a dying organisms and when it is unequivocally death, the carbon chemicals within the corps are essentially unchanged. The difference between life and death is that the water ceases to be the constraining element and it leaks away. The water loses its control. Trees are one of water's ways of getting around on land and into the air (Allen et al., 2001, p. 136).

This passage beautifully focuses on the crucial importance of water and the role that it (and its activity driven by dissipation of energy) plays in the functioning of terrestrial ecosystems. From this perspective, one can appreciate that the net primary productivity (NPP) of terrestrial ecosystems (the ability to use solar energy in photosynthesis to make chemical bonds) depends on the availability of a flow of energy of different natures (the ability to discharge entropy to the outer space, associated with the evapotranspiration of water). That is, the primary productivity of terrestrial ecosystems, which establishes a store of free energy in the form of chemical bonds in standing biomass, requires the availability of a different form of energy at a higher level. According to the seminal concept developed by Tsuchida, the identity of Gaia is guaranteed by an engine powered by the water cycle that is able to discharge entropy at an increasing rate (see Tsuchida and Murota, 1985). This power is required to stabilize favorable boundary conditions of the various terrestrial ecosystems operating on the planet.

When coming to agricultural production in agroecosystems, the situation is even more complicated. In fact, to have agricultural production, additional types of energy forms and conversions are required. At least three distinct types of energy flows (each of which implies a nonequivalent definition of identities for converters, components, wholes and admissible environments) are required for the stability of an agroecosystem:

1. **Natural processes of energy conversions powered by the sun and totally out of human control.** These can include, for example, heat transfer due to direct radiation, evapotranspiration of water, generation of chemical bonds via photosynthesis and interactions of organisms belonging to different species within a given community to stabilize existing food webs (i.e., for the reciprocal control predator–prey or plants–herbivores–carnivores–detritus feeders within nutrient loops in ecosystems).

2. **Natural processes of conversion of food energy within humans and domesticated plants and animals controlled by humans to generate useful power.** This metabolic energy is used to generate human work and animal power needed in farming activities, as well as plants and animal products (such as crops, fibers, meat, milk and eggs).

3. **Technology-driven conversions of fossil energy** (these conversions require the availability of technological devices — capital — and know-how, besides the availability of fossil energy stocks). Fossil energy inputs are used to boost the productivity of land and labor (e.g., for irrigation, fertilization, pest control, tilling of soil, harvesting). This input to agriculture is coming from stock depletion (mining of fossil energy deposits) and therefore implies a dangerous dependency of food security on nonrenewable resources.

These three types of energy flows have a different nature and therefore cannot be described within the same descriptive domain (not only the relative patterns are defined on nonreducible descriptive domains, but also their relative sizes, are too different). Therefore, it is important to be able to establish at least some sort of bridge among them. An integrated assessment has to deal with all of them, since they are required in parallel — and in the right range of values, intensive variable 3 — for sustaining a stable flow of agricultural production.

Relevant implications of this fact are:

1. When describing the process of agricultural production in terms of output/input energy ratios (using conventional energy analyses), the analyst tends to basically focus only on those activities and energy flows that have direct importance (in terms of costs and benefits) for humans. That is, the traditional accounting of output/input energy ratios in agricultural production refers to outputs directly used by humans (e.g., harvested biomass and useful by-products) and inputs directly provided by humans (e.g., application of fertilizers, irrigation, tilling of soil). That is, such an accounting refers to a perception of usefulness obtained from within the socioeconomic system (from within the black box). However, these two flows do not necessarily have to be relevant for the perspective of the ecosystem in which the agricultural production takes place. Actually, it should be noted that the two flows of energy considered in conventional energy assessments as input and output of the agricultural production are only a negligible fraction of the energy flowing in any agroecosystem. Any biomass production (both controlled by humans and naturally occurring) requires a very large amount of solar energy to keep favorable conditions for the process of self-organization of plants and animals. There are large-scale ecological processes occurring outside human control that are affecting both the supply of inputs and the stability of favorable boundary conditions to the process of agricultural production. That is, what is useful to stabilize the set of favorable conditions required by primary production of biomass in terrestrial ecosystems — the flow of useful activity of ecological processes — can be perceived and represented only by adopting a triadic reading of events on higher levels (level n, $n + 1$, or $n + 2$). This is a triadic reading different from that adopted to represent the process of agricultural production at the farm level (level $n - 1$, n, or $n + 1$). These mechanisms operating at higher levels are totally irrelevant in terms of short-term perception of utility for humans and tend not to be included in assessments based on monetary variables. A tentative list of ecological services required for the stability of primary productivity of terrestrial ecosystems and ignored by default by monetary accounting include (1) an adequate air temperature, (2) an adequate inflow of solar radiation, (3) an adequate supply of water and nutrients, (4) healthy soil that makes the available water and nutrients accessible to plants at the right moment

and (5) the presence of useful biota able to guarantee the various steps of reproductive cycles (e.g., seeds, insects for pollination).

2. The two flows of energy considered in conventional energy assessments as input and output of the agricultural production are referring to energy forms that require the use of different sets of identities for their assessment. The majority of energy inputs in modern agriculture belong to the type fossil energy used in converters, which in general are machines (what we before called exosomatic energy). The majority of energy output consists of the produced biomass, which belongs to the type food energy used in physiological converters (what we before called endosomatic energy). Therefore, this is a ratio between numbers that are reflecting logically independent assessments (they refer to two different autocatalytic loops of energy forms, as discussed in the two previous examples). This ratio divides apples by oranges — an operation that can be legitimately done to calculate indicators (e.g., dependency of food supply on disappearing stocks of fossil energy) or to benchmark (comparing environmental loading, or capital intensities of two systems of production), but not to study indices of performance about evolutionary trajectories of metabolic systems.

Just to provide an idea of the crucial dependency of the human food supply on the stability of existing biogeochemical cycles on this planet, it is helpful to use a few figures. The total amount of exosomatic energy controlled by humankind in 1999, for all its activities (agriculture, industry, transportation, military activities and residential), is around 11 TW (1 TW = 10^{12} J/sec), which is about 350×10^{18} J/year (BP-Amoco, 1999). For keeping just the water cycle, the natural processes of Earth are using 44,000 TW of solar energy (about $1,400,000 \times 10^{18}$ J/year), 4000 times the energy under human control. Coming to assessments of energy flows related to agricultural production, the amount of solar energy reaching the surface of the Earth per year is, on average, 58,600 GJ/ha (1 GJ = 10^9 J), which is equivalent to 186 W/m². This is almost 500 times the average output of the most productive crops (e.g., corn, around 120 GJ/ha/year). In the example of corn production, the amount of solar energy needed for water evapotranspiration is about 20,000 GJ/ha/year, which again is more than 150 times the crop output produced assessed in terms of chemical bonds stored in biomass. This assessment is based on the following assumptions: (1) 300 kg of water/kg of gross primary production (GPP), (2) 2.44 MJ of energy required/kg of evaporated water (1 MJ = 10^6 J), and (3) GPP = yield of grain $\times$ 2.62 (rest of plant biomass) $\times$ 1.3 (preharvest losses).

This deluge of numbers confirms completely the statement of Tim Allen reported earlier about the crucial role of water when compared to carbon chemistry in the stabilization of terrestrial ecosystems. Using again a metaphor of Professor Allen to explain the behaviors of terrestrial ecosystems, we should think of water as electric power, whereas carbon chemistry is the electronic part of controls.

The goal of this section full of figures is to make clear to the reader that several different output/input energy ratios can be calculated when describing agricultural production and the functioning of terrestrial ecosystems. Depending on what we decide to include among the accounted flows (different classes of energy forms) — as either output or input — we can generate a totally different picture of the relative importance of various energy flows or about the efficiency of the process of agricultural production. Conventional output/input energy analysis tends to focus only on those outputs and inputs that have a direct economic relevance (since they are linked to the short term and direct benefits and costs of the agricultural process). This choice, however, carries the risk of conveying a picture that neglects the importance of free inputs, which are provided by natural processes to agriculture. This picture ignores how the autocatalytic loop is seen from the outside (from the ecosystem point of view).

Without the supply of these free inputs (such as healthy soil, freshwater supply, useful biota, favorable climatic conditions), human technology would be completely incapable of guaranteeing food security. The idea that technology can (and will) be able to replace these natural services is simply ludicrous when analyzed under an energetic perspective as perceived by natural ecosystems. This is why we need an alternative view of the relations among the identity of parts and the whole of terrestrial ecosystems that reflect the internal relations between identity of parts and the whole, according to the mechanism of self-entailment of energy forms within ecological processes. The example presented below represents an attempt in this direction. The ILA rationale is applied to the analysis of a self-entailment of energy forms stabilizing the identity of terrestrial ecosystems.

7.2.4.2   An ILA of the Autocatalytic Loop of Energy Forms Shaping Terrestrial Ecosystems — Our impredicative loop analysis of the identity of terrestrial ecosystems tries to establish a relation between:

1. What is going on in them in terms of primary productivity (the making and consuming of chemical bonds) inside the black box (using the total amount of chemical bonds, extensive variable 1). This is information that can be linked to the analysis of agricultural activities.
2. The external power associated with the cycling of water linked to this primary productivity, which is a measure of the interaction of such a black box with the context (this measures the dependency of GPP on favorable conditions for the transpiration of water, extensive variable 2, according to the mechanism used to generate a mosaic effect across levels for dissipative systems, illustrated in Chapter 6).

Obviously, we cannot attempt to include the mechanisms occurring outside the investigated system to stabilize boundary conditions (the set of identities stabilizing the power associated with the cycling of water). By definition, there is always a level $n + 2$ that is labeled "environment" and therefore must remain outside the grasp of scientific representation within the given model. We just establish a set of reciprocal relations among key characteristics of the identity of terrestrial ecosystems (without considering the interference of humans). To do that, we benchmark the identities of various elements mapped using a specified form of energy (amount of chemical bonds, as extensive variable 1) against another form of energy (amount of water that is evapotranspired per unit of GPP, as extensive variable 2). If in this way we can find typologies of patterns that can be used to study the relations among characteristics and sizes of the parts and the whole of terrestrial ecosystems, then it becomes possible to study the effects of human alteration of terrestrial ecosystems associated with their colonization, in terms of distortion from the expected pattern. Applications of this analysis are discussed in Section 10.3.

An example of ILA applied to terrestrial ecosystems is given in Figure 7.6. The self-entailment among flows of different energy forms considered there refers to solar energy used for evapotranspiration, which is linked to generation and consumption of chemical bonds in the biomass. That is, the four angles of

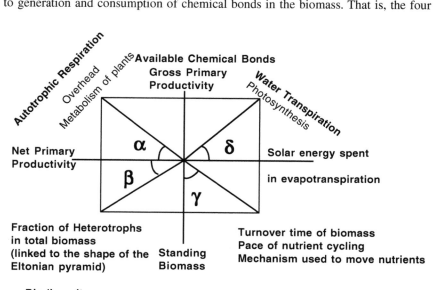

FIGURE 7.6 Terrestrial ecosystems.

Figure 7.6 refer to the forced congruence among two different forms of energy flowing in terrestrial ecosystems: (1) solar energy to power evaporation of water associated with photosynthesis, which is determining/determined by (2) biomass generated through photosynthesis, whose activity is used to organize and control the evaporation of water.

Put another way, the chicken–egg loop stabilizing terrestrial ecosystems is described in Figure 7.6 as an autocatalytic loop of two energy forms: (1) photosynthesis making biomass (storage of energy in the form of chemical bonds), which makes it possible to use solar energy through evapotranspiration of water and (2) solar energy invested in evapotranspiration of water, bringing nutrients to the leaves and making possible the photosynthetic reactions required for making biomass. Also in this case, it is possible to represent such a chicken–egg loop using a four-angle representation:

- **Angle α** — Due to the characteristic of the terrestrial ecosystem, a certain fraction of the energy made available to the ecosystem through photosynthesis (extensive variable 1) is used by the plants themselves. The fraction lost to autotrophic respiration — an overhead of the plants — defines the NPP of an ecosystem, given a level of GPP (internal loss at the level $n-1$).

- **Angle β** — The characteristics of the heterotrophic compartment of the terrestrial ecosystem defines the distribution of the total biomass among different subcompartments (the shape of the Eltonian pyramid, food web structure, or when adopting nonequivalent representations of ecosystems — e.g., network analysis, different graphs). The combined effect of this information will determine the ratio of SB /NPP (SB = standing biomass). This is still described using fractions of extensive variable 1.

- **Angle γ** — Due to the characteristic of the terrestrial ecosystem, there is a certain demand of water to be used in evapotranspiration per unit of total standing biomass (total standing biomass includes the biomass of heterotrophic and autotrophic organisms). This depends on the turnover of different types of biomass (with known identities in different compartments) and the availability of nutrients and way of transportation of them. This establishes a link between investment of extensive variable 1 and returns of extensive variable 2 at the level $n-1$ (over the autotrophic compartment).

- **Angle δ** — This angle represents the ratio between two nonequivalent flows of energy forms (two independent assessments) that can be used to establish a relation between:

  1. The size of the dissipative system terrestrial ecosystem seen from the inside (extensive variable 1), which is defined in size (total standing biomass and turnover time GPP/SB) using a chain of identities (energy carriers × transformers × whole), chemical bonds generated by photosynthesis making up flows of biomass across components — total size GPP. The internal currency expressed in GPP makes it possible to describe the profile of investments of it inside the system over lower-level compartments (e.g., autotrophs and heterotrophs).

  2. The size of the dissipative systems as seen from the context (extensive variable 2). This is the size of the energy gradient that is required from the context to stabilize the favorable boundary conditions associated with the given level of GPP (incoming solar radiation and thermal radiation into the outer space, which is supporting the process of water evapotrans-pirations, plus availability of sufficient water supply).

Obviously, we cannot fully forecast what are the most important limiting factors or what are the mechanisms more at risk in the future to stabilize such a power supply. But this is not a relevant issue here. Given a known typology of terrestrial ecosystems, we can study the relation between the relative flow of GPP and the solar energy for water transpiration associated with it, in terms of the relative characteristics (extensive and intensive variables) of parts and the whole, represented as stabilizing an impredicative loop of energy forms. The mapping of these two energy forms (extensive variables 1 and 2) can provide a reference value — a benchmark — against which to assess the size of the ecosystem (at level n) and the size of its relative components (at level $n-1$) in relation to the representation of events (intensive variables 3, associated with the identity of parts and lower-level elements) and the consumption and generation of energy carriers/chemical bonds (at level $n - 2$).

A detailed analysis of the self-entailment among characteristics of each one of the four angles (how the identity of lower-level components affects the whole and vice versa) is the focus of theoretical ecology applied to the issue of sustainability. Our claim is that ILA can provide a useful additional approach to study such an issue. We propose the theoretical discussion provided in the Section 7.5 and a few examples provided in Chapter 10 in Part 3 to support our claim. It is important to observe that studying the forced relations between the characteristics of identities of elements (and the size of the relative equivalence class) determining this impredicative loop in terrestrial ecosystems has to do with how to define concepts like ecosystem integrity and ecosystem health, and how to develop indicators of ecological stress. Even from this very simplified example, we can see that the concept of impredicative loop can help to better frame these elusive concepts (integrity or health of natural ecosystems) in terms of standard mechanisms of self-entailment among biological identities that are defining each other on different levels and on different scales. Integrity and health can be associated with the ability of maintaining harmony among the multiple identities expressed by ecological systems (the ability to respect the forced congruence among flows exchanged across metabolic components organized in nested hierarchies); see also last section of Chapter 6. Healthy ecosystems are those able to generate meaningful essences for their components (more on this in Chapter 8).

As in the previous example of the 100 people on a desert island, the perceived identity of terrestrial ecosystems is represented in Figure 7.6 in the form of an autocatalytic loop of energy forms that is related to the simultaneous perception (and definition) of identities of lower-level components, higher-level components and the congruence between functional relation and organized structures on the focal level. Again, looking at energy forms is just one of the possible ways to look at this system (whenever we use x-rays, we miss soft tissue). However, making explicit such a holarchic structure, in relation to a useful selection of formal identities, can be valuable for studying the effect of perturbations. For example, agriculture implies an alteration of the relations between key parameters determining the impredicative loop described in Figure 7.6. Monocultures, by definition, translate into a very high NPP with little standing biomass as averaged over the year, and a reduced fraction of heterotrophic respiration over the total GPP. An analysis of the stability of ILA applied to agroecosystems can be used to look for indicators of stress. A discussion of these points is given in Chapter 10 (e.g., Figure 10.13 through Figure 10.15). The main point to be driven home from the ILA approach now is that whenever we deal with parameters that are reciprocally entailed in a chicken–egg loop, we cannot imagine that it is possible to generate dramatic changes in just one of them, without generating important consequences over the whole loop. That is, whenever we decide to dramatically alter the holarchic structure of a terrestrial ecosystem, we have to expect nonlinearity in the resulting side effects (the breaking of its integrity). The use of impredicative loop analysis to study this problem should help in searching for mechanisms that can lead to catastrophic events across different scales that can be useful for such a search (for more information, see Chapter 10).

7.2.5 Parallel Consideration of Several Impredicative Loop Analyses

An overview of the three impredicative loop analyses presented in Figure 7.2, Figure 7.5 and Figure 7.6 is given in Figure 7.7. Actually, these three formalizations of impredicative loops refer to three possible ways of looking at energy forms relevant for the stability of agroecosystems. It is very important to note that these three formalizations cannot be directly linked to each other, since they are constructed using logically independent perceptions and characterizations of parts, the whole and contexts (nonequivalent descriptive domains). The meta-model used for the semantic problem structuring is the same, but it has been formalized (when putting numerical assessment in it) by referring to definitions of energy forms, and useful energy that is specific for the set of identities adopted to represent the autocatalytic loop. However, the three have some aspects in common, and this makes it possible to use them in an integrated way when discussing, for example, scenario analysis.

These three applications of the same meta-model, useful for catching different aspects of a given situation, required a tailoring of the general ILA on the specificity of a given situation. In this way it

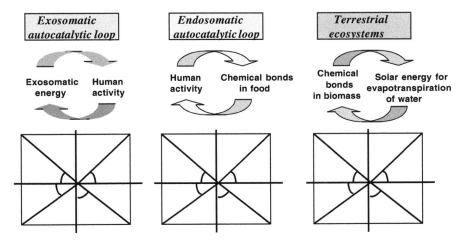

FIGURE 7.7 The nested hierarchy of energy forms self-entailing each other's identity.

becomes possible to build a set of integrated models reflecting different dimensions of analysis for a specific problem. That is, scientists that want to use this approach to deal with a specific issue of sustainability of an agroecosystem have to decide what are the relevant characteristics of the endosomatic autocatalytic loop (which is associated with physiological, demographic and social variables) and exosomatic autocatalytic loop (which is associated with both biophysical and socioeconomic variables), and critical factors affecting the self-entailment of energy forms in a specific terrestrial ecosystem (which makes it possible to establish a link with ecological analysis). The process through which scientists can decide how to make these choices has been discussed in Chapter 5. According to what was said there, we should always expect that different scientists asked to perform a process of multi-scale integrated analysis aimed at tailoring these three meta-models in relation to a specific situation (e.g., by collecting specific data about a given society perceived and represented at a given point in space and time as operating with a given terrestrial ecosystem) will come up with different variables and models to represent and simulate different aspects. The same can be expected with the direction of causality found in the analysis. Disciplinary bias (preanalytical ideological choice implied by disciplinary knowledge) is always at work.

Put another way, the predicament described in Figure 7.4 (What are the independent and dependent variables?), as well as all the other problems described in Chapter 3, can never be avoided, even when we explicitly introduce in our analysis multiple identities defined on multiple scales. What can be done when going for a multi-scale integrated analysis based on the parallel use of impredicative loop analysis is to take advantage of mosaic effects. If the analyst is smart enough, she or he can try to select variables that are shared by couples of ILAs. In this way, it becomes possible to look for bridges among nonequivalent descriptive domains. As illustrated in Figure 5.9, it becomes possible to generate integrated packages of quantitative models that are able to provide a coherent overview of different relevant aspects of a given problem. In particular, they can be used to filter out incoherent scenarios generated by simulations based on the *ceteris paribus* hypothesis. That is, they can be used to check the reliability of predictions based on reductionist models.

7.3 Basic Concepts Related to Impredicative Loop Analysis and Applications

7.3.1 Linking the Representation of the Identities of Parts to the Whole and Vice Versa

The examples of four-angle figures presented in the previous section (e.g., Figure 7.7) are representations of autocatalytic loops of energy forms obtained through an integrated use of a set of formal identities defined on different hierarchical levels (Figure 7.1).

To explain the nature of the link bridging the representation given in Figure 7.1 and the representation given in Figure 7.7, it is necessary to address key features associated with the analysis of the dynamic

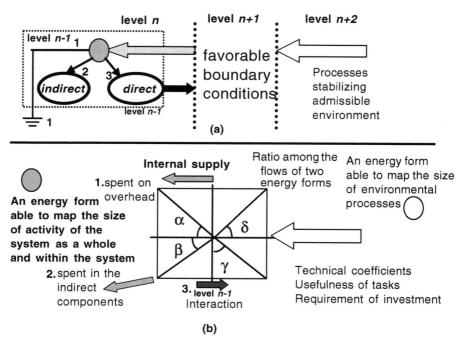

FIGURE 7.8
ILA: The rationale of the Meta Model.

energy budget of a dissipative system (Figure 7.8a). This alternative view provides yet another set of attributes that can be used to represent autocatalytic loop of energy forms in hierarchically organized dissipative systems. That is, the same network of elements represented in Figure 7.1 can be perceived and represented in a nonequivalent way by dividing the components described at the level $n - 1$ into two classes (Figure 7.8.a): (1) those that do not interact directly with the environment (aggregated in the compartment labeled "indirect") and (2) those that do interact directly with the environment, e.g., by gathering input from the environment (aggregated in the compartment labeled "direct"). In this view, the black box — seen as a whole (at level n) — can receive an adequate supply of required input thanks to the existence of favorable conditions (at level $n + 1$) and the work of the direct compartment (at level $n - 1$). This input feeding the whole can then be expressed in terms of an energy form, accounted for by using an extensive variable (which we called extensive variable 1 in Chapter 6). This variable is then used to assess how this total input is invested — within the black box — over its lower-level compartments. This variable measures the size of the whole in relation to its parts. Therefore, we can represent the total input, assessed using extensive variable 1, as dissipated within the black box in three distinct flows (indicated by the three arrows in Figure 7.8.a):

1. A given overhead for the functioning of the whole
2. For the operation of the compartment labeled "indirect"
3. For the operation of the compartment labeled "direct"

The favorable conditions perceived at the level $n + 1$, which make possible the stability of the environmental input consumed by the whole system, in turn are stabilized because of the existence of some favorable gradient generated elsewhere (level $n + 2$), which are not accounted for in this analysis. As noted, the very definition of environment is associated with the existence of a part of the descriptive domain about which we do not provide causal explanations with our model. Favorable gradients, however, must be available — to have the metabolism in the first place. These favorable gradients — assumed as granted — are exploited thanks to the tasks performed by the components belonging to the direct compartment (the representation of energy transformations occurring at the level $n - 1$). The return by the energy input made available to the whole system (at level n) per unit of useful energy invested by the direct compartment into the interaction with the environment (at level $n - 1$) will determine the strength of the autocatalytic loop of energy associated with the exploitation of a given set of resources.

This integrated use of nonequivalent representations of relations among energy transformations across levels is at the basis of the examples of impredicative loop analysis shown so far. The four-angle figures are examples of coherent representation of relations among formal identities of energy forms, which are generating an autocatalytic loop over five contiguous hierarchical levels (from level $n - 2$ to level $n + 2$). The transformation associated with the upper level (environment) is assumed by default.

The general template for performing this congruence check is shown in Figure 7.8.b. The four-angle figure combines intensive (angles) and extensive (segments) variables used to represent and bridge the characterization of metabolic processes across levels. The figure establishes a relation between a set of formal identities (given sets of variables) used to represent inputs to parts and the whole, and their interaction with the environment across scales.

The two angles on the left side (α and β) refer to the profile of distribution of the total available supply of energy carriers (or human activity or colonized land), indicated on the upper part of the vertical axis, over the three flows of internal consumption, according to the mapping provided by extensive variable 1. Angle α refers to the fraction of the total supply that is invested in overhead (e.g., for structural stability of lower-level components). Angle β refers to the profile of distribution of the fraction of the total left after the reduction, which is implied by angle α, between direct and indirect components. What is left of the original total — after the second reduction implied by angle β — for operating the direct compartment, at this point, is the value indicated on the lower part of the vertical axis. This represents the amount of extensive variable 1 (using still a mapping related to the internal perception of size) that is invested in the direct interaction with the environment.

The two angles on the right (γ and δ) are used for a characterization of the interaction of the system with the environment (the relation between the dark gray and black arrows in Figure 7.8a).

It is important to select a set of formal identities used to represent the autocatalytic loop (what variables have to be used in such a representation in terms of extensive 1 and extensive 2) that is able to fulfill the double task of making it possible to relate the perceptions and representations of relevant characteristics of parts in relation to the whole (what is going on inside the black box) with characteristics that are relevant for studying the stability of the environment (what is going on between the black box and the environment). Obviously, both extensive variables 1 and 2 have to be observable qualities (external referents have to be available to gather empirical information).

Therefore, the choice of identities to characterize an impredicative loop does not have as its goal the establishment of a direct link between the dynamics inside the black box and the dynamics in the environment. As has already been mentioned, this is simply not possible. The selection of two extensive variables (1 and 2) that can be related to each other simply makes it possible to establish bridges among nonequivalent representations of the identity or parts and wholes using variables that are relevant in different descriptive domains and in different disciplinary forms of knowledge. The logically independent ways of perceiving and presenting the reality, which are bridged in this way, must be relevant for a discussion of sustainability.

For example, the three impredicative loop analyses presented in Figure 7.7 reflect three logically independent ways of looking at an autocatalytic loop of energy forms according to the scheme presented in Figure 7.8b. These three formalizations cannot be directly linked to each other in terms of a common formal model, since they are constructed using logically independent perceptions and characterizations of identities across scales. However, they:

1. Share a meta-model used for the semantic problem structuring. This meta-model can be used to organize the discussion about how to tailor the selection of formal identities for parts, the whole and environment (when putting numerical assessments in it) to specific local situations.

2. Cover different aspects that are all relevant to a discussion of sustainability in relation to different dimensions of analysis (physiological and sociodemographic first, techno-economic second and ecological third).

3. Share some of the variables used for the characterization of the ILA.

7.3.2 An ILA Implies Handling in Parallel Data Referring to Nonequivalent Descriptive Domains

Before getting into the description of key features and possible uses of ILAs (the typology of four-angle figures introduced in the previous section), it is important to warn the reader about an important point. The four-angle figures presented so far all share the same features: (1) graphic congruence in relation to the extensive variables (rectangular shape of the four-angle figure) and (2) reasonable widths for all angles. These two features are obtained because the data represented across the four quadrants are not to scale; that is, the representation of angles and segments has been rescaled to keep the four-angle figure in a regular shape. It should be noted that the choice of rescaling the representation of data over an ILA is often an obliged one. In fact, if we want to compare the characteristics of parts to the characteristics of the whole, by using the same combination of intensive and extensive variables, we should expect to find big differences in the values found at different levels for (1) segments (in extensive terms, parts can be much smaller than wholes — the brain compared with the body) and (2) angles (in qualitative aspects of different specialized parts; a specialized part can have a value for an intensive variable that is much higher than the average value found for the whole — the brain compared with the body). In this situation, if we decide to keep the same scale of reference for the representation of both extensive (segments) and intensive (angles) variables used to characterize parts and the whole, we should expect to obtain graphs that are very difficult to read and use. An example of the difference between two four-angle figures based on a regular scale and a rescaled representation is given in Figure 7.9.

The two four-angle figures in Figure 7.9 reflect the situation of the hypothetical farming system — size of 100 ha (extensive variable 1 (EV1) referring to hectares of land) — described in the lower part of Figure 6.2. The two figures on the top represent a nonscaled graphical representation of the data set given in the lower part of Figure 6.2. Two dynamic budgets are considered: (1) the dynamic budget of food (EV2) (Figure 7.9a) and (2) the dynamic budget of money (EV2) (Figure 7.9b). The two figures on the bottom present the same couple of ILAs but after rescaling the values taken by the variables across quadrants. With this choice, the two reductions of the total available amount of extensive variable 1 divided among internal components, which is associated with the two angles on the left (α and β angles in Figure 7.8b), are reasonable: (1) a first overhead of 50% and (2) an allocation between direct and indirect of 10%. In this situation, it is still possible to follow the numerical values on the graph, keeping the same scale across quadrants. However, if we had used human activity as extensive variable 1 for studying the same two dynamic budgets, we would have found that the two reductions referring to the two angles on the left (α and β angles in Figure 7.8b) would have been (1) a first overhead of 90% and (2) an allocation between direct and indirect around 50%. This would have made it impossible to handle a useful graphic representation based on the representation of extensive variables using the same original scale.

A second qualitative difference that is relevant between the four figures shown in Figure 7.9 is between the two figures on the left (Figure 7.9a and c), in which there is no congruence between (1) the requirement of extensive variable 2 (food) consumed by the whole (at level n) and (2) the supply of extensive variable 2 produced by the compartment — land in production (at level $n - 1$). On the contrary, the two figures on the right, which are based on an extensive variable 2 of money (Figure 7.9b and d), are based on the assumption that what is consumed by the whole system (at level n) is actually produced by the compartment — land in production (at level $n - 1$). A few quick comments about these differences are:

- **Left side** — The budget related to food is not in congruence (this farming system is producing more food than it consumes). This can be used to classify this system in terms of a typology. For example, this pattern can be associated with an agricultural system net producer of food. The same ILA of land use (determination of a relation among identities of parts and the whole

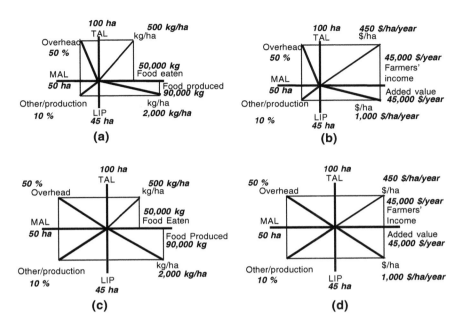

FIGURE 7.9 Different shapes of representations of ILA.

defined across levels in relation to spatial flow of food) could have applied to a city. In this case, the difference would have been very negative. This would have classified the system in the typology "urban system, heavy food importer." As discussed in the applications of Part 3, ILA can be used to define typologies (in terms of both pattern of land use and human time use). In the case of land use, this can help the characterization in quantitative terms of land use categories that can be associated with socioeconomic variables. This could help to integrate economic analysis to ecological analysis. Coming back to the example of ILA, given in Figure 7.9, according to (1) existing demographic pressure (food eaten per hectare), (2) respect of ecological processes (level of ecological overhead, which is very high in this example) and (3) available technique of production (technical coefficients expressed at the level $n - 1$), this farming system can be characterized as having a low level of productivity of food surplus (400 kg of food surplus produced per hectare of this typology of farming system). That is, from the perspective of a crowded country needing to feed a large urban population, this would not be a typology of farming system to sustain with *ad hoc* policies. Moreover, in this way, it is possible to individuate key factors determining this characteristics: (1) the small difference between angle γ expressed at the level $n - 1$ (the yield of 2000 kg/ha obtained in the land in production) and angle δ expressed at the level n (the level of consumption of the farming system, in terms of food consumption per hectare) and (2) the huge ecological overhead (the difference between total available land and managed available land). Obviously, we cannot ask too much of this very simplified example. This figure is useful only to indicate the ability of this approach to establish a relation among different dimensions of sustainability.

- **Right side** — The budget related to money flow is assumed to be in congruence. That is, the flow of added value considered in the compartment land in production at the level $n - 1$ (added value related to the value of the subsistence crops plus added value related to the gross return of the cash crop) has been used to estimate an average income for the farmers of this farming system at the level n. This is just one of many possible choices. A different selection of economic indicators (for example, using a combination of two indicators: net disposable cash for farmers and degree of subsistence) would have provided a quite different characterization of this farming system. In fact, only $15,000 in net disposable cash is generated in the simplified example considered in Figure 6.2, whereas in terms of income, the account should be $25,000 when

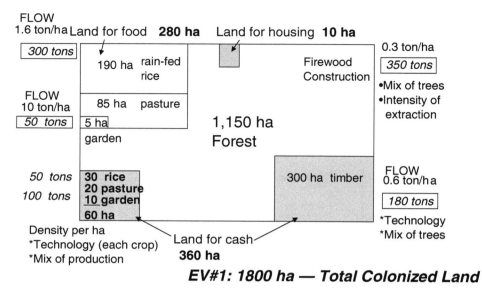

FIGURE 7.10 Examples of land use categories useful for characterizing a typology of farming system in highland Laos.

including also the value of subsistence crops. Changing hierarchical level would also imply a change in the mechanism of accounting. For example, the assessment of the net return of cash crops ($15,000) is obtained by subtracting the cost of production ($5000 paid for inputs). However, when considering the perspective of the socioeconomic system to which the farm belongs, this amount of money becomes part of the gross domestic product (GDP).

7.3.3 The Coupling of the Mosaic Effect to ILA

The example given in Figure 7.10 is related to a study aimed at characterizing typologies of farming systems in highland Laos (data from Schandl et al., 2003). The analysis refers to a farming system of shifting cultivation in the forest. The total size of the farming system (EV1) is expressed in this example in terms of hectares of colonized land (1800 ha). This area is then divided into lower-level compartments, keeping closure. Different categories used for the analysis are (1) land for food (280 ha), which is divided in three subcategories (rain-fed rice, pasture, garden); (2) land for housing (10 ha); (3) land for cash (360 ha), which is divided in two subcategories (cash crops and timber); and (4) forest (1150 ha). The identity assigned to these categories makes it possible to establish a technical coefficient (in terms of an intensive variable 3), establishing a given flow of biomass associated per hectare at the land use indicated. Examples are given, in which the yields (expressed in ton per hectare per year) of the various land uses associated with the production of useful biomass are reported.

At this point, it is possible to apply the ILA approach to such a system, as done in Figure 7.11. In the upper-right quadrant, three types of useful flows of biomass (rice, wood and vegetables) are indicated. The angles of this quadrant define for these flows a density in terms of production. The total amount can be obtained in terms of length of segments by multiplying the area of the various land use categories by the relative yields. However, the density of internal production does not coincide with the supply of useful biomass that is available for the socioeconomic context of this farming system. In fact, a certain fraction of this internal supply is reduced because of internal overhead of the system. A fraction of the internal supply of rice and vegetables, in fact, is eaten by the inhabitants of the farming system. This reduction due to the endosomatic metabolism of the farming system is indicated in the upper-left quadrant under the label "subsistence food." An additional reduction of the flow of biomass is related to the exosomatic metabolism of the farming system (the consumption of wood for firewood and construction). This additional internal overhead reduces the amount of biomass that can be exported per year from this

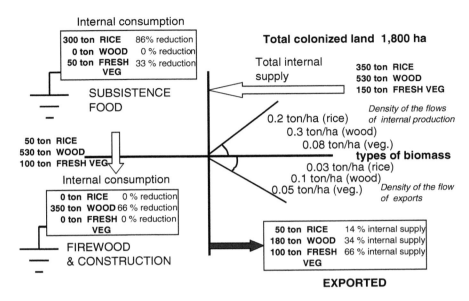

FIGURE 7.11 Defining the density of flows for the whole in relation to the definition of the parts (based on the data given in Figure 7.10).

farming system. In conclusion, the supply of useful biomass per hectare that the socioeconomic context of Laos can expect from this typology of farming system is indicated by the values of biomass flows expressed in the angles in the lower-right quadrant (e.g., in the quadrant labeled "exported": 0.03 ton/ha for rice, 0.1 ton/ha wood and 0.05 ton/ha for fresh vegetables).

Put another way, the amount of rice that the government of Laos can expect from this typology of farming system for feeding the cities does not depend only on the yield obtained per hectare in the relative category of land use (e.g., land in production of rice — 1.6 tons/ha). Additional and crucial factors are (1) the overhead of internal consumption (86% reduction of internal supply) and (2) the fraction of the total colonized land that is invested in the category rice production (220 ha of 1800). Again, we can reiterate the concept that when dealing with a metabolic system organized in nested compartments, the intensity of flows in individual compartments (or subcompartments) has always to be assessed in relation to the hierarchical structure of the whole in relation to the parts. The characteristics of a farming system refer to the whole (at level *n*), whereas technical coefficients refer to the level *n* – 1.

Two additional relevant points about this example are:

1. **Transparency** — As soon as this attempt to characterize this typology of farming system was discussed with the various researchers of the SEAtrans EC-Project, using both Figure 7.10 and Figure 7.11, every participant at the meeting jumped into it. In the following discussion, every term and assessment written in these figures was scrutinized in search for additional specifications. Then the selection of categories was questioned, with suggestions aimed at accommodating the various perspectives of different analysts (e.g., establishing a link between monetary variables and biophysical accounting). People coming from different countries and different disciplinary backgrounds were finally able to share meaning about how to perceive and represent the system under investigation.

2. **Distinction between types and realizations** — The definition of an extensive variable 1 defines a size for this metabolic system in terms of total area (expressed in this case in hectares of colonized land). However, when dealing with types, we have a size but not a specified boundary. In this type of analysis we can only deal with functional boundaries, since only individual realizations have a real specified boundary. By functional boundary we mean the boundary

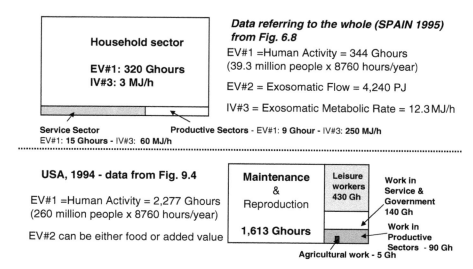

FIGURE 7.12 Examples of reduction (choice of EV1 and EV2) and classification (choice of categories providing closure) for representing societal metabolism.

determining the area required by the various land use categories included in the autocatalytic loop associated with the definition of total colonized land.

7.3.4   The Multiple Choices about How to Reduce and How to Classify

In the first part of the Faust of Goethe, Mephistopheles makes fun of the academic approach adopted at that time to study the phenomenon of life only in terms of *reduzieren* and *classifizieren*. On the other hand, as discussed in Part 1, when exploring the epistemological predicament faced by scientific analysis, when dealing with the representation of the complexity of reality, there are no alternatives to the search for useful categories and formal identities. Formal identities in turn must be associated with a closed and finite set of attributes. Crucial for getting out of this impasse is awareness that (1) any time we reduce and classify we are losing a part of the reality and (2) it is wise to always reduce and classify in parallel in several nonequivalent ways to increase the richness and reliability of the resulting integrated analysis.

Coming to technical aspects of this approach, we can look at the two examples given in Figure 7.12 that represent two nonequivalent ways of reducing and classifying countries (Spain in 1995 and the U.S. in 1994) when seen as metabolic systems organized over hierarchical levels. The main point of this example is that the same system admits and requires different choices about how to perceive and represent its identity in terms of parts and the whole, depending on the goal of the analysis.

In the example of the upper part of Figure 7.12 (based on data presented in Figure 6.8), the reduction and classification of the identity of the metabolic system has been obtained by the choice of the variables used for mapping the size of the whole and the parts. In this case, the EV1 is human activity, which is used to assess the size of the whole (total human activity in a year = population × 8760 h in a year) and the size of the lower-level compartment. The EV2 is exosomatic energy expressed in megajoules of oil equivalent (see the discussion in Chapter 6). In this way, using the concept of the mosaic effect, it becomes possible to associate the identity of lower-level elements (e.g., the typologies of patterns of consumption in the household sectors, the typologies of patterns of production in the economic sectors) with the average values of the whole. As discussed in Chapter 6, this is useful for studying how changes in technical coefficients at a given hierarchical level (e.g., in a subsector of the economy, which in this representation would be level $n - 2$) can be related to changes in the characteristics of the whole. Also in this case, as seen in Figure 7.11, changes in technical coefficients (more efficiency defined in terms of IV3 at the hierarchical level $n - 2$) are not necessarily translated into changes in the EMR of the whole (IV3 at the hierarchical level n). This is another way of looking at the effect of changes across levels that lead to the generation of Jevons' paradox, described in Chapter 1.

In the example given in the lower part of Figure 7.12 (based on data that are presented in Figure 9.4), we are still dealing with a definition of size (EV1) based on human activity, but this time the choice of categories is made with the goal of including in the analysis socioeconomic and demographic data. In particular, in this way it becomes possible to visualize the effect of socioeconomic pressure (a concept that will be introduced in Chapter 9), in terms of biophysical constraints on the density of flows associated with different elements. In relation to food production, we can see the seriousness of this biophysical constraint just by looking at Figure 7.12. That is, the requirement of food is associated with the size of the whole box determining the total human activity. In fact, as noted in Chapter 6, the endosomatic metabolism is related to the amount of human activity (2277 giga hours in a year), which is proportional to the population (260 million). On the other hand, the ability to provide an internal supply of food is related to the amount of working hours allocated in agriculture (the tiny box made up of 5 giga hours). To be self-sufficient in terms of food production, the U.S. in 1994 had to have an agricultural sector able to produce in a year, with 5 giga hours invested in production, the amount of food consumed to sustain 2277 giga hours of human activity. As illustrated in Chapter 9, the same approach can be applied to an analysis of the reciprocal density of flows of added value. The flow of added value associated with the production and consumption of goods and services in a year in the U.S. — related to total human activity, that is, 2277 giga hours — has to be produced by 235 giga hours of work, which are invested in the economic sectors generating added value (the categories selected to obtain closure in this case are services and government, productive sectors minus agriculture, and agriculture). Put another way, the density of the flow of added value per hour of human activity (the value of GDP per hour) can be related, with this method, to the economic labor productivity of the various economic sectors and subsectors. The economic labor productivity of an element in this approach is defined as the added value generated by the element divided by the working hours invested in that element.

Obviously, as noted in the previous example of the characterization of a typology of a farming system in Laos, whenever we attempt to do that with other people sitting around a table, there is a lot of explaining and discussions to be had. Different scientists coming from different geographic, social, ideological and disciplinary contexts will carry different but legitimate opinions about how to reduce and classify parts and the whole when representing flows and congruence constraints.

7.3.5   Examples of Applications of ILA

At this point we can try to resume the general feature of ILA. This approach requires considering various relevant types (used to represent parts and the whole), which are then characterized across contiguous levels using a common set of variables. In particular, the characterization of parts at the lower level and the whole at the focal level is obtained using a standard set of three variables: (1) extensive variable 1 (the common matrix that makes it possible to define compartments across levels while maintaining closure), (2) extensive variable 2 (characterizing the types in relation to the level of interaction with the context), and (3) intensive variable 3 (a combination of the previous two). By using this trick it becomes possible to generate mosaic effects across levels (when assuming a situation of congruence) or look for biophysical constraints associated with the particular role that a part is playing within the whole (when big differences in throughputs are studied over elements playing a different role in the system). Examples of applications are given in the sections below.

7.3.5.1   *The Bridging of Types across Different Levels* — Before getting into an analysis of examples of applications, it is important to illustrate the mechanism through which it is possible to establish a self-entailment over the formal identities used to represent types belonging to the same nested metabolic systems on two contiguous levels. Not only can such a mechanism be used to help scientists better discuss how to represent relevant aspects of the sustainability of an autocatalytic loop, but it also makes it possible to perform an operation of partial scaling in relation to the value taken by the variables shared by the two sets of types (the parts and the whole). To explain this point, let us use the example given in Figure 7.13.

In the example of ILA presented in Figure 7.13, the extensive variable 1 is hours of human activity and the extensive variable 2 is U.S. dollars of GDP. Intensive variable 3 is the flow of money of added

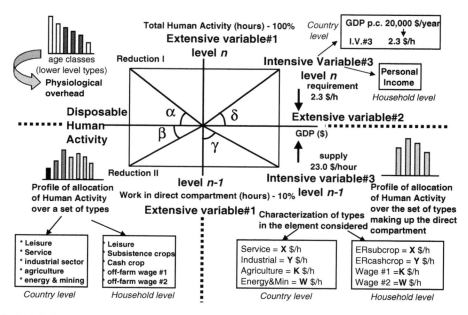

FIGURE 7.13 Relation among types in an impredicative loop.

value per hour of human activity (in either consumption or production, depending on the compartment and the hierarchical level considered). There is a strict analogy between this dynamic budget and that presented in Figure 7.2 about the society of 100 people on the desert island. An analysis by quadrants is given below:

- **Upper-right quadrant (angle δ)** — In this quadrant we have a characterization of the whole obtained by a combination of the three variables discussed before. The intensive variable 3 is the amount of added value (extensive variable 2, measured using a given currency referring to a given year) that is produced and consumed by the dissipative whole over a period of reference (a year) per unit of human activity (extensive variable 1). First imagine applying this analysis to a country. Then the intensive variable 3 will be equivalent to GDP p.c. Now imagine that we have a hypothetical value of $20,000/year. In this system of accounting, this would be expressed in terms of a level of added value of dissipation of $2.3/h of human activity (including sleeping, activities of children and the retired, and leisure of adults). To convert the value of GDP p.c. into the value of intensive variable 3, one has to divide GDP p.c. by the hours in a year — 8760 (box with white background).

- **Upper-left quadrant (angle α)** — This quadrant represents how the structural stability of lower-level compartments (i.e., body mass of humans) associated with the whole of human activity imposes an overhead on the amount of disposable human activity that can be used to generate and control useful energy. This concept has been discussed before; for a more detailed theoretical discussion, see Giampietro (1997). The important point in this discussion is that when dealing with the characterization of this quadrant, it is possible to:

  1. Define a known and predictable set of types (age classes) in which we will find humans, a set of types associated with a perception of events referring to the level $n-2$ (see the discussion about the mosaic effect across scales in Section 6.4.1, Figure 6.8 through Figure 6.10).

  2. Use this set of types to get the closure of the total size of the system (expressed in either number of individuals or kilograms of body mass).

  3. Use the profile of distribution of kilograms of body mass (or individuals) over the set of types to establish a link with a definition of compartments of human activity at the level $n-1$ (e.g., physiological overhead vs. disposable human activity). The assumptions used to calculate a

physiological overhead on total human activity, starting from the profile of distribution of individuals over age classes (or kilograms of body mass over age classes), have been discussed in the example of the 100 people on the remote island (Figure 7.2) in the previous section. This represents a first reduction of the original endowment of human activity.

- **Lower-left quadrant (angle β)** — This quadrant represents how the disposable amount of human activity — defined at the level $n - 1$ — is divided over different compartments. Also in this case, we have to start with a set of typologies of activities (the compartments in which we will divide investments of human activity) that provides closure. Additionally, a profile of allocation of human activity over the specified set of typologies will provide an assessment of the internal losses of this resource in providing control to internal compartments. In this example, an analysis referring to the country level, we select five typologies of human activity (box with white background): (1) leisure, (2) service, (3) industry, (4) agriculture and (5) energy and mining. At this point, depending on the choice of extensive variable 2 used in the ILA, only a fraction of these activities has to be included in the representation in the last quadrant (when defining the direct compartment). For example, dealing with added value, as in this case, we will include in the typologies characterizing the direct compartment all those economic sectors producing added value (service, industry, agriculture, energy and mining). Obviously, if the ILA were based on food as extensive variable 2 (as in the example of Figure 7.9c), we would have to include only the type agriculture in the right quadrant. In that case, we could split the agricultural sector into a subset of agricultural activities to better characterize the value taken by the intensive variable 3 in determining the angle γ. An example of this is given in Figure 7.15 discussed below.

- **Lower-right quadrant (angle γ)** — This quadrant characterizes relevant typologies determining the interaction of the direct compartment with the context. This representation deals with the ability to stabilize the flow of input required by the whole. To do that, this angle establishes a relation between the characteristics of the typologies making up the direct compartment and the characteristics of the whole. To do that, we need to have information related to (1) the level $n - 1$, the profile of investment of human activity (extensive variable 1) over the set of types used to represent the direct compartment and (2) the level $n - 2$, the characteristics of the intensive variable 3 over the selected set of types making up the direct compartment. In the example provided in Figure 7.13, the four compartments —service, industry, agriculture, energy and mining — are characterized in terms of their ability to generate added value per hour of labor invested in them. In this way, we can compare (1) the average level of consumption of the society (IV3, a given flow of added value ($2.3/h) that includes both the working and nonworking human activity) to (2) the capability of producing enough added value when working in those economic sectors producing profit ($23.0/h). Such a capacity obviously depends not only on the labor input, but also on the capital invested in these sectors and the availability and quality of natural resources. But this is not a relevant issue in ILA. The goal of this approach is just that of establishing a mechanism of accounting flows across levels and disciplinary fields. By combining the known profile of investment of hours of work in the various compartments and the known return of labor in each compartment, we can estimate an average return of labor (ARL) (in this example, $23.0/h) for the whole society (information referring to the interface level n/level $n - 1$). The values used to characterize the various returns of labor in different economic subsectors (ARL_i) are based on information referring to the interface level $n - 1$/level $n - 2$. The relation that can be used to bridge these two pieces of nonequivalent information has already been discussed in Section 6.4.1 and is $ARL = \Sigma (x_i \times ARL_i)$. The characteristics of ARL_i expressed in intensive variable 3 (e.g., X dollars per hour) for the hypothetical society considered in this example are given in the box with the orange background. Obviously the difference between the average rate of the flow of dollars per hour expressed at the level of the whole ($2.3/h, represented by angle δ) and the average rate of the flow of dollars per hour expressed at the level of the economic compartment generating added value (ARL = $23.0/h) must reflect the losses of total human activity over the two angles on the left (in this example, the working hours in the direct compartment are only 10% of total human activity). This is a very general feature found when performing ILAs of developed society. No matter what extensive variable 2 is considered in the analysis of the dynamic budget (food, exosomatic energy, added value, etc.), because of the very high value taken by the α angle, and the very high value taken by the β angle (linked to the profile of distribution of investments among various compartments at the level $n - 1$ — e.g., leisure, the various subsectors of the service sector, and the various subsectors of the productive sectors), in developed countries there is always a major reduction in size (measured in terms of hours of human activity) for specialized compartments. For example, agriculture in developed countries absorbs only 2% of the working time, which is already only 10% of the total human activity. This translates into a huge biophysical constraint in terms of productivity of labor, as discussed in Chapter 9. The same applies to the energy sector, where in general one can find that only 1% of the working time, which is already only 10% of the total human activity, is capable of handling all the energy conversions required to supply the huge amount of useful energy required by modern societies. This in turn requires massive investment of capital, technology and a large use of natural resources to boost the density of flows in these specialized compartments.

Before leaving this example, we can observe that the same type of analysis could have been applied at the level of a household operating within a given farming system. The meaning of the four angles would remain the same:

- **Angle δ** — In the analogy GDP p.c. would become the average income per capita (also in this case, using ILA, this should be expressed in dollars per hour — using a factor based on the household size and the hours over a year, indicated in the box with a gray background).
- **Angle α** — The profile of distribution of individuals over age classes can be used to assess a dependency ratio and a physiological overhead that indicate the amount of disposable human activity available for the household.
- **Angle β** — A set of typologies used to represent the range of human activities (obtaining closure) can be defined for this farming system (e.g., those indicated in the box with a gray background). The actual profile of allocation of the disposable human activity of the household over this set of activities will define the value taken by angle β.
- **Angle γ** — The combined use of two types of information: (1) the characterization of the typologies included in the set in terms of the value taken by the intensive variable 3 and (2) the profile of distribution of human activity in the direct compartment over the set of activities — makes it possible to calculate the value taken by the δ angle (or vice versa).

Examples with real data of this type of applications are discussed below.

7.3.5.2   *Mosaic Effect across Levels: Looking for Biophysical Constraints and for Useful Benchmarking* — In the previous example we made the point that an ILA can be seen as a systemic search for mosaic effect (self-entailment of identities of types defined on contiguous levels) over autocatalytic loops (representation of dynamic budgets) across levels. In this section we want to illustrate possible applications of this approach, in particular in relation to two possible goals:

1. The search for biophysical constraints and bottlenecks
2. Benchmarking, for comparisons and classifications of typologies in relation to different contexts

To do this we will use an example taken, this time, from a real application of ILA to farming system analysis. We will check the situation of a household type using two ILAs in parallel. The first is based on *land area* used as extensive variable 1 (Figure 7.14). The second is based on *human activity* as extensive variable 1 (Figure 7.15). In both examples, the extensive variable 2 is related to an assessment of a flow of added value (the variable here is Yuan, which is the Chinese currency). The data set is taken by the study presented in Pastore et al. (1999).

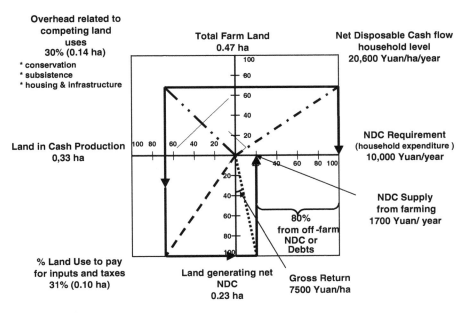

FIGURE 7.14 Application of ILA to farming system analysis. EV1, hectares of land; EV2, Yuan — Southeast China.

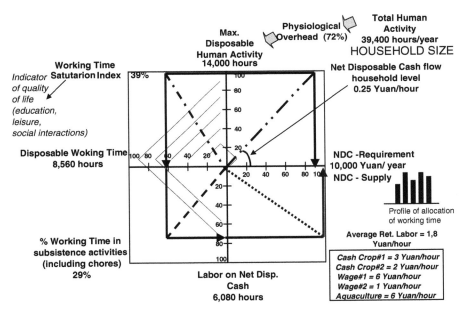

FIGURE 7.15 Application of ILA to farming system analysis. EV1, hours of human activity; EV2, Yuan — Southeast China.

The analysis presented in Figure 7.14 establishes a bridge between land use typologies and implications that an aggregate mix of these land use typologies has on the socioeconomic side. In particular, this analysis is an attempt to explain the option space seen by a particular household type found in the farming system considered in that analysis (a mainly off-farm household type).

- **Angle δ—** The intensive variable 2 in this example represents the level of net disposable cash spent by the household over the year (a different observable quality from income, which still uses Yuan per hour as IV3). This observable quality of the household can be associated with the level of economic interaction that the household manages to keep with its context in terms of economic transactions related to goods and services. Since this analysis is aimed at finding the bottleneck for this household type in relation to availability of land, the mapping of flows in this ILA is made against land area. That is, IV3 is Yuan per hectare per year. What is extremely clear from the beginning about this household type is that this dynamic budget is not working in a situation of congruence. That is, the amount of net disposable cash spent per hectare and per year at the level of the whole area occupied by the household is much higher than the amount of net disposable cash made available by the economic activities performed over the same period of time by the household in that area. To understand the reason for this deficit, we can go through the analysis of the other three angles.

- **Angle α**— The profile of distribution of land use over possible typologies can be used to assess what fraction of the available land is in cash production. This reduction is pretty high (30%), but this is due to the very limited amount of the total farm (0.47 ha) rather than to the choice of doing heavy investments in alternative land uses. That is, the total available to start with is so small that even a small amount of alternative uses (0.14 ha) represents 30% of the total.

- **Angle β**— In this analysis we decided to use the β angle to assess the reduction of EV1 related to net disposable cash due to the need to pay taxes and buy production inputs. This choice reflects the peculiarity of the farming system considered (taxes have to be paid in form of a certain amount of rice sold at a politically imposed price, much lower than market price; inputs are subsidized according to different mechanisms). For this reason, in this study, the amount of net disposable cash obtained from cash crops has been calculated by multiplying the value of gross return (7500 Yuan/ha) on a reduced amount of land in cash crop production (0.23 ha) to account for the costs, rather than using an assessment of net return (5200 Yuan/ha) over the actual area in production (0.33 ha). With this assumption, it was easier to discuss possible scenarios. Obviously, both options of how to account this reduction present pros and cons, and it is not possible to provide a substantive discussion of which of the two should be preferred.

- **Angle γ**— The final supply of net disposable cash from cash crop production depends on (1) the limited availability of land, (2) technical coefficients and economic variables and (3) the choices made by farmers, when modulating the value taken by the two angles on the left. By using an ILA we can have a look at possible combinations of this mix, separating what can be decided from what is a real constraint imposed on farmers' choices. It should be noted that in this example, this household type already invests a very limited amount of land in subsistence (together with housing and infrastructure — let alone ecological preservation — less than 0.15 ha). This implies that there is not much room for maneuvering. This household type must go for off-farm work, not only to get a decent economic condition, but also to cover its food security.

The existence of a clear biophysical constraint on the dynamic budget is evident. In this case, it is possible to study the possible effects of changes induced on intensive variables: in the price of the various crops, cost of the inputs, technical coefficients, and mix of crops to ease such a constraint. Obviously, when dealing with very severe biophysical constraints, not even the cultivation of very valuable crops, with heavy subsidies, can get this typology of household out of trouble. Off-farm jobs might be the only solution.

The analysis presented in Figure 7.15 establishes a bridge between different profiles of investment of human activity on work typologies and relative effects on the socioeconomic side. In particular, this a nonequivalent analysis of the option space seen by the same household type considered in the previous ILA (Figure 7.14).

- **Angle δ** — The intensive variable 2 represents the level of net disposable cash spent by the household over the year. The mapping of flows in this ILA against human activity implies that the intensive variable 3 is Yuan per hour per /year. During the study the system experienced a quasi-steady-state situation in relation to the dynamic budget considered in this example (the requirement and supply of net disposable cash were close enough to avoid big changes in the status of debts or savings). That is, the amount of net disposable cash spent per hour of household activity over the year was close to the amount of net disposable cash made available by the economic activities performed by the household in that period. To understand the mechanism generating this balance, we can go through an analysis of the other three angles. It should be noted that in this ILA we adopt a representation that keeps the same scale across the axes. To do that, however, we had to skip the large loss (reduction I) associated with physiological overhead (–72%). This makes it possible to better use the power of resolution of the four-angle figure. However, this trick is possible only when comparing household types operating within the same farming system (systems sharing the same level of physiological overhead).

- **Angle α** — This angle has been used to characterize the profile of distribution of disposable human activity in terms of two classes: nonworking and working. With this choice, the α angle can be directly used as an indicator of performance for the farming system. In fact, the level of saturation of the disposable human activity is a very powerful indicator of performance within the social dimension (Giampietro and Pastore, 2000). It reflects the level of attendance of children at school, the time dedicated to social interaction and reductions in workloads for adult workers.

- **Angle β** — In this analysis we decided to use the β angle to assess the reduction of working time due to the need of working in subsistence activities and chores. This choice reflects the peculiarity of the farming system considered (Southeast China), in which these two typologies of work absorb a consistent fraction of the available work.

- **Angle γ** — In this type of ILA the final supply of net disposable cash depends on (1) the identity of the option space (the set of possible work types considered, and their average return of labor — expressed in Yuan per hour) and (2) the profile of investment of available work over this set of possible works. It should be noted that we found a profile of investments of hours of work over all the options of the set. That is, the available human activity for work is not invested only in those work types that guarantee the highest average return per hour. This fact indicates the existence of different typologies of constraints preventing the maximization of economic profit. Those work types who have a very high return per hour in general cannot be performed for a large number of hours. This can be explained by the existence of biophysical constraints (e.g., it is not possible for a Chinese household sharing a pond with other households to invest 8000 h/year of human activity in the work type aquaculture) or by an economic mechanism (e.g., as soon as a very productive work type of activity is amplified, it reaches a scale at which the market value of the relative product drops).

An ILA makes it possible to analyze the dynamic budget of a household in terms of benchmarking. We can start by comparing the value of the δ angle — e.g., the average income per capita — against the analogous value for the country, or against the difference urban or rural household, or against the value reached by other households belonging to different farming systems in the same area or different typologies of households belonging to the same farming system within the same area. In the same way, we can compare the average return of labor associated with the various work typologies (wages for typologies of jobs, returns for typologies of crops) with analogous values found in different areas of the same province, same country or similar countries operating in different regional areas, but sharing the same farming system characteristics.

In all these cases, it is extremely important to have a tool able to characterize differences perceived at different levels and in relation to different relevant qualities. That is, the same income can be generated by a different combination of subsistence crops and net disposable cash. This can imply a different saturation index of disposable human activity or a different level of food security, reflected also in a different attendance of children at school, or a different level of surplus of food generated within the country to feed the cities, or a different ratio of income from on-farm and off-farm activities, or a different intensity in the use of inputs for boosting the agricultural production, or a different mix of typologies of land uses. It is important to build a network of analytical tools able to keep coherence in all these descriptions and to provide a multi-scale integrated analysis of changes, effects and trends.

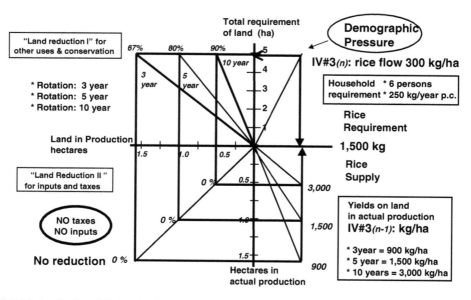

FIGURE 7.16 Application of ILA to farming system analysis. EV1, land area; EV2, food — shifting cultivation, Laos.

The last example of ILA analysis is about a bottleneck that seems to be operating on the societal side, which, however, is directly connected with biophysical realities. The case study illustrated in Figure 7.16 is based on a data set collected during a 3-year study in Laos (Schandl et al., 2003). The question investigated by this analysis is about the mechanism through which demographic pressure drives a reduction in the period of rotation adopted in shifting cultivation in Laos. We used field data reflecting average values referring to three typologies of shifting cultivation: (1) a 10-year rotation cycle, the old traditional method used when demographic pressure was low; (2) a 5-year rotation cycle, the temporary solution forced by demographic pressure in several areas of Laos; and (3) a 3-year rotation cycle, the solution adopted in those areas affected by high demographic pressure — a possible solution for obtaining stabilization of the farming activities in Laos, but a solution that also implies a few unsatisfactory features (low productivity of labor and high environmental impact). In this ILA we followed the same scheme illustrated in Figure 7.14.

- **Angle δ** — Assuming an average size for the household of six people, with a requirement of rice per capita of 250 kg/year, and assuming a demographic pressure that makes it possible to use 5 ha of land per household, we obtain a flow of rice to be stabilized per hectare (IV3 = 300 kg/ha).
- **Angle α** — This angle represents the first reduction associated with land conservation measures (ecological overhead). In this case, we can easily associate with this concept the fraction of land that is included in the rotation, but that is not directly used for production in a given year. With this assumption we obtain that the three typologies of cycles imply a reduction of (1) 90%, 10-year rotation; (2) 80%, 5-year rotation; and (3) 67%, 3-year rotation.
- **Angle β** — This second angle represents the second reduction associated with taxes and cost of inputs. In this case — shifting cultivation in Laos — these two factors are not relevant for the analysis (there are no taxes and no inputs in this typology of farming systems); this angle has no effect in determining the amount of land in production accounted for supply of food.
- **Angle γ** — This third angle describes how the characteristics of the part (the productivity of the small amount of land that is actually in production) will determine the average value for the whole (the flow of rice per hectare described by the angle γ). When using the values of IV3 characteristics of the three cycles of shifting cultivation, assessed at the level $n - 2$ (that is, at the field level), we got an unexpected result. The differences in densities of the yields at

the field level (IV3 at level $n - 2$) — 3000, 1500 and 900 kg/ha — are compensating exactly the differences associated with the reductions implied by the difference in the years of rotation. Put another way, in biophysical terms, the amount of land that has to be colonized by the three cycles over the time window of the full cycle to guarantee the same flow of rice is the same. The hectares required to get congruence (five) are the same for the three cycles.

This example clearly indicated that the problem with a 10-year rotation is not associated with an excessive demand of land for production. Rather, the problem is associated with an excessive demand of information and control to keep coherence in the societal mechanism of allocation of land to farmers. In this farming system, in fact, the various parcels used by farmers (summing up to 5 ha) are not contiguous. On the contrary, they tend to be scattered and mixed in a given area, depending on social events (wedding, deaths, moves). When dealing with a long rotation period, it is almost impossible to keep record at the societal level, by tracking the previous uses of a given piece of land. Moreover, the specific case of Laos is a case of a country that in the last decades has been through a lot of social perturbations (wars, revolutions, dramatic economic reforms). A typology of exploitation of natural resources, which requires the ability to express patterns of activities over a large area and over a large time window (e.g., the 10 years' cycle), can become impossible.

7.3.5.3 *Using ILA for Scenario Analysis: Exploring the Assumption* Ceteris

Paribus — Before concluding the presentation of possible applications of ILA analysis, we would like to mention a line of applications that has not been fully explored yet, but that in our opinion has huge potentialities. The overview provided in Figure 7.13 shows how to establish a relation among the characteristics of different types defined on different hierarchical levels. This mechanism establishing relations among types can be used to discuss the possibility of tracking the effect of changes occurring at one hierarchical level over a different hierarchical level. For example, we can study how the average value of the IV3 — at the level n (e.g., GDP p.c.) — will be affected by changes occurring at the level $n - 1$. Relevant changes for this analysis are:

1. Changes in the given characteristics of types. That is, given the same set of five possible activities (leisure, service, industry, agriculture, energy and mining), we can have improvements in the economic performance (e.g., higher returns per hour).
2. Changes in the profile of distribution over the type, that is, a change in the weight in the mix of activities determining the GDP of a country (less investments in agriculture and more in the industrial sector).
3. Change in the set of types considered to characterize sectors at the level $n - 1$. That is, some of the typologies of activities can be deleted because they became obsolete (e.g., charcoal making), or new typologies of activities can be included in the set (e.g., website designer).

Obviously, the same analysis can be made using biophysical variables; activities will be represented as performing biophysical transformations, qualitative changes will be interpreted as better technological coefficients, and emergent properties of the whole will be interpreted as a new set of definitions for useful tasks for the system (functions). In general, when dealing with emergence, we are dealing with the introduction of new typologies of activities, at the level $n - 1$, which are able to express new features at the level of the whole. These new features can be generated, at the beginning, by a stretch in the profile of distribution over the old set (by an anomalous amplification of just one of the specialized types) and by a quick change in technical coefficients in the particular typology under stress (the one undergoing a fast process of amplification). We can talk of real emergence when new relevant characteristics of the whole require the use of a new set of attributes (new observable qualities of the whole) that cannot be detected on lower-level elements (parts). Real emergence in evolution requires the development of new useful narratives for the complex observer–observed. Obviously, nobody can claim to be able to predict emergence and therefore what will happen in the future. This would require the possession of a reliable crystal ball.

However, when dealing with issues related to the future and to evolution, impredicative loop analysis can be used for qualitative trend analysis. For example, ILA can be used to characterize in quantitative terms the difference between growth and development. The four examples provided in Figure 7.5, comparing the situations of Spain and Ecuador at two points in time (1976 and 1996), can be used to explain what has generated the differences in the values of extensive and intensive variables. The difference between growth and development can be studied by looking at the relative pace of growth of the extensive variable (e.g., the increase in GDP vs. the increase in population size). Everybody knows that to study changes in the level of economic development of a country, one has to study changes in GDP per capita (an intensive variable) rather than changes in GDP in absolute terms. The GDP of a country can in fact increase due to a dramatic increase in population or to a slight decrease in the GDP per capita — when everybody is getting poorer. By performing in parallel several ILAs, based on different selections of extensive variables 1 and 2, and by using different definitions of direct and indirect compartments, it is possible to study this mechanism — how changes in intensive variables (vs. extensive variables) can be associated with evolution (vs. growth) — at different hierarchical levels of the system and in relation to different dimensions of the dynamic budget.

This approach also makes it possible to compare in quantitative terms trajectories of development. For example, at the level of individual economic sectors, an assessment of IV3 (throughput of megajoules of exosomatic energy per hour of human activity) has been used to analyze and compare the trajectory of development of Spain (Ramos-Martin, 2001) with those of other countries. In this case, average reference values for typologies of economic sectors found analyzing the trajectory of those belonging to the Organization for Economic Cooperation and Development (OECD) have been used for benchmarking. The set of reference values (e.g., 100 MJ/h for the service sector, 300 MJ/h for the productive sector and 4 MJ/h for the household sector) made it possible to individuate peculiarities of the Spanish situation. In this particular example, the power of resolution of this approach made it possible to detect a memory effect left in the Spanish system by the dictatorship of Franco. On the one hand, a certain compression of consumption under Franco's regime made it possible for Spain to have a quick capitalization of the economy. On the other hand, this left the household sector of Spain at a very low level of capitalization (the exosomatic metabolic rate — IV3 — the ability of human activity to consume exosomatic energy when out of work, can be used as a proxy for the level of capitalization of the compartment). In fact, the EMR of the household compartment (IV3 at the level $n - 1$) was 1.7 MJ/h in 1976. This was by far the lowest level in Europe in that year (including Greece and Portugal) (Ramos-Martin, 2001). At that time a fast growth of this parameter (almost doubled in 1996) could have been guessed, by looking at the European benchmark (around 4 MJ/h). Big gradients in the characteristics of analogous sectors tend to indicate top priorities in development strategy.

Another application of ILA to scenario analysis is related to the check for feasibility domains. By feasibility domain we mean the definition of admissible ranges of values taken by variables, in relation to the reciprocal relations imposed by the dynamic budget. This requires considering in parallel different dimensions of feasibility using nonequivalent ILAs. In fact, often models developed within a given disciplinary knowledge suggest the possibility of generating changes at one particular level (e.g., the possibility of improving technical coefficients in engineering, the possibility to improve the productivity of land by implementing economic policies, and the possibility to change unpleasant situations by implementing social policies). However, the effect of this change is envisioned within a given narrative. By using different ILAs based on the adoption of alternative narratives (different definitions of relevant typologies and expected relations between changes in extensive and intensive variables), we can increase the robustness of these scenarios.

Even when talking about systems, which still do not exist, the mosaic effect across scales and descriptive domains can be used to check the internal coherence of our hypotheses. As soon as we define technical coefficients, the relation among parts and the whole, and what has to be considered direct and indirect in relation to different typologies of the dynamic budget, it becomes possible to look for biophysical constraints and bottlenecks, as well as to look at the feasibility of becoming for the whole complex. A feasible process of becoming requires the ability of changing the identity of parts and the whole in parallel at different paces. This can be done for a while with smooth adjustments, but it should be expected that sooner or later a catastrophic rearrangement will occur. In terms of stability analysis, if we find congruence or lack of congruence on the axes representing a dynamic budget, we can conclude the degree of openness (the system is exporting inputs in surplus or importing inputs in shortage) or the degree of becoming (the system is reinvesting surplus to enlarge the size of its own metabolism). This enlargement of the size of the metabolism can be represented in terms of either development (the system is changing the profile of distribution of its activity over the given set of types by amplifying the scale of some of activities, while reducing in percentage the relevance of others) or growth (the system is simply increasing in size the same set of types by keeping the same profile of distribution).

Finally, a last possible application of ILA to understand deep changes associated with dramatic evolutionary processes is role analysis. A typical example (which will be discussed in Part 3) is the dramatic change in the role that the agricultural compartment plays in crowded developed societies. In the first half of 1900 this compartment was crucial not only in its obvious role of the specialized compartment guaranteeing the dynamic budget of food, but also in providing economic returns for both capital and labor invested there, which were higher than those obtained on average by the economy as a whole. In the last decades of that century, however, this economic role changed. In the dynamic budget of added value across levels and parts, investing capital and labor in agricultural activities is no longer raising the average of the economy. On the contrary, nowadays investing in that sector implies lowering the average at the level n. In spite of this fact, because of a cultural lock-in, the mechanism of controls used to regulate the agriculture still is based on an economic narrative.

7.4 Theoretical Foundations 1: Why Impredicative Loop Analysis? Learning from the Failure of Conventional Energy Analysis (*Technical Section*)

To better explore the meaning and possible use of impredicative loops in multiscale analysis, we will explore a well-known example of failure of reductionist science, when trying to deal with systems organized over multiple hierarchical levels: the failure of conventional energy analysis based on a characterization of dissipative systems in terms of a linear input/output approach. In this section, we claim that this failure can be explained by a systemic choice made by the analysts to deny the existence of chicken–egg processes. Our main point is that any attempt to deal with the representation and assessment of a set of energy flows and energy conversions across levels and scales in linear terms (in terms of input/output) can be equated to a statement that it is possible to define what energy is in a substantive way. That is, this implies believing that assessments of energy flows (and conversions) are independent from the choices made by the analyst on how to characterize a certain set of interactions over a given descriptive domain (rather than another).

The main message of complexity in this regard is clear. When dealing with the sustainability of complex adaptive systems, scientists must expect to deal with:

1. An unavoidable nonreducibility of models defined on different scales and reflecting different selections of relevant attributes. This is associated with an incomparability of perceptions and the relative nonequivalent representation of events. Scientists should acknowledge the impossibility of having a substantive and unique description of reality that can be assumed to be the "right" one.

2. Incommensurability among the priorities used by nonequivalent observers when deciding how to perceive and represent nested hierarchical systems operating across scales. That is, nonequivalent observers having different goals will provide logically independent definitions of what are the relevant attributes to be considered in the model. Scientists should acknowledge that it is impossible to have a substantive, agreed-upon definition of what should be considered the most useful narrative to be adopted when constructing a model.

7.4.1 Case Study: An Epistemological Analysis of the Failure of Conventional Energy Analysis

Attempts to apply energy analysis to human systems have a long history. Pioneering work was done by, among others, Podolinsky (1883), Jevons (1865), Ostwald (1907), Lotka (1922, 1956), White (1943, 1959) and Cottrel (1955). However, it was not until the 1970s that energy analysis became a fashionable scientific exercise, probably because of the oil crisis surging in that period. In the 1970s, energy input/output analysis was widely applied to farming systems and national economies and, more in general, to describe the interaction of humans with their environment (e.g., Odum, 1971, 1983; Rappaport, 1971; Georgescu-Roegen, 1971; Leach, 1976; Gilliland, 1978; Slesser, 1978; Pimentel and Pimentel, 1979; Morowitz, 1979; Costanza, 1980; Herendeen, 1981). At the International Federation of Institutes for Advanced Study workshop of 1974 (IFIAS, 1974), the term *energy analysis*, rather than *energy accounting*, was officially coined. The second energy crisis, in the 1980s, was echoed by the appearance of a new wave of interesting work by biophysical analysts (Costanza and Herendeen, 1984; Watt, 1989, 1992; Adams, 1988; Smil, 1991; Hall et al., 1986; Gever et al., 1991; Debeir et al., 1991) and a second elaboration of the original work by the "old guard" (Odum, 1996; Pimentel and Pimentel, 1996; Herendeen, 1998; Slesser and King, 2003). However, quite remarkably, after less than a decade or so, the interest in energy analysis quickly declined outside the original circle. Indeed, even the scientists of this field soon realized that using energy as a numeraire to describe and analyze changes in the characteristics of agricultural and socioeconomic systems proved to be more complicated than anticipated (Ulgiati et al., 1998).

We start our critical appraisal of the epistemological foundations of conventional energy analysis using one of the most well known case studies of this field: the attempt to develop a standardized tool kit for dealing with the energetics of human labor. This has been probably the largest fiasco of energy analysis due to the huge effort dedicated by the community of energy analysts to this subject — for an overview of issues, attempts and critical appraisal of results, see Fluck (1981, 1992), Giampietro and Pimentel (1990, 1991, 1991a and Giampietro et al. (1993).

Very quickly, looking at the vast literature on the energetics of human labor, one can find an agreement about the need to know at least three distinct pieces of information, which are required simultaneously, to characterize in useful terms indices of efficiency or efficacy. These are:

1. The requirement and availability of an adequate *energy input* needed to obtain the conversion of interest (an inflow of energy carriers — in the case of human labor, a flow of nutrients contained in food equals energy carriers compatible with human metabolism)

2. The ability of the considered converter to *transform the energy input into a flow of useful energy to fulfill a given set of tasks* (in this case, a system made up of humans has to be able to convert available food energy input into useful energy at a certain rate, depending on the assigned task)

3. The achievement obtained by the work done — *the results associated with the application of useful energy* to a given set of tasks (in this case, this has to do with the usefulness of the work done by human labor in the interaction with the context)

If we want to use indices based on energy analysis to formalize the concept of performance, we then have to link these three pieces of information to numerical assessments, based on observable qualities (Figure 7.17). For this operation we need at least four nonequivalent numerical assessments related to:

1. **Flow of a given energy input** (characterized and defined in relation to the given identity of the converter using it), **which is required and consumed by the converter.** In the case of the study of human labor, it is easy to define what should be considered food (energy carriers) for humans — something that can be digested and then transformed into input for muscles. If the converter were a diesel engine, food would no longer be considered energy input.

2. **The power level at which useful energy is generated by the converter.** This is a more elusive observable quality of the converter. Still, this information is crucial. As stated in a famous

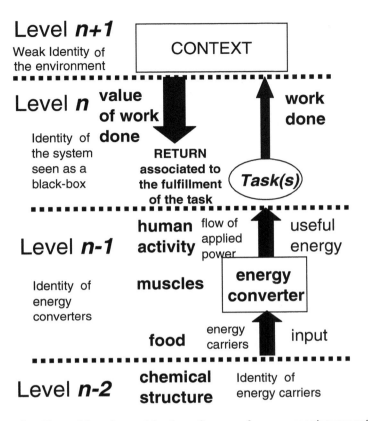

FIGURE 7.17 Energetics of human labor: characterizing the performance of energy conversions across hierarchical levels.

paper by Odum and Pinkerton (1955), when dealing with the characterization of energy converters, we have to always consider both the pace of the throughput (the power) and the output/input ratio. A higher power level tends to be associated with a lower output/input ratio (e.g., the faster you drive, the lower the mileage your car gets). It does not make any sense to compare two output/input ratios if they refer to different throughput rates. In fact, we cannot say that a truck is less efficient than a small motorbike on the basis of the information given by a single indicator, that is, by the simple assessment that the truck uses more gas per mile than the motorbike. However, after admitting that the power level is another piece of information that is required to assess the performance of an energy converter, it becomes very difficult to find a standard definition of power level applicable to complex energy converters (e.g., to human workers or subeconomic sectors). This is especially true when these systems are operating on multiple tasks. Obviously, this is reflected in the impossibility to define standard experimental settings to measure such a value. Moreover, power level (e.g., 1000 hp of a tractor vs. 0.1 hp of a human worker) does not map onto either energy input flows (how much energy is consumed by the converter over a given period of time used as a reference) or how much applied power is delivered (how much useful energy has been generated by a converter over a given period of time used as a reference). In fact, these two different pieces of information depend on how many hours the tractors or the human worker have worked during the reference period and how they have been operating.

3. **The flow of applied power generated by the conversion.** The numerical mapping of this quality clearly depends heavily on the previous choices about how to define and measure power levels and power supply. In fact, an assessment of the flow of applied power represents the

formalization of the semantic concept of useful energy. Therefore, this is the measured flow of energy generated by the converter, which is used to fulfill a specified task. However, the definition of such a task can only be given at the hierarchical level of the whole system to which the converter belongs. Put another way, such a task must be defined as useful by an observer that is operating at a hierarchical level higher than the level at which the converter is transforming energy input into useful energy. What a tractor produces has a value that is generated by the interaction with a larger context (e.g., the selling of products on the market). That is, the definition of usefulness refers to the interaction of the whole system — black box (to which the converter belongs as a component) with its context. This quality requires a descriptive domain different from that used to represent the conversion at the level of the converter (Giampietro, 2003). This introduces a first major epistemological complication: the definition of usefulness of a given task (based on the return that this task implies for the whole system) refers to a representation of events on a given hierarchical level (the interface level n/level $n + 1$) that is different from the hierarchical level used to describe and represent (assess indices of efficiency) the conversion of the energy input into useful energy (the interface level $n - 1$/level n).

4. **The work done by the flow of applied power (what is achieved by the physical effort generated by the converter).** Work is another very elusive quality that requires a lot of assumptions to be measured and quantified in biophysical terms. The only relevant issue here for the moment is that this represents a big problem with energy analysis. Even if assessments 3 and 4 use the same measurement unit (e.g., megajoules), they are different in terms of what are the relevant observable qualities of the system. That is, assessment 3 (applied power) and assessment 4 (work done) not only do not coincide in numerical terms, but also require a different definition of descriptive domain. In fact, the same amount of applied power can imply differences in achievement, because of differences in design of technology and in know-how when using it. The problem is that it is impossible to find a context-independent quality factor that can be used to explain these differences in substantive terms.

The overview provided in Figure 7.17 should already make clear (according to what was discussed in Chapters 2 and 3) that numerical assessments of energy input and output within a linear framework cannot escape the unavoidable ambiguity and arbitrariness implied by the hierarchical nature of complex systems. A linear characterization of input/output using the four assessments discussed so far requires the simultaneous use of at least two nonequivalent and nonreducible descriptive domains. Therefore, at least two of these four assessments would be logically independent. This opens the door to an unavoidable degree of arbitrariness in the problem structuring (root definitions of the system). Any definition or assessment of energy flows (as both input and output) will in fact depend on an arbitrary choice made by the analyst about what should be considered the focal level n. That is, what should be considered a converter, what should be considered an energy carrier, what should be considered the whole system to which the converter belongs, and what has to be included and excluded in the characterization of the environment, when checking the admissibility of boundary conditions and the usefulness of work done? A linear representation in energy analysis forces the analyst to decide from "scratch" in a total situation of arbitrariness about the set of formal identities (the finite set of variables used to describe changes in the various elements of interest) that are adopted in the model to represent energy flows and conversions. This choice of a set of formal identities for representing energy carriers, converters, the black box and the finite set of characteristics related to the definition of an admissible environment is then translated into the selection of the set of epistemic categories (variables) that will be used in the formal model (Figure 7.1). It is at this point that the capital sin of the assumption of linear representation becomes evident. No matter how smart is the analyst, any assessment of energy input (the embodied energy of the input) or energy output (what has been achieved by the work done) will be unavoidably biased by the arbitrary preanalytical choice of root definitions. In terms of hierarchy theory, we can describe this fact as follows: There is an unavoidable preliminary triadic filtering needed to obtain a meaningful representation of reality. That is, we have to select (1) the interface between the focal level n and the lower level $n - 1$ to represent the structural organization of the system and (2) the interface between the focal level n and the higher level $n + 1$ to represent the relational functions of the system. Ignoring this fact simply leads to a list of nonequivalent and nonreducible assessments of the same concepts. A self-explanatory example of this standard impasse in the field of energy analysis is given in Table 7.1, which reports several nonequivalent assessments of the energetic equivalent of 1 h of labor.

That is, every time we choose a particular hierarchical level of analysis for assessing an energy flow (e.g., an individual worker over a day), we are also selecting a space–time scale at which we will describe the process of energy conversion (over a day, a year, the entire life). This, in turn, implies a nonequivalent definition of what is the context (environment) and what is the lower level (where structural components are defined). Human workers can be seen as individuals operating over a 1-h time horizon (muscles are the converters in this case) or as households or entire countries (machines are the converters in this case). The definitions of identities of these elements must then be compatible with the identities of energy carriers (individuals eat food, developed countries eat fossil energy). This implies that whatever we choose as a model, the various identities of energy carriers, parts, the whole and the environment have to make sense in their reciprocal conditionings. Obviously, a different choice of hierarchical level as the focal level (e.g., the household to which the worker belongs) requires adopting a different system of accounting for inputs and outputs.

7.4.2   The Impasse Is More General: The Problematic Definitions of Energy, Work and Power in Physics

In the previous case study we defined as problematic a formal definition of energy, power and work when dealing with an energetic assessment of human labor. Since this is a point carrying quite heavy epistemological implications, we would like to elaborate a little bit more on it. That is, to be able to calculate within energy analysis substantive (absolute, of general validity outside specific situations) indices of performance based on concepts such as efficiency (maximization of an output/input ratio) and efficacy (maximization of an indicator of achievement in relation to an input), we should be able to define, first of all, three basic concepts — energy, work and power — in general (substantive) terms. That is, we should be able to agree on definitions that are independent from the special context and settings in which these three concepts are used. However, if we try to do that, we are getting into an even more embarrassing situation.

7.4.2.1   Energy — Much of the innate indeterminacy of energy analysis, especially when applied to complex systems, has its roots in the problematic definition of energy in physics. As Feynman et al. (1963, Chapter 4, p. 2) pointed out, "it is important to realize that in physics today, we have no knowledge of what energy *is*. … It is an abstract thing in that it does not tell us the mechanism or the *reasons* for the various formulas." In practice, energy is perceived and described in a large number of different forms: gravitational energy, kinetic energy, heat energy, elastic energy, electrical energy, chemical energy, radiant energy, nuclear energy, mass energy, etc. A general definition of energy, without getting into specific contexts and space scale- and timescale-dependent settings, is necessarily limited to a vague expression, such as "the potential to induce physical transformations." Note that the classic definition of energy found in conventional physics textbooks — "the potential to do work" — refers to the concept of free energy or exergy (which is another potential source of confusion), both of which still require (1) a previous formal definition of work and (2) a clear definition of operational settings to be applied.

7.4.2.2   Work — The ambiguous definition of energy (the ability to change something defined elsewhere) is also found when dealing with a general definition of work (a definition applicable to any specific space scale- and timescale-dependent settings). The classic definition found in the dictionary refers to the ideal world of elementary mechanics: "work is performed only when a force is exerted on a body while the body moves at the same time in such a way that the force has a component in the direction of the motion." Others express work in terms of equivalence to heat, as is done in thermodynamics: "the work performed by a system during a cyclic transformation is equal to the heat absorbed

TABLE 7.1

Nonequivalent Assessments of Energy Requirement for 1 h of Human Labor

Method[a] No.	Worker System Boundaries[b]	Energy Input (EI) per Hour of Labor		
		Food Energy Input[c] (MJ/h)	Exosomatic Energy Input[d] (MJ/h)	Other Energy Forms across Scales[e]
1	Man	0.5[f]	Ignored	Ignored
1	Woman	0.3[g]	Ignored	Ignored
1	Adult	0.4[h]	Ignored	Ignored
2	Man	0.8[f]	Ignored	Ignored
2	Woman	0.5[g]	Ignored	Ignored
2	Adult	0.6[h]	Ignored	Ignored
3	Man	1.6[f]	Ignored	Ignored
3	Woman	1.2[g]	Ignored	Ignored
3	Adult	1.3[h]	Ignored	Ignored
4	Man	2.5[f]	Ignored	Ignored
4	Woman	1.8[g]	Ignored	Ignored
4	Adult	2.1[h]	Ignored	Ignored
5	Household	3.9[i]	Ignored	Ignored
6	Society	4.2[j]	Ignored	Ignored
7a	Household	3.9[i]	39 (food system)[k]	Ignored
7b	Society	4.2[j]	42 (food system)[k]	Ignored
8	Society	Ignored	400 (society)[l]	Ignored
9	Society	Ignored	400 (society)[l]	2×10^{10} EMjoules[m]

[a] The nine methods considered are: (1) only extra metabolic energy due to the actual work (total energy consumption minus metabolic rate) in an hour; (2) total metabolic energy spent during actual work (including metabolic rate) in an hour; (3) metabolic energy spent in a typical workday divided by the hours worked in that day; (4) metabolic energy spent in a year divided by the hours worked in that year; (5) as method 4 but applied at hierarchical level of household; (6) as method 4 but applied at the level of society; (7) besides food energy, including also exosomatic energy spent in food system per year divided by the hours of work in that year (7a at household level, 7b at society level); (8) total exosomatic energy consumed by society in a year divided by the hours of work delivered in that society in that year; and (9) assessment of the EMjoules of solar energy equivalent to the amount of fossil energy assessed with method 8.

[b] The systems delivering work are typical adult man (man), typical adult woman (woman), average adult (adult), typical household, and an entire society. Definitions of typical are arbitrary and only serve to exemplify methods of calculation.

[c] Food energy input is approximated by the *metabolic energy requirement*. Given the nature of the diet and food losses during consumption, this flow can be translated into food energy consumption. Considering also postharvest food losses and preharvest crop losses, it can then be translated into different requirements of food (energy) production.

[d] We report here an assessment referring only to fossil energy input.

[e] Other energy forms, acting now or in the past, that are (were) relevant for the current stabilization of the described system, even if operating on space–time scales not detected by the actual definition of identity for the system.

[f] Based on a basal metabolic rate (BMR) for adult men of 0.0485 W + 3.67 MJ/day (W = weight = 70 kg) = 7.065 MJ/day = 0.294 MJ/h. Physical activity factor for moderate occupational work (classified as moderate) = 2.7 × BMR. Average daily physical activity factor = 1.78 × BMR (moderate occupational work). (Anonymous, 1985). Occupational workload: 8 h/day considering workdays only, and 5 h/day average workload over the entire year, including weekends, holidays, absence.

[g] Based on a basal metabolic rate for adult women of BMR = 0.0364 W + 3.47 (W = weight = 55 kg) = 5.472 MJ/day = 0.228 MJ/h. Physical activity factor for moderate occupational work = 2.2 × BMR. Average daily physical activity factor (based on moderate occupational work) = 1.64 × BMR. (Anonymous, 1985.) Occupational workload: 8 h/day (considering workdays only), or 5 h/day (average workload over the entire year, including weekends, holidays, absence).

[h] Assuming a 50% gender ratio.

[i] A typical household is arbitrarily assumed to consist of one adult male (70 kg, moderate occupational activity), one adult female (55 kg, moderate occupational activity), and two children (male of 12 years and female of 9 years).

[j] This assessment refers to the U.S. In 1993, the food energy requirement was 910,000 TJ/year and the work supply 215 billion hours (USBC, 1995).

[k] Assuming 10 MJ of fossil energy spent in the food system per 1 MJ of food consumed (Giampietro et al., 1994).

[l] Assuming a total primary energy supply in 1993 in the U.S. (including the energy sector) of 85,000,000 TJ divided by a work supply of 215 billion hours (USBC, 1995).

[m] Assuming a transformity ratio for fossil energy of 50,000,000 EMjoules (a method of accounting energy analysis introduced by H.T. Odum (1996)) /joule (Ulgiati et al., 1994),

by the system." This calorimetric equivalence offers the possibility of expressing assessments of work in the same units as energy, that is, in joules. However, the elaborate description of work in elementary mechanics and the calorimetric equivalence derived from classic thermodynamics are of little use in real-life situations. Very often, to characterize the performance of various types of work, we need *qualitative* characteristics that are impossible to quantify in terms of heat equivalent. For example, the work of a director of an orchestra standing on a podium cannot be described using the above definition of elementary mechanics, and it cannot be measured in terms of heat. An accounting of the joules related to how much the person sweats during the execution of a musical score will not tell anything about the quality of the performance. Sweating and directing well do not map onto each other. Indeed, it is virtually impossible to provide a physical formula or general model defining the quality (or value) of a given work (what has been achieved after having fulfilled a useful task) in numerical terms from within a descriptive domain useful to represent energy transformations based on a definition of identity for energy carriers and energy converters. In fact, formal measurements leading to energetic assessments refer to transformations occurring at lower levels (e.g., how much heat has been generated by muscles at the interface level n/level $n - 1$), whereas an assessment of performance — how much the direction has been appreciated by the public — refers to interactions and feedbacks occurring at the interface level n/level $n + 1$). This standard predicament applies to all real complex systems, especially to human societies (Giampietro and Pimentel, 1990, 1991a).

7.4.2.3   *Power* — The concept of power is related to the time rate of an energy transfer or, when adopting the theoretical formalization based on elementary mechanics, the time rate at which work is being done. Obviously this definition cannot be applied without a previous valid and formal mapping (in the form of a numerical indicator) of the transfer of energy or of doing the work. At this point, it should be obvious that the introduction of this "new" concept, based on the previous ones, does not get us out of the predicament experienced so far. Any formal definition of power will run into the same epistemological impasse as the previous two concepts discussed, since it depends on the availability of a valid definition in the first place.

It should be noted, however, that the concept of power introduces an important new qualitative aspect (a new attribute required to characterize the energy transformation) not present in the previous two. While the classic concepts of energy and work, as defined in physics, refer to quantities (assessments) of energy without taking into account the time required for the conversion process under analysis, the concept of power is, by definition, related to the rate at which events happen. This introduces a qualitative dimension that can be related to either the degree of organization of the dissipative system or the size of the system performing the conversion of energy in relation to the processes in the environment that guarantee the stability of boundary conditions. That is, to deliver power at a certain rate, a system must have two complementary but distinct relevant features: (1) an adequate, organized structure to match the given task (e.g., capability of doing work at a given rate — an individual cannot process and convert into power 100,000 kcal of food in a day) and (2) the capability of securing a sufficient supply of energy input for doing the task (e.g., to take advantage of the power of 100 soldiers for 1 year, you must be able to supply them with enough food). In short, gasoline without a car (case 1) or a car without gasoline (case 2) is of no use.

This is the point that exposes the inadequacy of the classic output/input approach — in the sense that in real dissipative systems, it is not thinkable to have an assessment of energy flows without addressing properly the set of relevant characteristics of the process of transformation (which implies addressing the existence of expected differences in power defined and measurable on different hierarchical levels and associated with the identity of the dissipative system in the first place). This means that the three theoretical concepts energy, work and power must be simultaneously defined and applied — within such a representation — in relation to the particular identity of a given typology of a dissipative system (whole, parts and expected associative context).

At this point, if we accept the option proposed in this book of using nonequivalent descriptive domains in parallel, we can establish a reciprocal entailment on the various definitions of identity of the various elements (converter, whole system, energy carriers, environment) used to characterize the set of energy transformations required to stabilize the metabolism of a dissipative system.

In this case, we have two couples of self-entailment among identities defined on different levels:

Self-entailment 1:
- The identities adopted for representing the set of various converters — what is transforming an energy input into a flow of useful energy (on the interface levels n and $n-1$). These identities are associated with the power level of the converter.

Define/are defined by:
- The identities adopted for representing the set of energy carriers — what is considered to be a material flow that is associated with the definition of an energy input (on the interface levels $n-1$ and $n-2$).

Self-entailment 2:
- The identities of the set of energy forms considered useful on the focal level — what are the energy forms considered useful — according to the organized structure of the system. This is a system-dependent characterization of the autocatalytic loop from *within* (on interface between level $n-1$ and level n). The usefulness of the whole organized structure and the combination of energy forms is decided in relation to their ability, validated in the past, to fulfill a given set of tasks.

Define/are defined by:
- The compatibility of the identities of the whole system in relation to its interaction with the larger context — what are the favorable boundary conditions related to the definition of identities of energy inputs, converters and the dissipative whole that are required to make the metabolism. This is a context-dependent characterization of the autocatalytic loop from *outside* (on the interface between level n and level $n+1$). This compatibility can be used to define the usefulness of tasks in relation to the availability in the environment of the required flow of input and sink capacity for wastes.

When using simultaneously these two self-entailing relations among identities, which depend on each other for their definitions, we can link nonequivalent characterizations of energy transformations across scales. An overview has been given in Figure 7.1 and Figure 7.8. This establishes two bridges:

- **Bridge 1** — Conversion rates represented on different levels must be compatible with each other; this implies a constraint of compatibility between the definition of the identity of the set of converters defined at the level $n-1$ (triadic reading, $n-2/n-1/n$) and the definition of the set of tasks for the whole defined at the level $n+1$ (triadic reading, $n-1/n/n+1$). This constraint addresses the ability of the various converters to generate useful energy (the right energy form applied in the specified setting) at a given rate, which must be admissible for the various tasks. This bridge deals with qualitative aspects of energy conversions (parts of parts vs. parts and parts vs. whole).

- **Bridge 2** — The flow of energy input from the environment and the sink capacity of the environment must be enough to cope with the rate of metabolism implied by the identity of the black box; this implies a constraint of compatibility between the size of the converters defined at the level $n-1$ (triadic reading, $n-2/n-1/n$) and the relative supply of energy carriers and sink capacity related to processes occurring at the level $n+1$. The set of identities of the inputs (from the environment to the converters) and wastes (from the converters to the environment) is referring to a representation of events valid also at the level $n-2$. In fact, these energy carriers will interact with internal elements of the converters to generate the flow of useful energy and will be turned out into waste by the process of conversions. However, the availability of an adequate supply of energy carriers and an adequate sink capacity is related to the existence of processes occurring in the environment, which are needed to maintain favorable conditions at the level $n + 1$. Put another way, the ability to maintain favorable conditions in face of a given level of dissipation can only be defined by considering level $n + 1$ as the focal one (triadic reading, $n/n + 1/n + 2$).

This consideration, however, implies the epistemological predicament discussed in Part 1. When dealing with quantitative and qualitative aspects of energy transformations over an autocatalytic loop of energy forms, we have to bridge at least five hierarchical levels (from level $n - 2$ to level $n + 2$).

By definition, the environment (processes determining the interface between level $n + 1$ and level $n + 2$) is something about which we do not know enough. Otherwise, it will become part of the modeled system. This means that when dealing with the stability of favorable boundary conditions, we can only hope that they will remain favorable as long as possible. On the other hand, the existence of favorable boundary conditions is a must for dissipative systems. That is, the environment is and must be in general assumed to be an admissible environment in all technical assessments of energy transformations.

If we accept this obvious point, we have also to accept that the existence of favorable boundary conditions — interface level $n + 1$/level $n + 2$ — is an assumption that is not directly related to a definition of usefulness of the tasks. The usefulness of tasks has been validated because of the admissibility of boundary conditions. That is, a technical definition of usefulness (efficiency, efficacy) does addresses only the effect of processes occurring on the interface level n/level $n + 1$, and therefore has limited relevance for discussing sustainability (recall here Jevons' paradox discussed in Chapter 1). The existing definition of the set of useful tasks at the level n simply reflects the fact that these tasks were perceived as useful in the past by those living inside the system. That is, the established set of useful tasks was able to sustain a network of activities compatible with boundary conditions (*ceteris paribus* at work). However, this definition of usefulness for these tasks (what is perceived as good at the level n according to the perceived favorable boundary conditions at the level $n + 1$) has nothing to do with a full evaluation of the ability or effect of these tasks in relation to the stabilization of boundary conditions in the future (in relation to processes occurring at level $n + 2$). For example, producing a given crop that provided an abundant profit last year does not necessarily imply that the same activity will remain useful next year. Existing favorable boundary conditions at the level $n + 1$ require the stability of the processes occurring at the level $n + 2$ (e.g., the demand for that crop remains high in face of a limited supply, as well as natural resources such as nutrients, water, soil and pollinating bees continuing to be available for the next year). This is information about which we do not know and cannot know enough in advance. This implies that analyses of efficiency and efficacy that are based on data referring to characterizations and representations relative to identities defined on the four levels — $n - 2$, $n - 1$, n, $n + 1$ (on the *ceteris paribus* hypothesis and reflecting what has been validated in the past) — are not very useful to study co-evolutionary trajectories of dissipative systems. In fact, (1) they do not address the full set of relevant processes determining the stability of favorable boundary conditions (they miss a certain number of relevant processes occurring at the level $n + 2$) and (2) they deal only with qualitative aspects (intensive variables referring to an old set of identities) — not quantitative aspects such as the relative size of components: How big is the requirement of the whole dissipative system (extensive variable assessing the size of the box from the inside) in relation to the unknown processes that stabilize the identity of its environment at the level $n + 2$. As observed earlier, the processes that are stabilizing the identity of the environment are not known by definition. This is why, when dealing with co-evolution, we have to address the issue of emergence. That is, we should expect the appearance of new relevant attributes (requiring the introduction of new epistemic categories in the model — new relevant qualities of the system so far ignored) to be considered as soon as the dissipative system (e.g., human society) discovers or learns new relevant information about those processes occurring at level $n + 2$ that were not known before.

In conclusion, an operational definition of the three concepts energy, work and power can be obtained only after adopting a structure of Chinese boxes (nonequivalent descriptive domains overlapping) in which the set of values taken by intensive and extensive variables used to represent (and assess) them are brought into congruence through a process of reciprocal entailment among definitions.

7.5 Theoretical Foundations 2: What Is Predicated by an Impredicative Loop? Getting Back to the Basic Fuzzy Definition of Holons Using Thermodynamic Reasoning (*Technical Section*)

The new paradigm associated with complexity, which is rocking the reductionist building, is the son of a big epistemological revolution started in the first half of the 19th century by classic thermodynamics (e.g., among others — Carnot and Clausius) and continued into the second half of the 20th century by nonequilibrium thermodynamics (e.g., by the ideas of Schroedinger and the work of Prigogine's school). Both revolutions used the concept of entropy as a banner. The equilibrium thermodynamics represented a first bifurcation from mechanistic epistemology by introducing new concepts such as irreversibility and symmetry breaking when describing real-world processes (e.g., unilateral directionality of real time). The nonequilibrium paradigm represents a final departure from reductionist epistemology since it implies the uncomfortable acknowledgment that scientists can only work with system-dependent and context-dependent definitions of entities within models. In particular, the concept of negative entropy — a concept introduced by Schroedinger (1967) to explain the existence of life — is not a substantive concept. Rather, this is a construction (an artifact) associated with the given identity of a dissipative system that is operating at a given point in space and time (within a particular setting of boundary and initiating conditions). The concept of negative entropy is crucial in our discussion, since this is a concept that imposes in scientific analysis that the perception and representation of quality for both energy inputs and energy transformations have to reflect a particular typology of metabolism (pattern of dissipation) of a specific dissipative system. That is, the concept of negentropy must refer to a given identity of a dissipative system that is assumed to operate within a given associative context (an admissible environment). According to this fact, food is an energy input for humans but not for cars. This fact has huge implications for energy analysts. In fact, saying that 1 kg of rice has an energy content of 14 MJ can be misleading, since this energy input is not an energy input for a car that is out of gas. In the same way, the definition of what should be considered useful energy depends on the goals of the system, which is expected to operate within a given associative context. Fish are expected to operate inside the water to get their food, whereas birds cannot fly to catch prey below the ground.

These can seem like trivial observations, but they are directly linked to a crucial statement that we want to make in this chapter. It is not possible to characterize, by using substantive formalisms (applicable to all conceivable dissipative systems operating in reality), qualitative aspects of energy forms. That is, it is impossible to define a quality index that is valid in relation to all the conceivable scales (from subatomic particles to galaxies) and when considering all the conceivable attributes that could be relevant for energy analysis (all possible observable qualities relevant in relation to the goal and the expected associative context). To characterize the behavior of dissipative systems, one has always to specify the preanalytical choices made by the analyst to perceive and represent energy transformations within a finite and closed information space. As noted earlier, this is impossible since these systems are (1) open (they are what they eat), and this blurs the distinction between system and environment across scales and (2) becoming in time (they are all special because of their history), and this requires a continuous updating of the set of typologies used to characterize them (why and how are they special and relevant for the observer).

In our view, this is why classic thermodynamics first, and nonequilibrium thermodynamics later, gave a fatal blow to the mechanist epistemology of Newtonian times. However, we cannot replace the hole left by the collapse of Newtonian mechanistic epistemology just by continuing to use a set of new terms derived from nonequilibrium thermodynamics (e.g., disorder, information, entropy, negentropy), as if they were substantive concepts (definable strictly in the physical sense as context independent). We should always recall the caveat presented by Bridgman:

> It is not easy to give a logically satisfying definition of what one would like to cover by "disorder."
> … Thermodynamics itself, I believe, must presuppose and can have meaning only in the context of a specified "universe of operations," and any of the special concepts of thermodynamics, such as entropy, must also presuppose the same universe of operations." (Bridgman, 1961)

The big problem with this point is that the approach suggested by Bridgman cannot be followed in a universe in which not only the observed system, but also the observer, is becoming something else in time. In fact, an observer with different goals, experiences, fears and knowledge will never perceive and represent energy forms, energy transformations and a relative set of indices of quality in the same way as a previous one (for more on this point, see Chapter 8).

This entails that concepts derived from thermodynamic analysis and energy indices of quality (such as output/input energy ratios, exergy-based indices, entropy-related concepts and embodied assessments associated with energy flows) can be seen as very powerful metaphors. However, like all metaphors, they always require semantic checks before their use. That is, all these concepts are very powerful to help in a discussion about the usefulness of alternative narratives when dealing with sustainability issues. They should not be used to provide normative indications about how to deal with sustainability predicaments in an algorithmic way (on the basis of the application of a set of given rules written in protocols). Put another way, it is not always certain that optimizing an index of efficiency or efficacy, reflecting one of the possible formalizations of a problem, is the right thing to do. There is no "magic" associated with thermodynamics that can provide analysts with an epistemological silver bullet, not even the concept of entropy.

7.5.1   A Short History of the Concept of Entropy

Originally the concept of entropy arose, even if in an implicit form, when dealing with qualitative aspects of energy transformations within thermal engines. Sadi Carnot (in *Reflections on the Motive Power of Fire*, 1824) proposed the existence of a set of predictable relations between the work produced by a steam engine and the characteristics of the heat transfer associated with its operation (Mendoza, 1960). Emile Clapeyron, in 1834, restated Carnot's principle in analytical form (Mendoza, 1960). That is, after framing the representation of a set of energy conversions within a predictable setting (thanks to the structural organization provided by the engine used for the experiments), it becomes possible to predict relations between losses, overheads and useful output. Clausius restated both the first (the energy of the universe is constant) and second (the entropy of the universe tends to a maximum) laws, assuming the universe as an isolated system (Clausius, 1867, p. 365). Details on this historical birth process of the term *entropy* are important since they show that the concept of entropy was alien to the prevailing mechanistic epistemology at that time (for more, see Mayumi and Giampietro, 2004).

Georgescu-Roegen introduced another crucial theoretical issue about entropy by proposing a *fourth law of thermodynamics*. This law refers to the impossibility of full matter recycling for a dissipative system. The point raised by Georgescu-Roegen is very important for the discussion of sustainability, since this is where the opinions of "prophets of doom" and "cornucopians" bifurcate. Georgescu-Roegen defines the idea of an economy that can recycle and substitute, using capital and technology, any limiting resource as an idea of a perpetual motion of the third kind. That is, he equates the idea of perpetual economic growth to the idea of having a closed thermodynamic system that can perform work at a constant rate forever or that can perform work between its subsystems forever. He then claims that perpetual motion of the "third kind" is impossible. This claim has generated an intense debate in the field of ecological economics (Bianciardi et al., 1993, 1996; Kümmel, 1994; Månsson, 1994; Converse, 1996, 1997; Ayres, 1999; Craig, 2001; Kåberger and Månsson, 2001). In fact, if the theoretical framework of thermodynamics is strictly followed, it is relatively easy to reach the following result, which contradicts Georgescu-Roegen's statement: it is *possible* to construct a *closed* engine that will work in a complete cycle and produce no effect except the raising of a weight, the cooling of a heat reservoir at a higher temperature, and the warming of a heat reservoir at a lower temperature (Mayumi, 1993). Actually and paradoxically, this closed system is nothing but a Carnot engine. The Carnot engine with fluid is indeed a closed system because heat can be exchanged during two isothermal processes (expansion and compression) through the base of the cylinder.

This is where the epistemological predicament implied by complexity enters into play. A Carnot engine is an ideal type of engine. In reality, each working engine is a special realization of such a type, which requires lower-level components guaranteeing structural stability. Because of this, it will have special differences from the general template used for its making. These differences, due to peculiar characteristics of lower-level components (the material structure) and stochastic events associated with the history of such an organized structure, imply that our knowledge of the type has limited applicability for dealing with individual special realizations. This is why material entropy is critically important. The main point of Georgescu-Roegen is that energy can be represented in models as a homogeneous substance. That is, energy representations of events are based on types. According to these energetic models, energy conversions from one energy form into another can be easily accomplished according to the laws included in the model. On the other hand, when looking at the same transformations in terms of matter (at a different level), we always find that material elements are highly heterogeneous, and every element has some unique physicochemical properties. This feature of matter explains the reason that the practical procedures for unmixing liquids or solids differ from case to case and consist of many complicated steps. Seemingly, the only possible way of reaching a quantitative measure of material entropy is to calculate indirectly the amounts of matter and energy needed to return to the initial state of matter in bulk in question given available technology.

Getting back to our historical overview, we can say that in the first century of its life, the concept of entropy was charged with various meanings by different users. Perhaps due to predominant addiction to formalism, and because of the obvious ambiguity of the term, the label entropy became associated with nonequivalent concepts such as irreversibility, arrow of time, expected trends toward disorder, expected directional changes in the quality of energy forms and even to quantitative assessments of information flows among communicating systems (for more, see Mayumi and Giampietro, 2004). In any case, there was a common connotation associated with the label entropy until the first half of the 20th century. The concept of entropy (no matter how defined) was always associated with a clear "prophet of doom" flavor. The universe is condemned to heat death, disorder will prevail, irreversibility and frictions are the unavoidable bad guys that are here to disturb the beautiful order of our universe and the work of scientists, etc.

The innate ambiguity associated with the concept of entropy, however, is so pervasive that it even made possible the overcoming of this negative connotation. A dramatic change in the perception of the role of entropy in the evolution of life came in the second half of the 20th century. The twist was primed by the ideas of surplus entropy disposal by Erwin Schroedinger (1967, in an added note to Chapter VI of *What Is Life?* written in 1945) and then by the work of the Prigogine school in nonequilibrium thermodynamics (Prigogine, 1961, 1978; Glansdorf and Prigogine, 1971; Nicolis and Prigogine, 1977; Prigogine and Stengers, 1984). With the introduction of the class of dissipative systems, the concept of entropy finally got out from the original outfit of villain. Self-organization and emergence are both strictly associated with the ability of exporting surplus entropy generated within the system into the environment. Schneider and Kay (1994) reformulated the second law and suggested that as systems are moved away from equilibrium, they will take advantage of all available means to resist externally applied gradients. Actually, looking at the evolution of biological systems, Brooks and Wiley (1988) arrived to see evolution as entropy (as stated in the title of their famous book). For an expert in thermodynamics, such a title can appear as an insult because entropy is a state function in classic thermodynamics. But this is the "magic" of entropy, as it were. First of all, in that book Brook and Wiley were using a definition of entropy derived from information theory. Second, they were exploring new frontiers (looking for new meanings) associated with the paradigm shift about how to perceive evolution. In this task, the ambiguity of the term might have been a blessing for them. In fact, after having accepted that irreversibility and friction are no longer the bad guys, the information entropy concept within their framework became the essential element that sustains and drives the evolution of the complex organization of dissipative systems.

Actually, at this point, it can be noted that the epistemological predicament associated with complexity (the impossibility of establishing a formal mapping when dealing with the identity of complex systems) is one of the main issues dealt with in recent thermodynamic discussions. For example, we can recall the point made by Bridgman (1961) about the clear fact that the definition of energy in thermodynamics is not necessarily logically equivalent to the definition of energy given in wave mechanics in quantum physics: "from this point of view it is therefore completely meaningless to attempt to talk about the energy of the entire universe" (p. 77). There is not a common and reducible set of definitions of energy that can be applied to all possible ways of perceiving and representing energy on different scales and in different contexts. The dilemma about the existence of nonreducible definitions of the identity of energetic systems and the relative assessment of energy forms directly recalls the historic impasse experienced in physics (e.g., by Boltzmann when attempting the unification of statistical representation and classic thermodynamics, and in quantum physics when attempting to handle the dual nature of particles). Such an issue has been directly investigated by Rosen (1985) in his *Anticipatory Systems*. Actually, this is the issue that led him to introduce the concept of complex time. In Chapter 4 of his book, he provides an overview of encodings of time, in which he shows that the formal definitions of time differentials within different representative frames (even when applied to conservative systems) are nonequivalent and nonreducible to each other. To prove this fact, he explores the various formal definitions of time differentials in Newtonian dynamics, thermodynamics and statistical analysis, probabilistic time, and time in general dynamical systems, where also the pace at which the observer can perform measurement matters. Concluding this chapter, Rosen (1985) says:

> We have abundantly seen that the quality we perceive as time is complex. It admits a multitude of different kinds of encoding. …Each of these capture some particular aspects of our time sense, at least as these aspects are manifested in particular kinds of situations. While we saw that certain formal relations could be established between these various kinds of time, none of them could be reduced to any of the other; nor does there appear to exist any more comprehensive encoding of time to which all of the kinds we have discussed can be reduced. (p. 271)

The impossibility of obtaining a substantive and formal definition of time is related to the impossibility of obtaining a substantive and formal definition of energy applicable across different scales to different situations.

This discussion supports the concern expressed by James Kay about the existing confusion in the definition and use of the concept of entropy in existing literature:

> I have not seen a good general treatment of the relationship between entropy change, entropy generated in a system, and environment and exergy change. There are a lot of examples of this being sorted out for specific cases, but not in general and not for biological systems (except for some specific cases). This is the reason that dissipation and degradation are used in sloppy ways as it is never quite clear if one is talking about entropy change, entropy generation, exergy change, gradient change, or heat transfer and if it is *for the system, or system plus environment*. (Kay, 2002, emphasis added)

In our view, the last statement is crucial, and we will discuss it more below.

7.5.2 Schroedinger's and Prigogine's Metaphor of Negative Entropy

We now want to briefly apply the rationale of hierarchical reading of energy transformations occurring within an autocatalytic loop of energy forms (the theoretical basis of ILA) to the famous scheme proposed by Prigogine to explain in entropic terms the biophysical feasibility of dissipative systems. This follows the intuition of Erwin Schroedinger (*What Is Life?*, 1945) that living systems can escape the curse of the second law thanks to their ability to feed on negentropy. Put another way, living systems are open systems that can preserve their identity thanks to a metabolic process that requires the compatibility of their identity with the identity of their context.

This original idea has been developed by the work of the school of Prigogine in nonequilibrium thermodynamics. They introduced the class of dissipative systems, in which the concept of entropy is associated with that of self-organization and emergence. In this way, it becomes possible to better characterize the concept of metabolism of dissipative systems. The ability of generating and preserving in time a given pattern of organization, which would be improbable according to the laws of classic equilibrium thermodynamics, depends on two self-entailing abilities: (1) the ability to generate the entropy associated with the energy transformations occurring within the system (those transformations required to generate the pattern) and (2) the ability to discharge this entropy into the environment at a rate that is proportional to that of internal generation. Put another way, the possibility of having life, self-organization and autocatalytic loops of energy forms is strictly linked to the ability of open dissipative systems of generating and discharging entropy into an admissible environment (Schneider and Kay, 1994). Actually, the more they can generate and discharge, the higher the complexity of patterns that can be sustained.

Using the vocabulary developed so far, we can say that open systems can maintain a level of entropy generation that is admissible in relation to their identity of a metabolic system (a given pattern of energy dissipation that is associated with an ordered structure of material flows and stocks characterized in relation to the expected favorable environment). Using the vocabulary developed by Schroedinger and Prigogine, this requires compensating the unavoidable generation of entropy associated with internal irreversibility (dS_i) with an adequate import of negentropy (dS_e) from the context. The famous scheme proposed by Prigogine to represent this idea is commonly written as

$$dS_T \Leftrightarrow dS_i + dS_e \qquad (7.1)$$

which implies that the identity of the system defined at the interface level n/level $n - 1$, associated with the two flows of entropy, internally generated (dS_i) and imported/exported (dS_e), must be congruent (or compatible) with the identity of the larger dissipative system in which they are embedded — the picture of the dissipative system as obtained from the level $n + 1$/level n (dS_T).

7.5.2.1 Applying This Rationale to Autocatalytic Loop across Hierarchical Levels —

Now visualize the application of this scheme to the representation of a nested hierarchical system made of metabolic systems, as illustrated in Figure 7.18. For example, imagine that the level n refers to the perception and representation of the metabolism of an organ (e.g., a liver) operating within an individual human being (level $n + 1$) that is operating within a household (level $n + 2$) that is operating within a given village (an environment with favorable characteristics). It should be noted that a structure such as the one represented in Figure 7.18 is mandatory according to the basic scheme proposed by Prigogine.

Several evident problems with this approach can be immediately detected even by a cursory look at this figure. The representations of the autocatalytic loop in Figure 7.1 and Figure 7.8 were based on two parallel nonequivalent mappings of matter and energy flows on different levels: (1) a view of the autocatalytic loop — from within — defined over three contiguous levels ($n - 2$/$n - 1$/n) used to explain how the black box is operating and (2) a view of the same autocatalytic loop — from outside — also defined over three contiguous levels (n/$n + 1$/$n + 2$), related to the compatibility of the behavior of the black box in relation to the characteristics of the environment. If we want to formalize (by assigning empirical data to) Equation 7.1, we have to use just one descriptive domain.

Now imagine that all this information (referred to views from inside and outside (Figure 7.1)) can be compressed thanks to the magic power of the entropy concept and as a result of a smart selection of three mappings: dS_T, dS_e, dS_i (letters referring to the white arrows), as done in Figure 7.18. These letters and arrows refer to a hypothetical formalization of Prigogine's scheme to characterize the effect of entropic processes related to level n. These three numerical assessments are referring to the interface level n/level $n + 1$, but they must include also information about mechanisms generating internal irreversibility (represented on the interface level $n - 2$/level $n - 1$). Because of the nested structure, these three mappings must be congruent with the other triplet of mappings: dS_T, dS_e, dS_i (letters referring to the gray arrows), this time used to characterize in numerical terms the effect of the arrows of entropies related to the application of Prigogine's scheme on level $n + 1$. These assessments are referring to the interface level $n + 1$/level $n + 2$, but they must also include information about mechanisms generating internal irreversibility (represented on the interface level $n - 1$/level n). Obviously, these three mappings must also be congruent with the other triplet of mappings: dS_T, dS_e, dS_i (letters referring to the black arrows), used to characterize in numerical terms the arrows related to the reading of the scheme on level $n + 2$. These assessments are referring to the interface level $n + 2$/level $n + 3$, but they must also include information about mechanisms generating internal irreversibility (represented on the interface level n/level $n + 1$).

To spare additional suffering to the reader, we stop here our stroll through this nested chain of representations of entropic processes associated with metabolic elements. In real situations, however, we have to expect a much longer journey across levels (when going from heat assessments associated with the movement of molecules within the body of consumers, to the processes generating the curves of demand and supply within economic systems, to the thermal engine of the water cycle discharging entropy for Gaia into outer space, as suggested by Tsuchida and Murota (1987)). It is obvious that the

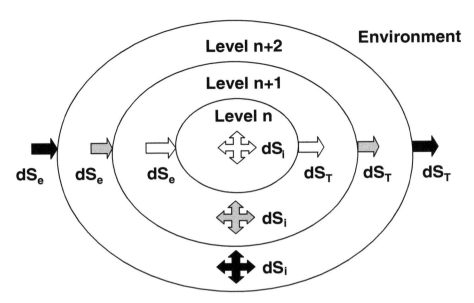

FIGURE 7.18 Hierarchical levels that should be considered for studying entropy exchanges according to Prigogine's scheme.

idea of being able to keep coherence in such a chain of formalizations across scales is ludicrous. This is why the epistemic triadic filtering (Chapters 2 and 3) is required.

The epistemological troubles related to the parallel representation of events that are occurring on different time scales are related to the impossibility of handling in formal way the complex nature of time (a la Rosen). That is, going back to Figure 7.1, we can see that over the three levels $n - 2/n - 1/n$, the mechanism of conversion of energy input (white arrow) into useful energy (gray arrow) requires assuming that the energy carriers associated with the representation of the white arrow (e.g., the food) enters into the converter (e.g., the farmers working in agriculture) *before* the gray arrow is generated. However, when dealing with the representation of the typology of the black box (when describing the interaction between the black box and its environment) — at the levels $n/n + 1/n + 2$ — it is the gray arrow that is considered to be the causal source of the arrival of the energy carriers into the black box. That is, it is the gray arrow (e.g., work in agriculture) that is generating the flow represented by the white arrow (e.g., the food harvested by the farmers). This applies also to an economic representation of the autocatalytic loop determining a farm:

> For example, an agricultural field is maintained as quite an improbable ecological community functioning in a particular context. Such an organized structure persists because the growth of crops generates enough money to allow the farmer to put in a new crop next year. If the market changes, the farmer shifts the realization by turning to a different crop. Thus the field persists as a production unit. The makings of the pattern reinforce themselves in a loop of structure feeding process, a loop that amounts to the whole process of ecological engineering. (Allen et al., 2003)

We saw in Chapter 3 that the identification of a given direction of causality in nested holarchies is impossible when dealing with systems operating on multiple scales and levels. For example, the concept of consumer democracy assumes that consumers when choosing goods on the market are determining what types of goods will be produced in the future (they provide an explanation to *why* certain goods remain and others disappear). On the other hand, when looking at *how* goods are produced (technical aspects of the productive process), the selection of goods on the market can only be done after these goods have been produced. Technical aspects provide an explanation to how certain goods arrive on the market.

Coming back to Figure 7.18, as soon as we look for the simultaneous validity of all the formalizations of the triplets of dS_T, dS_e, dS_i, referring to nonequivalent sets of representation of events perceived on different hierarchical levels, it becomes clear that this task would require the simultaneous adoption of assumptions about identities of elements and definition of space–time scales, which are inconsistent with each other. The metabolism of a liver can be represented in term of its "eating" glucose molecules as energy carrier (on a time horizon of hours), whereas the metabolism of a human being can be represented in term of its "eating" a variety of food products (on a time horizon of a year). Finally, the metabolism of a household can be represented in term of its "eating" a variety of energy carriers (electricity, gasoline, coal) over several years — associated with changes in age structure and the turnover of technical devices. Moreover, the typologies of energy carriers of a household will depend on the characteristics of its associative context (e.g., if it is operating in a subsistence society or in an industrialized country), whereas the energy carriers for a liver are the same in rich and poor countries.

In the situation represented in Figure 7.18, it is meaningless to look for a formal and substantive mechanism of accounting based on the concept of entropy, exergy or whatever other new thermodynamic function we want to introduce to address qualitative differences in energy flows. Qualitative differences of energy assessments are not substantive, but they always depend on the preliminary decision about how to perceive and characterize an autocatalytic loop across levels and scales. If we insist on looking at thermodynamic analysis to provide substantive quality assessment to energy forms, we simply keep trying to answer questions that cannot be answered. For example, what do we mean when we refer to the various triplets of entropies shown in the nested chain in Figure 7.18 with internal entropy production? What entropic assessment-related characterization of the system do we have in mind? For example, rate of change of entropy, rate of generation of entropy, rate of disposal of entropy. Moreover, such an assessment should be related to which substantive definition of system? For example, what is the system when dealing with a nested hierarchy of elements necessarily open on the top? How do we deal with the representation of the energy dissipation of individual elements in the middle? For example, should we be using a descriptive domain reflecting the perception and representation of events from within or a descriptive domain reflecting the perception and representation of events from its context, which in reality is another context? As noted earlier, both are needed and relevant, but they are not reducible to each other, since they are overlapping, but only when adopting nonequivalent definitions of time and space (definition of time and space differentials in relation to measurement scheme and the expected validity of the assumption of quasi-steady state). Finally, how do we decide what are the relevant forms of energy to be included or neglected from the accounting on different levels? Should we account for gravitational energy when accounting for energy consumptions of households living in coastal areas? Very few include such input (but the Odum school), even though tides can represent relevant agents determining the characteristics of the admissible environment of these households.

7.5.2.2   *Interpreting the Scheme Proposed by Prigogine in Metaphorical Terms* —
Now try to interpret in a metaphorical sense the scheme proposed by Prigogine to characterize autocatalytic loops of energy forms within nested metabolic systems:

1. dS_i refers to the representation of the mechanism of internal entropy production, which is associated with the irreversibility generated to preserve system identity. This assessment necessarily must be obtained using a representation of events related to a perception obtained within the black box — on the interface level $n – 2/n – 1/n$. This can be done by using a mapping of an energy form that makes it possible to represent how the total input used by the black box is then invested among the various parts. But this implies assuming a space–time scale for perceiving and representing events, which must be compatible with the identity of both energy carriers and energy converters. For example, this energy mapping can be related to the amount of chemical bonds made available through gross primary production to an ecosystem (and then divided within the various compartments of an ecosystem), as done in Figure 7.5, or the total amount of energy made available to organs operating within humans, as done in Figure 6.1.

2. dS_e refers to the representation of imported negative entropy. The term *negentropy* entails the adoption of a mechanism of mapping of energy that is directly related (it must be reducible) to the mechanism chosen when representing dS_i. In fact, the term *negentropy* requires establishing a bridge between the assessment of two forms of energy:

    a. One used to describe events within the black box in terms of dS_i (the mapping used to assess the effect of internal mechanisms associated with irreversibility) — e.g., extensive variable 1, such as chemical bonds obtained through photosynthesis or food energy eaten by people

    b. One used to describe the interaction of the black box with its context — e.g., extensive variable 2, such as solar radiation used to evapotranspirate the water associated with the photosynthesis generating the given amount of gross primary productivity (this would be the energy form associated with dS_e when dealing with gross primary productivity as the energy form associated with dS_i).

    The amount of entropy assessed when considering extensive variable 1 must be compatible with the room provided by extensive variable 2.

    That is, the concept of a sum of dS_i and dS_e can be referred to the coupling of two forms of energy within an autocatalytic loop. The first energy form is used to represent relevant mechanisms inside the black box (those associated with the stabilization of the identity of the box as a metabolic system). The second energy form is used to represent the interaction of the black box with its associative context (the energy form checking whether the hypothesis about admissibility of boundary conditions holds). It represents the view of the metabolism from the outside.

3. dS_T refers to the admissibility of the identity of a known typology of dissipative system (e.g., a human being, a dog, a car) with the actual context in which it is operating across the various levels, as it is required to represent the autocatalytic loop.

As noted earlier, the simultaneous check of this compatibility — a formalization in a substantive way of the relation among the three dS values — is not possible. However, the metaphorical message is the same as the one found when discussing ILA. An autocatalytic loop of different energy forms entails the congruence across levels of formal identities defined on different descriptive domains. This implies describing such an interaction in terms of a forced relation among known identities of energy forms defined over five contiguous hierarchical levels (from level $n - 2$ to level $n + 2$). If we structure our information space in a way that makes it possible to perform a series of congruence checks on the selected representation of the autocatalytic loop of energy forms, we can build a tool kit that can be used to look for biophysical constraints to the feasibility of this process.

However, this require accepting two negative side effects:

1. The same autocatalytic loop can be represented in different legitimate ways (by using different combinations of formal identities). By legitimate we mean a selection of identities that provides congruence among the representations of flows.

2. There is an unavoidable degree of ignorance associated with the representation. In fact, the very set of assumptions (e.g., about the future stability of the relation between investment and return, and admissibility of the environment on level $n + 2$) that make it possible to represent the system guarantee that such a representation is affected by uncertainty ("ceteris are never paribus" when coming to the representation of autocatalytic loops).

The concept of entropy, in this case, translates into a sort of Yin-Yang predicament. The triadic reading on contiguous levels, which is possible thanks to the self-entailment of identities across contiguous levels, makes it possible to perceive and represent an autocatalytic loop. On the other hand, it also implies that such a representation is just one among many alternatives and that it is affected by uncertainty and ignorance (a sure obsolescence).

7.6   Conclusion: The Peculiar Characteristics of Adaptive Metabolic Systems

Thermodynamic analysis, even when adopting mysterious and esoteric terms such as entropy, cannot be used to deal in substantive terms with the sustainability of complex adaptive holarchies. On the other hand, entropy provides a very powerful metaphor related to the peculiar characteristics of adaptive metabolic systems. The main semantic messages associated with this concept are:

1. Openness of these systems in terms of interaction with the context (therefore, one has to expect a fuzzy definition of what is the border between system and environment). There is a distinction between the functional boundary of types (associated with the expected domain of influence of a pattern) and a real boundary associated with a particular realization (which therefore is not particularly relevant in scientific terms).

2. Organization in nested hierarchies (therefore, one should expect to find different useful identities for the same system when observing it at different scales and when using different detectors).

3. The possibility to use types to perceive and represent characteristics of individuals (complex adaptive systems must be organized in equivalence classes of organized structures sharing the same template).

4. Awareness that our knowledge of types entails a certain degree of uncertainty, because all dissipative systems are special.

5. Unavoidable emergence of novelties (complex adaptive systems must become something else in time), which implies the existence of an unavoidable degree of ignorance in our forecasting of future states or events (it is impossible to predict future scenarios).

6. Unavoidable degree of arbitrariness in any formal representation (any representation of these systems reflects the perception associated with the peculiarity of an observer–observed complex), which cannot be considered substantive.

This is a crucial point that will be developed in the following chapter about the unavoidable arbitrariness of the choices made by scientists observing reality, which translates into the need to select useful narratives for surfing complex time. Scientists can only measure and study observer-dependent representations of complex adaptive systems, which are simplified versions of reality and which become obsolete in time. This translates into the existence of legitimate, but contrasting views about the usefulness of a given representation, which are directly related to legitimate, but contrasting interests of nonequivalent observers.

Looking at this list of semantic messages carried out by a metaphorical interpretation of the concept of entropy, we can easily understand why this concept is so popular in the debate over sustainability. In fact, if on one side entropy cannot provide us with a "magic bullet" made up of analytical tools that make it possible to formalize, measure and individuate in substantive terms the best course of action in relation to sustainability of complex systems, then on the other side we have to admit that such a concept has tremendous potential to help interdisciplinary researchers look at old problems by using different questions. The only caution is that one should always be aware that thermodynamics, in spite of its look as a very hard science, is just another possible narrative available to humans to make sense of their shared experience of reality (Funtowicz and O'Connor, 1999).

The same set of considerations suggesting that a substantive formalization of Prigogine's scheme is impossible (because of the epistemological predicament of complexity discussed so far) also provides a way out from the formal impasse. Scientists can take advantage of the robustness of mosaic effects of self-entailing identities of adaptive complex systems across scales. To this regard, the metaphorical message given by the triplet of entropic assessments proposed by Progogine is that it is possible to establish relations of congruence among nonequivalent definitions of formal identities across contiguous levels (impredicative loop analysis). This relation of congruence will impose constraints on the value taken by assessments of energy and matter flows in relation to the particular choice of formal identities assigned to the (1) system, (2) components, (3) energy carriers and (4) transformations in relation to what has be assumed to be an admissible environment.

Put another way, a known typology of metabolic system can be associated with a specific typology of autocatalytic loop of energy forms; we can expect an association between known identities and experienced patterns. This autocatalytic loop can be represented on contiguous hierarchical levels using formal identities of types in a way that imposes a set of reciprocal constraints on the value taken by the various assessments of energy flows associated with a given selection of identities (to a given choice of how to represent such a phenomenon). That is, we can expect to find the mosaic effect across levels.

These points are very relevant for developing analysis of biophysical constraints affecting the organization of nested dissipative systems. This is the reason why we believe that an integrated use of these concepts can be at the basis of useful procedures of multiple-scale integrated analysis of complex dissipative holarchies.

References

Adams, R.N. (1988), *The Eighth Day: Social Evolution as the Self-Organization of Energy*, University of Texas Press, Austin.

Allen, T.F.H., Havlicek, T. and Norman, J., (2001), Wind tunnel experiments to measure vegetation temperature to indicate complexity and functionality, in *Advances in Energy Studies: Exploring Supplies, Constraints and Strategies*, Ulgiati, S., Brown, M.T., Giampietro, M., Herendeen, R.A., and Mayumi, K., Eds., SGE Editoriali, Padova, Italy, pp. 135–145.

Allen, T.F.H., Giampietro, M. and Little, A.M., (2003), The distinct challenge of ecological engineering in contrast to other forms of engineering, *Ecol. Eng.*, in press.

Anonymous (1985), Energy and Protein Requirements. Report of a Joint FAO/WHO/UNU Expert Consultation. WHO Technical Report Series 724. World Health Organization, Geneva.

Ayres, R.U. (1999), The second law, the fourth law, recycling and limits to growth, *Ecol. Econ.*, 29, 473–483.

Bianciardi, C., Tiezzi, E. and Ulgiati, S., (1993), Complete recycling of matter in the frameworks of physics, biology and ecological economics, *Ecol. Econ.*, 8, 1–5.

Bianciardi, C., Tiezzi, E. and Ulgiati, S., (1996), The "recycle of matter" debate: physical principles versus practical impossibility, *Ecol. Econ.*, 19, 195–196.

BP-Amoco (2000), World-energy statistics, available at http://www.bpamoco.com/worldenergy/.

BBridgman, P. (1961), *The Nature of Thermodynamics*, Harper & Brothers, New York.

Brooks, D.R. and Wiley, E.O. (1988), *Evolution as Entropy*, The University of Chicago Press, Chicago.

Clausius, R. (1867), *Mechanical Theory of Heat, with Its Applications to the Steam-Engine and to the Physical Properties of Bodies*, Hirst, T.A., Ed., John van Voorst, London.

Converse, A.O. (1996), On complete recycling, *Ecol. Econ.*, 19, 193–194.

Converse, A.O. (1997), On complete recycling 2, *Ecol. Econ.*, 20, 1–2.

Costanza, R. (1980), Embodied energy and economic valuation, *Science*, 210, 1219–1224.

Costanza, R. and Herendeen, R. (1984), Embodied energy and economic value in the United States economy: 1963, 1967 and 1972, *Resources Energy*, 6, 129–163.

Costanza, R., Daly, H.E. and Bartholomew, J.A. (1991), Goals, agenda, and policy recommendations for ecological economics, in *Ecological Economics: The Science and Management of Sustainability*, Costanza, R., Ed., Columbia University Press, New York, pp. 1–20.

Cottrell, W.F., (1955) *Energy and Society: The Relation between Energy, Social Change, and Economic Development*, McGraw-Hill, New York.

Craig, P.P. (2001), Energy limits on recycling, *Ecol. Econ.*, 36, 373–384.

Debeir, J.-C. Deléage, J.-P., and Hèmery, D., (1991), *In the Servitude of Power: Energy and Civilization through the Ages*, Zed Books Ltd., Atlantic Highlands, NJ.

Feynman, R., Leighton, B. and Sands, M. (1963), *The Feynman Lectures on Physics: Mainly Mechanics, Radiation, and Heat*, Vol. I, Addison-Wesley Publishing Company, Menlo Park, CA.

Fluck, R.C. (1981), Net energy sequestered in agricultural labor, *Trans. Am. Soc. Agric. Eng.*, 24, 1449–1455.

Fluck, R.C. (1992), Energy of human labor, in *Energy in Farm Production*, Vol. 6 of Energy in World Agriculture, Fluck, R.C., Ed., Elsevier, Amsterdam, pp. 31–37.

Funtowicz, S. and O'Connor, M. (1999), The poetry of thermodynamics, in *Bioeconomics and Sustainability: Essays in Honor of Nicholas Georgescu-Roegen*, Mayumi, K. and Gowdy, J., Eds., Edward Elgar, Cheltenham, U.K.

Gell-Mann, M. (1994), *The Quark and the Jaguar*, Freeman, New York.

Georgescu-Roegen, N. (1971), *The Entropy Law and the Economic Process*, Harvard University Press, Cambridge, MA.

Gever, J., Kaufmann, R., Skole, D. and Vörösmarty, C. (1991), *Beyond Oil: The Threat to Food and Fuel in the Coming Decades*, University Press of Colorado, Niwot.

Giampietro, M. (1997), Linking technology, natural resources, and the socioeconomic structure of human society: a theoretical model, in *Advances in Human Ecology*, Vol. 6, Freese, L., Ed., JAI Press, Greenwich, CT, pp. 75–130.

Giampietro, M. (2003), Complexity and scales: the challenge for integrated assessment, in *Scaling Issues in Integrated Assessment*, Rotmans, J. and Rothman, D.S., Eds., Swets & Zeitlinger B.V., Lissen, Netherlands, pp. 293–327.

Giampietro, M., Bukkens, S.G.F. and Pimentel, D. (1994) Models of energy analysis to assess the performance of food systems. *Agric. Syst.* 45 (1): 19–41.

Giampietro, M., Bukkens, S.G.F., and Pimentel, D. (1993), Labor productivity: a biophysical definition and assessment, *Hum. Ecol.*, 21, 229–260.

Giampietro, M., Bukkens, S.G.F., and Pimentel, D. (1997), The link between resources, technology and standard of living: examples and applications, in *Advances in Human Ecology*, Vol. 6, Freese, L., Ed., JAI Press, Greenwich, CT, pp. 129–199.

Giampietro, M. and Mayumi, K. (2000a), Multiple-scale integrated assessment of societal metabolism: Introducing the approach, *Popul. Environ.*, 22, 109–153.

Giampietro, M. and Mayumi, K. (2000b), Multiple-scale integrated assessment of societal metabolism: integrating biophysical and economic representation across scales, *Popul. Environ.*, 22, 155–210.

Giampietro, M., Mayumi, K., and Bukkens S.G.F. (2001), Multiple-scale integrated assessment of societal metabolism: an analytical tool to study development and sustainability, *Environ. Dev. Sustain.*, 3, 275–307.

Giampietro, M. and Pastore, G. (2001), Operationalizing the concept of sustainability in agriculture: characterizing agroecosystems on a multi-criteria, multiple-scale performance space, in *Agroecosystem Sustainability-Developing Practical Strategies*, Gliessman, S.R., Ed., CRC Press, Boca Raton, FL, pp. 177–202.

Giampietro, M. and Pimentel, D. (1990), Assessment of the energetics of human labor, *Agric. Ecosyst. Environ.*, 32, 257–272.

Giampietro, M. and Pimentel, D. (1991a), Energy efficiency: assessing the interaction between humans and their environment, *Ecol. Econ.*, 4, 117–144.

Giampietro, M. and Pimentel, D. (1991b), Model of energy analysis to study the biophysical limits for human exploitation of natural processes, in *Ecological Physical Chemistry*, Rossi, C. and Tiezzi, E., Eds., Elsevier Publisher, Amsterdam, pp. 139–184.

Giampietro, M., Pimentel, D., and Cerretelli, G. (1992), Energy analysis of agricultural ecosystem management: human return and sustainability, *Agric. Ecosyst. Environ.*, 38, 219–244.

Gilliland, M.W., Ed. (1978), *Energy Analysis: A New Policy Tool*, Westview Press, Boulder, CO.

Glansdorff, P. and Prigogine, I. (1971), *Thermodynamics Theory of Structure, Stability and Fluctuations*, John Wiley & Sons, New York.

Hall, C.A.S., Cleveland, C.J., and Kaufmann, R. (1986), *Energy and Resource Quality*, John Wiley & Sons, New York.

Herendeen, R.A. (1981), Energy intensities in economic and ecological systems, *J. Theor. Biol.*, 91, 607–620.

Herendeen, R.A. (1998), *Ecological Numeracy: Quantitative Analysis of Environmental Issues*, John Wiley & Sons, New York.

IFIAS (International Federation of Institutes for Advanced Study) (1974), Energy Analysis, International Federation of Institutes for Advanced Study, Workshop on Methodology and Conventions, Report 6, IFIAS, Stockholm, p. 89.

Jevons, W.S. ([1865] 1965), *The Coal Question: An Inquiry Concerning the Progress of the Nation, and the Probable Exhaustion of Our Coal-Mines*. A. W. Flux (Ed.), 3rd ed. rev. Augustus M. Kelley, New York.

Kåberger, T. and Månsson, B. (2001), Entropy and economic processes: physics perspective, *Ecol. Econ.*, 36, 165–179.

Kay, J. (2002), Some Observations about the Second Law and Life, based on a dialogue with J.R. Minkel when he was writing the article on our work for *New Scientist*, by James J. Kay, available at http://www.jameskay.ca/musings/thermomusings.pdf.

Kleene, S.C. (1952), *Introduction to Metamathematics*, D. van Nostrand Co., New York.

Kümmel, R. (1994), Energy, entropy: economy, ecology, *Ecol. Econ.*, 9, 194–196.

Leach, G. (1976), *Energy and Food Production*, I.P.C. Science and Technology Press Ltd., Surrey, U.K.

Lotka, A.J. (1922), Contribution to the energetics of evolution, *Proc. Natl. Acad. Sci. U.S.A.*, 8, 36–48.

Lotka, A.J. (1956), *Elements of Mathematical Biology*, Dover Publications, New York.

Månsson, B. (1994), Recycling of matter, *Ecol. Econ.*, 9, 191–192.

Mayumi, K. (1993), Georgescu-Roegen's "fourth law of thermodynamics," the modern energetic dogma, and ecological salvation, in *Trends in Ecological Physical Chemistry*, Bonati, L., Cosentino, U., Lasagni, M., Moro, G., Pitea, D., and Schiraldi, A., Eds., Elsevier, Amsterdam, pp. 351–364.

Mayumi, K. and Giampietro M. (2004), Entropy in Ecological Economics In J. Proops and P. Safonov (Eds.) *Models in Ecological Economics*, Edward Elgar, Cheltenham, U.K. (in press).

Mendoza, E., Ed. (1960), *Reflections on the Motive Power of Fire by Sadi Carnot and Other Papers on the Second Law of Thermodynamics by É. Clapeyron and R. Clausius*, Dover Publications, Inc., New York.

Morowitz, H.J. (1979), *Energy Flow in Biology*, Ox Bow Press, Woodbridge, CT.

Nicolis, G. and Prigogine, I. (1977), *Self-Organization in Nonequilibrium Systems*, Wiley Interscience, New York.

Odum, H.T. (1971), *Environment, Power, and Society*. Wiley-Interscience, New York.

Odum, H.T. (1983), *Systems Ecology*. John Wiley, New York.

Odum, H.T. (1996), *Environmental Accounting: EMergy and Decision Making*, John Wiley, New York.

Odum, H.T. and Pinkerton, R.C. (1955), Time's speed regulator: the optimum efficiency for maximum power output in physical and biological systems, *Am. Sci.*, 43, 331–343.

Ostwald, W. (1907), The modern theory of energetics, *Monist*, 17, 481–515.

Pastore, G., Giampietro, M., and Li, J. (1999), Conventional and land-time budget analysis of rural villages in Hubei province, China, *Crit. Rev. Plant Sci.*, 18, 331–358.

Pimentel, D. and Pimentel, M. (1979), *Food, Energy, and Society*, Edward Arnold, London.

Pimentel, D. and Pimentel, M. (1996), *Food, Energy and Society*, rev. ed., University Press of Colorado, Niwot.

Podolinsky, S. (1883), Menschliche arbeit und einheit der kraft, *Die Neue Zeit* (Stuttgart), p. 413.

Prigogine, I. (1961), *Introduction to Thermodynamics of Irreversible Processes*, 2nd, rev. ed., John Wiley & Sons, New York.

Prigogine, I. (1978), *From Being to Becoming*, W.H. Freeman and Co., San Francisco.

Prigogine, I. and Stengers, I. (1984), *Order out of Chaos*, Bantam Books, New York.

Ramos-Martin, J. (2001), Historical analysis of energy intensity of Spain: from a "conventional view" to an "integrated assessment," *Popul. Environ.*, 22, 281–313.

Rappaport, R.A. (1971), The flow of energy in an agricultural society, *Sci. Am.*, 224, 117–133.

Rosen, R. (1985), Anticipatory Systems: Philosophical, Mathematical and Methodological Foundations. Pergamon Press, New York.

Rosen, R. (1991), Life Itself: A Comprehensive Inquiry into the Nature, Origin and Fabrication of Life. Columbia University Press, New York. 285 pp.

Rosen, R. (2000), Essays on Life Itself. Columbia University Press, New York. 361 pp.

Schandl, H., Grünbühel, C.M., Thongmanivong, S., Pathoumthong, B., and Schulz, N. (2003), Socio-Economic Transitions and Environmental Change in Lao PDR, final report of the Lao PDR country study of SEAtrans EC-Project I.F.F., Vienna.

Schneider, E.D. and Kay, J.J. (1994), Life as a manifestation of the second law of thermodynamics, *Math. Comput. Model.*, 19, 25–48.

Schroedinger, E. (1967), *What Is Life? Mind and Matter*, Cambridge University Press, London.

Slesser, M. (1978), *Energy in the Economy*, MacMillan, London.

Slesser, M. and King, J. (2003), *Not by Money Alone: Economics as Nature Intended*, Jon Carpenter Publishing, Charlbury, Oxon.

Smil, V. (1991), *General Energetics: Energy in the Biosphere and Civilization*, John Wiley & Sons, New York.

Tsuchida, A. and Murota, T. (1987), Fundamentals in the entropy theory of ecocycle and human economy, in *Environmental Economics*, Pillet, G. and Murota, T., Eds., R. Leimgruber, Geneva, pp. 11–35.

Ulgiati, S., Odum, H.T. and Bastianoni, S. (1994), EMergy use, environmental loading, and sustainability. An EMergy analysis of Italy. *Ecological Modelling* 73: 215–268.

USBC (United States Bureau of the Census) (1995), Statistical Abstract of the United States, 1990. U.S. Department of Commerce, Washington, DC.

Watt, K. (1989), Evidence of the role of energy resources in producing long waves in the US economy, *Ecol. Econ.*, 1, 181–195.

Watt, K. (1992), *Taming the Future*, Contextured Web Press, Davis, CA.
White, L.A. (1943), Energy and evolution of culture, *Am. Anthropol.*, 14, 335–356.
White, L.A. (1959), *The Evolution of Culture: The Development of Civilization to the Fall of Rome*, McGraw-Hill, New York.

8

Sustainability Requires the Ability to Generate Useful Narratives Capable of Surfing Complex Time*

This is the last chapter dealing with epistemological issues. Actually, after reading Chapters 6 and 7, in which the concepts of mosaic effects across levels and impredicative loop analysis were introduced, the reader fed up with epistemological discussions can skip this chapter and move directly to Part 3. The question answered by this chapter is: If we refuse the charge that the expression "sustainable development" is an oxymoron, then should we not be able to describe what it is that remains the same (sustainable) when the system becomes something else (development)? We understand that to some practitioners this question could appear too theoretical. However, the message proposed so far is that those analysts willing to deal with the issue of sustainability cannot just apply formal protocols. Complexity requires the adoption of flexible procedures of analysis that always imply an explicit semantic check. For this reason, we believe that those who are serious about developing analytical tools for dealing with sustainability should address first — as done in this chapter — the peculiarity of this predicament.

In this chapter, Section 8.1 introduces a few concepts that can be used to better frame the challenge implied by sustainability. The basic rationale proposed by Holling when representing evolutionary patterns (using the concepts of resilience, robustness and the cyclic movement among interrelated types — the adaptive cycle) is briefly introduced and translated into the narrative adopted so far in this book using the vocabulary presented in Part 1. Then Section 8.2, which is a technical section, deals with the concept of essence — something that cannot be formalized and that can be associated with the existence of multiple identities. The concept of essence requires a special discussion, since this is the elusive concept generating the epistemological predicament implied by complexity. In this section, first we provide several examples to show the relative unimportance of DNA in the definition of essences in biological systems. Then, using theoretical insights provided by the work of Rosen and Ulanowicz, we propose a mechanism that can be used to obtain a formal reading (images) of the unformalizable concept of essence within the analytical frame provided by network analysis. The final section, Section 8.3, deals with the definition of useful narratives in relation to the concept of complex time. Building on the concepts discussed in the previous two sections, we claim that useful narratives can only be defined by and in relation to a given complex observed–observer. Because of this, they have to be continuously updated during the never-ending process of evolution, which includes both the observer and observed system. In particular, the requirement of careful timing for updates becomes crucial when dealing with the reflexivity of human systems, that is, when observer and observed are at the same time observed by the observed and observing another observer, in a reciprocal process of interaction. This situation implies that both sides of the observer–observed complex can suddenly change their identity, implying that validated narratives can suddenly lose their usefulness.

8.1 What Remains the Same in a Process of Sustainable Development?

8.1.1 Dissipative Systems Must Be Becoming Systems

Dissipative systems are necessarily becoming systems (Prigogine, 1978), since they have to continuously negotiate their identity with their context in time. As discussed in the previous chapter, the very existence

_____
* Kozo Mayumi is co-author of this chapter.

(success in preserving its own identity) of a dissipative system implies the local destruction of favorable gradients on which its metabolism depends (a consequence of the second law of thermodynamics). Therefore, dissipative systems tend to destroy the expected stable associative context to which their current identity (type) is associated. Because of this, a reliable and predictable associative context must be a context that is stabilized by another process of dissipation (which requires in turn favorable boundary conditions on a higher level) occurring elsewhere (see Figure 7.18). This is the first mechanism that generates trouble in the representation of these systems. As noted before (Chapter 7), the stabilization of the identity of these systems can only be obtained through impredicative loops in which processes of dissipation occurring in parallel on different levels should be considered in terms of reciprocal entailment among identities defined on nonequivalent descriptive domains. When representing these systems, we have to select one (among many possible choices) window of three levels (triadic reading) and assume that (1) on the lower-lower level, structural stability is given and (2) on the higher level, favorable boundary conditions are stabilized by some benign process ignored by the model.

To make things more difficult, adaptive dissipative systems must use templates (e.g., DNA or social institutions) to guarantee the stability of their own identity (elements across levels). This implies the required stability of types expressed over a time window larger than the life span of individual components providing structural stability to the functions expressed by types. Realizations of an equivalence class have a shorter life span than that of the validity of the template used for making them. That is, organized structures sharing the same template undergo a process of turnover within a given set of expected types. Unfortunately, a mechanism of replication based on templates generates an additional problem of sustainability. Self-replicating dissipative systems are affected by an innate "Malthusian instability," the expression coined by Layzer (1988). As soon as a dissipative pattern associated with the existence of favorable boundary conditions finds a good niche (room for expansion), it tends immediately to expand its size by amplification (making more copies of the template). This means adding more individual organized structures belonging to the class sharing the characteristics of the type associated with the pattern. The sudden enlargement of the domain of activity of the pattern means jeopardizing the very survival of the mechanisms of replication. In fact, by making more copies of themselves (by making more of what seems to work under the existing perception of favorable boundary conditions), adaptive dissipative systems tend to amplify on a larger scale the rate of destruction of local favorable gradients. Probably a few readers have already recognized in this mechanism the ultimate driver generating the problems of sustainability of human affairs discussed in Chapter 1 (Jevons' paradox leading to the generation of various treadmills).

Any dissipative system that keeps growing in size just by amplifying the same basic process of dissipation will sooner or later get into trouble. We can recall here the story of Zhu Yuan-Chang's chessboard: if you put one kernel of rice on the first square, two on the second, four on the third, and keeping doubling the number for each square, there would be an astronomical number of kernels required for one position even before the 64th square is reached. This metaphor says it all. Using the expression proposed by Ulanowicz (1986), hypercycles (positive autocatalytic loops), when operating without a coupled process of control (and damping), do not survive for long — they just blow up. The expected trouble at one level (too much of an efficient type) implies that part of the surplus has to be invested in exploring new types (even if not efficient) able to diversify the set of relations expressed by the whole (on different levels).

Dissipative systems that use a template to replicate themselves, when taking advantage of existing favorable gradients, must reinvest a part of their energetic profit to become something else. This is the reason why mutations in DNA should not be considered errors, but a crucial mechanism associated with the ability of biological systems to evolve. We addressed this key feature of adaptive dissipative systems when representing (in Figure 7.8a) these systems as made up of two compartments: a direct compartment (where we can define the efficiency of the return on the investment) and an indirect compartment (where the system invests in adaptability). As noted in Section 3.6.3 (Figure 3.7), adaptive dissipative systems, to stabilize their own process of dissipation, have to balance their investments in efficiency (making stronger the actual set of identities) with investments in adaptability (expanding the option space of the set of virtual identities). This is what leads to the concept of sustainability dialectics. That is, it is not possible to formalize in a substantive representation of an optimizing function the expected trade-off between these two types of investments. Existing identities must not be too greedy; maximization of profit or efficiency implies reducing the option of expressing alternative virtual identities. The only certain point that can be driven home from the unavoidable process of becoming of complex adaptive system is that a strategy looking for a maximization of efficiency (obtained under the *ceteris paribus* hypothesis) is not the wise if one is concerned with the long-term stability of the system.

8.1.2 The Perception and Representation of Becoming Systems Require the Parallel Use of the Concepts of Identity and Individuality

In the 1970s Buzz Holling proposed a few concepts for the analysis of the sustainability of changes in ecological systems. These concepts were resilience, resistance (or robustness) and stability. The use of these concepts to represent the issue of sustainability of ecological systems has remained very popular among those trying to make formal analyses of sustainability for both human and ecological systems — an overview given by Holling himself about the use of these concepts is available in Holling and Gunderson (2002). It should be noted, however, that in spite of the large popularity of this narrative, and the crucial importance of these concepts for the understanding of the evolution and behavior of ecosystems, very little effort has been invested by those using these concepts in getting engaged in an epistemological discussion of them. If you were to ask different ecologists about the definition of these terms, you would get different answers. By looking at the literature in this field, one can find several definitions of resilience that are nonequivalent and nonreducible. Very often they are even listed as a set of interchangeable, optional definitions, without their mutual incompatibility and exclusiveness being addressed. It is obvious that the success of these terms is associated with their deep ambiguity, which can handle the different meanings that ecologists attach to them. A mathematician, on the other hand, would ask you to better specify the mathematical meaning of concepts like stability or resilience before getting into any discussion of their syntax. Obviously, this is not the way to advance in a critical epistemological appraisal of them. If we keep the terms too ambiguous, anyone can use them without problems, but in this way, one has to resort to a discussion about their semantics (what external referent should be used to share the meaning about them?). On the other hand, if the definition is too formal (as done by the mathematicians), everything is reduced to syntax. But exactly because of this, after having done that, it is no longer possible to discuss the semantic usefulness of the relative concept. Robert Rosen spent a large part of his academic career dealing with the epistemology of such a discussion. Therefore, this section has as its goal to share with the reader some of Rosen's insights.

Any theoretical discussion about the epistemology of these terms requires first answering the following question: When dealing with the analysis of the evolution of a given adaptive dissipative system, if we want to take measurements and make formal models about it, what remains the same when the system becomes something else?

Just to get our discussion started, let us try to describe the three concepts of resilience, resistance (or robustness) and stability. Two nonequivalent ways of defining these terms are listed below: (1) the definitions found in a dictionary (Merriam-Webster on-line) and (2) the semantic meanings conveyed by these terms according to a narrative and vocabulary taken from the work of Robert Rosen (1985, 1991, 2000). Obviously, we do not claim that what is posted below is the "right" interpretation of these terms. This is not the issue here. These definitions are needed for sharing with the reader the meaning assigned to these terms (to share a common understanding with the reader) in the rest of the chapter.

Resilience

> *From the dictionary*: The capability of a strained entity to recover its original condition (e.g., size, shape, structural characteristics) after a deformation caused by stress.

> *Narrative* (using Rosen terminology): Referring to the idea of multiple equilibrium states for a dynamical system. A given system has a certain identity. That system faces a perturbation (a nonadmissible environment) that makes its present state no longer viable (boundary conditions that are not compatible with the mechanism keeping the metabolism associated with a given type alive). The system can access alternative states (since it has multiple identities). In one of these alternative states the very same boundary conditions that were

not admissible for the previous identity become admissible. In this way, the system can preserve its individuality. This system must have the ability to switch among different viable states in relation to different definitions of admissible environment. Thus, it can preserve the ability to get back to the original state (type) when the perturbation is over. Examples are a tree branch bending under heavy wind (when the relevant state considered for defining its identity is only the position of the branch), bacteria forming a spore (relevant state considered is the original organizational structure of the bacteria, which comes back when the perturbation is over), and an ephemeral plant making seeds when the environment becomes too dry (as before).

Robustness (or Resistance)

From the dictionary: Having or showing firmness (firm = having a solid structure that resists stress, not subject to change or revision, not easily moved or disturbed).

Narrative: A given system has a certain identity. That system faces a perturbation (that would generate a nonadmissible environment). But the system can react to it, by fighting the process that is generating a hostile environment. This can be obtained by using a set of controls (a tool kit of alternative behaviors linked to anticipatory models based on previous experience of the same perturbation) — expressing behavior that is based on an anticipatory model "knowing" about the potential perturbation — or just having a size large enough or enough redundancy to overcome and dissipate the perturbation into an admissible noise. This requires also the ability to (1) expect possible perturbations and (2) control enough power (being able to express the dissipative pattern at a size large enough) to combat the exogenous perturbation. Examples are immune systems in mammals and storage of water in plants when facing a shortage of rain in the desert.

Stability

From the dictionary: The property of a body that causes it, when disturbed from a condition of equilibrium (or steady motion), to develop forces or movement that restore the original condition.

Narrative: If we want to translate this definition into a narrative based on Rosen's terminology, we should get into something very generic: the ability to retain your individuality in the face of perturbations no matter how you do it. This definition could be accepted as a variant of that of resilience or also as a variant of that of robustness. This is due to an open ambiguity in the definition of the terms used. To decide how to deal with this ambiguity, we should first be able to answer the following questions: What is the time threshold considered for retaining identity? What has to be considered a perturbation big enough to be distinct from normal noise? What defines a given individuality of a system that can change its identity in time? What defines a given type that is expressed by different individualities? Are we more interested in the preservation of types (same pattern stabilized by a turnover of lower-level structural elements) or individualities (path-dependent organized structures that changed their identity in time)?

The impossibility to answer in a general (substantive) way the above questions implies that often it is not possible to make a substantive distinction between the concepts of resilience, robustness and stability. Depending on what is the subject of our analysis (an individuality or a type), we can find different threshold values for assessing recover time and for defining the degree of perturbation and different useful strategies. To make things more difficult, the specification of these concepts is virtually impossible in nested hierarchical systems in which each of these concepts has to be defined on different scales (space–time domains), even though the resilience, robustness and stability of each level are affecting the others.

This deep epistemological ambiguity can explain why these concepts escape formalization. This also means that to better characterize this discussion in a different way, we have to introduce new epistemic categories. The introduction of new epistemic categories requires first of all the ability to share the meaning assigned to new labels and terms. This is the reason why this book invests a large part of its text in dealing with epistemological foundations and why, in the rest of this chapter, the reader will find a lot of pictures and examples taken from daily life experience, used to introduce concepts. Without introducing new concepts with examples familiar to everyone, it is impossible to share the meaning of new epistemic categories. On the other hand, without using new relevant concepts to be considered in analysis of sustainability (concepts that are ignored in reductionist science), it would be impossible to discuss how to conduct integrated assessments of agroecosystems in an innovative way.

Without a clear understanding of the differences in the meaning of concepts such as resilience, robustness and stability — or better, without having reached an agreement on the meaning that we want to assign to these labels or words in relation to the goals of our analysis — it is impossible to reach an agreement on how to represent the process of becoming (making analysis of sustainability), let alone to discuss strategies useful for improving the persistence of some of the characteristics of evolving systems that we (and who decides who is we?) would like to preserve.

To conclude this overview of the widespread confusion found in the field of analysis of sustainability of becoming systems, we can list additional concepts (variants of the previous ones) often used in literature that are associated with the ability to resist perturbation

1. **Redundancy/scale:** Because of this quality, the system can first resist and then even thrive on smaller-scale perturbations. It does so by incorporating them into the identity as functional activities, e.g., the use of wild fires by terrestrial ecosystems.
2. **Diversity:** Because of this quality, the system has the ability to work with multiple options —in terms of both possible behaviors and organizational states.

All the concepts listed so far are often confused, in their use, with each other, in the same way as the various strategies (redundancy, diversity and adaptability) are often ill-defined and used without a clear articulation of specific conditions and situations. Even worse is the situation with the term *adaptability*, which directly points to the process of becoming obtained by changing identity to preserve a given individuality. In a way, the concept of adaptability could also be associated with the concept of *persistence* (if only we were able to answer in formal terms the question: persistence of what?). Due to the relevance of the concept of adaptability (which implies a clear acknowledgment of the distinction and an innate tension between identity and individuality), we include below two nonequivalent definitions for adaptability and two metaphors useful for illustrating the concept:

Adaptability

From the dictionary: To make fit for a specific new use [goal] or new situation [context] often by modification.

Narrative: The ability to adjust our own identity to retain fitness in face of changing goals and changing constraints. Fitness means the ability to maintain congruence among (1) a set of goals, (2) the set of processes required to achieve them and (3) constraints imposed by boundary conditions. Since adaptive dissipative systems are history dependent, they preserve their individuality if they manage to remain alive in the process of becoming (the series of adjustments of their identity in time).

This definition can be confronted with the definition of sustainable development proposed in Section 4.2.2.

Useful metaphors about adaptability (from the *Bloomsbury Thematic Dictionary of Quotations*, available on the Internet):

- "If the hill will not come to Mahomet, Mahomet will go to the hill" (Francis Bacon).
- "President Rabbins was so well adjusted to his environment that sometimes you could not tell which was the environment and which was President Robbins" (Jarrel Randall).

The point to be driven home from all these examples of definitions is that the set of concepts proposed by Holling to deal with the evolution of adaptive systems entails an unavoidable severe epistemological

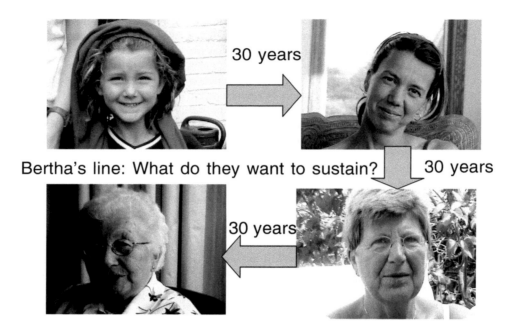

FIGURE 8.1 Sustainability of what? Sustainability for how long? Photos by Mario Giampietro.

challenge. Such a challenge is linked to the dilemma about (1) how to define the identity of the system, (2) how to define its context and (3) how to handle the fact that they change on different hierarchical levels at different paces. What is especially relevant in this discussion is the implicit constant requirement of both a syntactic and semantic appraisal of the terms used in these statements. When talking of adaptability and resilience, everything depends on:(1) what is considered to be the relevant set of characteristics used to determine (identify, perceive, represent) the identity of the system in the first place through observable qualities — type definitions and (2) what is considered to be the individuality of the system. The same individuality can remain — persist — even when its identity changes in time, as illustrated in the example of Figure 8.1. The four pictures given in Figure 8.1 can be imagined to be four views of Bertha, the old lady in the bottom-left picture, referring to four points in time of her life. As noted in Chapter 3, a peculiar way of expressing individuality of a holarchic system requires a preliminary choice made by the observer about an identity to be assigned to that individuality to make sense of the perceptions (signals carried by incoming data) referring to a given descriptive domain. The particular identity selected to organize our perceptions about a given individuality must be useful for the goal of the analysis. Differences in the choice of identity can be related to a different choice of scale or to a different choice of relevant attributes (as discussed in Chapter 3, e.g., Figure 3.1). In the case shown in Figure 8.1, we have an individuality (Bertha) that goes through a predictable trajectory of identities (types). Whenever the observer knows ahead of time that this will occur, she or he has to select the right set of observable qualities (epistemic categories) associated with the right type (the expected set of observable qualities useful to describe the individuality at a given historic moment).

This means that the characterization, perception, and representation of a given individuality of a becoming system over a large space–time domain (e.g., Bertha over her life span) requires the skillful handling of different identities. The same will occur if we want to study the multiple types that such an individuality could take (e.g., an overview of various members of different ages that are found in Bertha's family at a given point in time). Actually, when looking at the series of pictures given in Figure 8.1, we cannot know *a priori* if this series of pictures is representing the same person (individuality) at different points in time (e.g., taken at 30-year intervals) or if this series of pictures was taken in the same day looking at a genealogical line made up of a great-grandmother, her daughter, granddaughter and great-granddaughter. In both cases, we are dealing with a set of four types that are useful for describing a female human being. This implies that the selected type of identity used for the representation of a particular individual of female human being be appropriated to the goal of the analysis. This requirement translates into the need to use different models for representing and simulating the relative perception of changes associated with the selected type(s). Selecting just one among the possible relevant identities included in this set implies also selecting the relative appropriate model for simulating the expected behavior of the type. As already discussed in Part 1, formal models can refer to only one formal identity at a time. If we decide to represent Bertha when she is 90 years old, then we cannot imagine using a model that has been calibrated on the behavior of the type representing Bertha when she was 30 years old. In parallel, a model for simulating the behavior of a child cannot be used to simulate the behavior of an elderly person, even though they both represent women living in the Netherlands in the year 2000 (this is the homeland of Bertha).

That is, only after having specified one of the possible identities (i.e., the particular choice of triadic reading and the set of relevant attributes used to define the system), can we look for a model able to catch the set of expected causal relations used to predict expected changes in attributes. The scientist can attempt to make sense of experimental data only after having selected a given formal identity for the system and an inferential system able to simulate perceived changes in this formal identity (see Rosen, 1985, the chapter on modeling relations). The data set consists of different numerical values taken by a set of variables selected to encode changes in a set of relevant attributes, which are observable qualities associated with the choice of a measurement scheme, which are associated with the selection of a given formal identity. Because of this long chain of choices, all models are identity specific, and therefore they are bound to clash against complexity. Real natural systems are individualities operating on multiple scales or multiple types expressed simultaneously by a population of individualities. This is what entails the existence of multiple nonequivalent ways of mapping the same natural system when considering as relevant different sets of observable qualities (see Chapters 2, 3, 6 and 7).

As discussed in Part 1, the unavoidable existence of multiple valid models for the same reality is related not only to the complexity of the observed system, but also to the complexity of the observer. The existence of nonequivalent and nonreducible models for the same system is entailed by the simple fact that "life is the organized interaction of nonequivalent observers" (Rosen, 1985). In spite of being nonequivalent and nonreducible to each other, the various models used by nonequivalent observers can all be relevant for the study of the sustainability of becoming systems.

The main point made by Rosen (1985) about complex time is that any formalization of concepts such as resilience, robustness and adaptability into a mathematical system of inference has to deal with the existence of at least three relevant but distinct *time differentials*. The complexity of time in the process of making and using integrated set of models related to sustainability issues has to be contrasted with the simple time that is operating (only) within the simplified representation of reality obtained within reductionist models (formal systems of inference), when used one at a time. The three relevant time differentials are associated with the following processes:

1. The time differential selected for the dynamics simulated by the set of differential equations (in differential equations called dt).
2. The expiration date of the validity of the set of models used to simulate causality and the set of variables used to describe changes in the state space in relation to a given selection of relevant identities adopted in the problem structuring. When dealing with becoming systems, we have to explicitly address the unavoidable existence of a time horizon determining the reliability of the set of epistemic tools used to perceive, represent and simulate their behavior. The causal relation among observable qualities does change in time due to the process of becoming of these systems. This implies that functional forms and relations adopted in any given set of differential equations useful to simulate a becoming system at a given point in time should be updated sooner or later. The ability to observe and measure changes in observable qualities also evolves in time. That is, better proxies and better measurement schemes can become available to encode changes in relevant qualities of the system. This is another reason that can require changing and updating of the procedures adopted in the process of modeling. We call the time differential $d\tau$, at which the validity of the choice done in the process of modeling becomes obsolete.

3. The time horizon compatible with the validity of the problem structuring according to the *weltanschauung* of science and with the particular set of interests of the stakeholders in relation to a specific problem of sustainability. That is, any problem structuring implies a finite selection of (1) goals of the scientific analysis, (2) relevant qualities, (3) credible hypotheses about causal entailments, (4) observable qualities/selection of encoding variables, (5) related measurement protocols and data and (6) inferential systems — all of which must be compatible with each other. Out of a virtually infinite information space (including all the epistemological tools available to humans), scientists have to decide how to compress this intractable mass of information into a finite information space with which it is possible to do science (see Chapter 5). This process of compression of infinite to finite is called problem structuring, and it establishes an agreed-upon universe of discourse on which we apply our models to make sense of our potential actions. This choice will constrain what we perceive as happening in the world and what we decide to represent (actually what the scientists eventually represent) when defining the identity of the system to be investigated. As discussed at length in Chapter 5, this process of compression of infinite sets of identities, causal relations and goals into a finite set is in turn constrained by an underlying *weltanschauung* in which the scientific activity is performed and by the structure of power relations among the actors. The speed at which the basic *weltanschauung* is evolving (what the social consciousness defines as relevant issues and facts) can imply the obsolescence of some of the preanalytical choices associated with a given problem structuring. Changes at this level can imply important consequences on the speed at which the identity of the universe of discourse is evolving. This is especially clear in periods of paradigm shift. As noted in Chapter 4, the quality of the process generating a given problem structuring refers not just to the accuracy in the measurement and the calibration of models on data. The relevance of the set of qualities that should be included in the representation of system identity as well as the relevance of the set of causal relations that should be addressed by the model change in time. We call the pace of this process of evolution, in relation to the definition of complex time, a time differential, $d\theta$. When dealing with the perception and representation of sustainability, the relevance of this third time differential can become crucial.

In conclusion, we can define complex time as the parallel existence of nonequivalent relevant time differentials to be considered explicitly by the modelers (both inside and outside the model) when dealing with the implications of changes occurring in the observer–observed complex in relation to the validity of the model.

Why should a discussion about the existence of complex time be relevant for those reading this book? Because these concepts are crucial for discussing sustainability. It is very interesting to note that the distinction between identity (referring to the first two time differentials dt and $d\tau$) and individuality (referring to the second two time differentials $d\tau$ and $d\theta$) has been discussed by Rosen (1985, p. 403) using the metaphor of suicide. Suicide is a person terminating her or his individuality to resist the pressure of the context that would force a sudden change in her or his current identity. For example, there are people who take their life to avoiding aging, life without a loved one or because they are facing a failure. What is interesting in this case study is that when dealing with the complex observed–observer, the preservation of the current identity (just one among a set of possible ones) is obtained by eliminating (freezing) the observer (blocking the time differential $d\theta$), since nothing can be done about the changes on the ontological side (reality is forcing changes on the observed). The fact is that all becoming systems (biological and social entities) are history-dependent systems observing and making models of themselves. They must change their identity in time on both sides of the observation process (as both observed and observer). When the speed of the process of becoming pushes too close to the various time differentials (especially $d\tau$ and $d\theta$), then the predicament of postnormal science can become overwhelming. That is, the very identities of both the observed system and the observer system become fuzzy since they are affecting each other's definition at a speed that makes it impossible to have a robust validation. This can represent a serious problem of governance, related to a relatively new plague (widespread by mass media), which we can call the butterfly effect or pheromone attention syndrome, determined by the hypercyclic interaction observed–observer. Media focus on what is of concern for stakeholders, and stakeholders are concerned with what is focused on by media. The result is that what is on the spot of the public attention or in the debate about sustainability is often randomly generated by lower-level stochastic phenomena — what happened to be the initial problem structuring of a given problem given by media. Then this original input is amplified by lock-in effects (someone with a camera happened to be in a specific place catching a relevant fact). Nobody, however, can check how relevant is that particular fact, which is amplified by the spotlights, compared with other relevant facts ignored in the debate simply because they happened in the shadows.

8.1.3 The Impossible Use of Dynamical Systems Analysis to Catch the Process of Becoming

Adaptive holarchies can retain their individuality only if they are able to keep alive the mechanism generating coherence in the expression of their identity across the three time differentials defined in complex time. This implies the ability to keep harmony in the pace at which the various identities and individualities and their perceptions and representations are changing in time. This requires a deep interlocking of ontological and epistemological interactions (Chapter 2). The term *expression of an identity* refers to the concept of self-entailment between (1) establishing processes able to realize viable equivalence classes of organized structures sharing the same template at different levels (an ontological achievement) and (2) integrated processes across hierarchical levels able to determine essences in terms of the validity of mutual information used by interacting agents, which is associated with the perception, representation and running of anticipatory models at different levels (an epistemological achievement). This is a mechanism that cannot be fully represented using conventional formal systems of inference.

For example, the formalization of concepts such as resilience and stability is in general attempted from within the field of dynamical systems analysis. Actually, this field provides powerful images (e.g., basin of attractions) that are often used with semantic purposes. For example, the shape of the basin of attraction is a very popular metaphor. Resilient systems are depicted as having a shallow and large basin. Robust but fragile systems are associated with basins that are very deep and small in domain. An example of these two metaphors is given in Figure 8.2 (taken from Giampietro et al., 1997). These visualizations are certainly useful, but they do not avoid the original unsolved problem. Any formalization of resilience, robustness, stability or whatever label we want to use within the field of dynamical systems requires the previous definition of a given state space. By state space we mean a finite and closed (in operational terms) information space made up of variables, referring to observable characteristics of the system, that can be measured at a given point in space and time through a measurement scheme. The implications of this fact are huge. To represent a basin of attraction, you need numbers, which in turn requires assessments (measurement schemes), which in turn must refer to given typologies (types defined as a set of attributes), which are represented using a set of epistemic categories (variables). Put another way, if we plan to develop formal analytical tools to study the evolution of adaptive dissipative system, we need to measure key characteristics of them through an interaction within an experimental setting that makes it possible to encode observable qualities into numerical variables. These measurements are location specific. That is, they are and must be context and simple time dependent. Simple time is what is perceived from within the representation of reality (the model) obtained within a closed and finite information space, and what we generate within the artificial settings of an experimental scheme. Because of this, the dt of the model is reflecting (1) the choice of a triadic reading associated with our perceptions and (2) the filter on possible signals implied by the measurement scheme. That is, such a dt will reflect the preanalytical choices made when choosing the particular model.

The validity of simple models requires two assumptions related to the definition of identity for the system: (1) The existing associative context will remain valid (e.g., the environment is and will remain admissible also at a different point in space and time). That is, the validity of the model implies the absence of changes in relation to $d\tau$. (2) The choice of relevant attributes used to define the identity is agreed upon by all the observers (e.g., it is impossible to find a relevant user of this model who does not agree with its assumptions). That is, the validity of the model implies that the general agreement about its usefulness and relevance does not change in relation to $d\theta$.

However, considering these two assumptions valid — as required by dynamical systems analysis — puts the modeler in the unpleasant situation of defining concepts (resilience, robustness, stability)

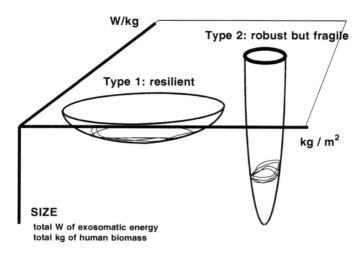

FIGURE 8.2 Shapes of basins of attraction.

associated with the identity of static dynamical systems perceived as operating out of complex time. These systems can have multiple attractors. They can even be able to switch from one attractor to another at command. They can jump; they can get chaotic and engage in any type of fancy mathematical behavior. But yet the identity of their information space does not evolve in time; see the work of Rosen (2000) and Kampis (1991) for a more elaborate discussion of this point. They are not alive, they are not becoming something else, and they are not adding new essences and new epistemic categories (emergence) to their original information space. Finally, and most important, they are not adding new meanings (for the observer) to their identity. Put another way, the real problem with complex systems is not that they are exhibiting nonlinear behavior. In fact, the technical feature of linearity or nonlinearity of dynamical systems refers only to changes occurring within the known state space and the simple time defined on dt.

Even when moving away from dynamical systems analysis to more advanced inferential systems based on the use of computers (e.g., cellular automata), the problem of a sound representation of the behavior of complex systems is not fully solved. These new mathematical objects can establish bridges between patterns and mechanisms operating on different levels, and this is a major step forward. However, also in this case, the mathematical tools only makes it possible to better clarify the mechanism associated with emergence. They can explain how a pattern expressed at one scale can be associated with patterns defined on different scales. We can find, using the output given by a computer, new properties that can be interpreted by the modeler in terms of additional insight provided by the algorithm. But the real challenge, in this case, remains that of finding the "right" set of external referents that can provide meaning to this analysis on multiple levels. In our view, there is a big risk associated with this new generation of sophisticated formalisms. Many practitioners tend to apply them to the analysis of sustainability, under the incorrect assumption that more complicated models and more powerful computers could handle the complexity predicament just by providing more syntactic entailment. Put another way, the risk that we see is that this new frontier of development of more powerful inferential systems can represent yet another excuse for denying a relatively simple and plain fact: becoming dissipative systems organized in holarchies have, and must have, a noncomputable and nonformalizable behavior to remain alive (Rosen, 2000). Modelers should just accept this fact.

8.1.4 The Nature of the Observer–Observed Complex and the Existence of Multiple Identities

Imagine that an extraterrestrial scientist belonging to an unknown alien form of life suddenly arrives on Earth to learn about the characteristics of holons — human beings. It would be confronted with the fact that humans can be classified in nonequivalent ways. These different ways could be seen as different attractor types using the vocabulary of dynamical system analysis, or different types associated with identities using the vocabulary developed in Chapters 2 and 3. For example, a given human being can be characterized as a system belonging to an equivalence class, which can be defined by adopting a set of observable qualities or attributes (temperature of the body and organs, pH of the blood, number of legs and arms, etc.). These characteristics have to be common to all the members of the equivalence class. In this case, a set of variables, which are proxies of the set of observable qualities associated with the class, must take a range of numerical values contained within a feasible domain of the class (individuals with 6 legs and individuals with 20 eyes are not included in the class of humans). These expected features and relations among variables associated with a given identity imply the definition of expected relations between numbers (if the specimen is human, it must have the expected number of arms, legs, eyes and ears) and a chain of tolerance ranges in the relative numerical assessments used to represent them. The value taken by the proxies used to assess each relevant quality must be included in a range (spread of the values around the average). The error bars associated with the measurement scheme must be compatible with the domain of feasibility for the variable. A very generic definition of identity for humans can be based on a set of attributes common to the majority of human types (e.g., temperature of the body and organs, pH of the blood, two legs and two arms, existence of typologies of organs).

Within this very generic definition of human beings, we can imagine a large set of possible typologies of realizations. Therefore, it is possible to define more specific typologies by constraining such a large domain with additional epistemic categories. For example, a human observer can combine the generic definition of human being (defined using the attributes listed before) with three relevant epistemic categories related to age (children, adults and elderly) and two additional relevant epistemic categories related to gender (male and female). In this way, it is possible to obtain the definition of six basic types for humans: boys, girls, men, women, old men and old ladies. This selection of types can be further expanded at will by adding new relevant categories (e.g., short, medium, tall; blond hair vs. black hair; dressed vs. undressed, etc.). The number of relevant human types found in this way will ultimately depend on the number of categories that are considered useful by the observers in organizing their perceptions.

Therefore, we cannot know the set of human types that alien observers would find. This would be determined by their selection of relevant characteristics sought in humans (are humans dangerous? are they good as food?). The consequent selection of the epistemic categories used by aliens for the definition of human types (e.g., level of presence of cobalt in their hair, amount of radioactive radiation emanating from the body) will reflect their choices about how to organize their perception about humans.

Depending on the size of the sample that the extraterrestrial will use, it could find very little variability in human types (e.g., a cluster of homogeneous human types found when sampling just 10 students in a classroom or 20 soldiers in a platoon) or great variability (e.g., when sampling the population of an entire continent). That is, by expanding the size of the sample and the diversity of detectors used to gather information about humans, the observers will change the universe of potential types. For example, alien observers can find:

- **A given set of attractors related to the existence of multiple equilibria (in terms of dynamical systems analysis) or multiple identities.** A massive project for studying humans on this planet based on a large sampling of humans at a given point in time would provide a set of well-defined categories to be used to characterize humans. These categories should be based as much as possible on the existence of equivalence classes occurring naturally in such a big sample. As noted in Chapters 2 and 3, the process of self-organization of dissipative holarchies naturally generates equivalence classes, types and essences. Therefore, smart aliens, to increase their anticipatory power when modeling humans, should be able to pick up a set of epistemic categories that would make it possible to maximize compression for their representation of the characteristics of humans. If this is true, we can only imagine that after a period of learning about humans, they probably could represent human types by using some of the same categories used by humans themselves. In this case, they will converge on the definition of a set of multiple identities existing in the holarchies making up humans (e.g., human organs, individual humans, baby girls, adult men, households). Depending on the available set of types used in pattern recognition (to categorize the individuals in the sample), an observation made at a particular point in time, but over a large space domain (e.g., on a particular day over an

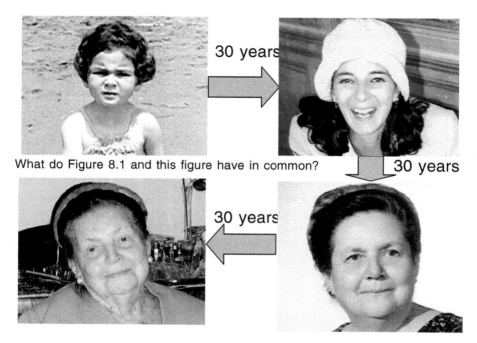

FIGURE 8.3 Gina's line represented as analogous to Bertha's line (Figure 8.1). (Photos courtesy M. Veneziani.)

entire continent) — a synchronic analysis — will provide a profile of distribution of individuals over the given set of possible types. That is, the system human being can be perceived and represented using a set of different identities at a given point in time.

- **A cyclical attractor (in terms of dynamical systems analysis) or a given trajectory across identities,** such as, for example, the set of identities represented in Figure 8.1 and Figure 8.3. The four pictures given in Figure 8.3 can be imagined to be four views of Gina, the old lady in the bottom-left picture. In analogy with Figure 8.1, these can be interpreted as four views of the same individuality represented at four points in time in her life. When comparing the four pictures given in Figure 8.1 with the four pictures given in Figure 8.3, we can note that even though we are dealing with two distinct individualities (Bertha and Gina), the set of four typologies through which these two individualities go in time and the sequence among the types (girl, adult woman, lady and old lady) are the same. If the project of investigation of the extraterrestrial expedition had followed a certain number of households over a long timescale — e.g., centuries — using the set of typologies adopted in Figure 8.1 and Figure 8.3, they would have found a predictable pattern in the order in time in which these typologies of identities appear in the life cycle of individual persons. That is, a diachronic analysis of human beings (e.g., history of a royal dynasty or important families such as the Kennedy or Bush families) makes it possible to look at a different set of typologies linked to a turnover of lower elements into a role. This can be seen as an emergent pattern, when considered at a higher hierarchical level (on a timescale larger than that related to the life span of an individual). The perception of this pattern, however, requires the adoption of a larger time horizon, which has to include the whole cycle of lower-level holons in the role defined at the higher level. This pattern overlooks the perspective of the individualities involved in its expression. When looking at this pattern (what the sequence of identities has in common in Figure 8.1 and Figure 8.3), we have to ignore the details, which are relevant to recognize either Bertha or Gina as individual persons. The process of aging described using types is nonequivalent to that used to describe the individuality of persons. Put another way, to see the turnover time of individual realizations within the relative type (babies becoming adults, adult becoming old, and dead people being replaced by newborn babies), we have to ignore information that is crucial when dealing with individual realizations. To make things more difficult, there are processes related to the structural stability of both types and individualities (the physiological processes keeping alive over a timescale of seconds each of the persons represented in the eight pictures in Figure 8.1 and Figure 8.3), which are defined at yet another scale (lower-lower level).

Two things are remarkable in this discussion: (1) the unavoidable arbitrariness in deciding what should be considered as a holon human being (especially when considering that holons are made of other holons made of other holons) and (2) coming to the problem of mapping and measurement, the only meaningful things that can be measured by an extraterrestrial expedition willing to know more about humans are the qualities of types, not the qualities of any special individual human being.

That is, when dealing with the perception and representation of learning adaptive holarchies, what is considered real by naive empiricists (special individual realizations materially defined in terms of structures) is not a relevant piece of information for scientific analysis and models. The input given by real entities to the process of measurement (extraction of data from the reality) is useful only when such input is processed in terms of typology within a valid interpretative scheme. In this case, the input is useful since it provides information about the characteristics of relevant types or essences of which the real entity is just a realization. Science deals with types (patterns defined over a space–time window that are useful to organize our perceptions in terms of epistemic categories). The space–time domain of validity of a definition of type is larger than that of individual realizations. On the other hand, data and measurement can only be referred to individual realizations seen and measured at a particular point in space and time. This is why we tend to see types as being out of time, since they refer to a standardized perception of a given relation type/associative context.

8.1.5   How to Interpret and Handle the Existence of Multiple Identities

The two series of pictures provided in Figure 8.1 and Figure 8.3 refer to the perception and representation of a given individuality going through a transition across a set of predictable identities. That is, different identities (types assiciated with equivalence class) defined as girl, adult woman, lady, and old lady are the expected states that a given individuality (person) will take during her expected trajectory of evolution in her lifetime. Obviously, to each of these types we must be able to apply the generic set of mappings defined for all human beings (temperature of the body and organs, pH of the blood, two legs and two arms, existence of organs). That is, to be a valid set of integrated identities, girls, adult women, ladies and old ladies must all belong to the class human beings in the first place.

In conclusion, when dealing with the representation of persons, we need three different pieces of information related to the perception and representation of the process of becoming shown in Figure 8.1 and Figure 8.3. These three types of information are:

1. A family of models able to represent the functioning of human beings (described in general terms) and that can be applied to each of the four identities. This would be, for example, the set of descriptions of physiological processes within the human body (e.g., those associated with respiration), which can be obtained by adopting a set of descriptive domains common to all four types. This requires a preliminary selection of relevant identities of lower-lower-level elements associated with the respiration of humans that refers to specific choices of triadic filtering, identification and representation of organized structures (e.g., alveoli, capillary, hemoglobin molecules). At this point, we can generate numerical assessments of values taken by variables and parameters. Examples of this type of information are given in Figure 8.4. The various dynamics represented Figure 8.4 are all expressed using a simple dt of general validity for the type supposed to operate by default in the right associative context (admissible environment, favorable boundary conditions).

2. A family of metaphors able to catch the similarity implied by the sequence of types into the cycle. That is, we should find a metaphorical knowledge able to tell us what all girls have in common when compared with adult women, ladies and old ladies in relation to the process of aging of a person. At the same time, the metaphor should also tell us what adult women have

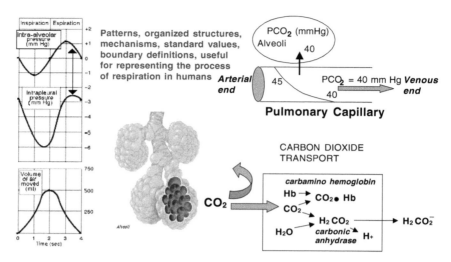

FIGURE 8.4 Respiration cycles within human cells. (from BIO 301 Human Physiology syllabus (eastern Kentucky University) -- Gary Ritchison. http://www.biology.eku.edu/RITCHISO/301notes6.html.)

in common when compared with girls, ladies and old ladies in relation to the process of aging of a given person, and so on. Obviously, in this case, we need a meta-model able to deal with the semantics of these relations. That is, the meaning of the relation among the four typologies included in the figure has to remain valid even when applied to different individualities (e.g., in this case, different persons) or a different type of essences (e.g., the process of aging of a dog). In this case, the problem is with the definition of the quality to be measured — the choice of attributes associated with the identity used to characterize a given equivalence class to which the individual realizations (the specimen under investigation) are supposed to belong. As discussed in the previous chapters, essences referring to adaptive metabolic systems (humans and biological/ecological systems) are always defined over a very large space–time window, since they require the simultaneous adjustment of the mechanisms determining the feasibility of the various equivalence classes on different hierarchical levels. This requirement of mutual information across scales implies that the set of qualities required to have sustainability in evolutionary terms has nothing to do with the type of information illustrated in Figure 8.4.

A very interesting example of metaphoric knowledge related to the cyclic sequence of types within a role — exactly what is shown in Figure 8.1 and Figure 8.3 — has been provided by Buzz Holling (1995; Gunderson and Holling, 2002), and it is shown in Figure 8.5. The metaphor proposed by Holling is interesting since it requires abandoning a formal representation based on exact models to move to a semantic description of events. This can be immediately realized by the fact that this metaphor was given several different names by different authors: adaptive cycle, cycle of creative destruction (recalling a similar idea of the economist Schumpeter), 4-box figure-8 adaptive cycle, or the lazy 8. This metaphor will be explored in detail in the next section.

3. Information about the history of the system that makes it possible to characterize the special individuality of this evolving system. All complex adaptive systems (learning holarchies) have and must have a history to be able to generate an integrated set of reliable identities. It is their special history that makes it possible to trace their individuality. However, it is exactly the keeping record of history that entails the development of narratives (selecting what relevant aspects should be included as records in the storage of information and what should be excluded). Keeping records means, in fact, selectively removing those details of a given history that are considered not to be relevant. This has important implications. This implies that the very decision of what represents the real individuality of a system, when this system has changed its identity in time, becomes an arbitrary decision. Defining what is the individuality of a becoming system is a matter that cannot be dealt in objective, substantive terms. This is

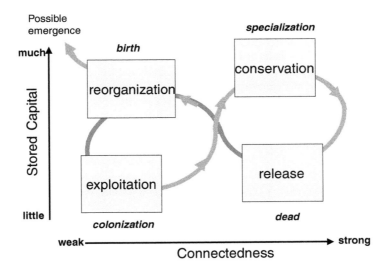

FIGURE 8.5 Holling's metaphor about adaptive cycles.

an operation that cannot be either described or performed from outside the complex observed–observer. For example, is a person that goes totally crazy (unable to retain awareness of her or his own identity) still the same person? For a citizen asked to vote for this person for president of the U.S., the answer is clearly not. For a mother asked to take care of this person, the answer is yes in most cases. The problem here is generated by the fact that a voter and a mother are generating different narratives about the history of (1) a public officer, candidate for a new term and (2) her own child, who had a car accident.

In fact, we can disclose to the reader that the set of pictures presented in Figure 8.1 are not referring to the same person, just as the set of pictures presented in Figure 8.3 are not. Rather, they are two nonequivalent combinations of mappings. Bertha's line (Figure 8.1) represents four generations of women: Bertha is the old lady, whereas Ria, her daughter, is the lady to her right. Sandra, the adult woman, is the daughter of Bertha's daughter Ria. Finally, Sofia is the granddaughter of her daughter Ria. On the contrary, Gina's line (Figure 8.3) presents only two persons: Gina is the lady on the lower level, shown in two pictures taken 30 years apart, whereas Marinella, the daughter of Gina, is on the upper level, in two pictures also taken 30 years apart. By giving this information to the reader, we changed the know-how of the reader/observer of these two figures. At this point, do you, the reader, still feel that the four pictures in each of these figures represent the same individuality? More specifically, at what point can the reader say that we are dealing with four individualities when looking at Figure 8.1 and only two individualities when looking at Figure 8.3? To make this statement, the reader must trust what has been written by these authors. Why should this distinction be relevant for someone who does not personally know the persons represented in these two series of pictures? For sure, there is something in common in these pictures that makes it possible to recognize the same individuality (same line) in each of the series in the figures. For the monarchy, the concept of individuality of the line is essential. If it is true that there is something in common between the four pictures within each of the two figures (a common line), it is also true that there is something in common among the typologies of the two lines — something that makes it possible to predict expected changes within an expected pattern of types for each of the two lines. In any case, predictions about the relation of multiple individualities, identities or essences found when observing a complex reality can be formalized only with great care and a deep awareness that alternative formalizations can be legitimate and valid. Moreover, uncertainty is always at work. Not even the most general characterization about the class human beings can be extrapolated to individual cases expecting full reliability. For example, an adult person could lose a leg because of a car accident, and therefore, not even a simple prediction (about the number of legs) in the next type (at the time t + 1), starting from the knowledge of the number of legs in the type (at the time t), is necessarily granted.

When we recognize that the same pattern of types referring to the same individuality is going through different stages, we have to expect that human systems operating at different stages of their life cycles will adopt different definitions of optimal strategies for sustainability. This is a crucial peculiarity that human holons have because of their reflexivity. This is what leads to the need to answer tough questions when dealing with the sustainability of human holons. Sustainability of what? Defined on which time horizon? (Are we sustaining an identity associated with a given type, an essence or an individuality?) Why should supporting an identity be more important than supporting an individuality (e.g., maximization of profit, which induces social stress)? Or why should the interest of individualities be more important than the preservation of essences (e.g., loss of cultural diversity because of widespread fast economic growth)? We are back to the unavoidable existence of contrasting indications and contrasting optimizing strategies (recall the nonequivalent explanations of death found in Figure 3.6) for agents belonging to contexts defined at different stages of the cycle. The optimal strategy for the young girl is that of growing and quickly obtaining the characteristics of older types (becoming a woman), whereas the optimal strategy for adult women is to avoid getting the characteristics of older types (Figure 1.1).

8.1.6   The Metaphor of Adaptive Cycle
Proposed by Holling about Evolution

To analyze the nature of the unavoidable ambiguity in determining a distinction between identity, essence and individuality, let us briefly go back to the discussion of the metaphor proposed by Buzz Holling under the name adaptive cycle. What is shown in Figure 8.5 is a very sophisticated application of this metaphor to the analysis of sustainability of an ecological system. A detailed description of this meta-model can be found in Holling and Gunderson (2002). A reading of this text (or more in general, the work of Holling in this direction) gives a clear idea of the powerful insights that can be gained by adopting it. We want to focus here only on the aspects related to the preliminary choice of a set of four identities (types) that have to be overimposed in time on a given individuality that the use of this metaphor entails. It should be noted that such a metaphor has very general applicability; the cyclic attractors can be used to explain various sequences of predictable states taken by an individuality. This individuality could be an ecosystem, and the four types can be the four seasons (in this case, the four pictures will be spring, summer, autumn and winter), using a small time window or the stages of development (in this case, the four pictures will be the stages that go from early colonization to senescence). Alternatively, the individuality could be a given person (or an organism), and the four types will be different stages of the life cycle (as in the examples given in Figure 8.1 and Figure 8.3).

For reasons that will be explained in the next section, we applied the representation of this cycle based on the use of an integrated set of four types to the description of the development of a car model (individuality). This view is shown in Figure 8.6. In fact, also the evolutionary cycle of a new model of a car can be expected to go through predictable stages according to Holling's scheme. Whenever there is an opening for a new model of car, a car maker can decide to go for it. The first template of such a model does not need to be very sophisticated. In step 1 what is needed is just something that is able to fill an empty niche. Whatever does the job is OK. We can recall here the story of the Ford Model T that was launched in 1908. At the beginning, the real issue for the U.S. buyers was having a car — getting out of the state of not having a car. No options were available. This is the basic reason that made it possible for Henry Ford to say the famous line "consumers can have it in all the colours they want as long as the colour is black." Since the filling of the niche was a success (buyers were buying more than the car maker was able to supply), the next problem was that of producing enough. That is, the next changes in the model were related to the improvements related to the process of fabrication of members of the equivalence class. By 1914, the moving assembly line enabled Ford to produce far more cars than any other company. However, a very large size of the niche (Ford built 15 million automobiles with the Model T engine) implied new problems:

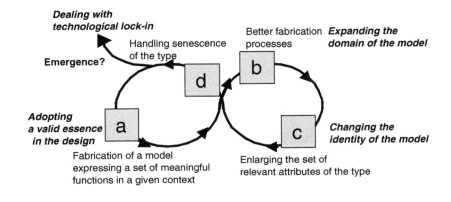

FIGURE 8.6 The metaphor of lazy 8 applied to a car. (Courtesy of FIAT spa and ARCHIVIO STORICO FIAT.)

1. Diversification of performance (since a large niche is geographically covering different expected associative contexts). Sooner or later the use of a huge amount of cars entails the requirement of performing different functions.
2. Fighting competition within the niche (since a large niche — many buyers willing to invest in cars — tends to attract competitors). This is the third stage, when the Ford Model T was made in different colors and versions.

Finally, we arrive to the final stage of maturity, when the basic structural organization of the template becomes obsolete. A new set of tasks and a new set of local associative contexts are now faced by the members of this equivalence class (cars). A different selective pressure is operating due to changes that occurred in the larger context. However, these changes in the larger context have been induced exactly because of the large success of the original model, which was able to amplify so much the domain of activity of this class. At this point, nobody would invest resources in building a new assembly line for making additional members of this obsolete equivalence class. On the other hand, as long as the production lines — existing capital — are still operating, it can pay to add a few possible gadgets to the template, to keep production alive. In the phase of senescence, car models tend to get into micro niches void of competitors, who would not invest to get there.

This very same cycle across different stages for a car model is shown in Figure 8.6. The model of car considered there is FIAT 500, of the Italian car maker FIAT: four different identities adopted by the same individuality over a predictable cycle that can be associated with differences in history, boundary conditions and goals. First, an idea is realized about a possible model filling an empty niche. Then when a positive experience confirms the validity of the original idea, it is time to patch the original process of realization according to operational problems (scaling up). In this way, it is possible to occupy as

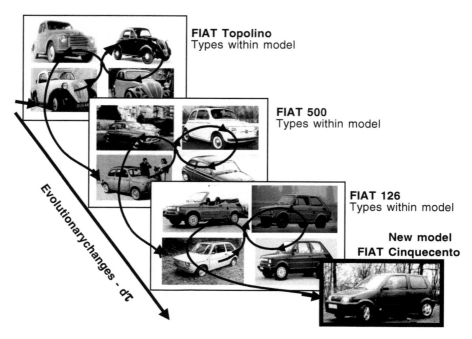

FIGURE 8.7 The lazy 8 in three dimensions. (Courtesy of FIAT spa and ARCHIVIO STORICO FIAT.)

much as possible the niche (take advantage of favorable boundary conditions to expand the domain of application of the type). When the size of the new pattern is large enough to guarantee enough protection against perturbations for the basic identity, it becomes possible to explore new functions and tasks that can be associated with complementing the original ones to expand the viability of the equivalence class in slightly different associative contexts. In fact, the large scale of operation of the original pattern tends to feed back in the form of a new definition of the context of the original essence. At this point, it is important to look for a different model (a new set of organized structures mapping onto a new set of tasks). However, because of the existing investment (lock-in), for a while it can be convenient to keep using the old process of fabrication of members of the obsolete equivalence class (for defining this situation, we can use expressions like Concorde syndrome or sunk cost).

We can gain a crucial insight from the metaphor of the lazy 8 of Holling if we add a third axis to the plane shown in Figure 8.5. This three-dimensional view is shown in Figure 8.7. As illustrated in both Figure 8.5 and Figure 8.6 during the adaptive cycle, after the phase of release and before the phase of reorganization (before spring, so to speak), there is the option for the process to become something different. That is, the phase of reorganization of a given type within a given associative context can lead to a phenomenon of emergence. In this case, the small changes accumulated at the level of the type and the small changes accumulated in the identity of the associative context can move the interacting type–associative context into a new self-entailment across identities. That is, we can look at the emergence of a new association between type and associative context across the various constraints operating at different levels. In this case, the self-organizing holarchy can jump into a different mechanism of self-entailment among identities across scales. The example given in Figure 8.7 shows first the cycle related to the model of the car FIAT Topolino, which reached its last stage in the 1950s, making possible the launch of a new model in the late 1950s. Then the FIAT 500 took over, going through the cycle to reach senescence in the late 1970s, when a new model, FIAT 126, took over (with a large engine — 700 cc of displacement — and better technical characteristics). But the larger engine was not enough to keep up with the changes occurring in the Italian socioeconomic context. This is what led to the definition of a new type (FIAT Cinquecento), with an even larger engine — 900 cc of displacement.

Two important observations are needed in relation to this example:

1. The three-dimensional representation given in Figure 8.7 shows the evolution in time of a car that is obtained by establishing congruence between processes occurring on the ontological side (the making and using of cars) and processes occurring on the epistemological sides (the decisions about producing and about the buying of cars) across hierarchical levels. The four different stages through which a given model of car (e.g., the FIAT Topolino) is expected to go through represent different ways of obtaining congruence among these two sets of processes. Therefore, in this figure we see the same pattern of movement across stages (the four types indicated in the adaptive cycle), which is repeated in time in relation to the trajectory of development of different individualities — models of car. In this example the three models of car are (1) FIAT Topolino, (2) FIAT 500 and (3) FIAT 126. That is, we can define the individuality as the model and the four different types as versions of this model.

   If we want to represent using numerical variables the changes in relevant system qualities associated with this process of evolution, we have to deal with the meaning of the third axis. That is, when describing evolutionary trajectories of models of cars, the identity of the information space — the variables used to represent the characteristics of the different models (the type of information similar to that given in Figure 8.4 about the respiration of persons) — has to be changed. That is, the movement from a model (an individuality) to another (from FIAT Topolino to FIAT 126) entails/requires a change in the identity of the descriptive domain (the set of numerical variables) used to represent the process of becoming over the four selected types. Recall here the example of the Benard cell discussed in Part 1. When a vortex is established (emergence of a pattern at a higher hierarchical level), this requires the use of new epistemic categories (an encoding variable associated with the category counter-clockwise) by the observer. To describe in useful terms the identity of the new system — a vortex — we have to use a new variable that was meaningless in the representation adopted in a molecular description. In the same way, we cannot use epistemic categories (e.g., number of wheels = 4; transparency of windows = yes; stopping by braking = yes) common to all three basic models — FIAT Topolino, FIAT 500 and FIAT 126 — to describe and compare changes associated with movements across the four types. To distinguish a FIAT 126 from a competitor in its niche (how and why the FIAT 126 is changing over the set of four stages), we have to use for its characterization a set of observable variables that were absent in the identity used to describe the FIAT Topolino.

2. The possibility of jumping into a new individuality during the stage of reorganization is related to the level in the holarchy at which the process is operating. This is a crucial point and can be related to the predicament of science for governance. The lazy 8 metaphor can be applied to different hierarchical levels found in a holarchy. By enlarging the scale of analysis (moving up in the levels), we can imagine applying the adaptive cycle to different models referring to the same essence of car, rather than to different versions of the same model (as done in Figure 8.6). This implies addressing what is the meaning of a car in a given context (the semantic definition of cars to which the various models refer). To clarify this concept, let us get back to the beginning of the car era, with the Ford Model T and the FIAT Topolino. At the beginning, the major task of the car industry was to make it possible for people to move around using a car. In this stage, comfort and safety were not very relevant. At the beginning of the auto industry, those pioneers that dared to use cars were expected to take chances. Therefore, the role of the FIAT 500 was to increase the ability to supply an increasing number of cars to those looking for them at the cheapest possible price. The FIAT 126 remains within the same basic definition of essence of the FIAT 500, but the new model had to include in its definition of minimum standard of quality for a car a new set of attributes of performance, at that point expected by vehicles circulating on modern roads (e.g., a decent cruise speed on the highway, new safety devices required by law). This is where the displacement of the engine had to be increased and several additional changes in the body of the car became necessary. At a different level, we can think of a different lazy 8 adaptive cycle based on the movement through different models, all referring to the same essence of car.

EXTINCTION

FIAT Cinquecento *model*   FIAT 500 *model*

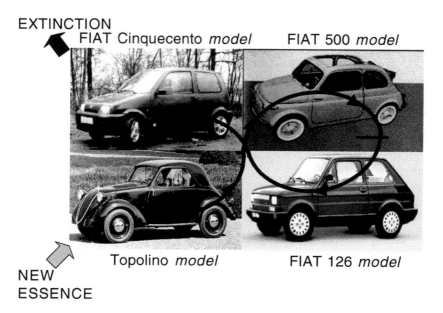

Topolino *model*   FIAT 126 *model*

NEW
ESSENCE

FIGURE 8.8 Evolution of models within the same essence. (Courtesy of FIAT spa and ARCHIVIO STORICO FIAT.)

That is, we can go for a higher level of perception and representation of the Holling metaphor, this time applied to models of cars, rather than to versions of the same model (Figure 8.8). If we look at the same process at a higher hierarchical level, we can see that the last emergent model at the end of the lazy 8 adaptive cycle — the FIAT Cinquecento — soon became extinct. This new model never managed to get into the four stages of different versions expected for a new model. The explanation for the extinction of this model can be found in the obsolescence of the relative essence, that is, a dramatic change in the role that the car started to play within the socioeconomic Italian context. At the end of the 1990s, the process of industrialization and the huge supply of cheap popular cars that FIAT provided in the previous 40 years implied a different meaning for small utilitarian vehicles (Figure 8.9). The old definition of role supporting the chain of four models illustrated in Figure 8.8 (a satisficing combination of attributes for those looking for a car) was split into two different roles for a car (Figure 8.9). The essence of a utilitarian car was no longer associated with the role of having a cheap car, but rather associated with a satisficing combination of attributes for a person owning already more than one car. Then the new meaning of a small car was that of a car to be used when getting into the heavy traffic of a city. This small car can get pretty expensive when doing that with a lot of comforts. Because of this change in the definition of performance (expected function) among the population of users, the old generation of utilitarian cars of the car maker FIAT is no longer mapping onto the expected functions of the Italian socioeconomic context. This is why FIAT is now trying to move the production of the FIAT Cinquecento into a different associative context (e.g., moving to East Europe). In that socioeconomic context the original semantic meaning of the car, associated with that model — the validity of the original niche (Figure 8.9 — generalist models, having a car vs. no car) — is still valid. Because of this, we can expect that within this typology of associative context (made up of buyers looking for their first car), new versions of this model and new models of utilitarian cars will be generated, following the expected adaptive lazy 8 metaphor shown in Figure 8.7.

The case of extinction of the FIAT Cinquecento model in Italy provides an example in which a large-scale change in the socioeconomic context made the semantic definition of essence for a given model of car (e.g., FIAT Cinquecento) no longer valid. Without a valid essence supporting the formal definition of types, the process of evolution of new types (versions of the model) stopped. This has nothing to do

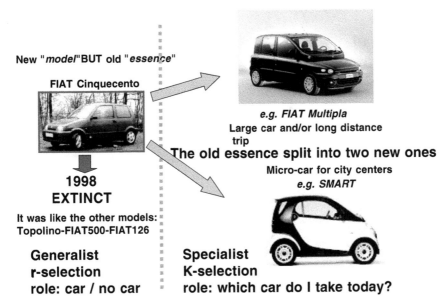

FIGURE 8.9 New model but old essence. (Courtesy of FIAT spa and ARCHIVIO STORICO FIAT.)

with the usefulness of blueprints for fabricating an equivalence class of cars belonging to the same model. In this phenomenon of extinction, the quality of the information contained in individual blueprints about the making of individual cars was quite irrelevant. Rather, it is the process of postindustrialization of the Italian economy that translated into the definition of new essences for cars, and therefore to the need to look for a new generation of car models.

As noted earlier, the phenomenon of emergence of new essences can be expected only on the top of the holarchy. Only on the top is it possible to establish new relevant attributes to the definition of identities in the interaction of five contiguous levels. This implies learning about how to better interact with the environment. On the contrary, on the bottom of the holarchy, the standard relation type–expected associative context, which is at the basis of the validity of the mapping between characteristics of the given organized structure and usefulness of the tasks and functions, is given and must remain given. A cell of our bones that gets old and that is replaced by a new cell must be organized according to the same template (type), reflecting the same essence over and over and over. The stability of the identity of lower-level structural holons is a must for the possibility of expressing new functions on the top of the holarchy. The structural stability of lower-level components must be assumed as given in dissipative holarchies. As noted in Part 1, when we apply the triadic reading, the requirement — on the lower level — of the stability of lower-level structural components is analogous to the requirement about the admissibility of the environment on the higher level.

For this reason, the last interface of the holarchy with its environment (the frontier of the holarchy on the triadic reading $n/n + 1/n + 2$) is the only place where the holarchy is able to generate new meaning about new forms of interaction with the context. This is where new essences are introduced into the information space (see the discussion about postnormal science in Figure 4.3).

8.2 Using the Concepts of Essence, Type and Equivalence Class When Making the Distinction between Identity and Individuality (*Technical Section*)

8.2.1 The Unimportance of DNA in the Definition of Essences in Biological Systems: The Blunder of Genetic Engineering

To introduce this discussion, let us start with an example dealing with the evolution of nonbiological essences. The two sets of identities given in Figure 8.10 show two trajectories of evolution referring to two

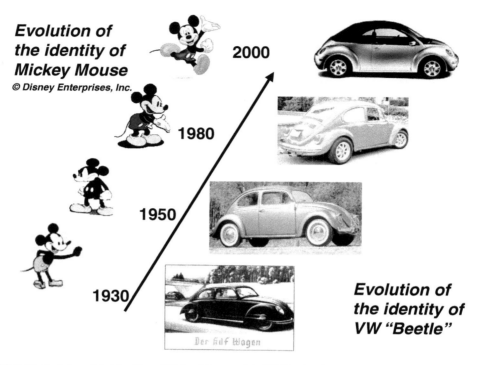

FIGURE 8.10 Evolution of the identities of Mickey Mouse and the VW Beetle. (Disney Enterprises, Inc. With permission.)

essences: (1) that of a famous species of cartoons, Mickey Mouse and (2) that of a famous species of cars, the Volkswagen Beetle. When conducting research on the Internet using a search engine and entering the key words *evolution* and *Mickey Mouse*, one can find, together with hundreds of sites that consider the evolution of Mickey Mouse as a fact, a couple of sites presenting teaching material in which the Mickey Mouse syndrome is proposed as a systemic error made by humans when attributing anthropomorphic characteristics (the ability to evolve in time) to unanimated objects. The basic argument of this reductionist analysis is that "Mickey Mouse is a dead object that cannot evolve since he does not have DNA. This reasoning is simple. Since the mechanism providing mutation to biological systems (assumed to be at the basis of the evolution of living systems) is associated with the existence of DNA, adaptive dissipative systems that do not have DNA cannot evolve in time. Therefore, if someone perceives as evolution a process of gradual changes in time in the identity of systems that do not have DNA, we are in the presence of a pathological phenomenon of transfer of anthropomorphic concepts to unanimated things.

In relation to this position, we happen to believe that, on the contrary, it is such a substantive association of DNA to both life and evolution that represents a pathological consequence of reductionism. Moreover, this overstatement of the role of DNA in determining both life and evolution is an important misunderstanding that is heavily affecting the efficacy of discussions about sustainability. This is why we believe that a general discussion about the mechanism driving the evolution of holons in terms of essence, types and equivalence class can be very useful in this regard.

The two concepts of type and individual are particularly useful for conceptualizing in semantic terms the nature of the process defining identities through impredicative loops. In fact, the concept of a type refers to that of a useful template used for the realization of an essence defined in the semantic realm. That is, a type is the representation of a set of qualities associated with a label used to refer to an equivalence class generated by the realization of a set of organized structures sharing a common template (e.g., a Volkswagen Beetle). The concept of individual refers to a special realization of a given type within a specific context (at a given point in space and time). Because of the particular path-dependent process (stochastic events accumulated in the history of each individual realization), this concept entails that we will always find special characteristics for each individual realization (e.g., my Volkswagen Beetle), even if they belong to a specified equivalence class.

The associated epistemological paradoxes illustrated in previous chapters entail that each specific realization of a given type (e.g., the fabrication of an organized structure) will be different from other realizations obtained adopting the same blueprint (e.g., we will never find two cars of the same model and brand that are exactly identical). In the same way, we will never find two homozygous twins that are exactly identical. However, in spite of this unavoidable existence of differences between individuals, it is possible to recognize and assign each of the realizations of a VW Beetle to the same type, which refers to the shared characteristics of the equivalence class to which the individual is supposed to belong. This means assuming the validity of the hypothesis that each specimen of Volkswagen Beetle is reflecting a common set of properties shared by all individual realizations. This assumption is based on the fact that all these realizations share (are mapping onto) the same blueprint.

The validation of these assumptions leads us back to the existence of an impredicative loop (a chicken–egg paradox) implied by this circular definition. We have to know the characteristics of the type to decide whether a particular individual realization belongs to an equivalence class, but we can know about the characteristics of the type only through direct interactions with a set of individual realizations (which we must already recognize as legitimate members of that class).

It is possible to get out of this impasse, because of the parallel existence of nonequivalent perceptions and representations of the reality on different levels (and descriptive domains). The robustness of a set of identities generated by an impredicative loop is based on the congruence (triangulation of information) between two nonequivalent sources of information about a given holon. These nonequivalent sources of information have to be generated by two nonequivalent external referents that exchange information about it. In the VW Beetle case, the two typologies of information are (1) the information used by the engineers when making an individual car following a given blueprint (validated by the top management of the company) and (2) the ability of consumers to recognize the type of a Volkswagen Beetle from certain characteristics — the stored image of the model in their memory (validated by social evaluation of such a model of car). This is what makes it possible for consumers to perceive individual realizations of such a model, when meeting it in traffic. The pattern recognition (identification of a particular realization as a legitimate member of a given equivalence class) at the level of the consumer is not necessarily obtained in the same way by different observers. For example, a blind person can recognize the model of this car by listening to the noise of the engine, whereas someone else will look at the design. Finally, children can recognize cars of the first model of the VW Beetle from their particular colors, some of which are unique among the cars running these days.

No matter how individual observers and potential buyers obtain their pattern recognition, humans will keep using the label Volkswagen Beetle (a mapping of an integrated set of qualities associated with a name) as long as it is useful for compressing the demand of computational capability to process information about their interaction with the reality. If the label Volkswagen Beetle is a valid one, then users will be able to apply to any individual member of the equivalence class (even if meeting it for the first time) the knowledge they have in relation to the type (e.g., where to find the switch for the lights or how to operate the gear box). That is, the knowledge about the type makes it possible to predict a given set of characteristics that are expected for each particular member of the class Volkswagen Beetle. This obviously requires that the mechanism of realization of the members of this equivalence class is capable of guaranteeing the predictability of their characteristics (according to a given common blueprint) — that the car maker guarantee a high level of reliability in the matching of expected standards for each car of a given model. From this example, we can say that the information stored in the blueprint used for making cars is the equivalent of the information stored in the DNA for making biological organisms.

We know that within the socioeconomic domain it is the economic mechanism that is in charge of keeping a correspondence between (1) the activities of engineers (those in charge of preserving the coherence of the characteristics of the equivalence class), which guarantee the validity of the identity of the type Volkswagen Beetle and (2) the mechanisms of recognition adopted by consumers (those that buy this model of car for its perceived better performance according to their personal expectations). This second activity guarantees the stability of the terms of reference (expected associative context) of the engineers working in the Volkswagen corporation producing additional cars of this class.

This is a crucial point, since even when we find a Volkswagen Beetle with a missing wheel, we are still able to consider it a member of that equivalence class. Knowing the identity of the type, it becomes possible to infer the existence of a discrepancy between the identity of the realization (what we perceive in our direct interaction in the reality) and the identity of the type (what we expect according to the information in the blueprint). This can be considered the sign of stochastic perturbations occurring during the process of realization (or during the consequent life of the realization). That is, even if a given specimen of Volkswagen Beetle does not fit totally with our expectations (e.g., since it shows signs of a recent crash), this will not affect our perception and representation of the type of its equivalence class. Whenever we have an identity for the type of an equivalence class that has been validated in our daily life, we will consider as deviant features of particular realizations or noise any information about special unexpected features found in individuals. We will judge such information as irrelevant for discussing the characteristics of types. This is why we tend to consider the characteristics of a type as out of time (we can recall the metaphor of Plato).

On the other hand, if we find a consistent number of three-wheel realizations of Volkswagen Beetles and these deviant realizations are relevant for our life (e.g., all three-wheel Volkswagen Beetles are used by police), we will have to quickly update our repertoire of useful typologies. That is, we will have to learn to make a distinction between three-wheel Volkswagen Beetles (labeled "Beetle police cars") and four-wheel Volkswagen Beetles (labeled "normal Beetle cars"). The number of wheels will be then included as a signal (a crucial attribute) to be used for quick pattern recognition. In this story we have an example of learning about the existence of (1) a new type (species?) of car (Beetle police cars) living in the jungle that we call urban traffic and (2) a new mechanism for pattern recognition (a new epistemic category — three-wheel vs. four-wheel), which has to be included by the observers in their repertoire of attributes associated with the detection of identities.

So far we used a few biologically related terms (species and jungle) to explore the similarity and differences between the roles of blueprints in the evolution of cars and DNA in the evolution of biological species. Can the distinction between realization and essence be used to better understand the difference between phenotype and genotype, as suggested by Rosen (2000)? The analogy between these concepts seems to be in a way already there, but it requires some additional comments.

8.2.1.1 *Phenotype: The Realization of a Type Associated with an Essence* — A phenotype can certainly be defined (at the organism level) as the realization of a particular organized structure referring to a given essence obtained at a given point in space. When defined this way, the concept of phenotype can be related to the ability to express agency through a member of an equivalence class. This agency refers only to the space–time domain to which the definition of essence is associated. Exactly because of this, the special individuality of agents operating on a smaller descriptive domain — e.g., the phenotype of an individual organism of a given species — is totally irrelevant in terms of information to be learned either by biological systems or by scientists about the role played by the species (to which the organism belongs) in a given ecosystem. The organism is operating at a hierarchical level too low to be relevant in terms of changes of types, let alone in terms of changes of essences. It is only when we consider the aggregate effect of all realizations (the various realizations of the phenotype), which are all mapping into the same semantic identity (the genetic information stored in the gene pool at the species level), that we reach an effect of aggregate agencies that has a scale large enough to be relevant for the definition of essences. That is, only when considering the aggregate effects of all the phenotypes realized in different points in space and time by a species (the result of the realization of different identities mapping onto the same pool of genes organized according to the species), do we have the possibility of having an exchange of information — over the same scale — between essence and realizations. Obviously, this implies adopting a definition of phenotype that is refers to a space–time domain much larger than that of the individual organisms (the set of populations operating in different associative contexts). This perception of phenotype is nonformalizable according to the structural organization of individual organisms. In fact, the set of all the populations of dogs cannot be expressed in relation to the identity of an individual organism belonging to that species (e.g., should we start with a cocker spaniel or a fox terrier?). As noted earlier, the set of all the phenotypes expressed in relation to the essence of a species can be expected to be determined by a component of stochastic events (noise)

that usually is averaged out on a large scale (e.g., gene flow). Alternatively, the set of all the phenotypes expressed by a given species can be seen as carrying useful information about the need to adjust the definition of essences (e.g., in determining the direction of speciation). Which one is the right interpretation? This cannot be known when looking at the information space associated with that individual species. This could be answered only when considering all the information spaces associated with the different species with which the first species is interacting. But this would imply considering simultaneously all the interactions at different scales, in different contexts (in different points in space and time). This is why we used before the expression "semantic identity of a species," which cannot be formalized.

Finally, if we accept along this reasoning that the definition of phenotype could also be applied to a hierarchical level higher than that of the individual organism (to arrive at a scale of operation associated with an impredicative loop, which is required for providing feedback on the identity of a species), we can no longer accept the central dogma of molecular biology requiring that the realization of an essence cannot imply any feedback on the definition of essence in the first place. The large-scale activity of all phenotypes (e.g., the ensemble of all individuals belonging to all populations of tigers) actually do feed back on the definition of the essence (the semantic identity) of the other species with which tigers are interacting. However, by doing that, they are obviously affecting the identity of their own environment (the expected associative context of their own type). The concept of essence, for adaptive dissipative systems, implies, in fact, that the environment of a given species is determined by the set of behaviors expressed by the other species interacting with it. This means that a change in the determination of the essence of other species will induce a loop through which the identity of the essence of the species tiger will also have to be updated. Recall here the example of the consequence of the huge production of FIAT 500 and other cheap utilitarian cars in Italy in the 1960s that led to a different definition of the role that a car had in the life of Italian car buyers in the year 2000 (see the comments for Figure 8.9).

8.2.1.2 Genotype: The Role Played by the Characteristics of the Template Shared by a Class of Realizations — As soon as one looks at the distinction between genotype and phenotype in relation to essences and realizations of equivalence classes of organized structures, one is struck by a noteworthy observation. In evolved life-forms (e.g., mammals) the material support used for the encoding information used for fabricating realizations (the sequence of DNA bases read during the process of ontogenesis) does not physically coincide with the material support transmitted to the offspring (the sequence of DNA bases used for encoding information used within the genic pool for transmitting and storing the identity associated with the type).

That is, the DNA that is actually read within an organism (the string of chemical bases used as a blueprint for the realization of a particular individual) represents just one of many formal identities that maps onto the same semantic identity. In fact, due to the crossing-over within the two sequences of bases, the final string passed to the offspring is not an exact copy of any of the two strings that were used to fabricate the parents. Put another way, even when dealing with the DNA of an individual organism — with the identity of genotype — we attend to a loss of a 1-to-1 mapping between the information used for the fabrication of realizations (used for the location-specific processes of ontogenesis) and the information used to store and transmit information about essence to the pool of the species (used for the larger-scale processes of handling mutual information). The dual nature of holons (a formal part related to structural organization at the local scale and a semantic part related to the expression of larger relational functions) is therefore reflected even at the molecular level, when dealing with the encoding of information in the DNA. The DNA string actually read for ontogenesis represents a formal identity that is physically and formally different from the DNA string passed to the rest of the species. Only in semantic terms is the genotype the same.

As a second observation, we can note that the information of the DNA transmitted to offspring (the DNA never read in the fabrication of the individual realization) contains not only information about the future realization of individuals of the same species (let us say a cat, organisms belonging to the species *Felis catus*), but also indirect information about the characteristics of the essence of other species with which the individual realizations will interact. Put another way, relevant characteristics of members of other equivalence classes (characteristics of expected features of other species) are included in the form of the expected associative context in which organisms belonging to the species *Felis catus* will interact. This second type of information can include, for a cat, indirect or direct information about expected features and behavior of the phenotype of mice, viruses, lizards, trees, poisonous plants, etc. Therefore, in a way, the ability of an individual realization of *Felis catus* to survive and reproduce in a given associative context (in a given point in space and time) represents a check for that species about the validity of two types of information:

1. The capability of fabricating realizations of individual organisms — other cats — which must be able to occupy the niche, that is, the viability of the ontogenesis in relation to the existence of favorable boundary conditions (how good is this information at making cats in an expected associative context).

2. The validity of the essence of "cat-ness," that is, the aggregate indirect information about relevant characteristics of other species with which the various phenotypes of the *Felis catus* will interact when operating within the expected associative context. This indirect information is embodied in the special characteristics of the organization of cats at different levels; e.g., it can refer to a better defense against possible microbial attacks, better efficacy when looking for specific preys, a better defense against larger-level predators. This implies that the usefulness of the template used to make individual cats can be associated with a validity check on how the characteristics of the essence of the *Felis catus* are compatible with the characteristics of the essence of the other species with which the cat will interact and that represent its associative context. That is, the validity of the essence of "cat-ness" is related to how good is the indirect information about the usefulness of various tasks that the organized structure is supposed to perform, in relation to the expected niche occupied by cats.

In a way, individual realizations of cats that are able to survive and reproduce are just experiments (or better probes) that the species sends around to confirm the validity of the relative information about the following two tasks:

1. Ability of realizing (bringing into existence) organized structures that can be considered members of the equivalence class *Felis catus* (within an admissible environment). Here the goal is not about discussing how to improve the identity of the essence. On the contrary, the identity of the essence must be given, to be able to solve the problem of how to make realizations of it.

2. Ability to reflect in a valid essence of *Felis catus* the constraints imposed on the species by crucial and relevant aspects of the essence of other interacting species. That is, how admissible is the environment in relation to what is expected according to the type (assumed to operate in a given associative context) associated with the template used in the process of fabrication?

Because of this, when an individual realization dies or thrives (at a given level n), it simply sends a message to the rest of the species (at the level $n + 1$) about the validity of the information it was carrying in relation to (1) blueprint and procedures of realization of the organized structure (triadic reading level $n - 2/n - 1/n$) — the fitness of structural stability and (2) usefulness of the information used to represent the essences of other species operating in the same environment, in the repertoire of models and detectors (triadic reading level $n/n + 1/n + 2$) — the fitness of relational functions.

Finally, we can get back to the discussion of the evolution of Mickey Mouse and the role that DNA plays in determining evolution. The analysis of biological systems presented so far can be perfectly applied to nonbiological species, that is, to species that do not store information about templates on strings of DNA (or RNA). First of all, what biological and nonbiological holarchies have in common is that they are evolving in time in terms of essences and not in terms of individual realizations. Organisms just express agency, reproduce and die. Realizations of specific types are just tools used to check whether the semantic definition of essence is still valid. Exactly as done by biological species, the people working at Volkswagen can be seen as doing an experiment based on the sending, every year, of a bunch of cars produced using a given a blueprint (i.e., VW Beetle) into a given socioeconomic system. The experiment is aimed at knowing whether the agents characterizing the context (potential buyers) are still willing to use this model (if the picture of an admissible environment embodied in the blueprint is still valid). If they are still willing to use this model of car, then they will pay for it. Then, if the consumers still pay for it, the validity of the type (which is associated with the characteristics of the given blueprint) is checked. In biological systems, which have to do everything by themselves (without an external design), the species gets back a semantic copy of the given blueprint, in sign of validation, every time the individual organism carrying it was able to reproduce. These two mechanisms are similar and provide the same result. The only difference is that in one case the blueprint is written on computer disks or paper (for the making of the VW Beetle) and in the other it is written using DNA bases (for the making of biological beetles). However, they represent two parallel validity checks determining whether a given essence used to define types and for the making of equivalence classes is still valid.

We can go back now to the evolution of the two essences shown in Figure 8.10. The possibility of generating equivalence classes of realizations of these two species is generated by the web of activities performed within the socioeconomic system in which they are expressed. This is why these two species changed in time and evolved. Over the considered time period, the same essence was expressed using different identities. Mickey Mouse can be seen as a sort of virus, which remains alive (is replicated by someone else) because of the energy dissipation paid for by those reading it. The readers make it possible to pay those that invest energy in drawing, printing, distributing and selling copies of it. The VW Beetle can be seen as a real organism that is dissipating on its own (it is eating gasoline). In this analogy, the VW Beetle can be seen as a domesticated species (e.g., milk cow), which depends on care and inputs from humans in exchange for a reliable supply of services. The only difference with cows is that the VW Beetle is not able to self-replicate. It should be noted, however, that this is a feature that is looked for in the latest development of high-tech genetic engineering (with varieties that cannot reproduce or when considering the terminator gene). This means that until the process generating the essence of these two species holds (the reciprocal entailment of expected behaviors of interacting agents, epistemological processes, makes possible the stabilization of the process of fabrication, ontological processes), it is possible to have smooth changes in the characteristics of the identity (in the templates used to define the type) used to realize the essence at different points in time. The set of characteristics shared by the members of the equivalence classes will change smoothly. The slight but continuous accumulation of small changes will be reflected in changes in the relative blueprint associated with these species, the representation of the template.

Remaining in the example of the evolution of species of cars, we can consider other examples of trajectories of evolution of car species. Two examples are illustrated in Figure 8.11. On the right side, we have the evolution of different species (models) defined in the market niche (essence) very cheap utilitarian cars of small size. This niche was filled at the beginning in Italy by the major Italian car maker with the FIAT Topolino model (from 1936). This model soon became extinct, and then FIAT launched in 1957 one of its most successful models ever: the FIAT 500. The FIAT 500 was characterized by a very limited supply of power, very severe driving conditions and a lot of noise, but still it was able to fulfill the task of supplying a car to Italian households during the transition toward industrialization. It was the car that made the transition from scooters to cars. Different versions of this basic model were generated during the years, trying to keep the demand high. However, the systemic weakness generated by the small size of its engine drove the model to extinction in 1975. The new model — the FIAT 126 — was built around a bigger engine and new basic technical solutions. The trajectory across these three models (FIAT Topolino → FIAT 500 → FIAT 126) has already been discussed in Figure 8.7. According to the metaphor proposed by Holling, these models of cars were following a natural pattern of evolution. The various models introduced were able to evolve through an expected sequence. The existence of this predictable sequence of stages for the type is the sign of an evolution in time of the relation type and associative context.

This means that the characteristics of the members of the equivalence class (species models, that is, typology of the car) were not only mapping onto a blueprint (the information used by those making the cars in the factory), but also were kept congruent with expectations about such a car expressed by the interaction of agents in the socioeconomic context. That is, the typology of a car associated with a given model was congruent with the mosaic of mutual information stored by those interacting with the realizations of this car in society. That is, Italy in those years was full of:

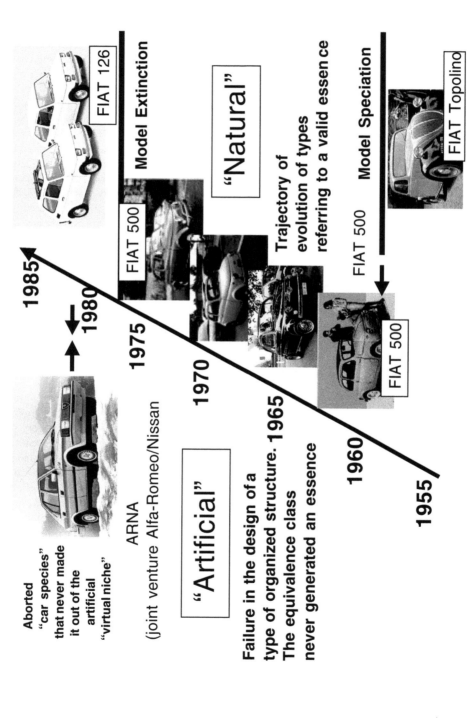

FIGURE 8.11 Failure in the design of a type of organized structure; the equivalence class never generated an essence. Trajectory of evolution of types referring to a valid essence. (Courtesy of FIAT spa and ARCHIVIO STORICO FIAT.)

1. Car buyers pleased with the performance of this given model
2. Mechanics able to repair this model
3. Small micro-factories making spare parts at a lower price than that of the official maker
4. Bankers willing to finance the purchasing of this model
5. Insurance companies willing to insure this model
6. Young women thrilled by the idea of dating an owner of such a car

Put another way, the realizations of FIAT 500 had an admissible environment (a niche in which operating) that was determined by the coherence between the information and the expected behaviors of all those agents interacting with individual realizations of this model of car. The model type was mapping onto a valid associative context. Completely different is the case of the model of car depicted in Figure 8.11 to the left of the axis indicating years. The model ARNA was generated by a joint venture of the Italian car maker Alfa-Romeo and the Japanese car maker NISSAN. However, the ARNA was only a pseudo-species car. This model was one of the biggest disasters in the history of these two car makers. This model did not reflect the existence of an essence in the real world. The expected niche imagined when projecting this model was simply not there. In this case, we have an example of (1) a useful blueprint, which was perfectly fit for making cars and (2) the existence of an equivalence class of individual realizations (based on a template), which were all sharing predictable characteristics (representing a potential type). The problem was that this set of characteristics never managed to generate a convergence of interest (they were never considered relevant) by the various agents operating within the socioeconomic environment. The mutation induced through design (the generation of a new valid blueprint), according to the prediction of the existence of a virtual niche for such a model, was validated by one of the external referents (the possibility of making functioning cars) at the local scale. However, such a process was not validated by the second external referent — the validity of the assumption about the existence of a benign associative context. Put another way, the information about how to fabricate a class of organized structures sharing the same set of characteristics was valid in syntactic terms (ARNAs were made and they were running fine). However, the type was not compatible in semantic terms with the information used by the other agents interacting with the members of this equivalence class — those determining the existence of a niche (admissibility of the environment) for the class of realizations of ARNAs.

The main point of this example is that we can learn a lot about the characteristics of Italian society in the 1960s by studying the characteristics of the FIAT 500 car model. Being a real species, the essence of the FIAT 500 carries information about its socioeconomic context. However, we cannot say anything about Italian society in the 1980s when looking at the characteristics of the ARNA. The shape and characteristics of the template used to make this car reflect only the personal opinions of the people that were in charge of its development.

This long discussion about the evolution of car models had the goal of making the following important point: When looking at the mechanism generating essences, we can conclude that in no way does the information stored in the blueprint (or in the DNA), which is useful for fabricating individual realizations of members of an equivalence class, have anything to do with either the definition of an essence or the process of evolution of individualities, through trajectories of types within the room provided by a valid essence.

The essence and the set of identities taken by the two species illustrated in Figure 8.10 (Mickey Mouse and VW Beetle), as well as the three species of car models illustrated in Figure 8.9 and Figure 8.11, are determined by the mutual information carried by adaptive systems interacting with each other on the basis of models they have of each other. The difference between biological systems and human artifacts is that in the latter case, it is the designer that uses models of the potential buyers to predict the future fitness of goods and services.

8.2.2 How to Obtain an Image of an Essence within the Frame of Network Analysis

To see an image of an essence within the frame of network analysis, it is important to acknowledge the existence of a few hidden assumptions, which are required to represent biological systems in terms of dissipative networks. Two old papers of Robert Rosen can be used to focus on these crucial, but neglected assumptions: "A Relational Theory of Biological Systems" (1958a) and "The Representation of Biological Systems from the Standpoint of the Theory of Categories" (1958b). In the second paper Rosen introduces his first ideas about category theory, which were developed later on (Rosen, 1991). The text of these two papers is quite obscure, but it still provides a few crucial points that are relevant for our goal.

In these two papers Rosen focuses on the characteristics that biological systems must have to be able to stabilize their own identities through a process of metabolism. First, Rosen starts his discussion by acknowledging that any decision about how to draw a graph representing interacting elements of a biological system (how to decide about boundaries) must be necessarily semantically driven. That is, this is a decision that cannot be formalized once and for all (in terms of a substantive protocol). In his own words, he says that a decomposition of a system into its components has to be obtained by some means. Then the dual nature of holons enters into play:

> Let us observe that the problem of determining the structure of a given system resolves itself naturally in two more or less distinct parts.
>
> In the first place, we can regard the system as a collection of simpler, fundamental sub-objects (which we shall term *components*), each of which also behaves in such a manner as to produce a defined set of output materials from a given set of input materials. These components are to be related to one another in the obvious manner; the output materials of a given component may serve as input materials to other components of the system. There will generally be other relations obtaining between the components of a given system; a complete enumeration of their relations will then determine the behavior of the system and will constitute a solution to what we may term the *coarse structure problem* of the theory of systems. Much of the information required for a solution of the coarse structure problem for a given system will be contained in the so-called *block diagram*, or flow chart, for the system. (Rosen, 1958a, p. 246)

This reflects the triadic reading level $n - 1$/level n/level $n + 1$.

> For the purposes of the block diagram, we see that each component is regarded as a structureless element (i.e., a "black box"). Once a coarse structure has been obtained (i.e., once we have decomposed our original system into a suitable set of components), we may begin to inquire about the actual physical realizations of these components; this aspect of the theory of systems may be termed the *fine structure problem*. Speaking very roughly then, the coarse and fine structure of systems bear to each other the same relation as do macroscopic and microscopic states of a thermodynamic system, as do physiology and anatomy, or, in Rashevsky's terminology (Rashevsky, 1954) as do relational and metrical aspects of biology. (Rosen, 1958a, pp. 246 and 247)

This reflects the triadic reading level n/level $n - 1$/level $n - 2$.

Then, if we want to represent biological systems in terms of interacting elements on a network, what is the list of characteristics that make this class of adaptive dissipative systems special?

1. The identity of a graph (level n) made up of lower-level elements (level $n - 1$) requires four pieces of information. Recall here the discussion about:
   a. What is the form of energy used for the metabolism of the whole (what flows of matter are associated with energy conversions operating at the interface between the black box and the environment). This defines an expected identity for energy carriers (admissible input in Rosen's terminology).
   b. What is the output/input ratio of specific transformations associated with each element of the network (at the level $n - 1$, as shown in Figure 8.12).
   c. What is the throughput ratio — the rate of energy dissipation of the converter and the pace of the input and output (at the level $n - 1$, as shown in Figure 8.12).

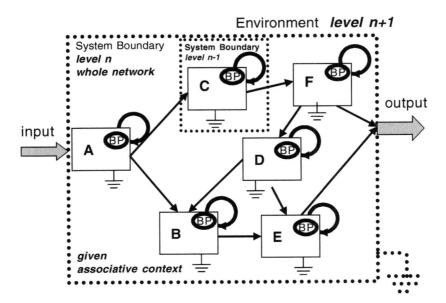

FIGURE 8.12 Rosen's view on the representation of a biological network.

    d. What is the relative position assigned to the element within the network with arrows indicating relations among components in the graph (defined at the level n, as in Figure 8.12).

2. All dissipative systems described at different levels (the whole system defined at the level n, individual elements defined at the level $n - 1$, and subcomponents making up the structure of components at the level $n - 2$) are assumed to operate (if they remain alive) in an expected associative context compatible with their identity. That is, their very identity of metabolizing systems entails the existence of favorable boundary conditions associated with the specific process of dissipation required by their metabolism. That is, the validity of the association of a type with an expected associative context is guaranteed by the activity of the black box for those components that are not interacting directly with the environment. The elements included in the indirect compartment have an associative context that is therefore associated with the survival of the black box. However, the assumption of favorable boundary conditions must be assumed valid by default for those components exchanging matter flows directly with the environment (those supported by environmental inputs and those generating environmental outputs). This different set of assumptions is at the basis of the distinction made among the components defined at the level $n - 1$ in two typologies:

    a. Those dealing with direct interactions with the environment (components A, F and E in Figure 8.12)

    b. Those not interacting directly with the environment operating within the boundary of the black box (internal components such as B, C and D in Figure 8.12).

The identity and the characteristics of these components are still affecting and affected by the environment, but in an indirect way. That is, these compartments affect the environment by affecting the characteristics of the components of the first group. This distinction has already been discussed in Figure 7.8a.

3. To be able to establish a stable relation among the various identities defined at different levels and scales, each element must be able to first guarantee the stability of its own characteristics. This implies the ability to reproduce itself using a template and express a set of behaviors useful for preserving both organized structure and functionality on a short timescale (e.g., repairing itself, avoiding dangerous situations, controlling the inflow of required inputs).

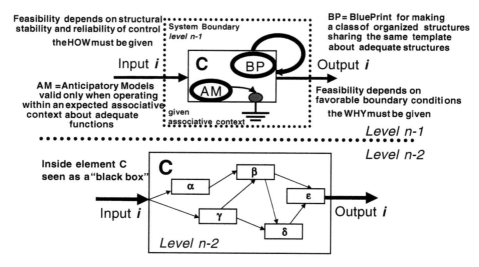

FIGURE 8.13 Different view of the identities of elements of biological networks when shifting the triadic reading to lower levels.

According to the specific characteristics of a particular element of the graph, we have to expect that each element not only must have a specified organized structure that shares a common set of characteristics (it belongs to an equivalence class), but also has an information system that is able to store and process two distinct types of information:

a. BP — A blueprint required to make additional copies/realizations of its own organized structure. This is required for preserving the identity; the reproduced component must be a member of the same equivalence class (to fulfill the task of reproduction)

b. AM — A set of anticipatory models required to express a set of behaviors useful for stabilizing the process of metabolism of the given dissipative structure in the short run within the expected associative context.

These two functions expressed at the level $n - 1$ and are guaranteed by lower-level mechanisms at the level $n - 2$; an overview is given in Figure 8.13.

4. After admitting that lower-level elements are able to preserve the given set of identities (the validity of points 1, 2 and 3) at both level $n - 2$ and level $n - 1$, then the particular configuration of the network (or graph structure or, as suggested by Rosen, flowchart or block diagram) at the level n assigns a functional identity to each of the internal components. This fact is represented in Figure 8.14 by the expression essence of B, which is indicated by the empty spot left when removing the component B from the graph (the same graph given in Figure 8.12). Put another way, after defining the identities over two different levels (level $n - 2$ and level $n - 1$), as specified in point 1, of all the other elements (A, D and E) linked to a given element (B) on a given graph, defined at both level $n - 1$ and level n, we can describe the image of an empty niche left there. The set of expected characteristics determining this empty niche is obtained using a mosaic effect of entailments across scales of the type discussed in Chapters 6 and 7.

   The crucial point of this discussion is that the information used to provide such a characterization of this network element is totally independent of the information required to make physical realizations (individual members of the equivalence class) occupying that position in the graph. This means that because of the identity of the other elements and the given position in the graph, the realizations of the typology of elements expected in position B must be able to express a common set of characteristics. These characteristics associated with legitimate members of the equivalence class, associated with the virtual element B, are expected by the network according to the mutual information carried out by the various identities of the other

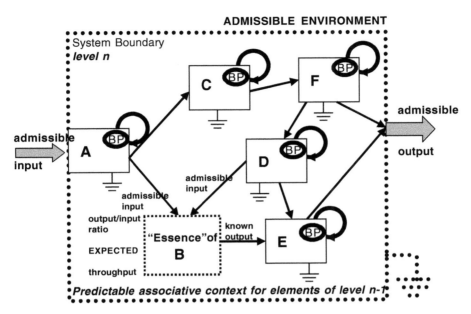

FIGURE 8.14 Definition of an image of an essence using a mosaic effect across scales.

elements of the holarchic dissipative network (the whole system, defined at the level *n*, and the other components, defined at the levels *n* − 1 and level *n* − 2). In this case, a given formalization of this dissipative network would make it possible to define an image of the essence of the particular virtual element B. Such an image is clearly just one of the possible images of that essence that will reflect the choices made by the analysts about how to represent the dissipative network in formal terms in the first place.

Different is the case of those elements (e.g., in Figure 8.12, elements A, E and F) that are directly interacting with the environment. In fact, the identity of these components is determined not only by the mutual information carried about them by other elements of the network, but also by the characteristics of what Rosen calls environmental inputs and environmental outputs. In this way, the reciprocal definition of identities is expanded to include, even if in partial terms, also the characteristics of the environment. The concept of metabolism, in fact, implies imposing a few assumptions — actually a sort of weak identity — on the environment. The environment of a dissipative system must be admissible by definition. This means that at the level *n* – 2, the definition of identity for energy carriers (admissible input for those components "eating" or depending on environmental inputs) must be compatible with the processes occurring in the environment at the level *n* + 2 (those processes that we do not know, but that must be there to stabilize the supply of input and the sink of wastes). The total input required by the system as a whole (at the level *n*) therefore coincides with the input required by the external components (at the level *n* – 1), and it must be compatible with the processes producing such an input in the environment at the level *n* + 2, to have an adequate supply of input at the level *n* + 1 (an admissible set of boundary conditions). Also in this case, a definition of an integrated set of identities for wholes and parts of the network translates into a requirement of compatibility in relation to the reciprocal conditioning of different processes determining admissible conditions on different scales and descriptive domains. We are back to impredicative loop analysis discussed in Chapter 7.

5. All the dissipative elements within socioeconomic systems and ecosystems (components of the graph, the whole network, and subcomponents of components) have a common (the same) reference state for their process of dissipation/self-organization. They all discharge heat as the final by-product of their metabolic process, meaning that the various epistemic categories used

to characterize their identities can be linked because in this case it is possible to establish a link between mapping of matter and energy flows. Recall here the various examples of impredicative loop analysis discussed in Chapter 7. The possibility of obtaining a reducible mapping between extensive variables 1 and 2 is based on the possibility of establishing an energetic equivalence between an energetic assessment and a mapping of a matter flow (tons of evapotranspirated water, kilograms of food, tons of oil equivalent). This is where the two mappings related to the representation of interaction — black box internal parts and black box environment — can be reduced to each other, even if accepting a certain degree of arbitrariness in this operation. Put another way, in the virtually infinite ladder of space–time scales that can be used to look for potential energy forms useful for perceiving and representing the reality, when dealing with the metabolism of terrestrial ecosystems and human societies, we can afford to focus only on a given space–time window. That is, when looking at biological systems or socioeconomic systems, we can ignore atomic and subatomic energy forms and also the energy flows associated with pulsars or black holes. Put in other terms, the metabolism of ecological processes and the metabolism of societal systems can be represented using nested elements having very strong identities. Therefore, we can perceive and represent their metabolism using only a finite set of relevant energy forms, transformations, and matter and energy flows. When additional categories become relevant (e.g., pollution by heavy metals), the analysis immediately becomes more difficult.

This eases selection of a finite representation/description of the metabolism of societal systems or ecosystems. Such a representation can be based on a selection of only a given finite and manageable set of components, energy forms, energy transformations and flows. Obviously, any representation will necessarily cover only a subset of all the possible components, energy forms, transformations and flows that could be used to perceive and represent terrestrial ecosystems or human societies. Actually, when nonequivalent observers (e.g., the interacting life-forms within a terrestrial ecosystem) use different detectors operating on different scales to observe the same ecosystem in relation to different goals, they are generating a quite large set of nonreducible valid narratives. Again, deciding which energy forms have to be included in a given analysis (when defining the identity of a network and their components) does not imply that those energy forms not included are not relevant in absolute terms.

8.2.3 The Difference between Self-Organization (Emergence of Essences and New Relevant Attributes) and Design (Increase in Fitness within a Given Essence = Set of Attributes/Associative Context)

Networks of nested dissipative elements are very important for the discussion of this chapter, since they can be associated with a peculiar set of characteristics that are typical of life. To introduce another relevant basic concept — centripetality — we quote here the work of one of the major experts in this field, Robert Ulanowicz:

> *Selective pressure that the overall autocatalytic form exerts upon its components.* ... Unlike Newtonian forces, which always act in equal and opposite directions, the selection pressure associated with autocatalysis is inherently asymmetric. ... They tend to ratchet all participants toward *ever greater levels of performance.* ... The same argument applies to every member of the loop, so that the overall effect is one of *centripetality*, to use a term coined by Sir Isaac Newton: the autocatalytic assemblage behaves as a focus upon which converge increasing amounts of exergy and material that the system draws unto itself. (Ulanowicz, 1997, p. 47)

A short description of the concept of centripetality, based on the text of his book, follows. By its very nature, autocatalysis is prone to induce competition and not merely among different properties of components. Its very material and mechanical constituents are themselves prone to replacement by the active agency of the larger system. For example, suppose that A, B, C and D are sequential elements comprising an autocatalytic loop as in Figure 8.15 and that some new element E appears by happenstance,

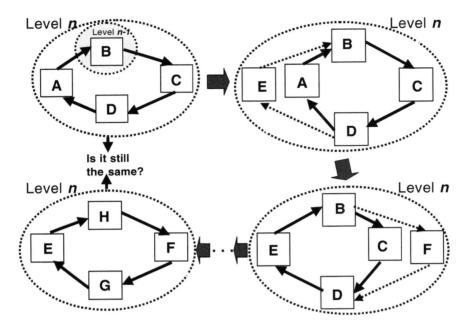

FIGURE 8.15 Ulanowicz's centripetality.

which is more sensitive to catalysis by D. This means that it provides greater enhancement to the activity of B than does A. Then E will grow to overshadow A's role in the loop or will displace it altogether. In like manner, one can argue that C could be replaced by some other component F (as in the lower-level graph on the right). Then D can be replaced by another component G, and the component B can be replaced in the same way by H. The final configuration of the sequential elements of the autocatalytic loop could become at that point E, H, F, G, which contains none of the original elements. (This is our summary of the text found on p. 47.)

> It is important to notice in this case that the characteristic time (duration) of the larger autocatalytic form is longer that that of its constituents. The persistence of active forms beyond present makeup is hardly an unusual phenomenon — one sees it in the survival of corporate bodies beyond the tenure of individual executives or workers, or in plays like those of Shakespeare, that endure beyond the lifetime of individual actors. (Ulanowicz, 1997, p. 48)

This refers to lower-level element turnover in the role defined at a higher level. If we try to apply the set of concepts developed so far to this representation of the process of evolution of an autocatalytic loop, we can say that what was described by Ulanowicz as centripetality refers to an improvement in the design of an autocatalytic loop. That is, we have a level n at which we are describing the network generating the autocatalytic loop. This network has a given role (set of tasks to fulfill) in relation to its interaction with the environment (defined at the interface $n/n + 1$). The ability to fulfill the required tasks translates into the ability to guarantee the stability of expected boundary conditions for the elements that are operating inside the network — at the level $n - 1$ — seen as a black box. Within this specification of roles across levels, we can define for lower-level components a formalization of efficiency. As specified by Ulanowicz, an improvement in relation to a given context-specific setting — (1) identities of the various elements at the level $n - 1$, (2) graph structure at the level n and (3) characteristics of the associative context (validity of the assumption of an admissible environment) — will feed back to the starting component as a reinforcement of this new behavior (over the whole cycle — at the level n). This is why more efficient components can replace obsolete ones in relation to the given set of relations and roles of the set of identities defined on the two levels n and $n + 1$ (black box/admissible environment). However, this has more to do with a natural implementation of the original design than with a real phenomenon of emergence. The substitution of an element A with an element E that does better in the niche of A is related to what in biology is called the survival of the fittest, when talking of natural selection. But the definition of fittest is already related to the definition of an essence for components operating within a specified associative context (determined by the identity of the larger whole characterized at the level n). This has nothing to do with emergence.

Emergence has to do with the generation of new essences — new bridges across levels in both ontological and epistemological terms. This implies considering, as discussed in Chapters 6 and 7, an integrated set of facts referring to and defined on two continuous triading readings: levels $n - 2/n - 1/n$ and levels $n/n + 1/n + 2$. The integration of these facts requires translating the definition of types across five levels in a way that implies a self-entailment of ontological and epistemological processes determining the feasibility of the resulting set of essences. That is, it requires the simultaneous compatibility on different scales of (1) the ontological process of realization of individuals within equivalence classes defined at different levels and (2) the epistemological process based on the information about experienced and expected facts, which is stored in the various systems of control regulating the predictability of patterns of dissipation at different levels.

The concepts of type, essence and realizations of a given equivalence class can be used to make an important point, which has been proposed to us by Tim Allen (personal communication). It is useful to make a distinction between two drivers in the process of evolution. The first of these two drivers can be associated with the concept of design, and the second with the concept of self-organization.

- We deal with design when we have the specification of a given role (a set of tasks to be fulfilled), and then different types of organized structures are changed in time to provide an improvement in relation to the established definition of that performance (this is the example of the centripetality illustrated in Figure 8.15).
- We deal with self-organization when an old set of organized structures (lower-lower-level structural elements ($n - 2$) organized in a structural pattern defined by the identity of lower-level types ($n - 1$) interact in a way that generates a new behavior at the level n. This new behavior makes possible the stabilization of a new set of functions on the interface black box–context — on the interface between level $n + 1$ and level $n + 2$. At this point we have an existing set of organized structures (on the lower triadic reading) that manages to express a new set of functions that makes sense in relation to processes detected on the higher triadic reading. A new set of categories is required to classify and represent this new set of functions/qualities.

In this regard, when discussing evolution and the role that mutations (changes in the template used to make classes of organized structures) can play in determining emergence, Popper says: "The main thing in my form of the theory is that mutations can succeed only if they fall within an already established behavioural pattern. That is to say, what comes before the mutation is a behavioural change, and the mutation comes afterwards" (Popper, 1993, p. 69). An established behavioral pattern, according to Chapters 6 and 7 of this book, implies a congruence of mutual information (reciprocally validated) on different levels by nonequivalent external referents. This term can be considered analogous to a valid essence. Therefore, we can translate the statement of Popper in our terms as: what comes before the mutation (a change in the template used to generate a class of organized structures) is the existence of a virtual new essence (a change in the self-entailing process of definition of valid essences), and then the mutation comes afterwards. This statement can be directly related to the discussion of the successful or aborted speciation of new models of cars described in the previous section (Figure 8.11).

We can now try to wrap up this discussion using the concepts introduced so far:

- Organization following design = new organized structures that are more useful according to the given problem structuring for increasing the perceived fitness (efficiency) of the system = making better types within a given essence.

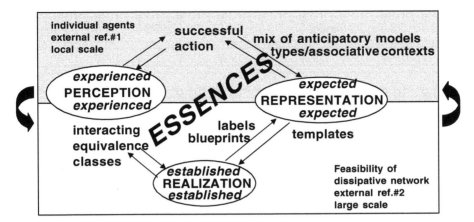

FIGURE 8.16 The generation of essences through a process of parallel validation at different scales (two nonequivalent external references).

- Self-organization = emergence = the making of new essences = the resultant need to introduce new attributes (variables) and new relational functions for the modeler.

8.2.4 Conclusions: So What? How Should We Deal with the Representation of Becoming Systems?

The challenge of describing the evolution of something that becomes something else while remaining the same cannot be confronted by using formalisms developed within the reductionist paradigm (models and formal definitions given once and for all). What we can do to face this predicament is try to understand as much as possible the implications of our arbitrary choices of what should be considered relevant (the types or the individualities) when representing this process.

We believe that concepts derived from complex systems thinking, such as essences, identity, types and individual realizations of members of an equivalence class, can be very useful for a better understanding of the epistemological implications of our choices in the step of representation. In particular, we can say that the unavoidable fuzzy characterization of holons and holarchies has to go through establishing a correspondence between an ontological and epistemological duality typical of adaptive dissipative systems. This duality is reflected in the following series of coupled terms:

A → Essence	Type	Niche	Role	Relational function
B → Realization	Individual	Population of a species	Incumbent	Organized structure

A valid essence reflects the ability to keep coherence/congruence throughout a process of autopoiesis within a holarchy between an epistemological side (row of definitions A) and an ontological side (row of definitions B). The two definitions obviously do not coincide in terms of descriptive domains. Therefore, they are related to each other in semantic terms, but they are nonequivalent and nonreducible in formal terms to each other. This is why a semantic check, based on a triangulation of nonequivalent external referents, is always required when dealing with the representation of complex systems. An example of this resonating process is given in Figure 8.16. This figure leads us back to the discussion of the Peircean triad given in Chapter 4. The parallel check done by two nonequivalent external referents in relation to two related, but not reducible, mechanisms of representations implies that such a process cannot be formalized. As already stated, it is impossible to obtain the "right" representation of complex systems, but only families of possible representations, which can be more or less useful depending on who will use such a representation.

When facing this predicament, it is wise to look for procedures that also include a discussion of the usefulness of a given representation in relation to a given situation and problem. In this context, impredicative loop analysis is useful, since it forces the analyst to adopt an integrated chain of choices that have to make sense in relation to a list of quality checks that are related to both (1) internal congruence of the scientific representation on different descriptive domains and (2) relevance of the choices of characteristics included in the definition of identities adopted in the models.

The adoption of two nonequivalent external referents in parallel (local and large-scale validation of a given identity in two nonequivalent ways) (Figure 8.16) has also been suggested by Tarsky (1944) as the only way out of the epistemological impasse implied by complexity. The famous line he uses to express this concept is: "'Snow is white' is true if and only if *snow is white*." The expression "snow is white" in quotation marks refers to a formal representation of the reality obtained within an object language (a formalized external referent), whereas the expression *snow is white* in italics refers to a validity check about the usefulness of the label (snow is white), which is obtained through a nonequivalent external referent (what he calls a meta-language).

Living systems routinely validate the integrated sets of multiple identities they use to make models of themselves for interacting with each other, throughout the establishment of impredicative loops (chicken–egg self-entailment between ontological processes and epistemological processes). They are making models of themselves all the time and use these models to generate integrated sets of predictable behaviors. The existence of these loops is perfectly explicable in scientific terms and very normal in reality. Actually, this mechanism of self-organization is much more general and likely to be found in nature than that of human design used by engineers for the making of machines. However, the process of self-organization of life seems to have a serious defect in the eyes of hard scientists. Such a process is not formalizable according to conventional differential equations and cannot be optimized in substantive way. As a result of this unpleasant characteristic, the existence of chicken–egg processes and complex mechanisms impossible to formalize is simply denied by rigorous scientists, when coming to the generation of models.

8.3 The Additional Predicament of Reflexivity: The Challenge of Keeping Useful Narratives within the Observed–Observer Complex

8.3.1 The Surfing of Narratives on Complex Time

We can start this section by recalling the implications of the concept of complex time for those attempting to generate useful models of the process of becoming of dissipative systems. As observed in Section 8.1, there are three time differentials that have to be considered:

1. dt — The time differential of the dynamics within the model (this is the time differential chosen within the observer–observed complex to represent events on a given descriptive domain).

2. $d\tau$ — The time differential at which models should evolve within a given problem structuring (this is the time horizon at which the validity of the model expires). This refers to the speed of change of the characteristics of the observed (e.g., is no longer behaving as before) and the observer (e.g., its ability of observing has changed).

3. $d\theta$ — The time differential at which the relevant features of the selected problem structuring change in relation to the characteristics of the observer–observed complex. This can imply changes in the original set of goals of the modeler, because of a change in power relations among stakeholders, because of the emergence of new relevant issues, or the acknowledgment of the inadequacy of previous problem structuring.

The definition and determination of these three time differentials is logically independent from the chain of choices made during the process of modeling.

The third timescale is the one on which it is more probable to have bifurcations, that is, disagreements about what should be considered the right set of relevant system qualities to be used to study a given problem. This translates into (1) potential troubles on the descriptive side and (2) expected troubles on the normative side. The formal identity of models (the identity of the system of differential equations used to describe movements in the state space according to the simple time defined by dt) should change over the time differential $d\tau$, whereas the semantic identity of useful metaphors (selection of a finite set of relevant relations among system qualities) should change over the time differential $d\theta$.

In this regard, we can recall the example of the standard failure of econometric analyses based on the *ceteris paribus* assumption. Whenever econometric models failed to be predictive, econometricians found a ready, yet self-defeating, excuse: "history has changed the parameters" (Georgescu-Roegen, 1976). Georgescu-Roegen, however, notes in this regard: if "history is so cunning, why persist in predicting it?" Rather than relying only on formal models based on a quasi-steady-state framing of the economic process (*ceteris paribus* hypothesis), it would be better to focus on understanding the basic mechanisms that are driving the evolution of the economic process (Georgescu-Roegen, 1976). As noted in Chapter 7, however, this is not easy, since it implies the need to use analyses referring to different hierarchical levels, and therefore in parallel nonequivalent descriptive domains.

The concept of complex time, which is associated with changes occurring in the observer–observed complex, introduces an additional problem for the analysts of sustainability. In fact, when dealing with the behavior of this complex, there is an additional and dangerous source of nonlinearity. This nonlinearity is not referring to the mechanisms of causality perceived in the natural system under investigation. Rather, this source of nonlinearity can be found on the semantic side. The semantic side is related to the process through which the observer assigns external referents to mathematical objects to provide them with meaning. Nonlinearity here refers to sudden changes in the interests of the observer (recall the example of the two logically independent definitions of size for London and Reykjavik, discussed in Section 3.1). When the observer is a social system (we are dealing with the *weltanschauung* in which the scientific activity is performed), nonlinearity can be generated by a sudden change in societal power relations or by a sudden discovery of a new typology of problems that entail a paradigm shift in scientific analysis.

A sudden change in the set of goals and organized perceptions in a society can induce a collapse in its sustainability, because it makes obsolete and inoperative the actual system of control. This can happen even when boundary conditions and biophysical processes remain the same. Let us recall the example of the fall of the Soviet empire, which was primed by a lower-level local collapse in the system of control operating at one of the gates of the Berlin Wall. Such a local failure quickly spread throughout the various systems of control up into the holarchic structure. In fact, nothing changed in terms of technological performance, in availability of natural resources for the Soviet economy in the following weeks and months. Still, the entire Soviet empire went through a process of implosion due to a quick loss of validity of reciprocal information stored in its hierarchical system of controls. The socioeconomic structure of the Soviet Union lost at an increasing speed the ability to keep alive the process of resonance between perception, representation and realization, required to validate the set of essences needed to guarantee the stability of types and equivalence classes of realizations (Figure 8.16). As a result, the Soviet system lost the ability to keep coherence in the behaviors of the realizations (members of the equivalence classes) associated with the various types across levels (from large state organizations to individual citizens). In the end, farmers stopped being farmers and policemen stopped being policemen.

The same can happen to modelers facing a sudden change in the definition of what is relevant in the problem structuring adopted in the definition of a particular dynamical system. A systemic loss of meaning can induce a sudden collapse in the validity of the model across levels. We can call this "the emperor has no clothes" effect.

8.3.2 The Moving in Time of Narrative

The various examples given in the set of figures seen so far enable us to make two important points about the relevance of changes that can take place on the side of the observer. This is a dimension characterized by the time differential $d\theta$. As discussed previously, this third dimension refers to the pace at which the observer is changing in time. This relevant rate of change can refer to both capability of observing (perceiving and representing) and the interests that are associated with the act of observing.

Let us start with an example of the first type of change in the observer. Imagine that we are watching an old tape of the U.S. Open tennis final on a TV. Looking at a tape on a TV set is like looking at the dynamic simulated by a mathematical model in a computer. There is a simple time determined by the representation given by the tape. Like in a mathematical model, there is a system (e.g., in this case, two interacting players that behave according to a specified set of rules) that is represented as operating in a given associative context. The potential behavior is determined *a priori*. As soon as a tennis match is recorded on a tape (as soon as a dynamic is formalized into a simple type), it becomes fully predictable (we can rewind the tape and then tell someone else exactly what will happen next). We can go backwards in time and even travel into the future (moving from the action recorded in the first set to the action recorded in the third set). This is possible because this is a finite representation of a real process, done in relation to a single descriptive domain. We are not dealing with a real event that is constrained by a set of parallel processes occurring on different scales and that can be observed in parallel on different levels. What we see on the screen of the TV now depends on the choices made by the directors and the technology available when the tape was recorded. We cannot look under the shoes of a player if this was not done by the director then. Imagine, for example, that we have a tape that is in black and white. The limits imposed by a black-and-white representation will remain a constraint when we run the tape on a modern color TV set.

The same problem (impossibility of perceiving enough relevant information from a formal representation of a modeled system) can occur in science. There are no pictures of Neanderthals (only reconstructions of their faces), because at that time humans could not use cameras for perceiving and representing each other. However, it should be noted that moving from a tape in black and white (e.g., differential equations) to a tape in color (e.g., using genetic algorithms to represent complex behaviors) simply makes richer our ability to perceive and represent reality. It does not change the fact that the tape (either in black and white or in color) will remain just a representation of a dynamic that occurs in simple time (it does not evolve). Even if we look at a color tape of the U.S. Open final between John MacEnroe and Bjorn Borg (in 1980), we can immediately perceive that the tape is old. That is, we can perceive a significant movement in relation to the time differential $d\tau$. In fact, we know that these two tennis players are no longer active in the tour. If we do not know this information, we can still observe that nobody in professional tennis still uses wooden rackets (used by them in the final). If we do not know this information, we can observe that the clothes and the hairstyles of those playing and those watching the match in this tape are different from what can be seen in the streets today. The consideration of all these clear signs of a movement of the narrative in time — in relation to $d\tau$ — tends unavoidably to indicate also a potential change in relation to the second type of change in the observer. Probably, if the observer of this tape watched the same match live in 1980, she or he was also dressed like the people in the stadium. A change in the way an observer is dressed or is arranging her or his hair is in general a sign of more important changes.

To deal with the effects of the second type of changes occurring in an observer, we can again use Figure 8.1 and Figure 8.3. We have here a practical example of the implications that a change in the interests of the observer can have on the representation of becoming systems. In particular, in relation to Figure 8.1, we can now inform the reader that Sandra is the wife of the first author of this chapter, whereas Marinella — the young woman represented in Figure 8.3 —was the companion of the first author 20 years ago. The bifurcation between two distinct complexes observer–observed (Mario–Sandra in Figure 8.1 vs. Mario–Marinella in Figure 8.3) can be seen as the explanation that caused the selection of two different representations of the process of aging in the two figures over two distinct individualities. The possibility of a sudden switch in how to perceive the same pattern is illustrated in Figure 8.17. Personal changes in the observer (in this example, Mario, considered at two different points in time — in the year 2000, when looking toward the left, and in 1980, when looking toward the right) have important implications on the process of representation (how aging is perceived and represented — what is shown in Figure 8.1 and Figure 8.3). In this example, the possibility of obtaining a different set of perceptions and representations depends not only on who is asked to make the observations (Mario vs. Kozo), but also on when this request is fulfilled (in 2000 or 1980).

It should also be noted that the new information given to the reader can change the explanations of what is represented in Figure 8.1 and Figure 8.3. At this point, after receiving this new information, the reader can infer that Sandra and Marinella are the relevant individualities (the real external referents assigning sense to the descriptions) that determined the choice of the set of representations used to

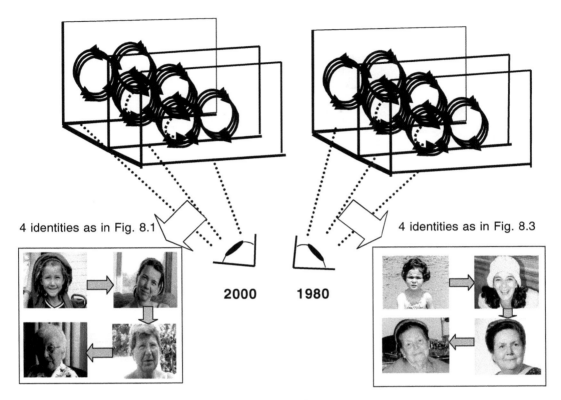

FIGURE 8.17 The relevant dθ at which the observer's interests change.

illustrate the various types associated with aging. In spite of the fact that Bertha and Gina were the individualities used to give the name to the two lines of four types, they were not the individualities relevant for the choice made by the observer. Characteristics of individualities per se can be irrelevant if perceived at the wrong moment in time. For example, a young male meeting Marilyn Monroe when she was 2 years old would not have considered her an exceptionally attractive woman. Readers have already learned about the process of aging illustrated in Figure 8.1 and Figure 8.3 well before looking at these figures to do that, but they used for it their own sets of relevant individualities.

8.3.3 In a Moment of Fast Transition, Moving Narratives in Space Can Imply Moving Them in Time

There is another important aspect of complex time that should be considered. When we discuss the existence of three time differentials that are relevant for the process of representation of reality, we refer only to the three time differentials that can be associated with a particular choice of a triadic reading of the reality — the meaning of a triadic reading that can be translated into the definition of a given narrative. Obviously, there are always several other potentially relevant time differentials besides these three. For example, Figure 8.18 shows that when studying the process of aging in primates and at the same time the process of evolution of primates, we should introduce yet another time differential, much larger than the previous three, that we can call dT.

If we do that, we have to put what is shown in Figure 8.1 and Figure 8.3 into a different perspective. Millions of years ago, an observer of a species of apes would have observed the aging of apes using a set of very coarse epistemic categories about which apes could share meaning (in the first box of Figure 8.18). A caveman observer would have observed the aging of cavemen using a set of epistemic categories about which cavemen could share meaning (in the second box of Figure 8.18) — exactly like a modern human being observes the aging of other human beings using a set of epistemic categories about which

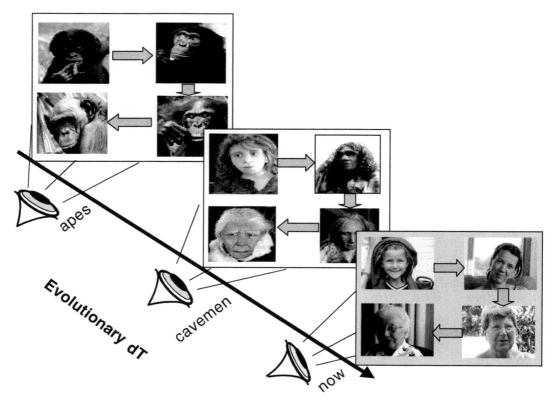

FIGURE 8.18 Evolutionary dT is too large to maintain useful narratives to be used within the observer–observed complex. Courtesy of EURELIOS – http://www.daynes.com.)

humans can share meaning (as in the last box of Figure 8.18). The problem is that a modern observer cannot get the meaning of the set of categories (signs? markers? labels?) that were used by nonequivalent observers in the past (either apes or cavemen). We do not know why and how their information space was useful for guiding their action. That is, we do not know what type of external referents their information space was referring to. A too large time lag implies the risk of losing the coherence within the observer–observed complex, and therefore losing track of useful narratives.

Useful narratives have a definite life span; after a while they simply lose sense, disappearing together with the complex observer–observed that generated them. This is why humans cannot put up with the concept of eternity. Our perception of time is related to the existence of narratives (coherent representations of perceptions of events). By definition, the simple time used in representation must be finite, since simple time just reflects the epistemological choices associated with a given narrative: (1) the perception or representation of a duration in relation to (2) another perception or representation of duration used as reference.

Therefore, whenever the complex observer–observed is forced to change at a pace that is too fast for the Peircean semiotic triad (Figure 4.2) to validate knowledge, there is a serious risk of destroying the coherence of useful narratives. In other words, we can say that when the process used to define the set of essences associated with an operational adaptive holarchy (Figure 8.16) does not have enough time to go through the necessary steps of validation, such a holarchy is under stress and is at risk of losing its internal coherence (individuality). This is the situation of human societies facing the predicament of postnormal science. This point is directly related to a serious threat to the sustainability of socioeconomic systems undergoing a fast period of transition.

This threat is particularly serious for developing countries, which are trying to catch up with the existing gap between their actual achievements and what already has been achieved by more developed

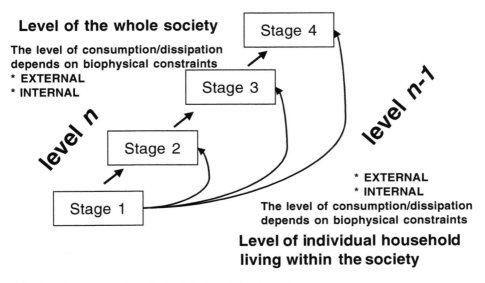

FIGURE 8.19 Different perspectives about a trajectory of development.

countries. An example of this problem is illustrated in Figure 8.19. Imagine that different countries (or different socioeconomic systems) are moving across a common trajectory of development as illustrated in Figure 8.19 along the diagonal line moving up toward right. The various stages of this trajectory are indicated in the figure by numbers. As will be discussed in Chapter 9, when representing socioeconomic systems as dissipative systems that stabilize their own identity through impredicative loops of energy forms, it is actually possible to individuate the existence of a common trajectory, which can be represented using predictable changes in some of the characteristics of societal metabolism.

Without getting into a technical description of such a phenomenon now, we can say that when considering a socioeconomic system as the focal level of analysis (level n), the possible changes in the characteristics of its metabolism are constrained by (1) internal constraints (the availability of enough accumulated capital, technology, infrastructure, know-how) and (2) external constraints (favorable boundary conditions related to the access to an adequate amount of natural resources and environmental services associated with a given level of metabolism). Obviously, if a country with a very low level of GDP (e.g., Laos) wants to change dramatically the characteristics of its metabolism to become more similar to a country with a very high level of GDP (e.g., the U.S.), there are a lot of internal and external constraints that have to be removed. On the side of internal constraints, it will take time to build infrastructures, to provide huge injection of capital and technology and to make a new generation of highly educated workers able to operate computerized technologies. On the side of external constraints, it will take time to become able to stabilize the required inflow of limiting natural resources (e.g., fossil energy and other strategic materials such as steel or metalloids needed for the making of microprocessors). This implies that the timing of evolution of the metabolism of the society as a whole at the level n — the speed at which the socioeconomic system can go through the different known stages of development — is affected by both types of constraint (internal and external), with the Liebig's law determining the final pace of evolution. The most limiting of these constraints — internal or external — will determine the overall speed of change.

On the other hand, individual households, which are operating at the level $n - 1$, have a different perception or definition of internal and external constraints. The external constraints for a household are given by the universe of accessible opportunities given by its local associative context (the socioeconomic interface within which it is operating). The internal constraints for a household are given by the definition of cultural identity of its members, which affects the definition of the set of activities that can and cannot be done, together with the overall perception and evaluation of costs and benefits derived from the interaction with the context.

Before the era of globalization, households (viewed as holons) were sharing with their context (the larger-level holon within which they were operating) the pace of evolution. Their pace of change as observers ($d\theta$) was in a way synchronized with the pace of evolution of the context (defining the need to update their know-how ($d\tau$)). The big problem induced by globalization is associated with the free circulation of (1) information (diffusion of mass media everywhere), (2) goods (huge spread of trade), and (3) people (huge migratory flows). At this point, the definition of sustainability at the household level is no longer univocally determined by what is going on in its direct context. The potential pace of change of a holon household is no longer determined by the most limiting factor between internal and external constraints affecting the pace of evolution of the holon community in which is embedded. Globalization transformed the definition of external constraints into an independent variable. In fact, in a globalized world individual households can choose (to a certain extent) their associative context. In this way, they can jump across typologies of observers. It is possible to move narratives across evolutionary time (like if a Neanderthal man could decide to perceive and represent the world using the same categories adopted by CNN). As soon as households of developing countries decide to do so, they can boost the pace of change of their own identity as much as possible. This is a standard solution whenever these new generations of observers find themselves not happy with what they get from their current associative context. As soon as this new typology of observer/household decides that it pays to change, it can no longer see as admissible the old context in which it used to operate (e.g., young people refusing to remain in rural areas to farm). In this case, we are in presence of an irreversible change in the identity of the observer that literally can no longer see, at a given point in space and time, the same reality seen by the elderly of the same village. The jump into a different typology of observer (an observer that is using a different set of epistemic categories to perceive and represent the costs and benefits of rural life and the related economic activities) requires also a dramatic change in the context within which such an observer can operate. To do that, first, these new observers must adjust (to a certain extent) their cultural identity to become compatible with a different definition of context (a different set of identities and essences).

This is where serious trouble starts. Very often in many developing countries, among young people, the new knowledge about how to perceive and represent reality is learned from television, that is, according to what is done in socioeconomic systems operating elsewhere at different stages of development. In this way, this generation of new observers tends to imitate mechanisms of perception and representation not particularly useful for operating in the local associative context of their household.

Reflexivity of humans implies that lower-level holons (e.g., households), even if belonging to a given focal-level holon (e.g., belonging to a given socioeconomic system), can decide to shift to another associative context by changing their identity as types. They can do that by taking advantage of the set of opportunities guaranteed by the different characteristics found in different socioeconomic systems. Because of this, as soon as they no longer accept the typology of household they happen to belong to, they tend to stop those activities required to interact with the original associative context. Often this implies the need to emigrate, which can become an obliged choice after the stop. This simply means that remaining in the old associative context becomes too costly in relation to the new identity assumed by the observer. This is what happened in the Soviet Union at the moment of the collapse, and this is what is happening in many rural areas of developing countries. Recall the example of the Chinese lady shown in Figure 5.1, who, after watching on television the possibility of operating in a different associative context, gave an assessment of the performance of her agroecosystem that was dramatically different from that given by Western scientists.

For this reason, it is crucial to be able to package the scientific information gathered in an integrated assessment of agroecosystems in a way that makes it possible to address explicitly the relevant attributes considered by lower-level agents (the households) when deciding whether it pays to sustain the socioeconomic context in which they are operating. It is impossible to sustain an agroecosystem in which farmers have lost their feeling of belonging to their particular identity as socioeconomic system (the set of essences and types associated with the impredicative loop that stabilizes their metabolism). A sound integrated analysis has to characterize the sustainability (feasibility) of agroecosystems in relation to the nonequivalent perceptions of performance of different agents operating at different hierarchical levels (this is illustrated in Chapter 11).

References

Georgescu-Roegen, N., (1976), *Energy and Economic Myths: Institutional and Analytical Economic Essays*, Pergamon Press, New York.

Giampietro, M., Bukkens S.G.F., and Pimentel, D., (1997), Linking technology, natural resources, and the socioeconomic structure of human society: examples and applications, in *Advances in Human Ecology*, Vol. 6, Freese, L., Ed., JAI Press, Greenwich, CT, pp. 131–200.

Holling, C.S., (1995), Biodiversity in the functioning of ecosystems: an ecological synthesis, in *Biodiversity Loss: Economic and Ecological Issues*, Perring, C., Maler, K.G., Folke, C., Holling, C.S., and Jansson, B.O., Eds., Cambridge University Press, Cambridge, pp. 44–83.

Holling, C.S. and Gunderson, L.H., (2002), Resilience and adaptive cycles, in *Panarchy: Understanding Transformation in Human and Natural Systems*, Gunderson, L.H. and Holling, C.S., Eds., Island Press, Washington, D.C., pp. 25–62.

Kampis, G., (1991), *Self-Modifying Systems in Biology and Cognitive Science: A New Framework for Dynamics, Information, and Complexity*, Pergamon Press, Oxford, 543 pp.

Layzer, D., (1988), Growth of order in the universe, in *Entropy, Information, and Evolution*, Weber, B.H., Depew, D.J., and Smith, J.D., Eds., MIT Press, Cambridge, MA, pp. 23–40.

Popper, K.R., (1993), *Knowledge and the Mind-Body Problem*, Notturno, M.A., Ed., Routledge, London.

Prigogine, I. (1978), From Being to Becoming. W.H. Freeman, San Francisco.

Rosen, R., (1958a), A relational theory of biological systems, *Bull. Math. Biophys.*, 20, 245–260.

Rosen, R., (1958b), The representation of biological systems from the standpoint of the theory of categories, *Bull. Math. Biophys.*, 20, 317–341.

Rosen, R. (1985), *Anticipatory Systems: Philosophical, Mathematical and Methodological Foundations*. Pergamon Press, New York.

Rosen, R. (1991), *Life Itself: A Comprehensive Inquiry into the Nature, Origin and Fabrication of Life*. Columbia University Press, New York 285 pp.

Rosen, R., (2000), *Essays on Life Itself*, Columbia University Press, New York, 361 pp.

Tarsky, A., (1944), The semantic conception of truth and the foundations of semantics, *Philos. Phenom. Res.* Vol. 4: 341–375.

Ulanowicz, R.E., (1986), *Growth and Development: Ecosystem Phenomenology*, Springer-Verlag, New York.

Ulanowicz, R.E., (1997), *Ecology, the Ascendent Perspective*, Columbia University Press, New York.

Part 3

Complex Systems Thinking in Action: Multi-Scale Integrated Analysis of Agroecosystems

Introduction to Part 3

What Is the Beef That Has Been Served in the First Two Parts of This Book?

After this long excursion through different issues and innovative concepts that has led us through very old philosophical debates and innovative scientific developments, it is time to get back to the original goal of this book. Why and how is the material presented and discussed so far in this book relevant for those willing to study the sustainability of agroecosystems? Part 3 provides examples of applications aimed at convincing the reader that the content of Parts 1 and 2 is relevant indeed to an analysis of the sustainability of agroecosystems. Before getting into such a presentation, however, it could be useful to have a quick wrap-up of the main points made so far:

1. Science deals not with the reality but with the representation of an agreed-upon perception of the reality. Any formalization provided by hard science starts from a given narrative about the reality. That is, any formalization requires a set of preanalytical choices about what should be considered relevant and on what time horizon. These preanalytical choices are value loaded and entail an unavoidable level of arbitrariness in the consequent representation. Substantive models of the sustainability of real systems do not exist.

2. To make things more difficult, science dealing with sustainability must address the process of becoming of both the observed system and the observer. This implies dealing with an unavoidable load of uncertainty and genuine ignorance, which is associated with the existence of legitimate nonequivalent perspectives found among interacting agents.

3. The process of generation of useful knowledge is therefore a continuous process of creative destruction. In his book *The Science of Culture*, White starts the first chapter, entitled "Science Is Sciencing," by saying: "Science in not merely a collection of facts and formulas. It is preeminently a way of dealing with experience. The word may be appropriately used as a verb: one sciences, i.e., deals with experience according to certain assumptions and with certain techniques" (1949, p. 3). Especially when dealing with science used for governance, it is easy to appreciate a sort of Yin-Yang tension in the process used by humans for dealing with their experience. The description of this tension by White says it all. There are two basic ways for dealing with the need to update our knowledge: one is science the other is art.

   The purpose of science and art is one: to render experience intelligible, i.e., to assist man to adjust himself to his environment in order that he may live. But although working toward the same goal, science and art approach it from opposite directions. Science deals with particulars in terms of universals: Uncle Tom disappears in the mass of Negro slaves. Art deals with universals in terms of particulars: the whole gamut of Negro slavery confronts us in the person of Uncle Tom. Art and science thus grasp a common experience of reality, by opposite but inseparable poles. (White, 1949, p. 3)

   We have at this point developed a new vocabulary to express this concept. To handle the growing mass of data associated with experience, humans must:

   a. Compress the requirement of computational capability needed to handle more sophisticated models and larger data sets. To do that they need science that uses types to describe equivalence classes of natural entities.

   b. Expand the information space used to make sense about the reality. This can only be done by adding new types and new categories about which it is possible to obtain a shared understanding.

This is where art enters into play. Art is needed to find out the existence of new relevant aspects of the reality, about which it is important to dedicate a new entry in our language or a new narrative about the meaning of reality. This leads to the idea that when dealing with science for governance, science cannot be taken from the shelf, as a repertoire of useful data and protocols. On the contrary, it is important to imagine science for governance as a set of procedures that can be used to do "sciencing."

4. There are already several attempts to develop procedures aimed at implementing the concept of sciencing. In Chapter 5 an example was given in relation to the soft system methodology proposed by Checkland. However, several other similar efforts in this direction can be found in the literature. The basic rationale is always the same. When dealing with a given perception of the existence of a problem, one has to start, necessarily, with a narrative. However, such a narrative should not be used directly, as such, to get into a scientific characterization. Rather, it is important to explore as many alternative narratives as possible to expand the possible useful perspectives, detectors, indicators and models to be used, later on, in the scientific problem structuring. Obviously, in the final choice of a given scientific problem structuring, the number of narratives, indicators and models used has to be compressed again. In a finite time, scientists can handle only a finite and limited information space. But exactly because of this, it is important to work on a semantic check of the validity of the narratives chosen as the basis for the analytical part.

5. If one agrees with the statements made in the previous four points, one is forced to conclude that when dealing with science for governance, there are two distinct tasks, which require a different type of expertise and a different approach. These two tasks, which imply facing a formidable epistemological challenge, should not be confused — as is done, unfortunately, by reductionist scientists. Task 1 is related to the ability to provide a useful and sound input on the descriptive side. This implies the ability to tailor the development of models, the selection of indicators and the gathering of data according to the specificity of the situation. Task 2 is related to the ability to handle the unavoidable existence of legitimate but contrasting values, fears and aspirations. This unavoidable existence of conflicts in terms of values will be reflected in the impossibility to determine in a substantive way (1) what should be considered the best problem structuring, (2) what should be considered the best set of alternatives to be evaluated, (2) what should be considered the best set of scenarios, (4) what should be considered the best alternative among those considered and (5) what is the best way for handling the unavoidable presence of uncertainty and ignorance in the problem structuring used in the process of decision making. Using the vocabulary adopted in Chapter 5, we can say that:

    *   Task 1 scientists should be able to provide a flexible input consisting of a multi-scale integrated analysis (generating a coherent but heterogeneous information space able to represent changes and dynamics at different hierarchical levels and in relation to different forms of scientific disciplinary knowledge).

    *   Task 2 has to be based on a process. That is, the issue of incommensurability and incomparability can only be handled in terms of societal multi-criteria evaluation. This concept implies forgetting about the approach proposed by reductionism. Different indicators should not be aggregated into one single aggregate function (e.g., as done in cost–benefit analysis). In this way, one loses track of the behavior of the individual indicators, meaning that their policy usefulness is very limited. The assumption of complete compensability should not be adopted, i.e., the possibility that a good score on one indicator can always compensate a very bad score on another indicator (money cannot compensate the loss of everything else). Any process of analysis and decision making has to be as transparent as possible to the general public.

From this perspective, we can define a reductionist approach as an approach based on the use of just one measurable indicator (e.g., a monetary output or a biophysical indicator of efficiency), one dimension (e.g., economic or biophysical definition of tasks), one scale of analysis (e.g., the farm or the country), one objective (e.g., the maximization of economic efficiency, the minimization of nitrogen leakage in the water table) and one time horizon (e.g., 1 year).

Reductionist analyses also imply a hidden claim about their ability to handle uncertainties and ignorance when they claim that a particular option (e.g., technique of production) is better than another one.

This is the reason why in multi-criteria evaluation it is claimed that what is really important is the decision process and not the final solution.

6. The set of innovative concepts presented in Part 2 can be used to organize a multi-scale integrated analysis of agroecosystems. These tools are required to organize conventional scientific analyses in a way that make explicit and transparent the chain of preanalytical choices made by the analyst. Actually, these decisions become an explicit object of discussion, since they are listed as required input to impredicative loop analysis.

In conclusion, what is presented in Part 3 is not an analytical approach aimed at finding the best course of action or indicating to the rest of society the right way to go to improve the sustainability of our agroecosystems. The text of Part 3 is just a series of examples of how the insight derived from complex systems theory can be used to organize scientific information to generate informed discussions about sustainability. To do that, the proposed approach generates useful information spaces made up of nonequivalent descriptive domains (integrated packages of nonreducible models) that can be tailored on the specific characteristics of relevant agents. The ultimate goal is that of structuring available data sets and models according to a selected set of narratives that have been defined as relevant for a given situation.

What Is the Beef That Is Served in Part 3?

If we do a quick overview of the literature dealing with sustainable agriculture, we will find a huge number of papers dealing with assessments and comparisons of either different farming techniques or different farming systems operating in different areas of the world. The vast majority of these papers are affected by a clear paradox:

1. Analyses of farming systems and assessments of the sustainability of agricultural techniques generally start with an introduction that makes an explicit or implicit reference to the following, quite obvious, two statements:

   a. What can be produced and what is produced in a farming system depends on the set of boundary conditions in which the farming system is operating (the characteristics of both the ecological and the socioeconomic interface of the farm). After conditioning what to produce, these characteristics also influence how to produce it (the choice of techniques of production and the choice of related technologies).

   b. Any assessment of the agricultural process obtained by considering only a particular perspective of farming (e.g., agronomic performance, economic return, social and cultural effects, ecological impact) necessarily misses other important information referring to other perspectives of the same process. To be meaningful, any evaluation of agricultural techniques should consider a plurality of perspectives through a holistic description of farming processes.

   So far, so good; the main message about the need for integrated analysis for complex systems seems to be clear to the majority of authors, at least when reading the introductory paragraphs. However, such wisdom tends to disappear in the rest of the paper.

2. Before entering into a discussion of case studies, comparisons of techniques of production or, more in general, analyses of sustainability of farming systems, authors omit providing in an explicit form *all* three pieces of information listed below:

   a. Characterization of boundary conditions with which the farming system is dealing:

   - According to the set of constraints coming from the socioeconomic side, how fast must be the throughput in the farming system? For example, what is the minimum level of productivity per hour of labor that is acceptable for farmers and the minimum level of productivity per hectare forced by demographic pressure, where applicable?

- According to the set of constraints coming from the ecological side — type of ecosystem exploited and intensity of withdrawal on primary productivity — what is the current level of environmental loading and what do we know about the eco-compatibility of such a throughput? That is, what room is left for intensification?

b. Characterization of the basic strategy affecting a farmer's choice:

- What is the optimizing strategy under which farmers are making decisions? For example, are they minimizing risk (farming system must be resilient since it is on its own in case of troubles), or are they maximizing return (the farming system is protected against risks such as crop failure by the rest of the society to which it belongs, as in developed countries)? Are there location-specific strategies affecting their choices?

- Are farmers sustaining the development of the rest of society (are farmers net tax payers), or are they subsidized by the rest of society (are farmers supported by subsidies)?

c. A critical appraisal about the limits of validity of the particular type of analysis performed on the farming system:

- Out of the many possible perspectives under which farming activities can be represented and assessed, any choice of a particular window of observation and a particular set of attributes to define the performance of farming (i.e., the one that was adopted in the study) implies missing other important views of the process. What consequences does it carry for the validity of the conclusions? For example, checking the agronomic performance and the ecological compatibility of different techniques does not say anything about the sustainability of these techniques.

  To discuss sustainability, we also need a parallel check on economic viability and on the compatibility of these techniques with cultural identity and aspirations of farmers that are supposed to adopt them.

- How possible is it to generalize the validity of the conclusions of this paper that are related to a location-specific analysis?

The three chapters of Part 3 have the goal of showing that it is possible to develop a tool kit for multi-scale integrated analysis of agroecosystems that makes it possible to:

1. Link the economic and biophysical reading of farming in relation to structural changes occurring in the larger socioeconomic system to which the farming system belongs during the process of development. This makes it possible to use an integrated set of indicators of development, able to represent the effects of changes on different hierarchical levels (from the country level to the household level) — Chapter 9.

2. Establish a bridge, which can be used to explain how changes occurring in the socioeconomic side are reflected in changes in the level of environmental impact associated with agriculture. The biophysical reading of these changes at the farm level makes it possible to explain the existing trends of increased environmental impact of agriculture to the existing trends of technical progress of agriculture — Chapter 10.

3. Represent agroecosystems in terms of holarchic systems. This makes it possible to study the reciprocal influence of the decisions of agents operating at different levels in the holarchy. In this case, indicators related to economic, social and ecological impacts can be integrated across levels to indicators of environmental impact based on changes in land use — Chapter 11.

Reference

White, L.A., (1949), *The Science of Culture*, Grove Press, Inc., New York, 444 pp.

9

Multi-Scale Integrated Analysis of Agroecosystems: Bridging Disciplinary Gaps and Hierarchical Levels

This chapter has the goal of illustrating examples of multi-scale integrated analysis of societal metabolism that are relevant for the analysis of the sustainability of agroecosystems. In particular, Section 9.1 illustrates the application of impredicative loop analysis (ILA) at the level of the whole country using in parallel different typologies of variables. In this way, one can visualize the existence of a set of reciprocal constraints affecting the dynamic equilibrium of societal metabolism. That is, feasible solutions for the dynamic budget represented using a four-angle figure can only be obtained by coordinated changes of the characteristics of parts in relation to the characteristics of the whole, and changes in the characteristics of the whole in relation to the characteristics of the parts. Section 9.2 provides the results of an empirical validation based on a data set covering more than 100 countries (including more than 90% of the world population) of this idea. In particular, such an analysis shows that an integrated set of indicators derived from ILA makes it possible to (1) establish a bridge between economic and biophysical readings of technical progress and (2) represent the effect of development in parallel on different hierarchical levels and scales. Section 9.3 deals with the link between changes occurring at the level of the whole country (society) and changes in the definition of feasibility for the agricultural sector. That is, socio-economic entities in charge of agricultural production must be compatible with their socioeconomic context. This implies the existence of a set of biophysical constraints on the intensity of the flow of produced output. Finally, Section 9.4 deals with trend analysis of technical changes in agriculture. Changes in the socioeconomic structure of a society translate into pressure for boosting the intensity of agricultural output in relation to both land (demographic pressure = increase in the output per hectare of land in production) and labor (bioeconomic pressure = increase in the output per hour of labor in agriculture). Indices assessing these two types of pressures can be used as benchmarks to frame an analysis of agroecosystems.

9.1 Applying ILA to the Study of the Feasibility of Societal Metabolism at Different Levels and in Relation to Different Dimensions of Sustainability

9.1.1 The Application of the Basic Rationale of ILA to Societal Metabolism

The general rationale of impredicative loop analysis, illustrated in Chapter 7, is applied here to the analysis of societal metabolism. The level considered as the level n is the level of the whole society (country). This requires:

1. A characterization of total requirement at the level n — this is a consumed flow assessed in relation to the whole. This is done by using an intensive variable 3 (IV3) mapping the level of dissipation (consumption of extensive variable 2) per unit of size of the whole (measured in terms of extensive variable 1).

2. A characterization of internal supply at the level $n - 1$ — this is a produced flow assessed in relation to a part of the whole. This is done by using an intensive variable 3 mapping the flow

of supply (measured using extensive variable 2) per unit of size of the part (measured in terms of extensive variable 1).

3. An analysis of the congruence over the loop of the reciprocal definition of identities of (1) the whole, (2) parts, (3) subparts and inputs and outputs of parts, and (4) the weak identity assigned to the environment (reflecting its admissibility).

The applications discussed below are based on the use of:

- **Two extensive variables 1** used to assess the size of the system, providing a common matrix representing its hierarchical structure. These two EV1 are human activity and land area.
- **Three extensive variables 2** used to assess the intensity of a flow, which can be associated with a certain level of production or consumption. These three EV2 are exosomatic energy dissipated, added value related to market transactions, and food. The definition of the size of parts (lower-level compartments), in terms of EV1, has to be done in a way that guarantees the closure of the assessments of the size of the whole across levels. The same applies to the distinction between the direct compartment generating the internal supply and the rest of society.

9.1.1.1   Step 1: Discussing Typologies

— Two possible choices considered here for extensive variable 1 are useful for addressing two main dimensions of sustainability: (1) Human time — when used as extensive variable 1 — is useful for checking the compatibility of a given solution within the socioeconomic dimension. (2) Land area — when used as extensive variable 1 — is useful for checking the ecological dimension of compatibility.

The first thing to do is therefore an analysis of possible types that can be used to establish an ILA according to the general scheme presented in Figure 9.1. When applying the scheme of Figure 9.1 to the analysis of the dynamic equilibrium of societal metabolism of a whole society using human activity as extensive variable 1, we are in a case that has been discussed on two occasions so far. There are different sets of types on different quadrants. The profile of distribution of individuals over the set of types will determine the value taken by the angle. For example, starting with the upper-left angle, we find that the level of physiological overhead on disposable human activity (DHA) can be expressed as generated by a set of types and a profile of distribution over it. This has been discussed in Figure 6.9 (profile of distribution of individuals over age classes) and Figure 6.10 (profile of distribution of kilograms

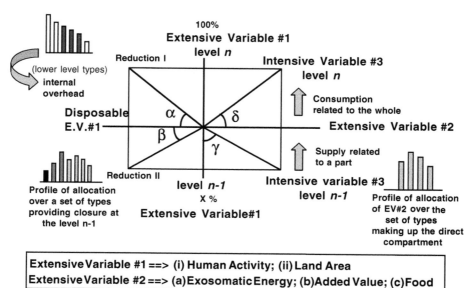

FIGURE 9.1 ILA: general relation among types in societal metabolism.

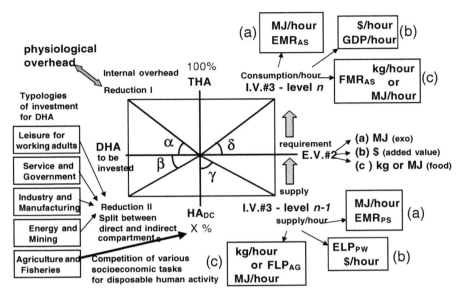

FIGURE 9.2 Choosing how to define and aggregate typologies over the ILA EV1: human activity.

of body mass over age classes). The effect of changes (either in the set of types — e.g., longer life span — or in the profile of distribution over the types), which can affect the physiological overhead, has been discussed in relation to Figure 7.2 and Figure 7.3 (when illustrating a simplified analysis of the dynamic budget of the societal metabolism — using food as extensive variable 2 — for a hypothetical society of 100 people on a remote island).

After subtracting from total human activity the physiological overhead, we obtain the amount of disposable human activity for the society — left side of Figure 9.2. This amount of disposable human activity is then invested in a set of possible activities. The various categories of human activities can be divided between work and leisure. Making this distinction always implies a certain degree of arbitrariness. This is why it is important to have (1) the constraint of closure across levels and (2) the possibility of making in parallel various ILAs based on a different selection of extensive variable 2. This is particularly important for the decision about the definition of the direct compartment, the compartment providing the internal supply, which is characterized in the lower-right quadrant. For example, we can decide to include the service sector among those lower-level parts making up the indirect compartment when studying the dynamic budget of exosomatic energy. That is, when making a four-angle figure with exosomatic energy as EV2, we can assume that the service sector does not produce either a direct supply of exosomatic energy or machines for using exosomatic energy. But when making a four-angle figure with added value as EV2, we have to include the service sector in the direct compartment. In fact, when considering the dynamic budget of added value, the service sector is among those sectors producing added value.

Obviously, the choice of the set of typologies used to obtain closure on disposable human activity is necessarily open. In this regard we can recall the crucial role of the category "other" to obtain closure (Figure 6.1). In this example, the difference between DHA and the sum of the various investments on working activities can be considered in this system of accounting as leisure. With this choice we can end up including into leisure investments of human activity typologies of work not included in the list of typologies.

The scheme of Figure 9.1 can also be applied to an analysis of the dynamic budget of societal metabolism, which uses land area as extensive variable 1. In this case, we start with a level of total available land defined as the area associated with the entity considered as the whole socioeconomic system (e.g., the border for a country or the area needed to stabilize a given flow). Also in this case, this scheme can be used to have a preliminary discussion of the standard typologies to be used for the analysis of land use. In general, a first list of land typologies is found when looking at data (e.g., desert, too

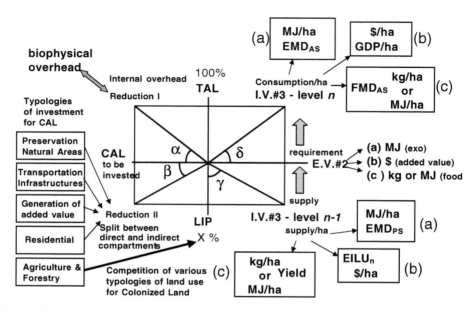

FIGURE 9.3 Choosing how to define and aggregate typologies over the ILA EV1: land area.

hilly, permanent ice, swamps, arable land, forest). The categories found in published data are not necessarily useful for a particular ILA. As soon as the analysts manage to obtain a set of useful typologies for the analysis, the profile of distribution of individuals hectares (unit used to assess the size according to extensive variable 1) over the set will define the level of biophysical overhead (reduction I) determining the colonized appropriated land (CAL) (see Figure 9.3). To indicate the process of permanent alteration of the identity of terrestrial ecosystems due to human interference on biological and ecological mechanisms of control, the group of the IFF of Vienna (Institute of Interdisciplinary Studies of Austrian Universities, see for example, Fischer-Kowalski and Haberl, 1993; Haberl and Schandl, 1999) suggests the term *colonization*. By adopting their suggestion we use the acronym CAL (colonized appropriated land).

At this point, we need a set of possible typologies of land use covering the entire colonized appropriate land to classify investments of human activity within this compartment. This is illustrated on the left in Figure 9.3. This is a very generic example, and depending on the type of problem considered, it requires an additional splitting of these coarse typologies into a more refined classification.

9.1.1.2 Step 2: Defining the Critical Elements of the Dynamic Budget — Depending on the EV2 that is chosen for the impredicative loop analysis and the specificity of the questions posed, it is necessary at this point to interpret the metaphorical message associated with Figure 9.2 and Figure 9.3. This requires that the analyst discuss how to formalize this rationale in relation to a specific situation, in terms of numerical assessments based on an available data set.

9.1.1.2.1 Example 1: Human Activity as EV1 and Food as EV2

Let us start with the example of an impredicative loop analysis referring to human activity (as extensive variable 1) and food (as extensive variable 2). This is a case that has already been discussed in the example of the 100 people on the remote island.

9.1.1.2.1.1 Assessing Total Requirement at the Level of the Whole: Level n — This is an assessment of total consumption associated with the metabolism of a given human system. In the case of food this flow can be written at the level n as

$$\text{Population} \times \text{Consumption p.c.} = \text{Total Food Requirement (EV2)} \qquad (9.1)$$

In Equation 9.1, total food requirement is expressed as a combination of the extensive variable 1 (size of the system — mapped here in terms of population) and the intensive variable 3 (consumption per capita (p.c.), which means a given level of dissipation per unit of size). Equation 9.1 can be easily transformed into

$$\text{THA} \times \text{FMR}_{AS} = \text{Total Food Requirement (EV2)} \qquad (9.2)$$

when considering that THA = population × 8760 = total amount of hours of human activity per year, and that consumption per capita represents an assessment of a given flow (e.g., megajoules of food or kilograms of food per year) that can be transformed into FMR_{AS} (food metabolic rate assessed as average of society) by dividing the relative value of consumption per capita (flowing in a year) by 8760. This provides the amount of flow of food consumed per hour of human activity. With this change we can write

$$\text{FMR}_{AS} = (\text{Consumption p.c.}/8760) = \text{IV3}_n \qquad (9.3)$$

9.1.1.2.1.2   Assessing Internal Supply: Level n − 1 — This is an assessment of the internal supply of input provided to the black box because of the activities performed within the direct compartment (HA_{AG}). This internal supply requires the conversion of energy input into useful energy able to fulfill the tasks. When mapping the effect of agricultural activities against human activity at the level $n - 1$ we can write

$$(\text{HA}_{AG} \times \text{BPL}_{AG}) = \text{Internal Food Supply (EV2)} \qquad (9.4)$$

The total supply assessed at the level n − 1 is expressed as a combination of extensive variable 1 (size of the lower-level compartment — HA_{AG} = human activity invested in the agricultural sector — the one labeled "direct" in the upper part of Figure 7.8) and intensive variable 3 (BPL_{AG} — biophysical productivity of labor in agriculture — which assesses the return of human activity invested in the set of tasks performed in the compartment labeled "direct"). BPL_{AG} measures the input of food taken from the land and delivered to the black box per unit of human activity invested in the direct compartment. This is the lower-level compartment in charge with the direct interaction with the context to get an adequate supply of input (see upper part of Figure 7.8).

$$\text{BPL}_{AG} = \text{Biophysical Productivity of Labor in Agriculture (IV3)}_{n-1} \qquad (9.5)$$

9.1.1.2.1.3   Checking the Congruence of the Required and Supplied Flows — At this point, by combining Equation 9.2 and Equation 9.4 we can look for the congruence among the two flows:

$$\text{THA} \times \text{FMR}_{AS} = \text{HA}_{AG} \times \text{BPL}_{AG} \qquad (9.6)$$

As noted before, these two flows do not necessarily have to coincide in either the short term (periods of accumulation and depletion of stocks) or long term (a society can be dependent on import for its metabolism or can be a regular exporter of food commodities).

Additional information can be added to the congruence check expressed by Equation 9.6. For example, recall the discussion given in Chapter 6 about the characterization of endosomatic flow in Spain across different levels (Figure 6.8). The characterization of the total food requirement can be expanded to include information referring to different hierarchical levels by substituting the term FMR_{AS} with three terms — in parentheses — as done in the following relation:

$$\text{THA} \times (\text{ABM} \times \text{MF} \times \text{QDM\&PHL}) = \text{HA}_{AG} \times \text{BPL}_{AG} \qquad (9.7)$$

In Equation 9.7 the total requirement of endosomatic energy, assessed at the level n, is expressed as a combination of extensive variable 1 (size of the system, mapped here in terms of total human activity, linked directly to the variable population) and three variables: (1) ABM (average body mass); (2) MF (metabolic flow), endosomatic metabolic rate per kilograms of human mass and unit of time; and (3) QDM&PHL, a factor accounting for quality of diet multiplier and postharvest losses. QDM&PHL accounts for the difference between the energy harvested in the form of produced food at the food system level (recall the assessment of embodied kilograms of grain vs. kilograms of grain consumed directly at the household level in Figure 3.1) and the endosomatic energy flowing within the population. QDM&PHL depends on (1) QDM, the degree of double conversion of crops into animal product (associated with the quality of the diet and the modality of production of animal products), plus other utilization of crops into the food system (seeds, industrial preparations associated with losses) and (2) PHL, direct losses due to pests, decay and damages in the steps of processing, handling, storage and distribution in the food system.

The congruence check suggested by Equation 9.7 is still related to (1) a requirement associated with the identity of the whole and (2) an internal supply associated with the identity of the direct compartment. However, the more elaborate characterization of the total food requirement makes it possible to consider a larger set of identities in the forced relation.

Before getting into other examples of impredicative loop analysis, it is useful to go through a few observations that can already be made after this first example.

When looking for closure in the representation of the black box (level n) on the lower level (level $n - 1$), we have to contrast, in the lower-right quadrant, the direct compartment with the rest of society. The size of the rest of society in this case is determined by:

1. Reduction I (expressed in terms of EV1) — associated with either physiological overhead (for human activity) or biophysical overhead (for land use).
2. Reduction II (expressed in terms of EV1) — associated with the fraction of investment of DHA or CAL, which is going to the indirect compartment. For example, in the system of accounting adopted in Equation 9.7 (Figure 9.2), the rest of society includes all the investments of human activity not included in the compartment agriculture.

It should be noted that the investments of human activity in the indirect compartment (see Figure 7.8) can be considered irrelevant in relation to the assessment of the specific mechanisms guaranteeing the supply of flows consumed by society — referring to a reading of this event at the level $n - 1$. However, when looking at events — at the level n — the size of HA_{RoS} (in the case of Equation 9.7, this would be all human activity not invested in agricultural work) becomes very relevant for two reasons: (1) because it participates in determining the total requirement of input at the level of the whole system and (2) because the indirect compartment includes different typologies of activities associated with different consumption levels. For example, even when considering activities belonging to leisure, the subcategory sleeping implies a much lower level of consumption than the subcategory running marathons. The higher the fraction of human activity invested in energy-intensive activities in the indirect compartment, the higher will be its share of total consumption. As a consequence, the higher will be the necessity for the fraction of human activity invested in the direct compartment to be productive.

To clarify this point, let us consider the profile of investments of human activity of a developed society such as the U.S., which is illustrated in Figure 9.4. Starting from a THA of 100%, we have a reduction I of 71% associated with the physiological overhead. Then leisure absorbs another 19% of THA. This implies that only 1 h of human activity of 10 is actually invested into typologies of work included in the class paid work. The internal competition among lower-level subcompartments of paid work implies that another 6% of THA goes in the sector service and government, leaving only 4% to the productive sectors (PSs) of the economy dealing with the stabilization of the endosomatic (food for people) and exosomatic (fossil energy for machines) metabolism. The vast majority of the work in the productive sectors goes to manufacturing and other activities related to energy and mining, leaving a very tiny fraction of work allocated to agriculture, which keeps shrinking in time. In 1994 (the year to which the profile of investments of human activity given in Figure 9.4 refers), the fraction of the workforce in agriculture was 2%. This means 2% of the 10% of paid work. At this point we can see that reduction II implies moving from the 29% of THA of disposable human activity, available after reduction I, to 0.2% of THA invested in the direct compartment agriculture. Put another way, after defining agriculture as the direct compartment in charge for producing the internal supply of food, we obtain that the size of the compartment rest of society is

$$\text{Rest of society} = \text{Red. I } (71.0\% \text{ THA}) + \text{Red. II } (28.8\% \text{ THA}) = 99.8\% \text{ THA} \qquad (9.8)$$

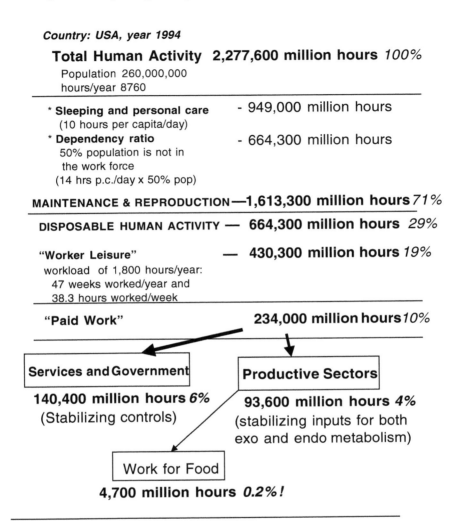

FIGURE 9.4 Profile of consumption × end uses of investments of human activity in the U.S., a developed country. (Giampietro, M. and Mayumi, K. (2000), Multiple-scale integrated assessment of societal metabolism: Introducing the approach. *Popul. Environ.* 22 (2): 109–153.)

The relation in size between the direct compartment (HA_{AG}) and the rest of society (which can be represented as $THA - HA_{AG}$) implies a constraint on the relative densities of the two flows (total requirement and internal supply) represented at different levels (n and $n - 1$) to obtain congruence. By recalling the definitions of IV3 at different hierarchical levels given by Equation 9.3 and Equation 9.5, and by using the equation of congruence (Equation 9.6), we can write

$$THA/HA_{AG} = BPL_{AG}/FMR_{AS} = 500 = 1/0.002 \qquad (9.9)$$

That is, the higher the difference between the size of the rest of society and the size of the direct compartment (according to extensive variable 1), the larger must be the ratio among the two intensities of the flows (IV3) assessed at the levels $n - 1$ and n. The assessment expressed in terms of intensive variable 3 obviously reflects the choice of an extensive variable 2 (in this case, food).

Reaching an agreement about the definition of what should be considered working and nonworking and about the correct assessment of the size of the resulting compartments (e.g., the profile of investments given in Figure 9.4) within a real society is anything but simple. Recall the example of the 100 people on the remote island discussed in Chapter 7. Any definition of labels for characterizing a typology of human activity is arbitrary. When dealing with the representation of human activity in relation to the metabolism of a country, a community or a household, nobody can provide a substantive characterization of what should be considered working (direct contribution to the stabilization of the input metabolized by society, which is taken from the context in the short term) and what should be considered nonworking. Any formalization of these concepts will depend on the timescale and on the selection of variables (epistemic categories) used to perceive and represent the mechanisms stabilizing the metabolism of the society in the first place. The working of a housewife preparing meals can be accounted as invested in the nonworking compartment (when characterizing the compartments using the categories household sector vs. paid work sector) or can be accounted as invested in the working compartment (when characterizing the compartments using the categories leisure vs. working and chores). In the same way, the service sector can be viewed as a sector producing added value in an economic accounting (as a part of the direct compartment in terms of production of added value), whereas it can be viewed as a net consumer of energy and goods in biophysical accounting. This means considering it as a part of the indirect compartment in terms of production of useful energy and material goods. This unavoidable arbitrariness, however, is no longer a problem, as soon as one accepts the use of nonequivalent representations in parallel, and as long as one addresses the technical aspects required to keep coherence in the nonequivalent sets of definitions (see Giampietro and Mayumi, 2000).

The various relations of congruence discussed so far are examples of impredicative loops, in which the definition of what are the activities included in the label "working in the direct compartment" will also define (1) the assessment of the IV3 (the output of work in the direct compartment) and (2) the definition of what has to be included under the label "rest of society." As soon as a particular system of accounting for assessing food requirement and supply is agreed upon, the relation among the identities expressed by the loop will become self-referential. That is, as long as the observer sticks to the definitions and the assumptions used when developing the specific system of accounting, impredicative loops can be used for looking at external referents that can provide mosaic effects to the integrated assessment.

9.1.1.2.2 *Example 2: Human Activity as EV1 and Added Value as EV2*

In this case, the congruence check over the dynamic budget is related to a characterization of the total requirement (on the left side of the relation) and to an internal supply (on the right side of the relation):

$$\text{THA} \times \text{GDP/hour} = \text{HA}_{\text{PW}} \times \text{ELP}_{\text{PW}} = [\text{THA} \times (\text{SOHA} + 1)] \times \text{ELP}_{\text{PW}} \tag{9.10}$$

The total requirement of added value, assessed at the level n, is expressed as a combination of an extensive variable 1 (size of the system — mapped here in terms of total human activity, linked directly to the variable population) and a well-known intensive variable 2 (the gross domestic product (GDP) per capita, expressed in dollars per hour). In this case, the GDP (or gross national product (GNP) depending on the selected procedure of accounting) is defined in terms of the sum of the expenditures of the various sectors. The only trivial transformation required by this system of accounting to make this variable compatible with the other nonequivalent readings is to divide the value of GDP per capita per year, by the hours of a year.

The internal supply of added value, assessed at the level $n - 1$, is expressed as a combination of an extensive variable 1 (size of the lower-level compartment, HA_{PW}), which considers all human activity invested in the generation of added value that is paid for (productive and service sectors including government), and an intensive variable 3 (ELP_{PW} — economic labor productivity of paid work).

An overview of the reciprocal entailment among the terms included in Equation 9.10 can be obtained using a four-angle figure, as shown in Figure 9.5. It should be noted again that ELP_{PW} has nothing to do with an economic assessment of how much added value is produced by the production factor labor. In fact, the assessment of ELP_{PW} refers to the combined effect of labor, capital, know-how and the availability and quality of natural resources used by a particular economy, sector, subsector, typology of activity or firm/farm.

$$ELP_{PW} = \frac{\text{Added value generated by an element } j \text{ over a given period of time}}{\text{Hours of human work in the element } j \text{ over the same period of time}} \tag{9.11}$$

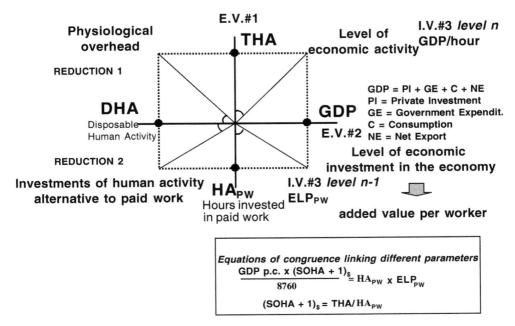

FIGURE 9.5 Example 2: ILA with an EV1 of human activity and an EV2 of added value. (Giampietro, M., Mayumi, K. and Bukkens S.G.F. (2001), Multiple-scale integrated assessment of societal metabolism: An analytical tool to study development and sustainability. *Environ., Develop. Sustain.*, Vol. 3 (4): 275–307.)

That is, we are dealing in this application only with a mechanism of accounting that has the goal of guaranteeing the congruence of nonequivalent systems of mapping providing an integrated analysis of the performance of a socioeconomic system. Put another way, ELP_{PW} is not used to study which particular combination of capital, labor, know-how and natural resources is generating a given flow of added value, to improve or optimize the mix. Rather, the only use of the assessment of ELP_{PW} is that of looking for the existence of constraints of congruence with nonequivalent, but related, assessments of flows, which can be obtained when looking at the same system, but on different hierarchical levels or using different definitions of identity for the elements.

To this ILA we can apply the same condition of congruence to the ratio between the intensities of the two flows of total requirement and internal supply seen in Equation 9.9:

$$THA/HA_{PW} = GDP_{hour}/ELP_{PW} = 10 \qquad (9.12)$$

In a developed society such as the U.S. the overhead over the investment of the resource human activity in the sector paid work is 10/1. This value reflects the combined effect of demographic structure and socioeconomic rules (high level of education, early retirement and light workloads for the economically active population). This translates into a requirement of a very high economic labor productivity (the average flow of added value produced in the economic sectors per hour of labor), which must be 10 times higher than the average level of consumption of added value per hour in the society.

9.1.1.2.3 *Example 3: Human Activity as EV1 and Exosomatic Energy as EV2*

At this point the reader can easily guess the basic mechanism of accounting for checking the congruence of the dynamic budget of exosomatic energy. Also in this case, the total requirement is characterized on the left and the internal supply on the right:

$$THA \times EMR_{AS} = HA_{PS} \times BPL_{PS} \qquad (9.13)$$

The total requirement of exosomatic energy, assessed at the level *n*, is expressed as a combination of an extensive variable 1 (size of the system — mapped here in terms of total human activity, linked directly to the variable population) and an intensive variable 3 (EMR_{AS} — which is the amount of primary energy consumed per unit of human activity as average by the society). In this case, we are accounting the total exosomatic throughput (TET) expressed using a quality factor for energy (e.g., converted into gigajoules or tons of oil equivalent), reflecting an appropriate procedure of accounting for the sum of the exosomatic energy expenditures of the various sectors. EMR_{AS} is the equivalent to what is usually defined in the literature as energy consumption per capita, and it is usually expressed in gigajoules of oil equivalent per year. Analogous with what was done with GDP p.c., this assessment given in gigajoules per year is converted into an assessment per hour (e.g., megajoules per hour). This is required to make possible the bridging of assessments at the level of individual sectors (level $n - 1$) and the whole system (level n).

In fact, the total supply of exosomatic energy, assessed at the level $n - 1$, is expressed as a combination of an extensive variable 1 (size of the lower-level compartment, HA_{PS}), that is, by considering the hours of human activity invested in those activities associated with the stabilization of the autocatalytic loop of exosomatic energy (Giampietro and Mayumi, 2000), and an intensive variable 3 (BLP_{PS}, biophysical labor productivity of the productive sector, assessed as the ratio between the flow of exosomatic energy consumed by society (TET) and the requirement of working hours in this sector ($BLP_{PS} = TET/HA_{PS}$)).

Due to the complete analogy with the two four-angle figures illustrated so far (Figure 9.2 and Figure 9.5), we can skip the representation of this congruence check using that scheme. It is time to move to a more elaborate analysis. In fact, the congruence check described in Equation 9.13 can also be written as

$$THA \times EMR_{AS} = HA_{PS} \times EMR_{PS} \times TET/ET_{PS} \tag{9.14}$$

In this relation BLP_{PS} has been replaced by $EMR_{PS} \times TET/ET_{PS}$. In this way, the use of an intensive variable 3 (EMR_{PS}) referring to the level $n - 1$ has been substituted by two terms, which imply the bridging of identities (establishing bridges among the values taken by variables) across different hierarchical levels.

In fact, the amount of exosomatic energy spent in the productive sector (called ET_{PS} in Chapter 6) can be written using the relation $ET_{PS} = HA_{PS} \times EMR_{PS}$. The ratio TET/ET_{PS}, however, has to respect the constraint $TET - ET_{PS} = ET_{RoS}$. That is, the difference between TET and the energy required to operate the PS, which is ET_{PS}, has to be enough to cover the required investments in the rest of society, which is ET_{RoS}. Therefore, the feasibility in relation to this constraint implies considering (depends on) a lot of additional parameters, for example:

1. The mix of tasks performed in the productive sectors
2. The mix of energy converters adopted in the productive sectors
3. The mix of energy forms dealt with in the energy sector
4. The mix of tasks performed in the various compartments of society — end uses
5. The mix of technologies adopted in the various compartments of society — end uses (with different degrees of efficiency)

Therefore, the application of Equation 9.14 requires a much more elaborate example of ILA. This is discussed in detail in the next two sections. Section 9.1.2 illustrates the possibility of establishing, in this way, bridges across an economic and a biophysical reading of the dynamic budget. Section 9.1.3 then illustrates the possibility of generating mosaic effects across levels.

9.1.2 Establishing Horizontal Bridges across Biophysical and Economic Readings

An overview of the relations between the terms used in Equation 9.14 is given in Figure 9.6. The reader can recognize immediately that this representation of the dynamic budget of exosomatic energy is different from the scheme used so far in Figure 9.2 and Figure 9.5. When applying the rationale implied by Equation 9.14, we obtain a four-angle loop figure that has been already illustrated in Chapter 7 (Figure 7.5). As promised then, we can now go into a detailed discussion about the selection of the set of parameters used over the loop.

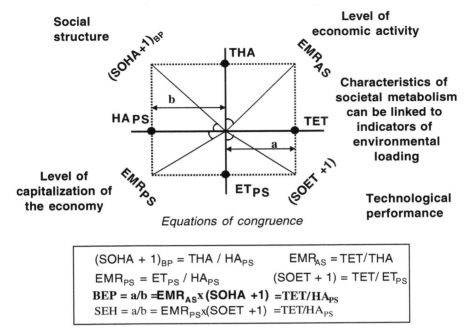

FIGURE 9.6 Example 3: ILA with an EV1 of human activity and an EV2 of exosomatic energy. (Giampietro, M., Mayumi, K. and Bukkens S.G.F. (2001), Multiple-scale integrated assessment of societal metabolism: An analytical tool to study development and sustainability. *Environ., Develop. Sustain.*, Vol. 3 (4): 275–307.)

Let us start with the total requirement of exosomatic energy — EV2 (TET) — which is expressed by using the three numerical assessments found on the northeast quadrant (upper right) (TET = THA × EMR_{AS}) where SOHA stands for societal overhead on human activity, THA is EV1 and EMR_{AS} is an IV3 assessed at the level *n*. On the left side, the total size of the system (expressed in terms of EV1) is reduced to the size of one of its lower-level elements considered the direct compartment (in this case, HA_{PS} is the investment of human activity in the compartment PS). This implies a first difference with the four-angle figures seen so far in this chapter. The northwest quadrant (upper-left quadrant) is used for representing the overall reduction (reduction I plus reduction II) related to the classification "rest of the society" × "direct compartment." In this example, the definition of direct compartment of productive sector) includes all the sectors stabilizing the autocatalytic loop of exosomatic energy. Such a reduction can be indicated as (SOHA + 1 = THA/HA_{PS}). The product (SOHA + 1) × THA therefore represents the size taken by the compartment "rest of society," which affects or is affected by the size of the direct compartment PS. For a representation based on real numbers, refer to Figure 7.5.

At this point, after having collapsed the two reductions in a single quadrant, there is an extra quadrant to be used. We can take advantage of this opportunity by using this extra quadrant (the lower right) to compare the size of the whole (assessed using extensive variable 2 at the level *n* (TET)) to the size of the direct compartment (assessed using extensive variable 2 at the level *n* − 1 (ET_{PS})). This relation is represented in the southeast quadrant (lower-right quadrant) under the label SOET + 1 = TET/ET_{PS}. This label has been chosen since the parameter societal overhead on exosomatic throughput (SOET) is the equivalent of SOHA in relation to EV2. That is, the shape of this angle will reflect/determine the relative size (expressed this time in EV2) of both the direct compartment PS and the rest of society.

The two profiles of investments for the two variables (EV2, expressed in fractions of TET; and EV1, expressed in fractions of THA) over the set of lower-level compartments are not the same. This is what generates differences in the value taken by IV3 on different compartments and on different levels.

As observed in the example of the parallel assessment of the metabolism of the human body and of its parts (Chapter 7), it is actually possible to associate the identity of a particular lower-level element

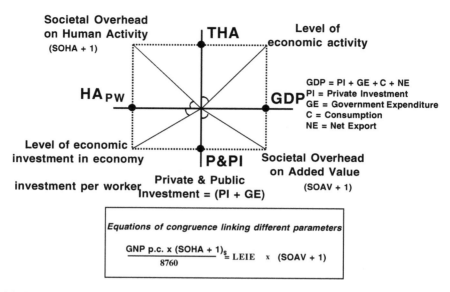

FIGURE 9.7 Example 4: Representation of ILA based on EV1 of human activity and an EV2 of added value different from that given in Figure 9.5. (Giampietro, M., Mayumi, K. and Bukkens S.G.F. (2001), Multiple-scale integrated assessment of societal metabolism: An analytical tool to study development and sustainability. *Environ., Develop. Sustain.*, Vol. 3 (4): 275–307.)

(e.g., the brain or the liver) with a specific rate of metabolism per kilogram, which is related to the very identity of its lower-lower-level elements. In metabolic systems, the given identity associated with the structural organization of lower-level elements represents nonequivalent external referents, which can be used to study the feasibility of the congruence in the representation of energy flows across levels. That is, we can associate with typologies of lower-level elements (e.g., urban households living in compact buildings or high-input agricultural sectors of a developed country) an expected level of intensity of flows. Put another way, it is possible to obtain experimental measurement schemes for both (1) the whole society at the level *n* (northeast, upper-right quadrant), associated with an external referent and (2) specific sectors at the level level *n* – 1 (southwest, lower-left quadrant), whose identity can be associated with the existence of a nonequivalent set of external referents.

Looking at the other quadrants in Figure 9.6, we can observe that:

- **Northwest, upper-left quadrant** (the reduction from THA to HA_{PS}) — This angle is related to the parameter SOHA, which can be associated with another set of external referents such as demographic variables, social rules or institutional settings, as discussed in Chapter 6.

- **Southeast, lower-right quadrant** (the ratio TET/ET_{PS}) — This angle is determined by technological efficiency and quality of natural resources used to guarantee the supply of required input. This has to do with determining what fraction of the total energy consumption goes into the household and into the service sectors (final consumption of exosomatic energy) and what fraction has to be invested just in the making of machines and in the extraction of energy carriers and material flows.

As soon as we represent the dynamic budget of exosomatic energy as in Figure 9.6, we discover that a very similar analysis can be obtained, for the same society, using flows of added value as extensive variable 2, rather than flows of exosomatic energy. An example of this parallel analysis is given in Figure 9.7. Technicalities linked to the calculation of these two four-angle figures are not relevant here (for a detailed discussion of this analogy and the mechanisms of accounting, see Giampietro and Mayumi (2000) and Giampietro et al. (2001)). What is important in the comparison of Figure 9.6 and Figure 9.7 is (1) the striking similarity in the characterization of the dynamic budget and (2) the fact that both types of extensive variable 2 (added value and fossil energy) are mapped against the same hierarchical structure in a matrix mapping the size of elements across levels provided by extensive variable 1. This means that *if*:

1. There is a relation (at the level n) between the values taken by the two IV3 variables, that is:
   a. EMR_{AS} (the exosomatic metabolic rate associated with the activity of producing and consuming goods and services in that society)
   b. GDP per hour (the added value metabolic rate, so to speak, which is associated with the activity of producing and consuming goods and services in that society)

and

2. There is a relation (at the level $n - 1$) between the values taken by the two IV3 variables, that is:
   a. EMR_{PS} (the exosomatic metabolic rate associated with an hour of human work in this sector as compared to EMR_{AS}), which is associated with the level of technical investments — exosomatic devices controlled by workers during their work, which is associated with their biophysical labor productivity
   b. ELP_{PW} (the amount of added value generated per hour of labor by workers in this sector compared to GDP per hour), which, in general, is associated with the level of economic investment per worker.

then, we can expect that

3. Changes in SOHA — the overhead of fixed investment of human activity required to have an hour of disposable human activity (defined in different ways according to the different identities assigned to the direct compartment associated with the choice of EV2)
4. Changes in SOET (the overhead of fixed investment of exosomatic energy required to have a megajoule of exosomatic energy in final consumption) and SOAV (societal overhead on added value, the overhead of fixed investment of added value required to have a dollar in final consumption) will be coordinated.

It is not the time to discuss the validity of assumptions 1 and 2 now. Section 9.2 is fully dedicated to the validation of this approach with an empirical data set. The important point to be driven home from the comparison of Figure 9.6 and Figure 9.7 is that when framing the analysis in this way, it is possible to establish a bridge among two different ways of looking at the dynamic budget associated with societal metabolism. One is based on biophysical variables, which can be compared with themselves across levels, and the other is based on economic variables, which can also be compared with themselves across levels.

Concluding this section, we can say that by using a representation of the metabolism of human systems based on the concept of impredicative loop analysis and using a set of parameters able to induce a mosaic effect across levels, it is possible to establish a relation between the representation of structural changes obtained when using economic variables and the representation of structural changes obtained when using biophysical variables. These two representations of structural changes using two nonequivalent descriptive domains can be linked because they are both mapped against the same nested structure of compartments used when adopting human activity as common extensive variable 1. This implies that we can expect that when going through structural readjustment of the whole in relation to its parts, even when adopting two nonequivalent descriptive domains to represent requirement and internal supply of flows (the economic one and the biophysical one), we should be able to find some common feature.

9.1.3 Establishing Vertical Bridges, Looking for Mosaic Effects across Scales (*Technical Section*)

There is another way to justify the name impredicative loop analysis for describing the typology of the four-angle figures presented in Figure 7.5, Figure 9.6 and Figure 9.7. Such a name is also justified by the fact that these figures represent the very same ratio between two variables TET/HA$_{PS}$, which is characterized simultaneously in two different ways. Let us discuss this fact, using again the example given in Figure 9.6:

1. The ratio TET/HA_{PS} can be viewed and defined as BEP (bioeconomic pressure) when looking at it from the requirement side (by considering the value taken by variables related to identities defined at levels n and $n-1$). In relation to Figure 9.6, we can write

$$BEP = a/b = EMR_{AS} \times (SOHA + 1) = TET/HA_{PS} = \Sigma x_i \, (EMR_i) \times (\Sigma HA_i)/HA_{PS} \qquad (9.15)$$

    The term $EMR_{AS} \times (SOHA + 1)$ can be viewed as the pace of dissipation of the whole (level n) per unit of human activity invested in the direct compartment (level $n-1$). Because of this, it can be expressed using the intensive variable 3. This assessment can be expressed using two focal-level characteristics $[EMR_{AS} \times (SOHA + 1) = TET/HA_{PS}]$. Alternatively, this ratio can be expressed using information gathered at the level $n-1$. After determining a set of identities for i components on the level $n-1$ that guarantee closure (e.g., imagine that we chose $i = 3$; productive sector, services and government, and household sector), we can write $THA = HA_{PS} + HA_{SG} + HA_{HH}$. Then we need information about the size and level of dissipation of each of these three lower-level elements. That is, we need the assessment of (1) the profile of investments of human activity HA_{PS}, HA_{SG} and HA_{HH} and (2) the level of dissipation of exosomatic energy per hour in these three compartments — EMR_{PS}, EMR_{SS} and EM_{HH} (or alternatively, the size of investments in exosomatic energy ET_{PS}, ET_{SS} and ET_{HH}) in these three sectors. At this point it is possible to express both EMR_{AS} $[= \Sigma x_i \, (EMR_i)]$ and $(SOHA + 1)$ $[= \Sigma HA_i/HA_{PS}]$ using only lower-level assessments (see Chapter 6).

    That is, the parameter BEP can be associated with a family of relations establishing a bridge between nonequivalent representations of events referring to levels n and $n-1$.

    The name *bioeconomic pressure*, which increases with the level of development of a society, indicates the need of developed countries for controlling a huge amount of energy in the productive sectors while reducing as much as possible the relative work requirement. Such a name was suggested by Franck-Dominique Vivien to refer to Georgescu-Roegen's (1971) ideas: increasing the intensity of the economic process to increase the enjoyment of life induces — as a biophysical side effect — an increase in the intensity of the throughputs of matter and energy in the productive sectors of the economy.

2. The ratio TET/HA_{PS} can be viewed and defined as the strength of the exosomatic hypercycle (SEH) when looking at it from the supply side (by considering the value taken by variables related to identities defined on the two interfaces level $n-2$/level $n-1$ and level n/level $n+1$ when representing the performance of the direct sector in guaranteeing the supply of the required input).

$$SEH = a/b = EMR_{PS} \times (SOET + 1) = TET/HA_{PS} \qquad (9.16)$$

    The last term on the right (TET/HA_{PS}) can be viewed as the characterization, in terms of intensive variables only (again the same unit as intensive variable 3) of the supply of energy delivered to the black box, megajoules of TET (level n assessment), per unit of investment of human activity in the lower-level component PS, hours of HA_{PS} (level $n-1$ assessment, productive sector). This characterization is based on variables referring to identities defined on the level n and the level $n-1$, and therefore compatible with what was done when determining BEP. However, we can express the term on the left side $[EMR_{PS} \times (SOET +1)]$ in relation to other variables that are reflecting characteristics defined and measurable only on different hierarchical levels. That is, the capability of the direct sector of generating enough supply of energy input for the whole is dependent on two conditions:

    a. Those working in the productive sectors must be able to control enough power (the level of EMR_{PS} per unit of human activity invested there) to fulfill the set of tasks required to guarantee an adequate supply. This condition is related to the value taken by the southwest angle (lower left) of Figure 9.6. The value of EMR_{PS} can be related to lower-level characteristics, the level of capitalization of the various subsectors making up the PS sector: [$EMR_{PS} = \Sigma x_j (EMR_j)$]. This analysis can be done by using the same approach discussed in Chapter 6 (dividing sector PS in lower-level compartments in terms of investments of HA and guaranteeing closure to the hierarchical structure used to aggregate lower-level elements into higher-level elements). The definition of a profile of values of EMR_j (reflecting the tasks to be performed in the various subsectors) will determine how much capital is required per worker in the various compartments defining the PS sector. The definition of ET_j, EMR_j and HA_j will make it possible to establish a relation between the characteristics of identities defined on level $n - 2$ and those defined on level $n - 1$.

b. The amount of power to be invested in fulfilling the set of tasks will depend on the return in the process of exploitation of natural resources (SOET + 1). This condition is related to the southeast angle (lower right) of Figure 9.6. That is, the lower the return on the investment to fulfill the tasks performed in the direct compartment, the higher will be the requirement of investment (expressed in terms of either ET_{PS} or HA_{PS}) in the direct compartment.

Given a high level of required ET_{PS}, it is possible to reduce the requirement of HA_{PS} (requirement of hours of working) by increasing the value of EMR_{PS} (requirement of technical capital per worker and exosomatic energy spent per working hour). Put another way, the constraints faced by the direct compartment to stabilize the flow of required input to the black box can be related to the two economic concepts of (1) level of capitalization (amount of exosomatic devices per worker), measured by the EMR of a given sector; (2) level of circulating capital, measured by the ET of a given sector; and (3) performance of technology [(SOET + 1) = TET/ET_{PS}]. This ratio, in fact, measures how much of the total energy used by society (TET) is consumed in the internal loop required for the metabolism of technical devices by the productive sectors for their own operation (ET_{PS}). The higher the fraction of TET used by technology, the lower is the relative performance.

The name *SEH* focuses on the fact that this ratio measures the return (the amount of spare input made available to the rest of society) obtained by investment of human activity in the sector labeled "direct" in the upper part of Figure 7.8. The ability to keep this ratio high is crucial in defining how much human time can be invested in activities not directly related to the stabilization of the flow of matter and energy required for the metabolism. Put another way, SEH determines the fraction of TET and THA that can be invested in final consumption (in adaptability, by exploring new activities and new behaviors).

At this point we can get back to Figure 9.6 to note that Equation 9.16 is determining the ratio between segments *a* and *b* going through the two lower angles of the four-angle figure. In doing so, it can be seen as the reciprocal of Equation 9.15, which links segments *a* and *b* going through the two upper angles. This means that in this four-angle figure, we are dealing with two nonequivalent representations of the same ratio TET/HA_{PS}, which are based on the reciprocal entailment of the identity of the elements of the loop. Such a ratio is characterized one time in terms of the total requirement using the terms included in Equation 9.15 and the other time in terms of internal supply using the terms included in Equation 9.16.

This impredicative loop requires two sets of external referents able to validate the representation of the same relation in two different ways. In Equation 9.15 the value of BEP can be calculated using data related to identities defined on only two hierarchical levels — the interface level $n - 1$/level n, whereas when dealing with the value of SEH, according to Equation 9.16, assessments of technical characteristics are related to both the interface level $n - 2$/level $n - 1$ (the conversion of an input into a specified flow of applied power to perform the set of tasks assigned to the direct compartment) and the return of a set of tasks defined on the interface level n/level $n + 1$. Recall the technical sections of Chapter 7. Moreover, the stability in time of this return (the stability of the supply of input gathered from the context to feed the black box, the stability of the quality of natural resources) is based on a hypothesis of admissibility for the context of the black box on level $n + 2$ (a hypothesis of future stability of boundary conditions), which is not granted. This is the hidden assumption implied by a representation of the steady state of dissipative systems, which entails defining a weak identity for the environment, as discussed in Chapter 7.

Any attempt to bring into congruence this four-angle figure in terms of a forced congruence between the two parameters BEP and SEH implies the challenge of bringing into coherence assessments referring to five different hierarchical levels. As noted earlier, when discussing holons and holarchies, it is impossible to do such an operation in formal terms (in the "correct" way). That is, we must expect that we will find different ways to formalize an impredicative loop (depending on the definitions and assumptions used for characterizing extensive and intensive variables over the four-angle figure, which will lead to a set of congruent assessments over the loop). The reader can recall the discussion of this problem in the example of the society of 100 people on the remote island given in Chapter 7 (Figure 7.4).

This implies that a model based on the application of the approach presented in Figure 9.6 will not represent the "right" representation of the mechanism determining the stability of the metabolism of a given society. Rather, it will be just one of the possible representations of one of the mechanisms that can be used to explain the stability of the investigated metabolism. Recall again the example of the 100 people on the island discussed in Figure 7.5. A very high return of food per hour of labor would not have guaranteed the long-term sustainability of such a human system, if all 100 people on the island were men. The analysis of the minimum number of fertile women as a potential constraint on the stability of a given societal metabolism would require the adoption of a totally different narrative.

The ability of impredicative loops to establish bridges among levels is based on the bridges across levels provided by intensive variable 3. As noted in Chapter 6, we can go through levels using a redundant definition of compartments across different hierarchical levels. For example, by starting with Equation 9.15 and substituting

$$EMR_{AS} = \Sigma x_i \, EMR_i \; = (MF \times ABM) \times Exo/Endo \tag{9.17}$$

$$(SOHA + 1) = THA/HA_{PS} \tag{9.18}$$

we can write

$$BEP = (MF \times ABM) \times Exo/Endo \times (THA/HA_{PS}) \tag{9.19}$$

Equation 9.19 establishes a reciprocal constraint on the set of values that can be taken by the three parameters on the right given a value of BEP. This is very important, since these three parameters happen to describe the characteristics of socioeconomic systems on different hierarchical levels and in relation to different descriptive domains. Examples of this parallel reading have been given in Chapter 6 (e.g., Figure 6.8 and Figure 6.9). In this case, the three parameters listed on the right side of Equation 6.19 are good indicators, describing changes in the metabolism of human societies on different hierarchical levels and in relation to different descriptive domains. For more details, see Pastore et al. (2000). In particular:

1. MF × ABM assesses the endosomatic metabolic rate (per capita per hour) of the population. This is an indicator of the average endosomatic flow per person (a value referring to an average assessed looking at the level of the whole society). This value refers to a descriptive domain related to physiological processes within the human body.

   - Metabolic flow is the endosomatic metabolic rate per kilogram of body mass of a given population — expressed in megajoules per hour per kilogram — determined by (1) the distribution of individuals over age classes and (2) the lifestyle of individuals belonging to each age class.

   - Average body mass is the average kilograms of body mass per capita of the population, determined by (1) the distribution of individuals over age classes and (2) the body size of the particular population at each age class.

   The higher the value of MF × ABM, the better are the physiological conditions of humans living in the society. According to the database presented in Pastore et al. (2000), the parameter MF × ABM has a minimum value of 0.33 MJ/h (short life expectancy at birth, small average body mass in very poor countries) and a maximum value of 0.43 MJ/h (long life expectancy at birth, large average body mass), which is a plateau reached in developed countries.

2. Exo/Endo is the exosomatic/endosomatic energy ratio between the exosomatic metabolism (megajoules per hour) and endosomatic metabolism (megajoules per hour). This is an indicator

of development valid at the socioeconomic hierarchical level (reflecting short-term efficiency (Giampietro, 1997a)). This ratio can be easily calculated by using available data on consumption of commercial energy of a country (assessing the exosomatic flow) and the assessment of endosomatic flow (food energy flow). The Exo/Endo energy ratio has a minimum value around 5 (when exosomatic energy is basically in the form of traditional biomass, such as fuels and animal power). The maximum value is around 100 (when exosomatic energy is basically in the form of machine power and electricity obtained by relying on fossil energy stocks). Exo/Endo is a good indicator of economic activity; it is strongly correlated to the GNP p.c. (see Section 9.2 for data). The higher the Exo/Endo, the more goods and services that are produced and consumed per capita.

3. $THA/HA_{PS} = SOHA + 1$; this is an indicator valid at the socioeconomic hierarchical level (reflecting long-term adaptability (Giampietro, 1997a)). It is the fraction of the total human activity available in the society per working time allocated in productive sectors of the economy. The ratio THA/HA_{PS} has a minimum value of 10 (crowded subsistence socioeconomic systems in which agriculture absorbs a large fraction of workforce). The maximum value is 45, in postindustrial societies with a large fraction of elderly and a large fraction of workforce absorbed by services. This indicator reflects social implications of development (longer education, larger fraction of nonworking elderly in the population, more leisure time for workers coupled with an increased demand for paid work in the services and government sector).

Concluding this section we can say that by using a representation of the metabolism of human systems based on the concept of impredicative loop analysis, it is possible to take advantage of the existence of mosaic effects to establish a relation between the representations of changes obtained using an integrated set of variables that refer to nonequivalent descriptive domains. That is, changes detected at one level using variables defined in a given descriptive domain (e.g., life expectancy, average body mass) can be linked to changes detected at a different hierarchical level, using variables defined on a descriptive domain that is nonequivalent and nonreducible to the first one (e.g., exosomatic energy consumption, GDP per capita, number of doctors per capita).

9.2 Validation of This Approach: Does It Work?

9.2.1 The Database Used for Validation

A validation of the analytical framework of multiple-scale integrated assessment of societal metabolism has been presented in Pastore et al. (2000). Data and figures presented in this section are taken from that source.

The analysis started with a database of 187 world countries, from which 55 countries with less than 2 million inhabitants were excluded because of their too small size (this excluded 0.6% of the total world population). For 25 of the remaining 132 countries (some countries from the former USSR, Yugoslavia, Czechoslovakia, plus South Africa, Libya, Algeria, Cambodia — which comprise 9% of the total world population) data are not available. Thus, the database includes 107 countries, comprising more than 90% of the world population. The database has been created using official data of the UN, FAO and World Bank statistics (specified in Pastore et al., 2000). BEP has been calculated according to Equation 9.19 as follows:

1. The term ABM × MF
   - ABM has been calculated by pondering the average weights (by age and sex classes) and the structure of population as reported by James and Schofield (1990) for all FAO countries. Data on the total population of 1992 are as reported by the World Tables published for the World Bank (1995b).
   - MF has been computed separately for each sex and age class following the indication given by James and Schofield (1990) and merged into national averages.

2. The term Exo × Endo

- The annual flow of exosomatic energy was evaluated according to United Nations (1995) statistics for commercial and traditional biomass consumption (expressed in tons of coal equivalent) in 1992, by using a conversion factor of 29.3076 terajoules per thousand metric tons of coal. However, a minimum value of 5/1 has been adopted for countries with a resulting value of Exo/Endo < 5. This is due to the fact that official statistics are mainly reflecting the use of commercial energy and therefore tend to underestimate, for rural communities, the contribution of animal power, biomass for cooking and building shelters (see Giampietro et al., 1993).

- The annual flow of endosomatic energy has been computed using the population size of 1992 as reported by the World Tables published for the World Bank (1995b), multiplied by the value of ABM × MF.

3. The term THA/HA$_{PS}$

- The fraction of the economically active population and the distribution of labor force in different sectors of the economy are both derived from United Nations (1995) statistics and refer to the latest available data in the period 1990–1993.

- In this analysis productive sectors of the economy include agriculture, hunting, forestry and fishing; mining and quarrying; manufacturing; electricity, gas and water; construction; and a fraction of transport. Transport (nonresidential) was in fact divided between productive sectors and the service sector, proportionally, for each country, according to the working time spent in the primary sectors and the working time spent in the service sectors (which include trade, restaurants and hotels; financing, insurance, real estate and business; community, social and personal services).

- Workload was estimated at a flat value of 1800 h/year when including vacations, absences and strikes (after Giampietro and Mayumi, 2000)..

The number of conventional indicators of material standard of living and development used in the analysis is 24. Such a selection of indicators basically reflected the selection found in the World Tables. The 24 indicators can be divided into three groups:

1. Eight indicators of nutritional status and physiological well-being:

    1 = Life expectancy

    2 = Energy intake as food

    3 = Fat intake

    4 = Protein intake

    5 = Average body mass index (BMI) adult

    6 = Prevalence of children malnutrition (weight/height < 2 z-score of U.S. National Center Health Statistics (NCHS) reference growth curve)

    7 = Infant mortality

    8 = Percent low birth weight

2. Seven indicators of economic and technological development:

    9 = GNP per capita

    10 = Percent GDP from agriculture

    11 = ELP$_{PW}$ — average added value per hour of paid work = GDP/(HA$_{SG}$ + HA$_{PS}$) (Note: This indicator has the label COL$_{AV}$ in the figures.)

    12 = Percent of labor force in agriculture

    13 = Percent of labor force in services

    14 = Energy consumption per capita

    15 = Percent of GDP expended for food

TABLE 9.1

Correlation of the Proposed Set of Integrated Indicators with Conventional Indicators of Development

	Ratio			
	log(BEP)	**log(Exo/Endo)**	**THA/HA$_{PS}$**	**ABM × MF**

Correlation between BEP, Exo/Endo Ratio, THA/HA$_{PS}$, ABM × MF and Some Major Indicators of Nutritional Status, Physiological Well-Being

Life expectancy	0.79	0.75	0.63	0.59
Energy intake	0.82	0.81	0.55	0.73
Fat intake	0.87	0.85	0.63	0.77
Protein intake	0.85	0.85	0.57	0.72
Children malnutrition	−0.71	−0.65	−0.63	−0.70
Infant mortality	−0.76	−0.74	−0.57	−0.58
Low birth weight	−0.65	−0.62	−0.49	−0.63

Correlation between BEP, Exo/Endo Ratio, THA/HA$_{PS}$, ABM × MF and Some Major Indicators of Economic and Technological Development

Log (GNP)	0.92	0.89	0.63	0.66
% Agriculture on GDP	−0.77	−0.73	−0.60	−0.54
Log (COL$_{AV}$)	0.92	0.87	0.71	0.63
% Laboratory force in agriculture	−0.90	−0.81	−0.72	−0.66
% Laboratory force in service	0.90	0.83	0.76	0.56
Energy constant/capacity	0.92	0.95	0.53	0.67
Expenditure for food	−0.86	−0.87	−0.69	−0.78

Source: Pastore, G., Giampietro, M. and Mayumi, K. (2000), Societal metabolism and multiple-scales integrated assessment: Empirical validation and examples of application. *Popul. Environ.* 22 (2): 211–254.

3. Nine indicators of social development:

   16 = Television/1000 people

   17 = Cars/1000 people

   18 = Newspaper/1000 people

   19 = Phones/100 people

   20 = Population/physician ratio

   21 = Population/hospital bed ratio

   22 = Pupil/teacher ratio

   23 = Illiteracy rate

   24 = Access to safe water (percent of population)

All data on these 24 indicators come from the FAO (1995), United Nations (1995) and World Bank (1995a), and each one of refers to the latest available year between 1991 and 1993. Data on prevalence of malnutrition in children come from ACC/SCN (1993).

9.2.2 The Representation of Development According to Economic Variables Can Be Linked to Structural Changes in Societal Metabolism Represented Using Biophysical Variables: The Correlation of BEP with the Chosen Set of Indicators of Development

The analysis of Pastore et al. (2000) indicates that BEP is strongly correlated with:

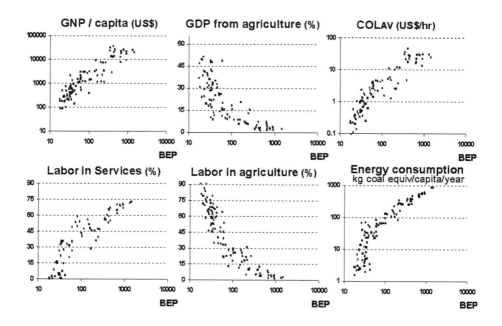

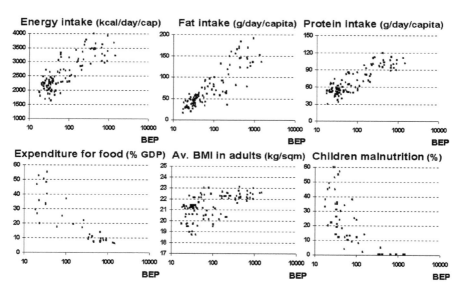

FIGURE 9.8 Conventional indicators of development vs. BEP. (Pastore, G., Giampietro, M. and Mayumi, K. (2000), Societal metabolism and multiple-scales integrated assessment: Empirical validation and examples of application. *Popul. Environ.* 22 (2): 211–254.)

1. All classic economic indicators of development. See gray column of Table 9.1 — average value of r = 0.88 (ranging from 0.77 to 0.92) — and upper part of Figure 9.8 for graphic representation.

2. All nutritional status and physiological well-being indicators. See the gray column of Table 9.1 — average value of r = 0.78 (ranging from 0.65 to 0.87) — and the lower part of Figure 9.8 for a graphic representation.

3. All health indicators (Figure 9.9, upper part) and social development indicators (Figure 9.9, lower part) — average value of r = 0.76 (ranging from 0.44 to 0.89). See the gray column of Table 9.2.

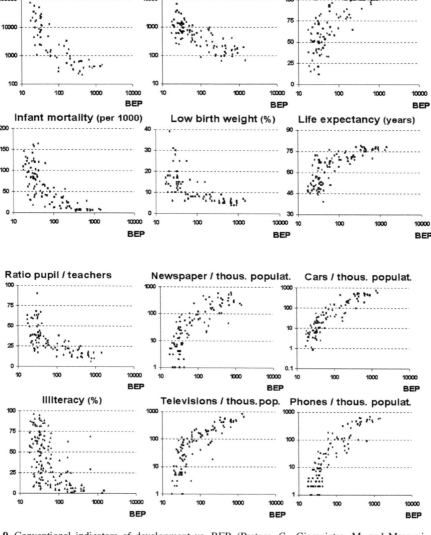

FIGURE 9.9 Conventional indicators of development vs. BEP. (Pastore, G., Giampietro, M. and Mayumi, K. (2000), Societal metabolism and multiple-scales integrated assessment: Empirical validation and examples of application. *Popul. Environ.* 22 (2): 211–254.)

9.2.3 Changes Associated with Economic Development Can Be Represented Using an Integrated Set of Indicators on Different Levels and Descriptive Domains: Assessing Changes Related to Development on More Hierarchical Levels

The ability to see changes coupled with development on more hierarchical levels has been verified by studying the correlation of each one of the three terms composing BEP with conventional indicators of development in the fields of (1) nutritional status and physiological well-being (by using only ABM × MF); (2) economic development at the level of society (by using only Exo/Endo); and (3) socioeconomic level (by using only the ratio THA/HA_{PS}).

Since we are no longer dealing with a parameter such as BEP that is the product of three factors, the graphs illustrating the correlation with each of these individual parameters are no longer based on a logarithmic scale. This makes more evident the existence of threshold values for the various parameters

TABLE 9.2

Correlation of the Proposed Set of Integrated Indicators with Conventional
Indicators of Development

	Ratio			
	log(BEP)	log(Exo/Endo)	THA/HA$_{PS}$	ABM × MF

Correlation between BEP, Exo/Endo Ratio, THA/HA$_{PS}$, ABM × MF and Some Major Indicators of Social Development

Television/inhabitant	0.89	0.89	0.62	0.72
Cars/inhabitant	0.88	0.91	0.59	0.72
Newspaper/inhabitant	0.77	0.80	0.47	0.60
Phones/inhabitant	0.87	0.88	0.61	0.71
log(population/physician)	−0.81	−0.76	−0.60	−0.67
log(population/hospital bed)	−0.77	−0.78	−0.51	−0.70
Pupil/teacher	−0.77	−0.76	−0.51	−0.62
Illiteracy rate	−0.61	−0.58	−0.42	−0.44
Primary school enrollment	0.44	0.39	0.38	0.36
Access to safe water	0.78	0.77	0.53	0.59

(Pastore, G., Giampietro, M. and Mayumi, K. (2000), Societal metabolism and multiple-scales integrated assessment: Empirical validation and examples of application. *Popul. Environ.* 22 (2): 211–254.)

determining the characteristics of the energy budget (BEP and SEH), when considered in an evolutionary trajectory; see, for example, Figure 9.10. Above a given value (the same for the entire sample of correlation with the 24 indicators), all countries seem to converge toward an attractor value.

9.2.3.1 *Physiological/Nutrition: Individual Hierarchical Level* — ABM × MF proved to be a good indicator for assessing changes coupled to the process of development at the physiological level. Actually, the example of correlation with six indicators provided in the upper part of Figure 9.11 shows quite clearly that it is possible to individuate a threshold value of ABM × MF around 9 MJ/day (0.4 MJ/h), above which socioeconomic systems tend to converge on similar values.

9.2.3.2 *Economic Development: Societal Hierarchical Level (Steady-State View)* —
Exo/Endo is an excellent indicator to assess changes coupled to the process of development at the societal level (Figure 9.10, upper and lower parts). Besides the obvious correlation with energy consumption and GNP (not reported in these graphs, since it is a perfect diagonal with a light dispersion due to the noise typical of the data), Exo/Endo shows an extraordinary ability to detect sharp structural changes of the socioeconomic system, which can be related to the process called demographic transition (Giampietro, 1998). The graphs reported in Figure 9.10 clearly show a threshold value of Exo/Endo (about 25/1). This indicates a change in the path of expansion of the activity of self-organization (measured by TET). After this threshold point, socioeconomic systems stop expanding by increasing in human mass. Further increases in size of economic activity (TET) are obtained by increasing the Exo/Endo energy ratio (the EMR per unit of HA). Such a change, in turn, affects the value taken by the THA/HA$_{PS}$ ratio (which keeps growing). This has important consequences in evolutionary terms. In fact, this can be seen as an increase in the fraction of exosomatic energy and human activity that are invested in long-term adaptability (allocated in social roles not strictly encoded, such as leisure time, and in job positions in the services sector) instead of in short-term efficiency (Giampietro, 1997a; Giampietro et al., 1997).

9.2.3.3 *Socioeconomic Level: Societal Hierarchical Level (Evolutionary View)* —
THA/HA$_{PS}$ also correlates with traditional indicators of economic development (Figure 9.11, lower part). However, such a correlation is not as strong as the one found with Exo/Endo (Table 9.1 and Table 9.2). The power of resolution of THA/HA$_{PS}$ increases when it is compared to indicators of social development. Also in this case, we can see a threshold value (at about 30/1) that could be interpreted as an indication for a switch to a new form of meta-stable equilibrium of the dynamic energy budget (for more, see

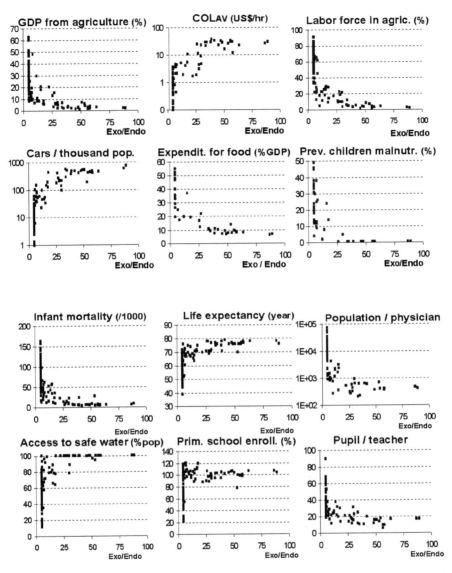

FIGURE 9.10 Conventional indicators of development vs. Exo/Endo. (Pastore, G., Giampietro, M. and Mayumi, K. (2000), Societal metabolism and multiple-scales integrated assessment: Empirical validation and examples of application. *Popul. Environ.* 22 (2): 211–254.)

Giampietro, 1998). The increase in THA/HA_{PS} is possible only when SEH and BEP both increase in a coordinated way (Giampietro et al., 1997). However, there is an inertia toward changes in socioeconomic parameters (especially those under direct cultural control, when dealing with profiles of allocation of human activity) determining a lag time in adjustments of THA/HA_{PS}. This inertia is generated by the existing social and cultural identity, which is related to the history of the socioeconomic system. This could explain the lower correlation of THA/HA_{PS} with traditional indicators when compared with the correlation factors of the other parameters of BEP: Exo/Endo and ABM × MF. The possible role of cultural identity in slowing down the changes induced by development can also be seen when looking at the graph BEP vs. illiteracy in the lower part of Figure 9.9. Such a relation is much looser that the ones found in the graphs referring to economic indicators. For example, in many Islamic countries the illiteracy rate of women is higher than the value expected in countries having the same level of GDP/THA.

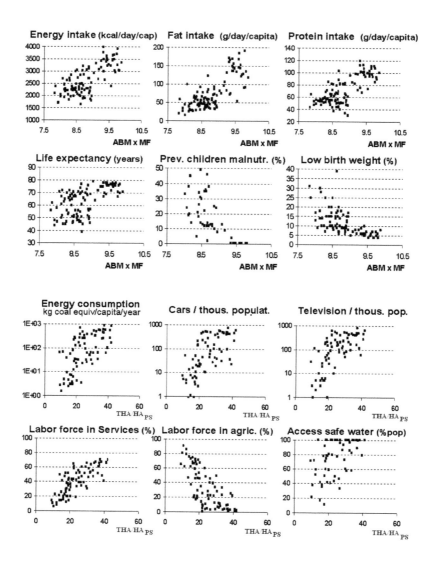

FIGURE 9.11 Conventional indicators vs. ABM × MF and THA/HA$_{PS}$. (Pastore, G., Giampietro, M. and Mayumi, K. (2000), Societal metabolism and multiple-scales integrated assessment: Empirical validation and examples of application. *Popul. Environ.* 22 (2): 211–254.)

9.3   Applications to the Analysis of the Role of Agricultural Systems

9.3.1   The Particular Identity of the Metabolism of a Socioeconomic System Implies Minimum Thresholds on the Pace of Throughputs in the Various Components

Conventional analyses of the agricultural sector are based on the generic (unspecified) assumption that this sector is in charge of the generation of the supply of two types of flows: a flow of food and a flow of added value.

However, as soon as we consider different entities related to agriculture at different hierarchical levels (e.g., farmers, rural villages, agricultural sectors), we discover that in relation to these two flows, different entities can belong to one of three categories:

In relation to food production:
1. Net producer of food surplus for the rest of society
2. Self-sufficient in terms of food production/consumption (pure subsistence)
3. Covering only a fraction of the internal requirement of food (worst-case scenario)

In relation to added value production
1. Generating surplus of added value for the rest of society
2. Managing to remain economically viable by breaking even in economic terms
3. Covering only a fraction of their consumption of added value (worst-case scenario)

As noted in the introduction to Part 3, an integrated analysis of the performance of an agroecosystem has to be structured in a completely different way, depending on which one of the three above options applies to the particular farming system and agroecosystem under analysis. Therefore, before getting into a discussion on how to formalize an integrated analysis of a given agroecosystem, it is crucial for the analyst to individuate critical goals, and constraints have to be addressed in the relative problem structuring. This requires individuating the particular role that the agricultural holon (viewed as a lower-level element operating at the level $n - 1$) is playing within the larger agricultural holon (viewed as the context at the level n, to which the first holon belongs).

To do that, one can explore the hierarchical relation that the two elements (e.g., the agricultural sector and the whole society) have in relation to their ability to generate and consume both flows: food and added value. There are cases in which the agricultural sector is a net producer of both food and added value for the rest of society (e.g., in some countries at low demographic density, such as Argentina). In this case, the agricultural sector belongs to the direct sector when mapping both types of flows. It has to be considered direct since it generates a positive return on the investment in relation to both food energy and added value. There are other cases in which the agricultural sector is a net producer of food, but a consumer of added value (e.g., in some crowded developed countries in Europe). In this case, the agricultural sector is the direct sector when mapping flows of food (it generates positive return on the investment of human activity) and a net consumer of added value (it generates a return on the economic investment that is lower than the national average). Finally, there are cases in which poor subsistence farming systems tend to collapse, becoming a humanitarian problem (they become net consumers of both food and added value flows).

This discussion can be generalized by saying that when studying the stability of the metabolism of a system organized in nested elements, the contribution (and the role) of a particular element (in this case, the agricultural sector) is determined by the relative value of average densities of flows in the whole and in the parts. Any assessment of these densities, however, is not absolute or substantive; rather, it reflects the choices made by the analyst on how to account for quantities included in the flow (again the reader can recall the example of the four different assessments of the consumption of cereal per capita within the U.S. — Figure 3.1). The applicability of the numerical indication will therefore depend on the agreement of the users of this information in relation to the choice of the narrative used to formalize the analysis.

In this section, we apply the methodological approach presented so far to compare the density of flows within the agricultural sector (viewed as level $n - 1$) and the density of the same flows characterized at the level of the whole society (viewed as level n).

In Chapter 11, the same rationale is applied to the study of lower-level interfaces (to do benchmarking within the holarchy). There, the performance of an agroecosystem, perceived and represented at the community level (viewed as level $m - 1$) is framed in relation to the perception and representation of the performance of the same agroecosystem viewed at the larger level (e.g., the province) to which it belongs (viewed as level m). Within the same frame, it is possible to bridge the representation of the typology of individual households (viewed as level $m - 2$) in relation to the representation of the performance of a community (viewed as level $m - 1$) to which the selected type of household belongs.

This process of comparison across levels requires looking at different definitions of constraints on the throughputs (IV3), which are characterized using variables relevant for the analysis (e.g., kilograms of food per hour and per hectare, dollars of added value per hour and per hectare, megajoules of exosomatic energy per hour and per hectare). The next section looks at the existence of constraints related to the biophysical productivity (density of food produced) against human activity (per hour of work in the agricultural sector) and against land in production (per hectare of land in production). In the same way, one can look at the existence of constraints related to the economic productivity (density of added value produced) against human activity (per hour of work in the agricultural sector) and against land in production (per hectare of land in production).

9.3.2   Determination of Minimum Thresholds for Congruence over the Loop

9.3.2.1   *Minimum Throughput of Food per Hour of Labor in the Agricultural Sector —*

The congruence check for the agricultural element can be done by applying Equation 9.7; we can write a food metabolic requirement (FMR) at the level of the food system:

$$FMR_{FS} = THA \times (ABM \times MF \times QDM\&PHL) = HA_{AG} \times BPL_{AG}$$

Just to give an example of the application of Equation 9.7 based on a very simple system of mapping, we can represent the requirement of endosomatic energy per capita (left side of the equation) using an assessment based on kilograms of grain-equivalent per capita per year. A more elaborate analysis of energy consumption in relation to the quality of the diet and the double conversions within the food system is given in the last section of Chapter 10.

Assuming an average level of consumption of 250 kg of grain-equivalent per person, and a society of 1 million people, we get a final requirement of a flow of 250 million kg of grain per year. Depending on the quality of the diet and the characteristics of the food system, we can apply to this initial flow of endosomatic energy a coefficient QDM&PHL of 3/1 for developed countries, and a coefficient QDM&PHL of 1.5/1 for poor developing countries.

This will provide the following assessment of the flow of grain required in production (for a society of 1 million people):

$$FMR_{FS} = 750 \text{ million kg for a society with a food system typical of developed countries}$$

$$FMR_{FS} = 375 \text{ million kg for a society with a food system typical of developing countries}$$

Let us check the congruence threshold on the pace of the relative throughput in production. This check can be obtained by using Equation 9.9:

$$THA/HA_{AG} = BPL_{AG}/FMR_{FS} = 500 = 1/0.002 \text{ (for the U.S.)}$$

These values are obtained by applying to our hypothetical society of 1 million people the values found for the profile of allocation of human activity shown in Figure 9.4:

$$\text{Using extensive variable 2: THA} = 8.76 \text{ billion; } HA_{AG} = 17.5 \text{ million}$$

$$\text{Using fractions of reduction of THA: } 1/(SOHA + 1) = 0.1; HA_{AG}/HA_W = 0.02$$

To be self-sufficient, the agricultural sector of this hypothetical society should operate above a minimum threshold of labor productivity (assuming that the whole agricultural sector is totally dedicated to grain production) of 40 kg of grain per hour of labor. The assessment of this threshold can be obtained using extensive variables, by dividing the total requirement FMR_{FS} (750 million kg) by the available supply of working hours HA_{AG} (17.5 million hours). Alternatively, the assessment of this threshold can be obtained using intensive variables. According to Equation 9.9, we can assess BPL_{AG} as the product of $FMR_{FS} \times 500$, starting with a FMR_{FS} of 750 kg of grain per year (250×3 for a developed country). This assessment has to be divided by 8760 (to get the assessment of kilograms of food per hour), and then multiplied by 500 according to the congruence constraint. Both methods, obviously, provide the same answer: 40 kg of grain per hour of labor as a minimum threshold of labor productivity for being self-sufficient.

This threshold, however, is never relevant in developed countries. In fact, in the agricultural sector of a developed society, grain production is just one of the tasks to be performed in agriculture. Actually, it is exactly because of the ability to reach levels of biophysical labor productivity much higher than this minimal threshold (in the order of hundreds of kilograms of grain per hour of labor) that it is possible for the agricultural sector of developed countries to produce a lot of animal products (based on a double conversion of cereal) and a diversified abundant supply of fresh vegetables.

It should be noted, however, that a biophysical labor productivity of 40 kg of grain per hour is completely out of the range of technical coefficients found in subsistence farming systems.

9.3.2.2 *Minimum Throughput of Food per Hectare of Land in the Agricultural*

Sector — Let us now look for the constraint on the throughput of endosomatic energy flows based on the adoption of land area as extensive variable 1. This time the threshold can be calculated in relation to availability of land to be invested in production. Using the four-angle figure given in Figure 9.3 we can see that total available land (TAL) is first reduced to colonized available land (CAL), which is further reduced to land in production (LIP). At this point, we can use a variation of Equation 9.9 to assess the minimum thresholds of land.

We can look for an expression based on intensive variables, using the concepts of:

- **BOAL** (biophysical overhead on available land) — a variation of SOHA. This represents the ratio NAL/CAL (natural available land over colonized available land). By adopting the same system of accounting, we can write (BOAL + 1) = TAL/CAL.

- **YLP$_{AG}$** (yield of land in production in the agricultural sector) — This is the yield (or the average of aggregated yields) perceived within the element considered in the analysis. It can be the direct yield (e.g., kilograms of rice per hectare assessed in a defined field), an average value for a farming system (e.g., kilograms of total output per hectare assessed in a defined year), or the aggregate output of crop energy for a food system of a given country. Obviously, depending on the choices made by the analyst, it is crucial to remain consistent in the system of accounting (selection of variables and measuring schemes) across hierarchical levels.

The finding of a threshold value on YLP$_{AG}$ based on Equation 9.9 is a trivial task. The rationale of this analysis has already been illustrated in the various examples seen so far. Starting with the total requirement of food (e.g., FMR$_{FS}$ = 750 million kg per year calculated for the hypothetical developed society of 1 million people) and given the overall ratio TAL/LIP, it is possible to calculate the minimum threshold YLP$_{AG}$. Depending of the goal of the analysis, we can decide to keep disaggregating the lower-level element LIP by considering a mix of different techniques of production within LIP. Depending on additional categorizations, we can split the area allocated to agricultural production LIP into agricultural production for subsistence and agricultural production for cash. Within the area in production for subsistence, we can then divide cereal from vegetables, and so on.

It is always important to define the set of categories used to classify the identities on level $m - 1$ in a way that provides closure on level m. It is at this point that the category "other uses" enters into play. A discussion of this mechanism of accounting based on aggregation through levels has been given in Figure 6.1 and Figure 6.2.

Also in this case, it should be noted that the minimum threshold found when dividing FMR$_{FS}$ by LIP is a minimum threshold of throughput density that would be required to keep the system self-sufficient. As noted earlier, such information (a minimum throughput to be self-sufficient) can be useful or not. We are living in a market economy in which trade plays a crucial role in allocating investments of labor and capital. If a given social element is not self-sufficient — like a city — it can always import food from elsewhere.

Two quick points can be made to rebut this objection: (1) When talking of agriculture, the degree of self-sufficiency of food systems remains in any case a relevant criterion of performance, a criterion that is logically independent from economic considerations. (2) An element can be "buy food from its context" only when it generates enough surplus of added value to be able to do so. Very often, in poor rural areas of developing countries, the following three issues are deeply mixed: (1) minimization of risk and self-sufficiency in food production; (2) lack of purchasing power to guarantee food security, whenever the biophysical constraints prevent the system from accessing an adequate flow of food; and (3) insufficient generation of added value to increase the income of the family in terms of net disposable cash for nonfood expenditures.

As noted earlier, it is important to be able to characterize first of all what is the situation of the particular farming system considered (e.g., net producer/consumer of food, net producer/consumer of added value, a household with characteristics above or below the average values found in the country).

When discussing the minimum threshold of labor productivity, we found that the food systems tend to avoid operating with values of throughput too close to such a threshold. In fact, this would imply a very dangerous situation (possibility of collapses in case of perturbations). In this case, a food system that is not able to generate enough endosomatic energy input is forced to produce a very limited variety of crop outputs. When in trouble with the balancing of the dynamic budget, only those tasks (crops) that provide a maximum in throughput and return are amplified in the profile of investments over the set of potential tasks. This implies that other tasks (crops) with lower throughputs are neglected when selecting the final profile of investments of land and human work.

Unfortunately, as will be discussed in Chapter 11, this predicament (food systems operating too close to viability thresholds) is affecting more and more agroecosystems of both developing and developed countries. In developing countries the problem is generated by the increasing severity of biophysical constraints. That is, many subsistence societies can no longer achieve the biophysical viability of the autocatalytic loop: food ↔ human activity ↔ food. This is pushing the food systems of these societies toward a monotonous diet geared around a cereal (or a starchy root where possible). This key crop, whose production is amplified to boost efficiency, is the crop that according to boundary conditions, provides a maximum throughput both per hour and per hectare. Put another way, the existence of a strong external constraint affecting the feasibility of the metabolism of the social system is indicated by a very skewed distribution — at the lower level — of the profile of investments of both human activity and land in production over the set of potential tasks (in this case, over the set of cultivated crops). This has been suggested to be a general feature in the organizational pattern of human societies (Bailey, 1990). To boost the efficiency of the system, investments of labor, suitable land or technology tend to be concentrated only on the crop(s) providing the highest throughput both per hour of labor and per hectare of land in production.

A similar skewed distribution of investments over possible productive tasks in agriculture is found in the food system of developed countries. In this case, however, it is a constraint on economic viability that generates the set of relevant constraints. The food system of rich countries tends to focus only on crops, which are easy to mechanize (e.g., cereal) to boost labor productivity, and on animal products. Animal products have two major advantages compared with conventional crops: (1) they make it possible to reduce the space requirement of the production system through feed imports (externalization of the requirement of space to other systems — this avoids the constraint of space on the generation of flow of added value) and (2) they make it possible to better use the large investment of capital they require. In fact, contrary to what happens in crop production, the expensive equipment required for animal production is used every day for the entire year.

In conclusion, farmers in developed countries are forced, by the economic constraint on their viability, to focus only on those productions that maximize the flow of added value per hour of labor and that maximize the economic return of economic investments.

9.3.2.3 *Minimum Throughput of Flows of Added Value per Hour of Labor in the Agricultural Sector* — We can apply the same approach used so far to detect a minimum threshold of flow density related to added value, viewed as EV2. However, before getting into the discussion of how to assess a minimum threshold on the throughput of added value in different socio-economic entities (the economy of a country, an economic sector or subsector, a firm, a farm, or a household), one has to first answer the obvious question: In which sense can we speak of a flow of added value? To answer this question, I do not intend to get into a technical discussion of foundations of economic theory, but just to mention a few points about the method generally used to assess the gross domestic product of an economy (the mechanism of accounting for national income).

First, the analyst has to define a space–time domain (the system — an aggregate of interacting actors within a boundary defined in a given area and space). This already leads to a distinction between GDP (the accounting is referring to geographical location of the flows of added value) and GNP (the accounting is referring to the nationality of those getting revenues).

The assessment of GDP is based on the required congruence of three nonequivalent systems of accounting of the same system quality defined and assessed three times in nonequivalent ways. For the sake of simplicity, I do not include the effect of imports and exports in the description of the three methods given below:

1. GDP as the sum of the assessments of added value produced by the various sectors. That is, the assessment of GDP average for a society is

$$GDP_{AS} = \Sigma x_i \, GDP_i \qquad (9.20)$$

The values added generated in sectorial GDPs (GDP_i) are calculates as final values of goods and services produced by the sector minus cost. How to formalize such an assessment in specific situations, obviously, is an open question. But this problem is found with each one of the assessments discussed here. Equation 9.20 implies the ability to express the characteristics of level n (the whole) as a function of assessments referring to characteristics of elements defined at the level $n - 1$. After selecting a particular protocol for accounting, we will find out that not all the sectors of the economy produce added value at the same rate or produce added value at all (e.g., the household sector).

2. The flow of added value (GDP average for a society) can be assessed as the sum of the assessments referring to the revenues received by the various production factors of the economy.

$$GDP_{AS} = WFW \times HA_{PW} + \Sigma(PF_i \times RPF_i) + GS \times TFS \qquad (9.21)$$

GDP_{AS} is determined by three terms:
a. Wages [WFW (wages for work) × HA_{PW} (working time)]
b. Profits (including rent) for those owning capital and other production factors besides labor [PF_i (production factors) × RPF_i (return of production factors)]
c. Taxes for the government and other administrative entities, which can be expressed as [GS (governmental services) × TFS (taxes for services)]

Equation 9.21 can be used to explain once again that ELP (economic labor productivity) expressed as the ratio GDP/HA_{PW} has nothing to do with either the wages received by the worker or the economic return of labor as a factor of production. In fact, from Equation 9.21 it is clear that other elements enter into play in the determination of GDP besides HA_{PW}.

3. The flow of added value (GDP average for a society) can be assessed as the sum of the assessments referring to the expenditures of two different compartments: private sector and public sector. We can write

$$GDP_{AS} = [(CGS \times PGS) + (IGS \times CIGS)]_{PRIVATE} + [(CGS \times PGS) + (IGS \times CIGS)]_{PUBLIC} \qquad (9.22)$$

The expenditures of the two sectors, private and public, can be assessed using the following formula:

[CGS (consumption of goods and services) × PGS (price of goods and services)] and
[IGS (investment in goods and services) × CIGS (cost of the investment in goods and services)]

This represents the profile of investments within the black box of the available input of added value.

Practical aspects of the definitions of the various acronyms used in Equation 9.20 through Equation 9.22 are beside the point here. Rather, the relevant point is that the formalization of each one of these three assessments is anything but simple (even when asking the professional consultancy of expert economists). The problem, which has been already discussed over and over about impredicative loop analysis, is that the various acronyms written in the various equations can be formalized in different ways depending on a lot of arbitrary choices of the analyst. For example, this can start with very basic questions:

- What is value added? (Should we account for goods produced but not yet sold? How should we account for potential changes in the values of stocks of goods already available due to potential future changes in price?)
- How should we calculate revenues? (How should we deal with the variation of the value of fixed capital, which could move suddenly up or down? Depreciation is not only related to physical obsolescence, but also to functional obsolescence, which is much more difficult to deal with. A computer perfectly working can suddenly lose its value, because of the introduction on the market of a new model.)
- How should we distinguish between consumption and investment when having to choose between different systems of accounting? (Should an expensive wine chosen to impress a potential business partner be accounted for as a business investment? What about the private purchase of a car? A car in modern times is for leisure — or is it a necessary investment for getting a decent job?)

We have not even mentioned the problem of deciding how to deal with the simultaneous discounting of different forms of capital on different time horizons.

It should be obvious at this point that the mechanism used to define what should be considered an assessment of GDP is very similar to what has been proposed in the previous chapter for the biophysical accounting of societal metabolism (TET assessed in relation to BEP and SEH, which in turn are assessed in relation to the characteristics of lower-level elements). A lot of semantic definitions of system qualities, which are impossible to formalize, are brought in congruence by using mosaic effects and impredicative loop analysis (defining the same thing in nonequivalent ways, looking for different external referents determining assessments on nonreducible descriptive domains). Only in this way does it become possible to arrive at a coherent representation of these concepts through convergence on an arbitrary number that — on the other hand — must match the reciprocal constraints imposed by the set of definitions selected by the analyst.

Another important observation associated with this mechanism of accounting is related to the fact that assessments of GDP are related to the integrated process of producing and consuming flows of added value across compartments over the whole system. The overall assessment has to be congruent through the set of assessments performed at the level $n - 1$ and the whole level n. This means that if we define

$$\text{ELP}_{AS} = \text{GDP}_{AS}/\text{HA}_{PW} \tag{9.23}$$

then we obtain an average value of generation of added value per hour of human activity invested in working for the direct compartment (the average economic labor productivity for a given economy in which the direct compartment assessed in terms of investment of human activity is defined as paid work). The value of ELP_{AS} can be assumed to be a sort of cost opportunity of labor in that society. It should be recalled, in fact, that at the hierarchical level n of the whole society, working time is already a cost for the society, due to the societal overhead on human activity. In fact, ELP_{AS} represents the amount of dollars of GDP generated at the hierarchical level of the whole society (level n) in a defined year, per hour of work supply delivered in the direct compartment (paid work) at the level $n - 1$.

When dealing with economic mechanisms of regulation, things are in reality more complex since debts can be used to buffer the difference between requirement and supply of added value at a particular moment in time. However, let us remain here in a basic theoretical analysis. We can write, at level $n - 1$, Equation 9.23 j times, applied to j economic sectors (with $1 < i < j$):

$$\text{ELP}_i = \text{GDP}_i/\text{HA}_i \tag{9.24}$$

Combining Equation 9.23 and Equation 9.24, we obtain

$$\text{GDP}_{AS} = \text{HA}_{PW} \times \text{ELP}_{AS} = \Sigma \, (\text{HA}_i \times \text{ELP}_i) \tag{9.25}$$

When a particular economic sector (or activity) w has an average return of added value per unit of labor lower than the one achieved at the societal level ($ELP_w < ELP_{AS}$), then the hours of work supply allocated to that particular sector (activity) become a sort of economic cost for the society. In fact, in the presence of working time allocated in sectors with an ELP_i lower than the average of the society, to maintain the same societal average ELP_{AS}, it is necessary to allocate work in other economic sectors with a higher ELP_i ($ELP_h > ELP_{AS}$). In particular, the surplus of added value generated in these sectors (e.g., assume that all the other sectors are included in K) has to be equivalent to the deficit generated by the hours of work allocated in sector w:

$$\text{Cost opportunity of work in } w = HA_w \times (ELP_{AS} - ELP_w) \tag{9.26}$$

Equation 9.26 establishes a new form of constraint on the amount of working time that can be allocated to activities that have an average return of added value lower than the average return at the societal level ARL_{AS}. A society can afford to allocate hours of labor to these economic sectors (or activities) while remaining at its original level of ELP_{AS} only if the remaining hours of work supply allocated to other economic sectors (or activities) are able to generate enough surplus to pay for them. Recalling Equation 9.26, this means

$$HA_w \times (ELP_{AS} - ELP_w) \times HA_k \times (ELP_k - ELP_{AS}) \tag{9.27}$$

In any particular case, the higher the difference ($ELP_{AS} - ELP_w$), the higher will be the pressure at the level of society to reduce the investments of working time allocated at the societal level on activity w.

If we apply this rationale to the role that agriculture plays within a given country, we can write Equation 9.27 as

$$GDP_{AS} = HA_{PW} \times ELP_{AS} = (HA_{AG} \times ELP_{AG}) + (HA_{OS} \times ELP_{OS}) \tag{9.28}$$

where:
AS = average society
AG = agriculture
PW = paid work
OS = other sectors

9.3.2.4 *Minimum Throughput Dollars per Hectare of Land in the Agricultural*

Sector — The ability to generate a given amount of added value per hectare can be used to detect the existence of constraints on the aggregate value of added value, which can be associated with a given mix of land uses over a given land area. Depending on the yield (YLP_{AG}) expressed in biophysical variables (e.g., kilograms of crops produced per year and per hectare), we can associate with the various typologies of crop production a relative flow of added value. Such a flow will reflect the relative difference between revenues (determined by the price of the produced crop) and costs (determined by the expenditures associated with the production — e.g., cost due to the remuneration of production factors). Again, the procedure to obtain such an assessment is difficult to generalize. When looking at different farming systems around the world, one can find several strange definitions for costs and revenues to be considered when looking for this assessment. In any case, no matter how we decide to formalize this assessment, it will always be possible to find a constraint related to the availability of productive land, which affects the supply of added value. Such a constraint can be expressed as

$$GDP_{AG} = LIP \times YLP_{AG} \times AVP_{AG} \tag{9.29}$$

where:
LIP = land in production
YLP_{AG} = yield on the land in production
AVP_{AG} = the amount of added value associated with the agricultural production assessed in biophysical terms

A more elaborate discussion of the implications of this equation is provided in Chapter 11, when this constraint is explored in relation to the choices available to the farmer. Given a limited amount of land to be used in production (LIP), it is possible to increase the value of the yield (YLP_{AG}) by increasing the use of technical inputs. However, this implies sooner or later reaching a plateau in the average production of added value associated with this higher biophysical productivity (because of a higher fraction of production costs). This is a predicament that is well known and studied in agricultural economics. However, when analyzing this mechanism within an approach of integrated analysis, it is possible to complement the reading of this trade-off with a parallel analysis of other incommensurable trade-offs (e.g., minimization of risk for subsistence, material standard of living associated with the choice of production techniques, ecological impact of selected techniques).

9.4   Demographic Pressure and Bioeconomic Pressure

9.4.1   Introducing These Two Basic Concepts

9.4.1.1   *Demographic Pressure* — The concept of demographic pressure is traditionally related to the ratio population size/area occupied by the society. Such a parameter is an important factor affecting the choice of techniques of agricultural production (e.g., Boserup, 1981). Besides scientific analyses, common sense suggests that high demographic pressure in a society tends to cause the selection of farming techniques with a high yield of food per unit of area.

In relation to the intensity of such pressure we can calculate the following two parameters:

- AP_{DP} (agricultural productivity due to demographic pressure) — the level of productivity of land (yield of food energy per hectare) that would be required to obtain self-sufficiency (to match the aggregate food demand) given (1) current population and (2) current availability of land. Such an indicator is obtained by considering the food system under analysis as closed. Then the aggregate demand of food (which depends on population, current characteristics of the diet, postharvest losses) is divided by the amount of land used for generating food supply (which depends on population size, endowment of land, characteristics of the available land, existence of alternative land uses implied by socioeconomic organization).

- AP_{ha} (actual agricultural productivity of land) — the level of productivity of land (yield of food energy per hectare) actually achieved by a country. Such an indicator is obtained by dividing the assessment of aggregate internal production of food (which depends on the mix of cultivation and yields of different crops) by the assessment of land used in food production.

The goal of self-sufficiency would imply reducing the difference between these two parameters ($AP_{DP} - AP_{ha} \rightarrow 0$), even though this solution is very seldom reached by societal systems (at all levels — entire food systems, provinces, individual villages or households). In any case, an increase in demographic pressure tends to select a mix of productions and production techniques that increase the output per hectare of land in cultivation. The alternative is the expansion of food production on marginal land, by reducing the ratio TAL/LIP. This can imply the reduction of the fraction of area of terrestrial ecosystems not colonized by humans and alternative land uses to LIP.

9.4.1.1.1   *Indicators of the Level of Demographic Pressure on Agriculture*

To characterize this pressure, we can use three extensive variables 2 (exosomatic energy, added value and food) to characterize societal metabolism in relation to land area, in this case, assessment of local supply or requirement of added value (dollars per hectare) or food (kilograms per hectare) calculated at different levels (e.g., whole country, individual economic sector, subsectors and individual firms/farms). These indicators can be used to characterize the performance of agroecosystems in relation to other socioeconomic characteristics. An assessment of the exosomatic energy applied per hectare in agricultural production can be used to characterize the performance in relation to ecological impact.

9.4.1.2 Bioeconomic Pressure

9.4.1.2 Bioeconomic Pressure — In parallel with the demographic pressure, we can define bioeconomic pressure as determined by the ratio total human activity in a society (THA)/hours of human activity invested in work in agriculture (HA_{AG}). Since food demand is proportional to total human activity, whereas internal food supply is proportional to the amount of work in agriculture, the value of the fraction THT/HA_{AG} will affect the productivity of an hour of labor in the agricultural sector (Giampietro et al., 1997). That is, analogous with what is seen with demographic pressure, we can expect that high socioeconomic pressure tends to cause the selection of farming techniques that generate a large quantity of food produced per unit of labor delivered in the agricultural sector.

To calculate an indicator assessing the intensity of such pressure, let us first calculate the following two parameters:

- AP_{BEP} (agricultural productivity due to bioeconomic pressure) — the level of productivity of labor (yield of food energy per hour) that would be required to obtain self-sufficiency at the societal level (to match the aggregate food demand). Such an indicator is obtained by considering the food system under analysis as closed, by dividing (1) the aggregate demand of food (which depends on population, current characteristics of the diet, postharvest losses) by (2) the labor available within the country for generating food supply (which depends on population size, fraction of population that is in the working age, unemployment, workload for the working population, fraction of workforce absorbed by nonagricultural sectors). AP_{BEP} at the level of the whole country is the equivalent of BPL_{AG} described in Equation 9.6 and Equation 9.9. An economy such as that of the U.S., which allocates only 2% of its workers to agriculture, has a level of AP_{BEP} of at least 270 MJ of food energy per hour. Such an AP_{BEP} is well out of the range of productivity of farmers in all developing countries (e.g., lower than 4 MJ/h in China) and is not even reached by farmers in the European Union (EU) (lower than 100 MJ/h) (Table 9.3).

- AP_{hour} (agricultural productivity per hour of labor) — the level of productivity of labor (throughput of food energy per hour of work supply in the agricultural sector) actually achieved by a country. Such an indicator is obtained by dividing (1) the aggregate internal production of food (which depends on the mix of cultivation and yields of different crops) by the (2) amount of working time allocated in food production.

The goal of self-sufficiency implies reducing the difference between these two parameters ($AP_{BEP} - AP_{hour} \to 0$), even though this solution is very seldom reached by societal systems (at different levels — entire food systems, provinces, individual villages or households). In general, economic considerations are determinant in driving changes in techniques of production in agriculture. In particular, it is the difference between the economic labor productivity in agriculture and the economic labor productivity averaged over the various economic compartments that tends to select a mix of productions and production techniques that increase the output per hour of labor in agriculture. However, to boost economic labor productivity, it is often necessary to subsidize human labor with huge injections of fossil energy in the form of technical inputs.

9.4.1.2.1 Indicators of the Level of Bioeconomic Pressure on Agriculture

To characterize this pressure, we can use three extensive variables 2 (exosomatic energy, added value and food) to characterize societal metabolism in relation to human activity. In this case, assessments of local supply or requirement of added value (dollars per hour) or food (kilograms per hour) can be calculated at different levels (e.g., whole country, individual economic sector, subsectors and individual firms/farms). These indicators can be used to characterize the performance of agroecosystems in relation to other socioeconomic characteristics. An assessment of the exosomatic energy associated with an hour of labor in agricultural production can be used to characterize the level of capitalization of this sector.

TABLE 9.3

Parameters Defining Demographic and Bioeconomic Pressure in a Sample of 60 Countries (1991)

Country	GNP p.c. U.S.$[a]	% Labor in Agriculture[b]	% GNP from Agriculture[c]	ELP_{AS}[d]	THA/HA_w[e]	HA_{AG}/THA[f]	ha/Capita[g]	ha/Farm[h]
Algeria	1991	24.0	14.0	4.3	21.3	0.011	0.14	2.55
Argentina	3966	10.0	8.1	11.9	14.5	0.007	0.38	10.50
Australia	17,068	5.0	3.3	22.0	11.0	0.004	1.35	57.53
Bangladesh	205	69.0	36.8	0.4	16.3	0.041	0.04	0.19
Brazil	2920	24.0	10.0	4.6	13.8	0.016	0.16	1.85
Burkina Faso	290	84.0	44.0	0.4	11.4	0.073	0.19	0.44
Burundi	218	91.0	55.0	0.2	10.3	0.088	0.10	0.21
Cambodia	202	70.0	48.9	0.3	13.1	0.053	0.13	0.45
Cameroon	858	61.0	23.0	1.3	13.9	0.042	0.25	1.10
Canada	20,740	3.0	3.0	24.8	10.9	0.003	0.85	53.30
Central African Republic	407	63.0	41.0	0.5	11.2	0.053	0.31	1.09
Chad	212	75.0	43.0	0.4	14.8	0.048	0.28	1.12
China	364	67.0	28.4	0.5	8.6	0.076	0.04	0.10
Colombia	1254	27.0	16.1	2.7	16.6	0.015	0.06	0.68
Congo	1060	60.0	13.2	1.6	15.1	0.039	0.03	0.14
Costa Rica	1841	24.0	15.8	3.8	15.4	0.014	0.05	0.57
Ecuador	1010	30.0	13.4	2.2	15.9	0.018	0.08	0.81
Egypt	611	41.0	18.0	1.4	18.5	0.021	0.02	0.19
El Salvador	1084	37.0	11.2	1.8	12.0	0.029	0.05	0.46
Ethiopia	123	75.0	47.0	0.1	11.4	0.062	0.13	0.42
EU	17,393	6.3	3.0	24.9	11.9	0.005	0.10	3.90
Finlandia	24,089	8.0	6.0	0.6	15.6	0.004	0.25	6.27
Gambia	367	81.0	28.5	2.2	17.9	0.051	0.10	0.28
Guatemala	944	51.0	25.7	0.6	10.0	0.028	0.08	0.51
Honduras	587	55.0	20.0	1.1	15.6	0.034	0.16	0.91
India	330	66.0	31.0	0.5	13.6	0.048	0.10	0.38
Indonesia	592	48.0	21.4	1.1	12.7	0.036	0.04	0.23
Iran	2274	28.0	21.0	4.8	19.8	0.013	0.14	1.87
Jamaica	1446	27.0	5.0	1.5	9.4	0.027	0.03	0.24
Japan	26,824	6.0	2.5	36.6	10.2	0.005	0.02	0.52
Jordan	935	6.0	7.0	2.6	18.9	0.003	0.04	3.21
Kenya	350	77.0	27.0	0.4	11.8	0.064	0.04	0.13
Korea (South)	6227	25.0	9.0	10.1	11.6	0.019	0.02	0.21
Madagascar	207	77.0	33.0	0.3	13.4	0.056	0.11	0.32
Malawi	200	75.0	35.0	0.3	14.4	0.050	0.08	0.28
Mali	251	81.0	42.1	0.5	16.4	0.048	0.11	0.44
Mauritania	500	64.0	22.0	0.9	15.0	0.042	0.05	0.24
Mexico	2971	30.0	8.0	5.8	14.0	0.020	0.14	1.31
Morocco	1033	37.0	16.8	1.9	15.7	0.022	0.18	1.57
Mozambique	84	82.0	64.0	0.1	8.7	0.092	0.10	0.24
New Zealand	12,301	9.0	8.4	16.0	11.1	0.008	0.06	1.44
Nicaragua	283	37.0	31.1	0.6	15.7	0.022	0.15	1.39
Niger	303	87.0	34.8	0.3	11.3	0.076	0.23	0.52
Nigeria	305	65.0	37.0	0.4	12.0	0.053	0.14	0.55
Pakistan	383	50.0	26.0	0.8	17.3	0.028	0.09	0.58
Paraguay	1266	46.0	27.8	2.7	15.9	0.029	0.25	1.56
Philippines	728	47.0	22.1	1.3	13.9	0.033	0.04	0.25
Senegal	736	78.0	20.3	1.0	12.1	0.064	0.16	0.47
Sri Lanka	495	52.0	27.0	1.0	13.9	0.037	0.03	0.14
Sweden	25,254	4.0	3.0	29.1	10.3	0.003	0.16	8.51
Switzerland	33,850	4.0	3.6	46.6	11.4	0.003	0.03	1.45
Tanzania	95	81.0	61.0	0.1	10.5	0.075	0.05	0.14
Thailand	1697	64.0	12.7	2.3	9.7	0.064	0.16	0.45

TABLE 9.3 (Continued)

Parameters Defining Demographic and Bioeconomic Pressure in a Sample of 60 Countries (1991)

Country	GNP p.c. U.S.$[a]	% Labor in Agriculture[b]	% GNP from Agriculture[c]	ELP$_{AS}$[d]	THA/HA$_w$[e]	HA$_{AG}$/THA[f]	ha/Capita[g]	ha/Farm[h]
Tunisia	1504	24.0	18.0	3.1	15.9	0.013	0.18	2.24
Turkey	1793	48.0	18.0	4.2	12.3	0.037	0.22	1.07
U.S.	22,356	2.0	2.0	0.2	10.5	0.020	0.37	32.85
Uganda	163	81.0	51.0	30.2	10.7	0.075	0.14	0.39
Uruguay	2883	14.0	11.3	4.4	13.0	0.010	0.20	3.90
Venezuela	2728	11.0	5.4	4.7	14.5	0.007	0.08	2.15
Zimbabwe	641	68.0	20.0	0.8	13.0	0.051	0.13	0.50

[a] GNP per capita expressed in U.S. dollars (1991). From WRI (World Resources Institute) (1994), *World Resources 1994-95*. Oxford University Press, New York.

[b] Percentage of labor force in agriculture. From WRI, 1994.

[c] Percentage of GDP in agriculture. From WRI, 1994.

[d] ELP$_{AS}$, U.S. dollars/hour, economic labor productivity average for the society (obtained by dividing GDP by the amount of hours worked in that year, HA$_w$, which is the economically active population × 2000 h).

[e] THA/HA$_w$, ratio of total human activity/human activity allocated to working (THA = population × 8760; HA$_w$ as above). From WRI, 1994.

[f] HA$_{AG}$/THA, ratio of human activity in agriculture/total human activity (HA$_{AG}$ = HA$_w$ × percent of labor force in agriculture).

[g] Hectares of arable land per capita. From WRI, 1994.

[h] Hectares of arable land per farmer — total arable land divided by HA$_{AG}$/2000.

Source: Giampietro, M. (1997a), The link between resources, technology and standard of living: A theoretical model. In: L. Freese (Ed.), *Advances in Human Ecology*, Vol. 6. JAI Press, Greenwich, CT. pp. 73–128.

9.4.2 The Effect of the Quality of the Diet, Trade and Market

Several practical problems make it difficult to formalize (in terms of substantive numerical assessments) these basic concepts. For example:

1. A different quality of the diet due to a different mix of food products will determine different space and labor demands for the same amount of megajoules of food.

2. The same applies for a different profile of postharvest losses for different mixes of products.

3. The definition of working time in the food system is quite variable. In general terms, when considering a food system, producing food is only the first step of a long chain of activities. A lot of postharvest tasks are also required for storing, transporting, processing and preparing meals. Moreover, activities related to food security are difficult to quantify in terms of work demand and are often performed, at least in part, by children or elderly not included in the workforce.

4. The calculation of the area required for food production is also quite tricky (e.g., in the case of shifting cultivation, when an integrated use of the landscape implies that agricultural production is incorporated by other activities, such as hunting and gathering food from wild ecosystems).

All these difficulties have already been discussed when the concept of impredicative loop analysis was introduced. Definitions can be variable, but after agreeing on one particular definition (accepting the consequent approximations), we can write individual — definition-specific — equations of balance. For example, we can decide to check the existence of biophysical constraints only in relation to a particular food input about which it is easier to gather data (e.g., check only on the supply of the main staple food produced in the subsistence farming system). Alternatively, we can decide to focus only on a particular step in the food system (e.g., considering only agricultural production at the field level). Additional help is represented by the fact that when looking for benchmarking in the representation of the performance of an agroecosystem, when comparing the severity of socioeconomic constraints affecting farmers in developed countries with those affecting pure subsistence farming systems, differences are huge. Therefore, whatever approximation we use to calculate constraints on the intensity of through-puts, we will still obtain significant results for characterizing different systems of production. For example, for cereal production, the level of productivity of labor in subsistence societies is in the order of 1 or a few kg/h, whereas in developed countries it is in the order of hundreds of kilograms per hour.

Another objection to the idea of using assessments of minimum threshold for the value of the throughput is related to the roles of market and trade. That is, when a society is based on trade and market, the amount of food made available to society per hour of labor is no longer determined by the amount of food produced within the society. In fact, market and trade can change dramatically the requirement of congruence over the four-angle picture. In a modern country, rather than biophysical quantities of food commodities produced within the society, it is the amount of added value generated per hour of labor that, through the international price of commodities, defines the availability of food for internal consumption (assuming the existence of an adequate supply on the international market). Put another way, the possibility of importing food could reduce the relevance of the constraint of congruence over the dynamic budget imposed by food security.

However, when looking at data describing the performance of actual agricultural sectors, we find the opposite. Because of the effect of economic variables, the bioeconomic pressure becomes more relevant in pushing up the actual levels of productivity of labor in different countries. The bioeconomic pressure, in fact, acts on two hierarchical levels in parallel:

1. At the hierarchical level of the whole society — by reducing the amount of work allocated in those economic sectors that have a productivity of added value lower than ELP_{AS}. In fact, in this way the society can increase its economic performance.

2. At lower hierarchical levels (according to the perspective of a region, village or farmer living within such a society) — by the effort of individual lower-level elements (rural regions, rural villages, farmers) to achieve an income similar to the average income enjoyed at the level of society by other regions, villages and citizens. Agricultural entities (holons) are forced to keep up with the economic development of their context to maintain a standard of living comparable to that enjoyed by the rest of society.

9.4.3   How Useful Are These Two Concepts When Looking at Data?

Data in this section are taken from Giampietro (1997b) and Conforti and Giampietro (1997); they refer to the year 1990–1991 over a sample of 60 countries (see Table 9.3). The sample of 60 countries is representative of different combinations of demographic pressure, economic development and geographic location.

9.4.3.1   *Technological Development, GDP vs. Fraction of GDP from the Agricultural Sector, and Employment in Agriculture* — The analysis of this data set indicates three main points, which are well known within the field of traditional economic analyses:

1. When, in a country, the fraction of labor force in agriculture is high, the GDP per capita is low. That is, agricultural labor alone, without a significant support of technological inputs provided by a strong manufacturing sector, generates a flow of added value per hour of labor that is much lower than the one generated in developed countries. This point is well illustrated by Figure 9.12, showing that all the countries with a GDP per capita higher than $10,000/year have less than 7% of the working force in agriculture.

2. In developed countries (where GDP p.c. > $10,000/year),

    % of labor force in agriculture ≥ % of GDP from agriculture < 10%

    The higher the GDP p.c., the lower the percent of labor force in agriculture and GDP. This point is illustrated by Figure 9.12, which shows a general trend over the entire sample of 60 countries (covering the entire range of GNP p.c.).

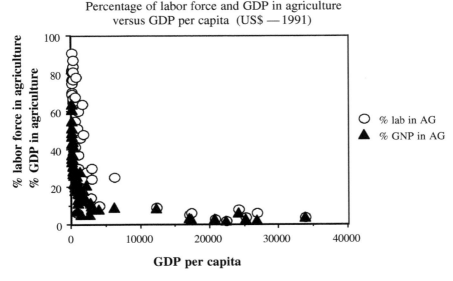

FIGURE 9.12 If the workforce is just producing its own food, the society will never become rich. (Giampietro, M. (1997b), Socioeconomic pressure, demographic pressure, environmental loading and technological changes in agriculture. *Agric. Ecosyst. Environ.* 65: 219–229.)

3. In poor developing countries (where GDP p.c. < $1000/year),

    % of labor force in agriculture > 20% > % of GDP from agriculture

    This point is illustrated by Figure 9.13, which presents in detail the situation of developing countries (it includes only the countries of the sample with a GDP p.c. below $5000/year). In this situation, developing countries would increase their GDP per capita if they were able to allocate a larger fraction of their work supply to nonagricultural activities.

These three points can be summarized into two general trends associated with technical and economic development of a country:

1. A continuous decrease of the fraction of labor force in agriculture, to arrive at a level below 5% (when the GNP p.c. is over $15,000)
2. A tendency to reduce percent of labor force and percent of GNP from agriculture toward single-digit percentage value

9.4.3.2 The Effect of Demographic and Bioeconomic Pressure

— Within modern societies trade plays a significant role in stabilizing the equilibrium between food demand and food supply. Therefore, as already discussed above, the two indicators AP_{DP} and AP_{BEP} (the density of biophysical throughput per hour and per hectare needed in agriculture to achieve self-sufficiency) do not indicate mandatory threshold values for the stability of a given food system. Rather, these are values toward which the system tends to operate. When $AP_{DP} > AP_{ha}$ or $AP_{BEP} > AP_{hour}$, society has to rely on imports to offset existing differences between internal demand and internal supply of food (on both a land basis and agricultural labor basis). This solution has three basic negative implications: (1) dependency on foreign countries for food security, (2) risk of economic shocks in the case of price fluctuations and (3) economic burden to guarantee the needed flow of imports. The degree of economic development and the size in population of a country will determine the importance of these factors. Obviously, there are also cases in which farming systems are operating in countries that are net food exporters (using previous indicators when $AP_{DP} < AP_{ha}$ or $AP_{BEP} < AP_{hour}$).

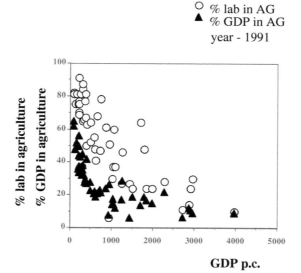

(countries of the sample with a GDP p.c. < 5,000 US$)

FIGURE 9.13 In developing countries, investments of human activity in agriculture contribute less to the GDP than investments in other sectors. (Giampietro, M. (1997b), Socioeconomic pressure, demographic pressure, environmental loading and technological changes in agriculture. *Agric. Ecosyst. Environ.* 65: 219–229.)

Data reported in Figure 9.14 and Figure 9.15 refer to a significant sample of world countries (the same considered in Figure 9.12 and Figure 9.13). From this comparison it is possible to see the existence of a direct link between (1) characteristics of a socioeconomic system (determining the values of AP_{DP} and AP_{BEP}) and (2) characteristics of techniques of agricultural production adopted in a defined society (determining the values of AP_{ha} and AP_{hour}). In particular:

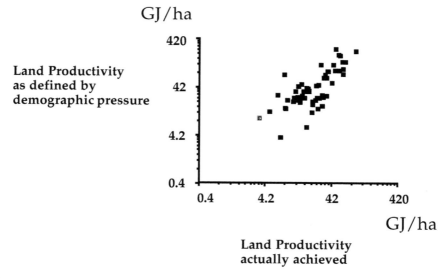

FIGURE 9.14 Demographic pressure: the higher the population density, the higher the productivity of agricultural land. (Giampietro, M. (1997b), Socioeconomic pressure, demographic pressure, environmental loading and technological changes in agriculture. *Agric. Ecosyst. Environ.* 65: 219–229.)

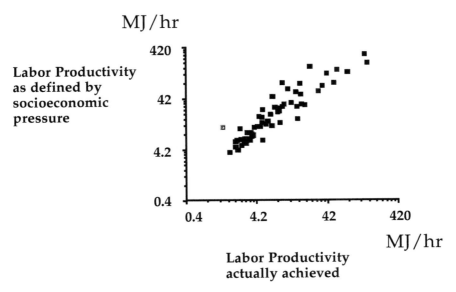

FIGURE 9.15 Bioeconomic pressure: the higher the societal overhead on human activity in agriculture, the higher must be the productivity of labor. (Giampietro, M. (1997b), Socioeconomic pressure, demographic pressure, environmental loading and technological changes in agriculture. *Agric. Ecosyst. Environ.* 65: 219–229.)

- The higher the value of AP_{DP} in a country, the more probable it becomes to also find a higher AP_{ha} (throughput of food energy produced by the agricultural sector per hectare) — Figure 9.14. That is, current technological performance in terms of yield per hectare is affected by existing demographic pressure.
- The higher the value of AP_{BEP} in a country, the more probable it becomes to find a higher AP_{hour} (throughput of food energy produced by the agricultural sector per hour of work delivered) — Figure 9.15. That is, current technological performance in terms of productivity of labor is affected by existing socioeconomic pressure.

For the poorest countries of the sample, where the openness of the food system is minor and the aversion to risk higher, the constraint given by the goal of self-sufficiency in both land and labor terms is obviously stricter (that is, $AP_{ha} \leftrightarrow AP_{DP}$ and $AP_{hour} \leftrightarrow AP_{SEP}$).

Developed countries, however, can heavily rely on fossil energy to boost internal supply of agricultural products both per hectare and per hour. Where an adequate amount of arable land per farmer is available, injection of machinery can boost the value of AP_{hour} well above the value of AP_{BEP} (e.g., U.S., Canada, Australia).

However, it should be observed that the two levels of productivity (per hectare of arable land and per hour of labor) are not independent. As Hayami and Ruttan (1985) show, we can write the productivity of labor as a function of yield per hectare:

$$BPL_{AG} = \text{output (kg)/hour} = \text{output (kg)/ha} \times \text{ha managed/hour} \qquad (9.30)$$

Therefore, it is not possible to achieve a very high level of kilograms per hour (average referring to the whole year) when producing at a very low level of kilograms per hectare (e.g., 1 ton/ha). Two main problems would occur in this case: (1) land required to achieve targeted labor productivity could exceed the endowment of hectares available per farmer and (2) too much technical investment would be required to harvest an increasing area with a limited amount of labor. The same applies to the case in which it is the second term (hectares available to be managed) that becomes limiting. For example, when the value of AP_{BEP} is in the order of hundreds of kilograms of grain per hour and the arable land per farmer is in the order of 1 ha (e.g., Japan), there is no feasible technical solution to match such a challenge. At that point, a limiting constraint on land entails two consequences: (1) crowded and developed countries have to depend on imports for their food security (if the selected diet entails a high value for QDM&PHL) and (2) their farmers have to depend on subsidies to get an income similar to the average income typical of other economic sectors.

This brings us to the third indicator of pressure, ELP_{AS}, the amount of added value related to a particular agricultural production, which depends on revenues (on yield and outputs' price) and the structure of costs at the farm level (depending on inputs' cost). All these factors are heavily influenced by international agreements, governmental policies and regulations. At a particular moment in time, given the set of current boundary conditions (prices on the international market, cost of inputs, existing laws, rules and policies affecting the market), a given farmer is affected by pressure to achieve a certain economic return, determined by the characteristics of the society to which she or he belongs. In turn, a set of biophysical constraints is also affecting a particular farmer operating at a defined point in time and space (shortage of land or water, climatic conditions). Given all this information, to break even, a farmer should be able to produce a determined crop pattern at determined levels of throughput per hour of labor and per hectare. Translating economic variables into biophysical variables and vice versa (establishing a relation between BPL and ELP) is especially useful when comparing farming systems that are operating under widely different levels of demographic and bioeconomic pressure. In this way, by an integrated use of these numerical indicators, we can characterize the influence of different socioeconomic contexts on technical and economic performance of farming. For example, Japanese farmers are operating under high bioeconomic pressure ($THT/WS_{AG} = 0.005$; Table 9.3) and very high demographic pressure (about 0.04 ha of arable land per capita; Table 9.3). In terms of added value, Japan was characterized by an ELP_{AS} of about \$37/h in 1991 (Table 9.3). According to the range of technical coefficients achievable in agriculture, the task for Japanese farmers can become impossible (very small endowment of land —less than 1 ha per farmer (Table 9.3) — and the need to reach levels of labor productivity of hundreds of kilograms of rice per hour). The characteristics of their socioeconomic context place them at a disadvantage when competing with both:

1. Farmers operating under much lower bioeconomic pressure — e.g., Chinese rice producers with a negligible opportunity cost of labor (a ELP_{AS} of about \$0.5/h in 1991; Table 9.3). Chinese farmers can afford to produce rice at a level of productivity of labor in the order of tens of kilograms per hour and still be economically viable.

2. Farmers operating under much lower demographic pressure — e.g., U.S. farmers can boost their labor productivity by using an endowment of arable land that is more than 30 ha per farmer (Table 9.3).

The integrated use of these indicators makes it possible to link economic analyses to the effect of demographic changes. According to actual trends, we can expect that a general economic development of the various countries of our planet will determine a continuous increase in both demographic and bioeconomic pressure in the next decades. What consequences can we expect for the environment from this trend? This is the subject of the next chapter.

References

ACC/SCN, (1993), *Second Report on the World Nutritional Situation*, FAO, Rome.

Bailey, K.D., (1990), *Social Entropy Theory*, State University of New York Press, Albany.

Boserup, E., (1981), *Population and Technological Change*, University of Chicago Press, Chicago.

Conforti, P. and Giampietro, M., (1997), Fossil energy use in agriculture: an international comparison, *Agric. Ecosyst. Environ.*, 65, 231–243.

FAO (Food and Agriculture Organization), (1995), *Production Yearbook 1994*, FAO Statistic Series 125, FAO, Rome.

Fischer-Kowalski, M. and Haberl, H., (1993), Metabolism and colonization: modes of production and the physical exchange between societies and nature, *Innovation Soc. Sci. Res.*, 6, 415–442.

Giampietro, M., (1997), The link between resources, technology and standard of living: a theoretical model, in *Advances in Human Ecology*, Vol. 6, Freese, L., Ed., JAI Press, Greenwich, CT, pp. 73–128.

Giampietro, M., (1998), Energy budget and demographic changes in socioeconomic systems, in *Life Science Dimensions of Ecological Economics and Sustainable Use*, Ganslösser, U. and O'Connor, M., Eds., Filander Verlag, Fürth, Germany, pp. 327–354.

Georgescu-Roegen, N. (1971), *The Entropy Law and the Economic Process*. Harvard University Press, Cambridge, MA.

Giampietro, M., Bukkens, S.G.F., and Pimentel, D., (1993), Labor productivity: a biophysical definition and assessment, *Hum. Ecol.*, 21, 229–260.

Giampietro, M., Bukkens S.G.F., and Pimentel, D., (1997), Linking technology, natural resources, and the socioeconomic structure of human society: examples and applications, in *Advances in Human Ecology*, Vol. 6, Freese, L., Ed., JAI Press, Greenwich, CT, pp. 131–200.

Giampietro, M. and Mayumi, K., (2000), Multiple-scale integrated assessment of societal metabolism: introducing the approach, *Popul. Environ.*, 2, 109–153.

Giampietro, M., Mayumi, K., and Bukkens, S.G.F., (2001), Multiple-scale integrated assessment of societal metabolism: an analytical tool to study development and sustainability, *Environ. Dev. Sustain.*, 3, 275–307.

Haberl, H. and Schandl, H., (1999), Indicators of sustainable land use: concepts for the analysis of society-nature interrelations and implications for sustainable development, *Environ. Manage. Health*, 10, 177–190.

Hayami, Y., and Ruttan, V. (1985), *Agricultural Development: An International Perspective*, 2nd ed. Johns Hopkins Univ. Press, Baltimore.

James, W.P.T. and Schofield, E.C., (1990), *Human Energy Requirement*, Oxford University Press, Oxford.

Pastore, G., Giampietro, M., and Mayumi, K., (2000), Societal metabolism and multiple-scales integrated assessment: empirical validation and examples of application, *Popul. Environ.*, 22, 211–254.

United Nations. (1995), Statistical Yearbook. 1993. U.N. Department for Economic and Social Information and Policy Analysis; Statistical Division, New York.

World Bank, (1995a), *Social Indicators of Development 1995*, Johns Hopkins University Press, Baltimore.

World Bank, (1995b), *World Tables 1995*, Johns Hopkins University Press, Baltimore.

WRI (World Resources Institute) (1994), *World Resources 1994-95*. Oxford University Press, New York.

10

Multi-Scale Integrated Analysis of Agroecosystems: Technological Changes and Ecological Compatibility

According to the analysis presented in the previous chapter, a general increase of both the demographic and bioeconomic pressure on our planet is the main driver of intensification of agricultural production at the farming system level. In turn, a dramatic intensification of agricultural production can be associated with a stronger interference on the natural mechanisms of regulation of terrestrial ecosystems — that is, to a reduced ecological compatibility of the relative techniques of agricultural production. To deal with this problem, it is important to first understand the mechanisms through which changes in the socioeconomic structure are translated into a larger interference on terrestrial ecosystems. This is the topic of this chapter. Section 10.1 studies the interface socioeconomic context–farming system. At the farm level, in fact, the selection of production techniques is affected by the typology of boundary conditions faced by the farm. In particular, this section focuses on the different mix of technical inputs adopted when operating in different typologies of socioeconomic context. Section 10.2 deals with the nature of the interference on terrestrial ecosystems associated with agricultural production. A few concepts introduced in Part 2 are used to discuss the possible development of indicators. The interference generated by agriculture can be studied by looking at the intensity of the throughput of appropriated biomass per unit of land area. Changing the metabolic rate of a holarchic system (such as a terrestrial ecosystem) requires (1) a readjustment of the relative size of its interacting parts, (2) a redefinition of the relation among interacting parts and (3) changing the degree of internal congruence between produced and consumed flows associated with its metabolism. When the external interference is too large, we can expect a total collapse of the original system of controls used to guarantee the original identity of the ecosystem. Finally, Section 10.3 looks at the big picture presenting an analysis, at the world level, of food production. This analysis explicitly addresses the effect of the double conversion associated with animal products (plants produced to feed animals). After examining technical coefficients and the use of technical inputs related to existing patterns of consumption in developed and developing countries, the analysis discusses the implications for the future in terms of expected requirement of land and labor for agricultural production.

10.1 Studying the Interface Socioeconomic Systems–Farming Systems: The Relation between Throughput Intensities

10.1.1 Introduction

After agreeing that technological choices in agriculture are affected by (1) characteristics of the socioeconomic system to which the farming system belongs, (2) characteristics of the ecosystem managed for agricultural production and (3) farmers' feelings and aspirations, it is important to develop models of integrated analysis that can be used to establish bridges among these three different perspectives. This requires defining in nonequivalent ways the performance of an agroecosystem in relation to (1) socioeconomic processes, (2) ecological processes and (3) livelihood of households making up a given farming system.

The link between economic growth and the increases in the intensity of the throughput per hour of labor and per hectare at the societal level (due to increasing bioeconomic and demographic pressure) has been explored in Chapter 9. That is, that chapter addresses the link related to the first point of the previous list. This chapter explores the implications of the trend of intensification of agricultural production in relation to ecological compatibility — it addresses the link implied by the second point of the list. An integrated analysis reflecting the perspective of farmers seen as agents in relation to the handling of these contrasting pressures at the farming system level — the link implied by the third point of the list — is proposed in Chapter 11.

The need to preserve the integrity of ecological systems — the ecological dimension of sustainability — in effect can be seen as an alternative pressure coming from the outside of human systems, which is contrasting the joint effect of demographic and bioeconomic pressure, a pressure for growth coming from the inside. That is, whereas human aspirations for a better quality of life and freedom of reproduction push for increasing the intensity of the throughputs within the agricultural sector, a more holistic view of the process of co-evolution of humans with their natural context provides an opposite view, pushing for keeping as low as possible the intensity of throughput of flows controlled by humans within agroecosystems. As noted in Part 1, the sustainability predicament is generated by the fact that these two contrasting pressures are operating at different hierarchical levels, on different scales, and this makes it difficult to interlock the relative mechanisms of control.

At the level of individual farms, at the level of villages, at the level of rural areas, at the level of whole countries and at the supranational level, different rules, habits, allocating processes, laws and cultural values are operating for enforcing the two views. However, an overall tuning of this complex system of contrasting goals is anything but easy — especially when considering that humankind is living in a fast period of transition, which implies the existence of huge gradients among socioeconomic systems (very rich and very poor) operating on different points of the evolutionary trajectory.

This implies that human agents at different levels, at the moment of technological choices, must decide the acceptability of compromises (at the local, medium or large scale) in relation to the contrasting implications of these two pressures. This chapter obviously does not claim to be able to solve this Yin-Yang predicament. Rather, the goal is to show that it is possible to use the pace of the agricultural throughput to establish a bridge between the perception and representation of benefits and constraints coming from the societal context (when using the throughput per hour of labor) and the perception and representation of benefits and constraints referring to the ecological context (when using the throughput per hectare) of a farm.

To make informed choices, it is important to have a good understanding of the mechanisms linking the two types of pressures: (1) the internal asking for a higher level of dissipation and therefore for an expansion into the context and (2) the external reminding that a larger level of dissipations entails higher stress on boundary conditions and therefore a shorter life expectancy for the existing identity of the socioeconomic system generating the ecological stress. The debate over sustainability, in reality, means discussing the implications of human choices when looking for compromise solutions between these two pressures.

The analysis described in Section 9.4 (Figure 9.12 through Figure 9.15) indicates the existence of a clear link between the values taken by:

1. Relevant characteristics of the food system defined at the hierarchical level of society (using the two IV3: AP_{DP} and AP_{BEP}), which can be characterized by a set of variables such as gross national product (GNP) and density of produced flow, which can be related to other relevant system qualities such as age structure, life span of citizens, profile of labor distribution over economic sectors, and workload (as discussed in Chapter 9). These variables refer to the societal system seen as a whole, without any reference to the farming system level.

2. Relevant characteristics of the food system defined at the hierarchical level of the farming system (using the two IV3: AP_{ha} and AP_{hour}), which are determined by a set of biophysical constraints such as technical coefficients, technical inputs and climatic conditions, and location-specific socioeconomic constraints such as local prices and costs and local laws. These characteristics, for example, refer to the horizon seen by farmers when making their living. The variables used to represent these system qualities are well known to the agronomist, agricultural

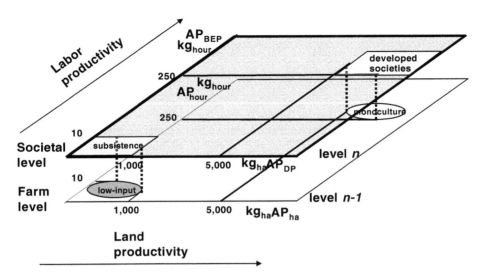

FIGURE 10.1 The link between two assessments based on two definitions of IV3. AP_{hour} and AP_{ha} at level $n - 1$ and AP_{BEP} – AP_{DP} at level n. (Giampietro, M., (1997a), Socioeconomic pressure, demographic pressure, environmental loading and technological changes in agriculture, *Agric. Ecosyst. Environ.*, 65, 219–229.)

economists and agroecologists (technical coefficients, economic parameters characterizing the economic performance of the farm, local indicators of environmental stress).

This link among two different hierarchical levels — society as a whole (level n) and individual farming system (level $n - 1$) — can be visualized by using a plane describing the agricultural throughput according to two IV3: (1) agricultural throughput per hectare (when using human activity as EV2) and (2) agricultural throughput per hour (when using land area as EV2). In this way, the technical performance of a farming system can be described in parallel on two levels (Figure 10.1):

- On the level n, society as a whole, by considering values of AP_{DP} and AP_{BEP} (which are two types of IV3$_n$) assessed by using societal characteristics. These values must be compatible with the constraints coming from the socioeconomic structure associated with the particular typology of societal metabolism.
- On the level $n - 1$, individual farming system, by considering values of AP_{ha} and AP_{hour} (which are two types of IV3$_{n-1}$). These values must be feasible according to local economic and biophysical constraints and available technology.

In this way, the characteristics of an agricultural throughput can be seen as determined by (1) the set of constraints coming from the context (societal level) and (2) the set of constraints operating at the farming system level.

On the upper plan of Figure 10.1 (with the axes x and y represented by values of AP_{DP} and AP_{BEP}, respectively) it is possible to define areas of feasibility for agricultural throughputs according to socioeconomic characteristics. As noted earlier, developed countries require agricultural throughputs above 5000 kg of grain per hectare and above 250 kg of grain per hour of labor, when talking of cereal cultivation. On the lower plan (with the axes x and y represented by values of $AP_{ha} \leftrightarrow kg_{ha}$ and $AP_{hour} \leftrightarrow kg_{hour}$, respectively) it is possible to define areas of feasibility for agricultural throughputs according to farm-level constraints and characteristics of techniques of production. For example, subsistence societies that do not have access to technical inputs cannot achieve land and labor productivity higher than 1000 kg of grain per hectare and 10 kg of grain per hour (clearly, these values are general indications and are not always applicable to special cases — e.g., delta of rivers). As noted earlier, we can expect that farming systems belonging to a particular socioeconomic system tend to adopt techniques of production described by a combination of values of AP_{ha} and AP_{hour} that keep them as much as possible close to the area determined by socioeconomic constraints.

In conclusion, when describing technological development in agriculture on a plane $AP_{DP} - AP_{BEP}$, we can expect that:

- Farming systems operating within different socioeconomic contexts (in societies described by different combinations of $AP_{DP} - AP_{BEP}$) tend to operate in range of land and labor productivity ($AP_{ha} - AP_{hour}$) close to the values defined by socioeconomic constraints. As noted in Chapter 9, whenever a biophysical constraint on land imposes an $AP_{hour} \ll AP_{BEP}$ (in developed countries), imports (market and trade) must be available to cover the difference. Getting into an economic reading, in a situation in which $ELP_{AG} \ll ELP_{PW}$, farmers require protection from international competition and direct subsidies, to keep a level of income similar to that achieved by workers making a living in other economic sectors. This requires the availability of financial resources (surplus of added value), at the country level, which can be allocated to subsidize the agricultural sector.

- Changes in demographic and socioeconomic pressure ($AP_{DP} - AP_{BEP}$) will be reflected, sooner or later, in changes of technical coefficients of farming techniques ($AP_{ha} - AP_{hour}$). As soon as economic growth (parallel increase in GNP per capita (p.c.) and population size) translates into a parallel increase of demographic and socioeconomic pressure, technical progress is coupled to changes in socioeconomic characteristics that require techniques of agricultural production characterized by high values of AP_{hour}. The same link between economic development and increases in labor productivity in agriculture is found when adopting a more conventional economic reading of technological development of agriculture (Hayami and Ruttan, 1985).

According to this integrated analysis, we should be able to represent general trends in the evolution of food production techniques for different types of socioeconomic systems on the two-dimensional plane (made using IV3): productivity of land (kilograms per hectare) and productivity of labor (kilograms per hour), as illustrated in Figure 10.2. For the sake of simplicity, the plane describes productivity of land and labor mapped in terms of kilograms of grain. Four main types of socioeconomic systems, having different combinations of demographic and bioeconomic pressure, are represented there:

1. Socioeconomic systems with low demographic and low bioeconomic pressure. This situation is characterized by more than 0.5 ha of arable land per capita (this value depends on available productive land and population size) and less than $1000 per year of GNP per capita (depending on economic performance). This type of socioeconomic system includes several African countries, such as Burundi.

2. Socioeconomic systems with low demographic and high bioeconomic pressure. This situation is characterized by more than 0.5 ha of productive land per capita and more than $10,000 per year of GNP per capita. This type of socioeconomic system includes countries such as the U.S., Canada, and Australia.

3. Socioeconomic systems with high demographic and low bioeconomic pressure. This situation is characterized by less than 0.2 ha of arable land per capita and less than $1000 per year of GNP per capita. This type of socioeconomic system includes countries such as China and Egypt.

4. Socioeconomic systems with high demographic and high bioeconomic pressure. This situation is characterized by less than 0.2 ha of arable land per capita and more than $10,000 per year of GNP per capita. This type of socioeconomic system includes several countries of the European Union and Japan, among others.

According to existing trends in population growth and economic development for these four different types of socioeconomic systems, we can expect the following movements in the plane (see Figure 10.2):

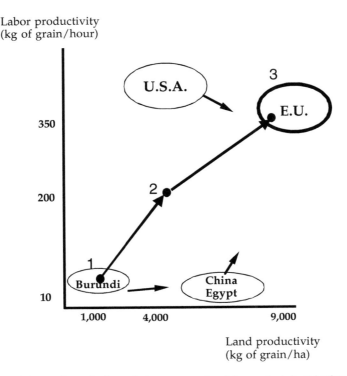

FIGURE 10.2 Trends and changes in production techniques over a plane labor productivity × land productivity. (Giampietro, M., (1997a), Socioeconomic pressure, demographic pressure, environmental loading and technological changes in agriculture, *Agric. Ecosyst. Environ.*, 65, 219–229.)

1. Societies with low demographic and bioeconomic pressure (e.g., some African countries): The population is growing faster than the GNP per capita, which means that AP_{DP} will grow faster than AP_{BEP}. Hence, they will move toward a situation typical of China.

2. Societies with low demographic and high bioeconomic pressure (e.g., Canada, U.S.): Economic development is expected to be maintained (GNP per capita will remain high) and population growth will be relatively slow but steady (medium or low internal fertility but high immigration rate). On the plane, this means a slow movement toward higher values of AP_{DP}.

3. Societies with high demographic and low bioeconomic pressure (e.g., China): These societies look for a quick economic growth (increasing GNP per capita) and they are expected to maintain if not expand their already huge population size. At a national level, an increasing GNP per capita will result in an accelerated absorption of the labor force currently engaged in agriculture (e.g., 60% in China at present) by other sectors (primary and service sectors) of the economy. This will inevitably require a dramatic increase in agricultural labor productivity (AP_{hour}) to maintain food security. Hence, a movement toward the West European conditions of agricultural production is to be expected.

4. Societies with high demographic and bioeconomic pressure (e.g., The Netherlands, Japan): These societies have no alternative but to try to maintain a high material standard of living and keep population growth to a minimum. This means a more or less stable and high level of AP_{BEP} and a very slowly increasing value of AP_{DP} (mainly due to the strong pressure of immigrants). For these societies, trying to reduce the environmental impact of their food production becomes a major factor.

Note that food imports from the international market, a must for countries where biophysical or economic constraints determine a value of $AP_{DP} > AP_{ha}$ or $AP_{BEP} > AP_{hour}$, are based on the existence of surpluses produced by countries where the relation between these parameters is inverse. Countries producing big surpluses in relation to both types of pressures are scarce. In 1992, the U.S., Canada, Australia and Argentina combined produced over 80% of the net export of cereal on the world market (WRI, 1994), but at their present rate of population growth (including immigration) and because of an increasing concern for the environment (policies for setting aside and developing low-input agriculture), this surplus might be eroded in the near future. For instance, the U.S. is expected to double its population in 60 years (USBC, 1994). However, the situation is aggravated when including in this analysis the legitimate criteria of respect of ecological integrity. This criterion is already leading to a push for a less intensive agriculture all over the developed world (slowdown, at the farming level, of the rate of increase in AP_{ha}). This combined effect could play against the production of food surpluses in those countries that could do so. In conclusion, at the world level, demographic and bioeconomic pressures are certainly expected to increase, forcing the countries most affected by those two pressures to rely on imports for their food security.

It is often overlooked that at the world level, there is no option to import food from elsewhere. When increases in demographic and bioeconomic pressure are not matched by an adequate increase in productivity of land and labor in agriculture, food imports of the rich will be based on starvation of the poor. This simple observation points at the unavoidable question of the severity of biophysical constraints affecting the future of food security for humankind. How do these trends fit the sustainability of food production at the global level?

The rest of this section focuses on the changes in techniques of production (in particular in the pattern of use of technical inputs) that can be associated with changes in demographic and bioeconomic pressure as perceived and represented from the lower level (changes in techniques of production at the farm level). Section 10.2 deals with the relation between changes in techniques of production associated with changes in demographic and bioeconomic pressure as perceived and represented from the higher level — the aggregate effect that these changes have on the integrity of terrestrial ecosystems. This is where the ecological dimension of sustainability becomes crystal clear. Agricultural production, in fact, depends on the stability of boundary conditions for the productivity of agroecosystems. Finally, Section 10.3 explores the relation between qualitative changes in the diet — the implications of increasing the fraction of animal products and fresh vegetables and fruits (changes in the factor QDM (quality of diet multiplier)) — and changes in the profile of use of technical inputs in perspective and at the world level. Increasing the amount of animal products in the diet requires a double conversion of food energy (energy input to crops and crops to animals). In the same way, increasing the amount of fresh vegetables in the diet requires a mix of crop production associated with a much higher investment of human labor per unit of food energy produced and a reduced supply of food energy per hectare. Both changes (typical of the diet of developed countries) represent an additional boost to the problems associated with higher demographic and bioeconomic pressure.

10.1.2 Technical Progress in Agriculture and Changes in the Use of Technical Inputs

The classic analysis of Hayami and Ruttan (1985) indicates that two forces are driving technological development of agriculture:

1. The need to continuously increase the productivity of labor of farmers; this is related to the need of:
   a. Increasing income and standard of living of farmers
   b. At the societal level, making more labor available for the development of other economic sectors during the process of industrialization.
2. The need to continuously increase the productivity of agricultural land. This is related to the growing of population size, which requires guaranteeing an adequate coverage of internal food supply using a shrinking amount of agricultural land per capita.

It is important to understand the mechanism through which bioeconomic and demographic pressure push for a higher use of technical inputs in agriculture. In fact, the effect of these forces is not the same in developed and developing countries. In developed countries, the increasing use of fossil energy had mainly the goal of boosting labor productivity in agriculture to enable the process of industrialization. This made possible a massive move of the workforce into industrial sectors, increasing at the same time the income of farmers. For example, in West Europe the percentage of the active population employed in agriculture fell from 75% before the industrial revolution (around the year 1750) to less than 5% today. In the U.S., this figure fell from 80% around the year 1800 to only 2% today. As observed in Figure 9.12, none of the countries considered in that study with a GNP per capita higher than $10,000 per year has a percent of workforce in agriculture above the 5% mark. In the same way, none of the countries with more than 65% of the workforce in agriculture has a GNP per capita higher than $1000 per year of GNP. The supply of human activity allocated in work (HA_{PS}) is barely capable of producing the food consumed by society; there is no room left for the development of other activities of production and consumption of goods and services not related to food security.

In developing countries, the growing use of fossil energy has been, up to now, mainly related to the need to prevent starvation (just producing the required food) rather than to increase the standard of living of farmers and others. Concluding his analysis of the link between population growth and the supply of nitrogen fertilizers, Smil (1991, p. 593) beautifully makes this point:

> The image is counterintuitive but true: survival of peasants in the rice fields of Hunan or Guadong — with their timeless clod-breaking hoes, docile buffaloes, and rice-cutting sickles — is now much more dependent on fossil fuels and modern chemical syntheses than the physical well-being of American city dwellers sustained by Iowa and Nebraska farmers cultivating sprawling grain fields with giant tractors. These farmers inject ammonia into soil to maximize operating profits and to grow enough feed for extraordinarily meaty diets; but half of all peasants in Southern China are alive because of the urea cast or ladled onto tiny fields — and very few of their children could be born and survive without spreading more of it in the years and decades ahead.

The profile of use of technical inputs can be traced more or less directly to these two different goals. Machinery and fuels are basically used to boost labor productivity, whereas fertilizers and irrigation are more directly related to the need to boost land productivity.

The data presented in this section are taken from a study of Giampietro et al. (1999). Twenty countries were included in the sample to represent different combinations of socioeconomic development (as measured by GNP) and availability of arable land (population density). Developed countries with low population density are represented by the U.S., Canada and Australia. Developed countries with high population density include France (net food exporter), the Netherlands, Italy, Germany, the U.K. and Japan (net food importers). Countries with an intermediate GNP include Argentina (with abundant arable land), Mexico and Costa Rica. Countries with a low GNP and little arable land per capita include the People's Republic of China, Bangladesh, India and Egypt. Other countries with low GNP include Uganda, Zimbabwe (net food exporters), Burundi and Ghana. The data on input use refer to the years 1989 and 1990. Technical details can be found in that paper.

The relation between the use of irrigation and the amount of available arable land per capita over this sample of world countries is shown in Figure 10.3. The upper graph clearly indicates that the different intensities in the use of this input reflect differences in demographic pressure (the curve is smooth in the upper graph — Figure 10.3a) more than differences in economic development. If we use the same data of irrigation use vs. an indicator of bioeconomic pressure (e.g., GNP p.c.), we find that crowded countries, either developed or developing, tend to use more irrigation than less crowded countries, with very little relevance of gradients of GNP (Figure 10.3b).

It is remarkable that exactly the same pattern is found when considering the use of synthetic fertilization over the same sample of countries (Figure 10.4). The upper and lower graphs of Figure 10.4 are analogous to those presented in Figure 10.3 for irrigation. The only difference is that they are obtained with data referring to nitrogen fertilizer. The similarity between the two sets of figures (Figure 10.4a and Figure 10.3a vs. Figure 10.4b and Figure 10.3b) is self-explanatory. Demographic pressure seems to be the main driver of the use of nitrogen and irrigation.

Completely different is the picture for another class of technical inputs: machinery (tractors and harvesters in the Food and Agriculture Organization (FAO) database used in the study of Giampietro et

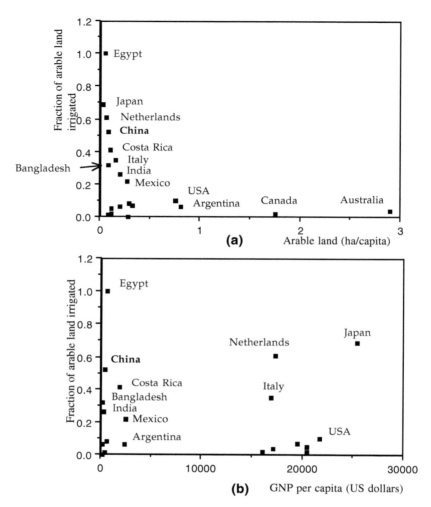

FIGURE 10.3 (a) Irrigation and demographic pressure. (b) Irrigation and bioeconomic pressure. (Giampietro, M., Bukkens, S., Pimentel, D., (1999), General trends of technological change in agriculture, *Crit. Rev. Plant Sci.*, 18, 261–282.)

al. (1999)). The two graphs in Figure 10.5 indicate that machinery for the moment is basically an option of developed countries (Figure 10.5b). Within developed countries, huge investments in machinery can be associated with large availability of land in production. This is perfectly consistent with what is discussed in Chapter 9. To reach a huge productivity of labor, at a given level of yields, it is necessary to increase the amount of hectares managed per worker. This requires both plenty of land in production and an adequate amount of exosomatic devices (technical capital) to boost human ability to manage large amounts of cropped land per worker. This rationale is confirmed by the set of data represented in the graph of Figure 10.6a. Over the sample considered in the analysis of Giampietro et al. (1999), the highest levels of labor productivity are found in the agricultures that have available the largest endowment of land in production per worker.

Finally, it should be noted that there is a difference between agricultural land per capita (land in production divided by population) and agricultural land per farmer (land in production divided by workers in agriculture). In fact, a reduction of the workforce in agriculture (e.g., by reducing the fraction of the workforce in agriculture from 80 to 2%) can increase the amount of land per farmer at a given level of demographic pressure. However, this reduction of the workforce in agriculture can only have a limited effect in expanding the land in production per farmer. An economically active population is only half

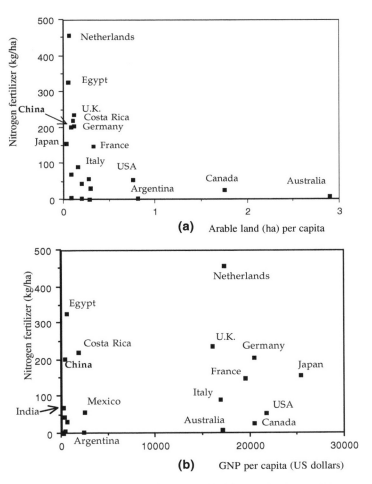

FIGURE 10.4 (a) Nitrogen fertilizer and demographic pressure. (b) Nitrogen fertilizer and bioeconomic pressure. (Giampietro, M., Bukkens, S., Pimentel, D., (1999), General trends of technological change in agriculture, *Crit. Rev. Plant Sci.*, 18, 261–282.)

of the total population, and when the accounting is done in hours of human activity, rather than in people, we find that the effect on AP_{BEP} is even more limited, since $HA_{Working}$ is only 10% of total human activity (THA). When looking at the existing levels of demographic pressure and the existing gradients between developed and developing countries (Figure 10.6b), it is easy to guess that such a reduction, associated with the process of industrialization, will not even be able to make up for the increase in the requirement of primary crop production associated with the higher quality of the diet (higher quality of diet mix and postharvest losses), which industrialization tends to carry with it (more on this in Section 10.3).

10.1.2.1 The Biophysical Cost of an Increasing Demographic and Bioeconomic Pressure: The Output/Input Energy Ratio of Agricultural Production —

The output/input energy ratio of agricultural production is an indicator that gained extreme popularity after the oil crisis in the early 1970s to assess the energy efficiency of food production. Assessments of this ratio are obtained by comparing the amount of endosomatic energy contained in the produced agricultural output to the amount of exosomatic energy embodied in agricultural inputs used in the process of production. Being based on accounting of energy flows, such an assessment is generally controversial (see technical section in Chapter 7). The two most famous problems are (1) the truncation problem on the definitions of an energetic equivalent for each one of the inputs (Hall et al., 1986) (as

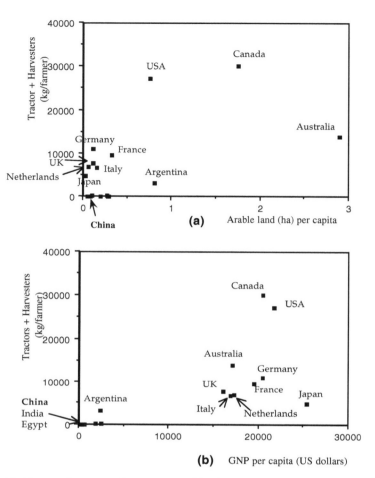

FIGURE 10.5 (a) Machinery and demographic pressure. (b) Machinery and bioeconomic pressure. (Giampietro, M., Bukkens, S., Pimentel, D., (1999), General trends of technological change in agriculture, *Crit. Rev. Plant Sci.*, 18, 261–282.)

noted in Chapter 7, this has to do with the hierarchical nature of nested dissipative systems) and (2) the summing of apples and oranges — in particular the most controversial assessment of energy input is that related to human labor (especially the summing done by some analysts of endosomatic and exosomatic energy) (Fluck, 1992). As noted in Chapter 7, this has to do with the unavoidable arbitrariness of energy assessments that, rather than linear analysis, would require the adoption of impredicative loop analysis (ILA). Methodological details are, however, not relevant here.

This ratio is generally assessed by considering (1) the output in terms of an assessment of an amount of endosomatic energy that is supplied to the society (e.g., the energy content of crop output) and (2) the input in terms of an assessment of an amount of exosomatic energy consumed in production. To obtain this assessment, it is necessary to agree on a standardized procedure (e.g., on how to calculate the amount of fossil energy embodied in the various inputs adopted in production).

If the analysis focuses only on the embodied requirement of fossil energy in the assessment of the input, then the resulting ratio (the amount of fossil energy consumed per unit of agricultural output) can be used as an indicator of biophysical cost of food. In fact, it measures the amount of exosomatic energy (one of the possible EV2 — fossil energy — that can be used for the analysis of the dynamic budget of societal metabolism) that society has to extract, process, distribute and convert into useful power to produce a unit of food energy.

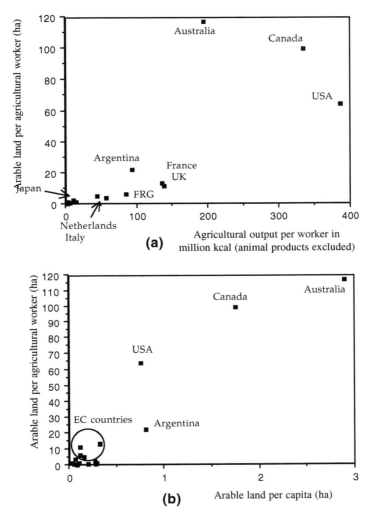

FIGURE 10.6 (a) Biophysical productivity of labor vs. arable land per worker. (b) Arable land per worker vs. arable land per capita. (Giampietro, M., Bukkens, S., Pimentel, D., (1999), General trends of technological change in agriculture, *Crit. Rev. Plant Sci.*, 18, 261–282.)

Thus, by using this ratio we can study the relation between (1) biophysical cost of food production, (2) level of socioeconomic development (when using as an indicator either the fraction of working force in agriculture, GNP p.c. or AP_{BEP}) and (3) level of demographic pressure (by using as the indicator a measure of agricultural resource per capita, such as arable land). An overview of the relation between these factors — represented on a 2×2 matrix — is provided in Figure 10.7. All these indicators are obtained by applying the analysis of the dynamic budget of societal metabolism using a different combination of extensive variables 1 (human activity and land area) and extensive variables 2 (exosomatic energy, added value and food).

Where the combination of the two pressures is high/high, we have societies that have the lowest values of output/input energy ratios in agriculture; for a more detailed analysis, see Conforti and Giampietro (1997). Therefore, these societies face the highest biophysical cost of one unit of food produced. On the other hand, Figure 10.7 shows the importance of performing an integrated assessment of agricultural performance based on nonequivalent indicators reflecting different dimensions. In fact, simple observation of the values presented in the 2×2 matrix makes it easy to realize that the output/input energy ratio of agricultural production should not be considered a good optimizing parameter. Very high values

Indicator of environmental loading: output (food energy): input (fossil energy)

		SOCIO-ECONOMIC PRESSURE		
		HIGH	**LOW**	
DEMOGRAPHIC PRESSURE	HIGH	e.g. Netherlands **Very low** **< 1/1**	e.g. China **Medium** **~ 3/1**	*Arable land per capita* *< 0.20 ha*
	LOW	e.g. U.S.A. **Medium** **~ 3/1**	e.g. Traditional Subsistence **Very high** **> 20/1**	*Arable land per capita* *> 0.75 ha*

```
              * Work force:        * Work force:
                < 7% in agriculture   > 50% in agriculture
              * Exo/Endo > 30/1     * Exo/Endo < 10/1
              * Labor Productivity: * Labor Productivity:
                > 200 kg cereal/hour < 10 kg cereal/hour
```

(output:input energy ratio of agriculture shown in bold)

FIGURE 10.7 Combined effect of demographic and socioeconomic pressure on technical performance of agricultural production. (Giampietro, M., Bukkens, S., Pimentel, D., (1999), General trends of technological change in agriculture, *Crit. Rev. Plant Sci.*, 18, 261–282.)

of output/input are found in those agricultural systems in which the throughput is very low. This situation is generally associated with very poor farmers and a low level of societal development. The goal of keeping the biophysical cost of food low — assessed in terms of a fossil energy price — is in conflict with the goal of keeping the material standard of living high.

10.2 The Effect of the Internal Bioeconomic Pressure of Society on Terrestrial Ecosystems

10.2.1 Agriculture and the Alteration of Terrestrial Ecosystems

Three simple observations make evident the crucial link between food security and the alteration of terrestrial ecosystems worldwide:

1. More than 99% of food consumed by humans comes from terrestrial ecosystems, and this percentage is increasing (FAO food statistics).
2. More than 90% of this food is produced by using only 15 plant and 8 animal species, while estimates of the existing number of species on Earth are in the millions (Pimentel et al., 1995).
3. Worldwide, land in production per capita is about 0.24 ha (FAO food statistics) and is expected to continue to shrink because of population growth.

In addition, arable land is being lost. During the past 40 years nearly one third of the world's cropland (1.5×10^9 ha) has been abandoned because of soil erosion and other types of degradation (Pimentel et al., 1995). Most of the added land (about 60%) that replaces this loss has come from marginal land made available mainly by deforestation (Pimentel et al., 1995). High productivity per hectare on marginal lands requires large amounts of fossil energy-based inputs. This occurs at the very same time that the economic growth of many developing countries is dramatically increasing the demand

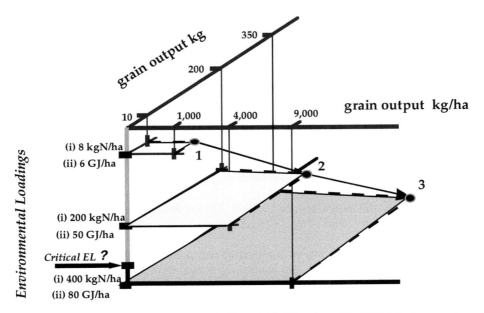

FIGURE 10.8 Adding a third axis to the plane shown in Figure 10.2. (Giampietro, M., (1997a), Socioeconomic pressure, demographic pressure, environmental loading and technological changes in agriculture, *Agric. Ecosyst. Environ.*, 65, 219–229.)

of oil for alternative uses (e.g., construction of industrial infrastructures, manufacturing and household consumption).

Agriculture can be defined as a human activity that exploits natural processes and natural resources to obtain food and other products considered useful by society (e.g., fibers and stimulants). The verb *exploits* suggests that we deal with an alteration of natural patterns, which is disturbance. Indeed, within a defined area, humans alter the natural distribution of both animal and plant populations to selectively increase (or reduce) the density of certain flows of biomass that they consider more (or less) useful for the socioeconomic system.

When framing things in this way, it becomes possible to establish a link between two types of costs and benefits that refer to two logically independent (nonequivalent) processes of self-organization, which we can perceive and represent as two impredicative loops (referring to the definition within nested hierarchies of identities and essences). On the side of human systems we can look for impredicative loops based on endosomatic and exosomatic energy in which the definition of identities of socioeconomic entities is related to biophysical, social and economic variables (examples have been given in Figure 7.7). In this way, the throughput in agriculture can be related to the characteristics of human holons on different hierarchical levels, as illustrated in Figure 10.1. This makes it possible, for example, to guess — when using a graph such as the one described in Figure 10.2 — that a given technique of production characterized by point 1 on the plane can be adopted in Africa but not in the U.S., whereas a technique of production characterized by point 3 can be adopted in Europe but not in China. It is interesting to observe that by adopting this analysis we can find out that an intermediate technology, for example, a technique characterized by point 2, is not necessarily a wise solution to look for. Such an intermediate solution can be unsuitable for any of the agroecosystems considered. That is, the parallel definition of compatibility at two levels can imply that something that is technically feasible is not compatible in socioeconomic terms, whereas something looked for in socioeconomic terms cannot be realized for technical or ecological reasons.

If we characterize the effect of three different techniques of production — the same three solutions indicated in Figure 10.2 using three different points — in relation to their impact on terrestrial ecosystems, we have to add new epistemic categories to our information space. That is, we have to add a new dimension to our representation of performance. Just to provide a trivial example, this requires adding a third axis to the plane, in which the third axis is used as an indicator of environmental impact. This is done in Figure 10.8 in which a vertical axis called environmental loading has been added to the representation of Figure 10.2.

By adding an additional attribute used to characterize the performance of the system, we can obtain a richer biophysical characterization of technical solutions. That is, we can check (1) the socioeconomic compatibility, using the plane labor and land productivity and (2) the ecological compatibility, looking at the level of environmental loading associated with the technique of production. In the example of Figure 10.8 two proxies/variables (kilograms of synthetic nitrogen fertilizer per hectare and total amount of fossil energy embodied in technical inputs per hectare) are proposed on the vertical axis as possible indicators of environmental loading. After having established a mapping referring to the degree of environmental loading, it is possible to assess the situation against a given critical threshold of environmental loading that is assumed to be the level at which damages to the structure of ecological systems become serious (we can call such a threshold value CEL). The distance between the current level of EL and the critical threshold (the value of the difference $|EL_i - CEL|$ assuming that $EL_i < CEL$) can be used as an indicator of stress.

Even in this very simplified mechanism of integrated representation of the performance of agriculture it is possible to detect the existence of two nonequivalent optimizing dimensions. The best solution in relation to a socioeconomic reading (solution 3 when considering only the information given in Figure 10.2), not only is the worst when considering the degree of environmental impact, but also could be nonsustainable (not feasible) according to the constraint imposed by the ecological dimension ($EL_3 >$ CEL). That is, according to the identity of the particular ecosystem considered (determining the value of CEL), the given technical solution defined by point 3 as an optimal solution in relation to socioeconomic considerations should be considered not ecologically compatible, when considering the process of self-organization of terrestrial ecosystems.

The indicator used in Figure 10.8 — the amount of fossil energy associated with the management of agroecosystems per hectare — is a very versatile indicator. In fact, it not only tells us the degree of dependency of food security on the depletion of stocks of fossil energy, but also indicates how much useful energy has been invested by humans in altering the natural impredicative loop of energy forms associated with the identity of terrestrial ecosystems. Such interference is obtained by injecting into this loop a new set of energy forms, which are not included in the original set of ecological essences and equivalence classes of organized structures.

We can represent the use of fossil energy to perform such an alteration using a 2×2 matrix (Figure 10.9) that is very similar to that given in Figure 10.7. Also in this case, we can observe that the different intensity of use of fossil energy (to power the application of technical inputs) is heavily affected by the characteristics of the socioeconomic context. The only difference between the two matrices is that rather than the ratio output/input, the variable used to characterize typologies of societal context is the total throughput of fossil energy that is required to alter the natural identity of the terrestrial ecosystem within the agricultural sector. The cluster of types of countries obtained in the matrix of Figure 10.9 is the same as that found in the matrix of Figure 10.7. This could have been expected when considering the message of the maximum power principle (technical sections of Chapter 7: the output/input ratio of a conversion is inversely correlated to the pace of the throughput). Again, this observation can be used to warn those considering efficiency an optimizing factor in sustainable agriculture. Increasing the efficiency of a given process, in general, entails (1) a lower throughput and (2) less flexibility in terms of regulation of flows.

As noted in the theoretical discussion in Chapter 7, when dealing with metabolic systems that base the preservation of their identity on the stabilization of a given flow, it is impossible to discuss the effect of a change in a output/input ratio or, more in general, the effect of a change in efficiency of a particular transformation if we do not specify first the relation between the particular identity of the system and the admissible range of values for its specific throughput. Increasing the output/input by decreasing the throughput is not always a wise choice. This trade-off requires careful consideration of what is gained with the higher output/input and what is lost with the lower pace of throughput. This is particularly evident when discussing flows occurring within the food system.

Indicator of environmental loading: GJ of fossil energy input/ha cultivated land

		SOCIOECONOMIC PRESSURE		
		HIGH	**LOW**	
DEMOGRAPHIC PRESSURE	**HIGH**	e.g. Netherlands **Very high** **> 80 GJ/ha**	e.g. China **Medium/High** **~ 50 GJ/ha**	*Arable land per capita* *< 0.20 ha*
	LOW	e.g. U.S.A. **Medium/Low** **~ 30 GJ/ha**	e.g. Traditional Subsistence **Very Low** **~ 0 GJ/ha**	*Arable land per capita* *> 0.75 ha*

* Work force: * Work force:
  < 7% in agriculture > 50% in agriculture
* Exo/Endo > 30/1 * Exo/Endo < 10/1
* Labor Productivity: * Labor Productivity:
  > 200 kg cereal/hour < 10 kg cereal/hour

Fossil energy equivalent of technical input application shown in bold

FIGURE 10.9 Combined effect of demographic and socioeconomic pressure on the environmental loading of agriculture. (Giampietro, M., (1997a), Socioeconomic pressure, demographic pressure, environmental loading and technological changes in agriculture, *Agric. Ecosyst. Environ.*, 65, 219–229.)

10.2.2 The Food System Cycle: Combining the Two Interfaces of Agriculture

The overlapping within the agricultural sector of flows of energy and matter that refer to the set of multiple identities found in both terrestrial ecosystems and human societies can be studied by tracking our representation of inputs and outputs through the four main steps of the food system cycle (Giampietro et al., 1994). This approach is illustrated in Figure 10.10. The four steps considered in that representation are:

1. Producing food — where food is defined as forms of energy and matter compatible with human metabolism
2. Making the food accessible to consumers — where accessible food is defined as meals (nutrient carriers) ready to be consumed according to defined consumption patterns of typologies of households
3. Consuming food and generating wastes — where wastes are defined as forms of energy and matter no longer compatible with human needs
4. Recycling wastes to agricultural inputs — where agricultural inputs are defined as forms of energy and matter compatible with the productive process of the agroecosystem

The scheme in Figure 10.10 shows that it is misleading to assess inputs and outputs or to define conversion efficiency by considering only a single step of the cycle. All steps are interconnected, and hence the definition of a flow as an input, available resource, accessible resource or waste is arbitrary. First, the definition of the role of a flow depends on the point of view from which the system is analyzed. For example, the introduction of trees in a given agroecosystem can lead to increased evapotranspiration. This can be negative in terms of less accessible water in the soil, but positive in terms of more available water in the form of rain clouds. Second, the definition of the role of a flow depends on the compatibility of the throughput density with the processes regulating the particular step in question. For example, night soil of a small Chinese village is a valuable input for agriculture (recall here Figure 5.1), but the sewage of a big city is a major pollution problem — same flow but different density in relation to the

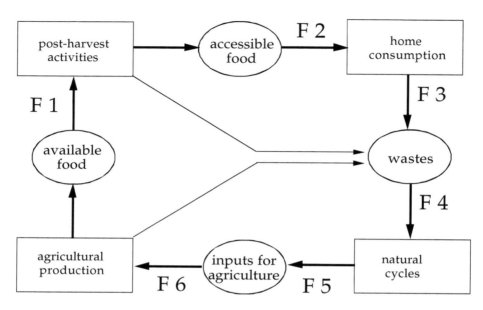

FIGURE 10.10 The food system cycle. Boxes represent the components of the food systems. Ellipsoids describe the nature of flows. The arrows marked by F and numbers indicate: F1, food crops available after agricultural production and harvest; F2, food accessible to consumer after processing, packaging, distribution and home preparation; F3, food required for food security; F4, Wastes and pollutants generated by the food system; F5, wastes and pollutants degraded and recycled by the ecosystem; F6, nutrients consumed by agriculture. (Giampietro, M., Bukkens, S.G.F., and Pimentel, D., (1994), Models of energy analysis to assess the performance of food systems, *Agric. Syst.*, 45, 19–41.)

capability of processing it for a potential user. Whether the speed of a throughput at any particular step is compatible with the system as a whole depends on the internal organization of the system. Feeding a person 30,000 kcal of food per day, about 10 times the normal amount, would represent too much of a good thing, meaning that person would not remain alive for a long period of time. Why then do many believe it to be possible to increase the productivity of agroecosystems several times without generating any negative side effects on agroecosystem health?

Technical progress sooner or later implies a switch from low-input to high-input agriculture:

- Low-input agriculture, which is based on nutrient *cycling* within the agroecosystem (Figure 10.11a). In this case, the relative size of the various equivalence classes of organisms (populations) that are associated with the various types and components of the network (ecological essences — see the discussion about Figure 8.14) is related to their role in guaranteeing nutrient cycling.
- High-input agriculture, in which the throughput density of harvested biomass is directly controlled and maintained at elevated levels through reliance on external inputs that provide *linear* flows of both nutrients and energy (Figure 10.11b).

In low-input agriculture the harvested flow of biomass reflects the range of values associated with a natural turnover of populations making up a given community. That is, the relative size of populations of organized structures mapping in the same type (species), has to make sense in relation to the job done by that species within the network — the essence associated with the role of the species. In this situation the activity of agricultural production interferes only to a limited extent with the ecological system of controls regulating matter and energy flows in the ecosystem. This form of agriculture requires that several distinct species are used in the process of production (e.g., shifting gardening, multi-cropping with fallows) to maintain the internal cycling of natural inputs as a pillar of the agricultural production process. For humans (the socioeconomic system), this implies poor control over the flow of produced biomass in the agroecosystem because of the low productivity per hectare when assessing the yield of a particular crop at the time. Especially serious is the problem associated with low-input agriculture,

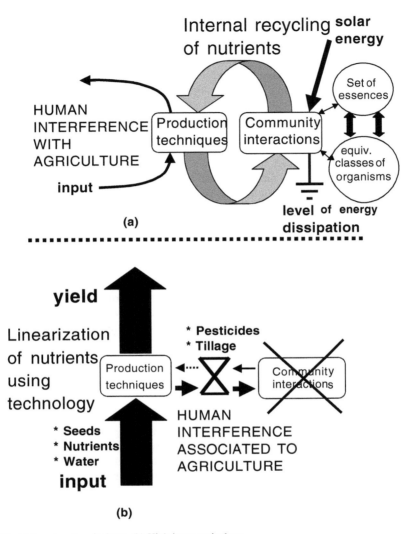

FIGURE 10.11 (a) Low-input agriculture. (b) High-input agriculture.

when facing a dramatic increase in demographic pressure. This type of farming cannot operate properly when the socioeconomic context would require levels of throughput per hectare too high (e.g., 5000 kg/ha/year of grain). On the positive side, with low-input agriculture the direct and indirect biophysical costs of production are very low. Few technical inputs are required per unit of output produced, and when population pressure is not too high, the environmental impact of this form of agricultural production can remain modest.

The contrary is true for high-input agriculture. In this case, the harvested flow of biomass is well out of the range of values that is compatible with regulation processes typical of natural ecosystems (when considering the natural expected yields of an individual species at the time). Harvesting 8 tons of grain every year (bringing away the nutrients from the agroecosystem) would not be possible without putting back the missing nutrients in the form of human-made fertilizer. In high-input agriculture, human management is based on an eradication of the natural structure of controls in the ecosystem. In fact, when several tons of grain have to be produced per hectare and hundreds of kilograms of grain per hour of agricultural labor, natural rates of nutrient cycling and a natural structure of biological communities are unacceptable. In high-input agriculture, not even the genetic material used for agricultural production is related to the original terrestrial ecosystem in which production takes place. Seeds are produced by transnational corporations and sold to the farmer. In this way, humans keep the flow of produced biomass tightly under control. Humans can adjust yields to match increasing demographic and socioeconomic pressure (e.g., green revolution). However, this control is paid for with a high energetic cost of food production. When adopting this solution, humans must continuously (1) provide artificial regulation in the agroecosystem in the form of inputs and (2) defend the valuable harvest against undesired, competitive species. Therefore, the environmental impact of high-input food production is necessarily large and involves a dramatic reduction of biodiversity in the altered area — that is, the destruction of the entire set of mechanisms that regulated ecosystem functioning before alteration (e.g., predator–prey dynamics and positive and negative feedback in the web of water and nutrients cycling). In addition, high-input agriculture has several negative side effects such as on-site and off-site pesticide and fertilizer pollution, soil erosion and salinization.

In conclusion, we can expect that under heavy demographic and socioeconomic pressure agriculture will experience a drive toward a dramatic simplification of natural ecosystems in the form of linearization of matter flows and use of monocultures. How serious is this problem in terms of long-term ecological sustainability? Can we individuate reliable critical thresholds of environmental loading that can be used in decision making? Even if we find out that a certain level of human interference over the impredicative loop determining the identity of a terrestrial ecosystem can be associated with an irreversible loss of its individuality, can we use this indication in normative terms? For example, can we use in optimizing models the fact that the environmental stress associated with an environmental loading equal to 80% of the value of critical environmental loading is the double of the environmental stress associated with an environmental loading equal to 40% of the critical threshold? If we try to get into this quantitative reasoning, how important are the issues of uncertainty, ignorance, nonlinearity and hysteretic cycles?

10.2.3   Dealing with the Informalizable Concept of Integrity of Terrestrial Ecosystems

In Chapter 8 the integrity of ecosystems was associated with their ability to preserve the validity of a set of interacting ecological essences. In that analysis the concept of essence was not associated with a material entity, but rather with a system property defined as the ability to preserve in parallel the reciprocal validity of (1) nonequivalent mechanisms of mapping representing a type (a template of organized structure) that is supposed to perform a set of functions in an expected associative context and (2) the actual realization of equivalence classes of organized structures (determined by the typology of the template used in their making). This validity check is associated with the feasibility of both the process of fabrication and the metabolism associated with agency of these organized structures that are operating as interacting nonequivalent observers at different levels and across scales.

The main concepts presented in both Chapter 7 and Chapter 8 point to the possibility of associating a particular throughput (used to characterize a specific form of metabolism) of learning holarchies with a set of identities used to characterize the nested hierarchical structure of their elements. In particular, the reader can recall here both Figure 8.13 and Figure 8.14 referring to the possible use of network analysis to generate images of essences and to study relations among identities. This rationale has been utilized on the right side of Figure 10.11a to represent the characterization of a given community in relation to an expected throughput of nutrients. A given level of dissipation of solar energy, required to stabilize the cycling of nutrients at a given rate, can be associated with the existence of a given set of essences (a valid definition of identities of ecological elements and their relation over a network), which are, in fact, realized and acting in an actual area.

The concept of a profile of distribution of an extensive variable over a set of possible types providing closure to express characteristics of parts in relation to characteristics of the whole can be used to have a different look at the mechanism regulating both (1) the rate of input/output of energy carriers in ecological networks and (2) change in relative size of the various components of the network. In particular, it is possible to apply the concepts of age classes and profile of distribution of body mass over age classes (used to study changes in the socioeconomic structures — see Figure 6.10) to the analysis of changes in turnover time of biomass over populations (considered as elements of a network).

In his discussion of the mechanism governing population growth, Lotka (1956, p. 129, emphasis mine) downplays the importance of fertility and mortality rates in defining the dynamics of growth of a particular species:

> Birth rate does not play so unqualifiedly a dominant role in determining the rate of growth of a species as might appear on cursory reflection. ...Incautiously construed it might be taken to imply that growth of an aggregate of living organisms takes place by births of new individuals into the aggregate. This, of course, is not the case. The new material enters the aggregate another way, namely in the form of food consumed by the existing organisms. Births and the preliminaries of procreation do not in themselves add anything to the aggregate, but are merely of directing or catalyzing influences, initiating growth, and guiding material into so many avenues of entrance (mouths) of the aggregate, *provided that the requisite food supplies are presented.*

The same point is made by Lascaux (1921, p. 33, my translation): "Both for humans and other biological species, the density is proportional to the flow of needed resources that the species has available."

Placing this argument within the frame of hierarchy theory, we can say that fertility and mortality are relevant parameters to explain population growth only when human society is analyzed at the hierarchical level of individual human beings (Giampietro, 1998). When different hierarchical levels of analysis are adopted, for example, when studying the mechanisms regulating the demographic transition, a different level of analysis is required. A study of the relation of changes of the size of parts and wholes (e.g., ILA) can be much more useful to individuate key issues. For instance, when human society is described as a black box (society as a whole) interacting with its environment, we clearly see that its survival is related to the strength of the dynamic budget associated with its societal metabolism. Hence, such a system can expand in size (increase its population at a certain level of consumption per capita or expand the level of consumption per capita at a fixed size of population or a combined increase of consumption per capita and population size) only if able to amplify its current pattern of interaction with the environment on a larger scale (Giampietro, 1998). Exactly the same reasoning can be applied to the size of a population operating in a given ecosystem.

As observed by Lotka (1956), within an ecosystem the total amount of biomass of a given population (the amount of biomass included in all the realizations of organisms belonging to the given species) increases because the population of organized structures is able to increase the rate at which energy carriers are brought into the species compared to the rate at which energy carriers (for other species) are taken out. When looking at things in this way (Figure 10.12) the total amount of food utilized by a given species to sustain the activity of the various realizations of organisms can be represented using (1) the set of typologies of organisms (e.g., age classes or types found in the life cycle of a species) and (2) a profile of distribution of biomass over this set of types. Obviously, biomass tends naturally to move from one age class (or from one type) to the next during the years, whereas there is a set of natural mechanisms of regulation determining the input and output from each age class or type (e.g., natural causes of death and selective predation). For an example of a formalization of the analysis of the movement across population cohorts, see Hannon ad Ruth (1994, Section 4.1.1).

Therefore, the size of a given population in a situation of steady state can be associated with a given profile of distribution of biomass over the different typologies associated with the set of possible types. As noted for socioeconomic systems (e.g., Figures 6.9 and 6.10), changes in such a profile of distribution can be associated with transitional periods in which the size of the whole is either growing or shrinking.

When applying these concepts to agroecosystems we can say that the more the amount of agricultural biomass harvested from a defined population per unit of time and area (considering a single species at a time) differs in density from the natural flow of biomass per unit of time and area (size × turnover time) of a similar species in naturally occurring ecosystems, the larger can be expected the level of interference that humans are determining in the agroecosystem. Because of this larger alteration of the natural mechanisms of control, we can expect that the energetic (biophysical) cost of this agricultural production will be higher. Indeed, to maintain an artificially high density of energy and matter flow only in a selected typology of a population of a certain species (amplification of an equivalence class associated with a type, well outside the value that the ecological role associated with the relative essence would

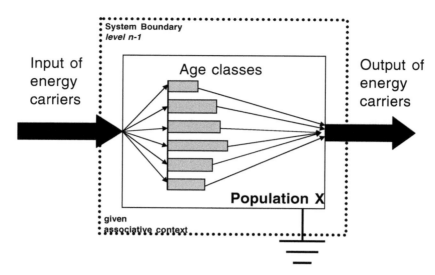

FIGURE 10.12 The distribution of input/output of energy carrier to/from the biomass of population X over age classes.

imply), humans must interfere with the impredicative loop of energy forms determining the identity of terrestrial ecosystems.

In this way, humans end up amplifying the genetic information of the selected populations (species) and suppressing genetic information of competing populations (species). Whenever, this process is amplified on a large scale, this systemic interference carries out the possibility of destroying the web of mutual information determining the set of essences at the level of the web of interaction. As noted, this interference in the dissipative network (altering the profile of admissible inputs and outputs over the set of constituent elements) must be backed up by a supply of alien energy forms. In modern agriculture these alien energy forms are imposed by human activity, which is amplified by exosomatic power and an adequate amount of material inputs (external fertilizers and irrigation). Extremely high densities of agricultural throughput (per hectare or per hour) necessarily require production techniques that ignore the functional mechanism of natural ecosystems, that is, the nutrient cycles powered by a linear flow of solar energy. Clearly, individual species, taken one at a time, cannot generate cycles in matter since they perform only a defined job (occupy a certain niche associated with a given essence) in the ecosystem.

Following the scheme of ecosystem structure proposed by the brothers H.T. Odum (1983), we can represent a natural ecosystem as a network of matter and energy flows in which nutrients are mainly recycled within the set of organized structures composing the system and solar energy is used to sustain this cycling (Figure 10.13a). Within this characterization we can see that the amount of solar energy used for self-organization by the ecosystem is proportional to the size of its matter cycles. In turn, matter cycles must reflect in the distribution of flows and stocks the relative characteristics of nodes in the networks and the structure of the graph (Figure 8.14). As noted in Chapter 7, in terrestrial ecosystems this has to do with availability and circulation of water, which makes it possible to generate an autocatalytic loop between (1) solar energy dissipated for evapotranspiration of water required for gross primary productivity and (2) solar energy stored in living biomass in the form of chemical bonds through photosynthesis, which is required to prime the evapotraspiration.

The interference provided by agriculture on terrestrial ecosystems consists in boosting only those matter and energy flows in the network that humans consider beneficial and eliminating or reducing the flows that they consider detrimental to their purposes. Depending on the amount of harvested biomass, such a process of alteration can have serious consequences for an ecosystem's structure (Figure 10.13b). This has been discussed before when describing the effects of high-input agriculture associated with linearization of nutrient flows (Figure 10.11). When going for high-input agriculture, humans (1) look for those crop species and varieties that better fit human-managed conditions (this is the mechanism

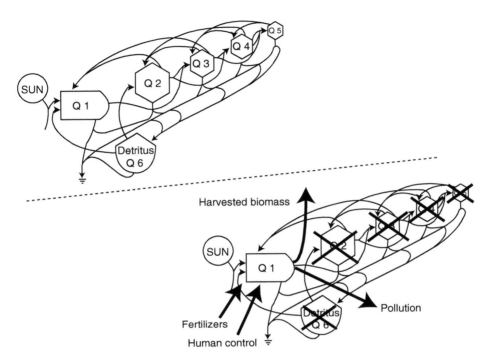

FIGURE 10.13 (a) H.T. Odum graphs: a natural ecosystem without heavy human interference. (b) H.T. Odum graphs: the effects of the interference associated with high-input agriculture. (Giampietro, M., (1997b), Socioeconomic constraints to farming with biodiversity, *Agric. Ecosyst. Environ.*, 62, 145–167.)

generating the reduction of cultivated species and the erosion of crop diversity within cultivated species) and (2) tend to adopt monocultures to synchronize the operations on the field (substitution of machine power for human power). This translates into a skewed distribution of the profile of individuals over age classes described in Figure 10.12.

10.2.4 Looking at Human Interference on Terrestrial Ecosystems Using ILA

The autocatalytic loop of energy forms stabilizing the identity and activity of a terrestrial ecosystem has been described in the form of a four-angle representation of an impredicative loop in Figure 7.6. To discuss the implications of that figure in more detail, a nonequivalent representation of the relation among the parameters considered in that four-angle figure is given in Figure 10.14. This representation of the relation among the key parameters is based on the use of an economic narrative.

The ecosystem starts with a certain level of capital (the amount of standing biomass (SB) of the terrestrial ecosystem), which is used to generate a flow of added value (gross primary productivity (GPP), that is, a given amount of chemical bonds, which are considered energy carriers within the food web represented by the ecosystem). To do that, it must take advantage of an external form of energy (solar energy associated with water flow, generating the profit keeping alive the process). A certain fraction of this GPP is not fully disposable since it is used by the compartment in charge for the photosynthesis (autotrophic respiration of primary producers). This means that the remaining flow of chemical bonds, which is available for the rest of the terrestrial ecosystem — net primary productivity (NPP) — can be used either for final consumption (by the heterotrophs) or by replacing or increasing the original capital (standing biomass). What is important in this narrative is the possibility of establishing a relation among certain system qualities. That is:

1. To have a high level of GPP, an ecosystem must have a large value of SB. In the analogy with the economic narrative, this would imply that to generate a lot of added value, an economic system must have a lot of capital.

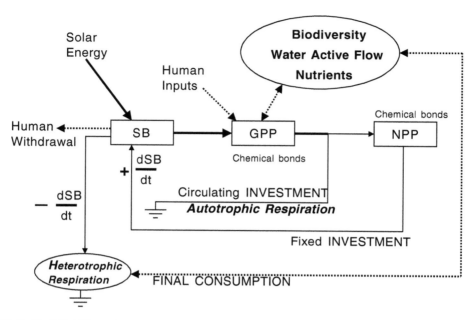

FIGURE 10.14 GPP, NPP and human appropriation in terrestrial ecosystems (a different view of the impredicative loop analysis described in Figure 7.6).

2. To keep a high degree of biodiversity, it is important to invest a reasonable fraction of NPP in the heterotrophic compartment. In the analogy with the economic narrative, this would imply that to have a large variety of activities in the system (i.e., biodiversity), it is important to boost the resources allocated in final consumption (e.g., postindustrialization of the economy). As noted by Zipf (1941), an adequate supply of leisure time associated with an adequate diversity of behaviors of final consumers can become a limiting factor for the expansion of the economies after the process of industrialization. To be able to produce more, an economy must learn how to consume more.

3. Human withdrawal, in this representation, represents a reduction of the capital available to the system.

4. Human interference with the natural profile of redistribution of available chemical bonds among the set of natural essences expected in the ecosystem implies an additional compression of final consumption. Heterotrophs, which are usually called within the agricultural vocabulary pests, are within this analogy those in charge for final consumption in terrestrial ecosystems — those that would be required to boost primary productivity according to Zipf, and that in reality are the big losers in the new profile of distribution of NPP imposed by humans in high-input agroecosystems.

There is an important implication that can be associated with the use of the concept of capital for describing the standing biomass of terrestrial ecosystems. When dealing with metabolic systems there is a qualitative aspect of biomass that cannot be considered only in terms of assessment of mass. That is, the size of a metabolic system is not only related to its mass in kilograms, but also to its overall level of dissipation of a given form of energy, which implies a coupling with its context. This is why an ILA represents a nonequivalent mechanism of mapping of size that is obtained by using in parallel two extensive variables (EV1 and EV2), able to represent such an interaction from within and from outside, in relation to different perceptions and representations of it. Getting back to the example of biomass of different compartments of different ecosystems, 1 kg of ecosystem biomass in the tundra has a very high level of dissipation (the ratio of energy dissipated to maintain a kilogram of organized structure in its expected associative context). This high level of dissipation can be explained by its ability to survive

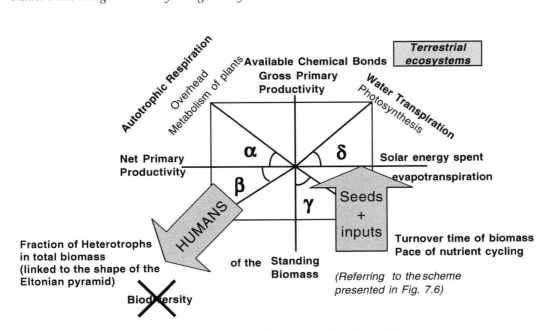

FIGURE 10.15 Terrestrial ecosystems, referring to the scheme presented in Figure 7.6.

in a difficult environment, that is, to maintain its ability of transforming solar energy into biological activity, despite severe boundary conditions (Giampietro et al., 1992). The same concept of capital can be applied to study the activity of terrestrial ecosystems in different areas of the biosphere. Not all hectares of terrestrial ecosystems are the same. Tropical areas are very active in the process of sustaining biosphere structure (a high level of utilization of solar energy per square meter as average, meaning more biophysical activity per unit of area and therefore more support for bio-geochemical cycles). On the other hand, because of the low level of redundancy in the genetic information stored in tropical ecosystems (K-selection means a lot of essences — a spread distribution of the total capital over different types — which translates into many species with a low number of individuals per each species), these systems are very fragile when altered for human exploitation (Margalef, 1968).

Getting back to the representation of the impredicative loop of energy forms defining the identity of terrestrial ecosystems (Figure 7.6), we can try to represent the catastrophic event associated with the process of linearization of nutrients due to high-input agriculture and monocultures (Figure 10.15). Two major violations of the constraints of congruence over the loop are generated by human interference and are related to the parameters and factors determining the two lower-level angles:

1. Humans appropriate almost entirely the available amount of net primary productivity. This does not leave enough capital in the ecosystem to guarantee again a high level of GPP in the next year.
2. Humans interfere with the natural profile of distribution of GPP among the various lower-level elements of the ecosystem (e.g., a distortion of the shape of Eltonian pyramids).

10.2.4.1 *Can We Use This Approach to Represent Different Degrees of Alteration?* — Giampietro et al. (1992) proposed a method of accounting based on biophysical indicators that can be used to describe the effect of changes induced by human alteration of terrestrial ecosystems. This system of representation is based on the use of thermodynamic variables such as watts per kilogram and kilograms per square meter (and their combination — watts per square meter), which assumes that terrestrial ecosystems are represented as dissipative systems, that is, a combination of EV2 and EV1 variables. The rationale of this analysis is that human intervention implies sustaining

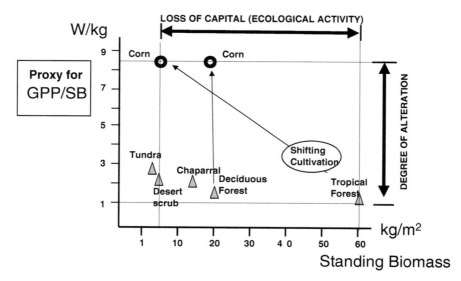

FIGURE 10.16 Plane to represent the alteration of terrestrial ecosystems (adopting a thermodynamic rationale).

an agroecosystem that is improbable according to the natural process of self-organization. Human intervention is therefore viewed as an interference preventing the most probable state of a dissipative system from a thermodynamic point of view. This makes it possible to study the degree of alteration by measuring the increase in the level of energy dissipation per kilograms of standing biomass. "The effect of human intervention can be better detected by the change in the level of energy dissipation of cornfield biomass. After human intervention the biomass of a cornfield in the USA dissipates 9.0 W/kg, compared with 0.5 W/kg dissipated by the wild ecosystem biomass that was replaced" (Giampietro et al., 1992).

An example of this type of representation is given in Figure 10.16. On the horizontal axis the variable used to represent the state of terrestrial ecosystem is the amount of standing biomass (expressed in kilograms of dry mass) averaged over the whole year. This requires the averaging of the amount of standing biomass assessed per week (or per month) over the 52 weeks (or 12 months) making up a year. On the vertical axis, we have a variable that represents a proxy for the ratio GPP/SB (in this case the total flow of solar energy for water transpiration associated with GPP, per unit of biomass available), which is assessed in watts per kilogram. The assessment of solar energy is obtained by calculating the plant active water flow (PAWF), which is the amount of energy required to evaporate the water used to lift nutrients from the roots to the leaves, which must be associated with the relative amount of GPP. This calculation is based on a fixed amount of water required to bring nutrients from the root to the place where photosynthesis occurs (more details in Giampietro et al., 1992). Technical aspects of the choice of the mechanism of mapping of PAWF, however, are not relevant here. What is important is that it is possible to establish a mechanism of accounting that establishes a bridge between the solar energy used for evapotranspiration of water, which can be directly linked to the amount of chemical bonds (GPP — the internal supply of energy input for the terrestrial ecosystem), and the amount of ecological capital (the amount of standing biomass) available to the terrestrial ecosystem per year.

Possible configurations for terrestrial ecosystems are represented in Figure 10.16 (data from Giampietro et al., 1992). According to external constraints (i.e., different boundary conditions — soil, slope and climatic characteristics) different types of ecosystems can reach different levels of activity, measured by this graph by the value reached by the parameter watts per square meter (a combination of the values taken by the two variables watts per kilogram and kilograms per square meter). This should be considered a sort of level of technical development reached by the ecosystem. By adopting a variation of what was proposed for the analysis of socioeconomic systems, we can define the equivalence (1) exosomatic energy for human societies = solar energy spent in evapotranspiration by terrestrial ecosystems) and (2) endosomatic energy for human societies = energy in the form of chemical bonds — GPP in terrestrial ecosystems). At this point, we can say that those terrestrial ecosystems that are able to dissipate a larger fraction of solar energy associated with GPP per square meter (thanks to the use of more water) are able to express more biological activity and store more information (define more identities and essences) than those ecosystems that are operating at a lower level of dissipation.

An increase in the level of dissipation can be obtained by stocking more biomass (more kilograms) at a lower level of dissipation per kilograms (at lower watts per kilograms) or vice versa. As noted earlier, the parallel validity of the maximum power principle and the minimum entropy generation principle pushes for the first hypothesis (a larger amount of structures at a lower level of dissipation). This is perfectly consistent with the discussions presented in Chapters 1 and 7 about the physical root of Jevons' paradox. An improvement assessed when adopting an intensive variable (lower ratio of watts per kilogram) tends to be used by dissipative adaptive systems to expand their capability to handle information in terms of an expansion of an extensive variable (kilogram of biomass stored in the same environment). Terrestrial ecosystems can increase their stability and the strength of their identity (at least in the medium term) by increasing the quantities of reciprocal interactions and mechanisms of control operating in a given ecosystem, while keeping the relative negentropic cost (watts per kilogram) as low as possible.

Going back to Figure 10.16, the points of the plane indicated by triangles represent the most probable states in which we can find the typologies of terrestrial ecosystems indicated by the various labels. The essences and relative identities of these systems have been determined by millions of years of evolution and reflect the biological and ecological knowledge accumulated in impredicative loops of energy forms stabilized by the information encoded in elements participating in self-entailing processes of self-organization.

The point indicated by circles represents improbable states for natural terrestrial ecosystems. However, as noted before, the improbable configuration of a cornfield in which a single species (the monoculture crop) enjoys a stable dominance over other potential competitors is maintained because of the existence of an alien system of control, reacting to different signals. That is, it is the profit looked for by farmers, which makes it possible to buy the biophysical inputs and the seed required to obtain another cornfield in the next season. This system of control is completely unrelated to the need for closing matter cycles within the ecosystem. On the contrary, when adopting an economic strategy of optimization, the more linear are the flows, the higher is the pace of the throughput, and therefore the more compatible is perceived the agricultural production with human needs. The ecological improbability of a cornfield (in terms of profile of distribution of GPP over the potential set of types of biota) is indicated by the very high ratio GPP/SB, which in our thermodynamic reading is indicated by a very high level of energy dissipation (watts per kilogram of solar energy used to evaporate water associated with photosynthesis (PAWF), per unit of standing biomass, averaged over the year). This also requires that the needed inputs must arrive at the right moment in time (e.g., fertilizers and irrigation have to arrive when they are needed — in the growing season — and not as average flows during the year).

It is interesting to observe that by adopting the representation of the characteristics of the impredicative loop associated with the identity of terrestrial ecosystems, it becomes possible to study the interference induced by humans using two variables rather than one (Figure 10.16). On the vertical axis, we can detect a quantitative assessment related to the degree of alteration, that is, how much human alteration is increasing the negentopic cost associated with the energy budget of the produced biomass. On the horizontal axis, we can detect the level of destruction of capital implied by the management of a given area generally associated with a given typology of ecosystem. That is, the management of a tropical forest implies the destruction of a huge quantity of biophysical capital, and therefore we can expect that this will have a huge impact on biodiversity (see also Figure 10.15). As shown by Figure 10.16, a cornfield dissipates much less energy and sustains much less standing biomass (on average over the year) than a tropical forest. This type of assessment is completely missed if we describe the performance of these two systems in terms of net primary productivity (an indicator often proposed to describe the effects of human alteration of terrestrial ecosystems). In fact, a monoculture such as corn or sugar cane can have very high level of NPP — similar to that found in forests. This can induce confusion when assessing their ecological impact. By adopting this two-variable mechanism of representation of the alteration of a terrestrial ecosystem for crop production, we can see that managing a temperate forest to produce corn is much better than managing a tropical forest. Corn produced by displacing a temperate forest provides a large supply of food for humans and destroys less natural capital than corn produced by displacing tropical forests.

A last consideration about the representation of human alteration is given in Figure 10.15. According to what is represented in that figure, the total appropriation of the available NPP by humans implies a complete reshuffling of the profile of investment of the resource chemical bonds available to the ecosystem over the lower-level component (either to boost final consumption or to increase the level of capitalization of the system — increasing SB). This, in Figure 10.15 at the level of angle β, translates into a major distortion of the natural profile of the Eltonian pyramid of the disturbed ecosystem. An analysis of this rationale, this time applied to aquaculture, has been performed by Gomiero et al. (1997). The main results are given in Figure 10.17. Given the natural profile of distribution of energy flows in the Eltonian pyramid of a natural freshwater ecosystem (on the left of the figure), we can see that the two forms of aquaculture — low-input aquaculture in developing countries (e.g., China) (upper graph) and high-input aquaculture in developed countries (e.g., Italy) (lower graph) — can be characterized as following two totally different strategies:

> The Chinese system is based on polyculture, which means rearing several species of fish in the same water body. As different species have different ecological niches (they feed on different sources), a balanced polycultural system has the potential to reach a very high efficiency, in terms of the use and recovery of potential resources. Even if the pond is an artificial setting in which inputs are imported, the internal characteristics of the managed system still play a funda-mental role in the regulation of matter and energy flows. In the Chinese polycultural system as many as eight or even nine fish species can be reared in the same pond in a balanced combination of size and number. ... Italian intensive monocultural systems rely on carnivorous fish species for 85% of their production. (Gomiero et al., 1997, pp. 173–174)

In practical terms, this means that Chinese farmers replace the role that top carnivores play in natural freshwater ecosystems. At the same time, they provide adequate input able to keep the population associated with the types of lower-level components of the Eltonian pyramids in the right relation between size and number (they preserve the meaning of the essences required to determine the interaction of the food web on lower levels). On the contrary, the producers in developed countries keep the top carnivores in the managed ecosystems and replace all the rest of the ecosystem. That is, humans have to provide the expected associative context that is required to provide the stability of top carnivores. In other terms, to do that, high-input aquaculture has to guarantee a huge supply of feed and the capability of absorbing the relatively huge flow of wastes. An idea of the difference in requirement of environmental services associated with these two different strategies can be obtained by looking at the embodied consumption of environmental services per kilogram of produced fish in these two systems.

The high-tech aquaculture system has a very high level of throughput:

1. IV3 (hour) = 20/80 kg of fish/hour of labor
2. IV3 (hectare) = 80,000 kg of fish/ha/year

whereas the Chinese system operates with a much lower throughput:

1. IV3 (hour) = about 1 kg of fish/hour of labor
2. IV3 (hectare) = 2400 kg of fish/ha/year

Because of its high productivity, the Italian system operates at a ratio of 30 kg of imported nitrogen in the feed per kilogram of nitrogen in the produced fish. It consumes 200,000 kg of freshwater per kilogram of produced fish and is generating 30,000 kg of nitrogen pollution per hectare of water body. The Chinese system operates with a much higher efficiency ratio (2.5 kg of nitrogen in the feed per kilogram of nitrogen in the produced fish) and a much lower environmental loading (only 7000 kg of freshwater per kilogram of produced fish and a negligible load of pollution of nitrogen from the ponds) (Gomiero et al., 1997). Actually, weeds are often planted by Chinese farmers around crop fields to

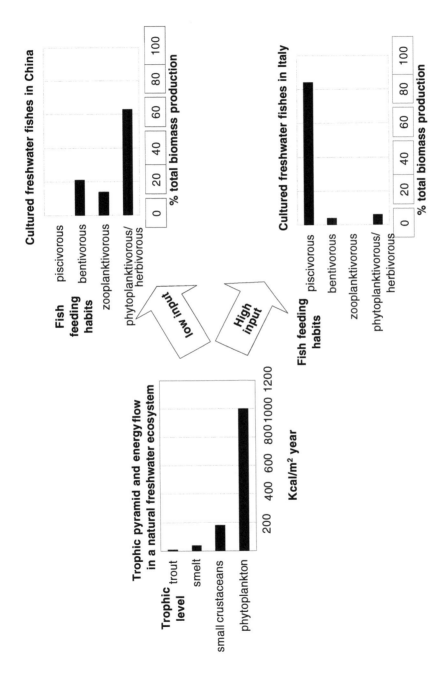

FIGURE 10.17 Trophic pyramids in natural and human-managed freshwater systems. (Modified from Gomiero, T. et al., *Agric. Ecosyst. Environ.*, 62, 169–185.)

recover nitrogen runoff from cropped areas, and then these weeds are fed to the herbivores fish reared in the pond, with an overall effect of reducing the pollution from nitrogen in the agricultural landscape. We have here another example of the same activity — aquaculture — that can be used (1) to reduce nitrogen pollution, while producing at the same time valuable output (animal protein) in rural China and (2) to produce added value, at the cost of a pollution of nitrogen and a taxation on other aquatic ecosystems (industrial pellets used for high-input aquaculture are made up mainly of fish meal obtained from marine catch). Also in this case, an approach based on the definition of (1) a set of types and (2) a profile of distribution of an EV1 over the set can be used to study changes induced from an external perturbation on a natural organization of a metabolic system organized over different hierarchical levels.

In conclusion, the examples discussed in this section seem to confirm the potentiality of impredicative loop analysis to study the representation of the interference that humans induce with agriculture on natural ecosystems. By framing the analysis in this way, it is possible to study and assess which energy forms are associated with ecological mechanisms of control (expressed by the set of identities typical of natural ecosystems) and which energy forms are associated with the mechanisms of control expressed by human society (technology and fossil energy in developed societies).

10.3 Animal Products in the Diet and the Use of Technical Inputs

10.3.1 Introduction

Before concluding this chapter dealing with the consequences that the drivers of technological progress in agriculture can imply on the overall ecological compatibility of food production, it is important to have an idea of the effect of an additional key factor: the quality of the diet adopted by humans in a given food system. As mentioned several times before, the same amount of food calories consumed per capita at the household level can imply a dramatic difference in the requirement of food production, depending on the fraction of animal products in the diet. As a matter of fact, levels of consumption per capita of more than 600 kg of cereal per year can only be obtained because of animal production.

To address the implications of this point, this section provides an overview of food production at the world level, accounting for both vegetal and animal products included in the diet of developed and developing countries. The analysis starts by examining technical coefficients and the requirement of land, labor and technical inputs related to existing patterns of consumption across different continents. Then the relative data set is used to discuss implications for future scenarios. The following analysis is taken from Giampietro (2001).

10.3.2 Technical Coefficients: Material and Methods Used to Generate the Database

As already discussed in Chapter 7, any assessment of energy use in agriculture is difficult and, in a way, arbitrary. Reading the literature of energy analysis, we can find a variety of energy assessments, which almost reflects the variety of authors generating them. For example, very often, standard statistics of energy consumption assign to agriculture a negligible share of total consumption of a country (e.g., less than 3% of total energy consumed in developed countries). These assessments refer to the fraction of energy directly consumed in this sector (e.g., fuels, electricity). That is, they do not include the large amount of fossil energy spent in the making of fertilizers and machinery used in modern farming (this energy consumption is recorded as fossil energy spent in industrial sectors). In this study, indirect inputs to agricultural production are included in the total assessment. However, afterwards, where should such an accounting stop? (Should we include the fossil energy spent at the household level by rural populations?) Again, the decision regarding what should be considered fossil energy spent in agriculture remains necessarily dependent on the logic and goals of the analyst. There is not a unique "right" way to make this accounting.

For this reason, a large part of this section is dedicated to explaining the assumptions and data used to generate the assessments of fossil energy input in plant and animal production at the world level, presented in the tables. In this way, those willing to use a different perspective for the analysis, or a different boundary definition for agricultural production, or a different set of energy conversion factors for the inputs considered, will still be able to use and compare the set of data presented here with others. Additionally, any attempt to assess fossil energy use in developing and developed countries, both for crop and animal production, is unavoidably affected by additional practical problems, such as the reliability of the statistical data and the availability of appropriate conversion factors to be used in the calculations.

The assessments given in this study are based on:

1. A common data set (taken from FAO Agricultural Statistics, available on Internet, 1997 data). This database covers different aspects of agricultural production:
   a. Means of production — e.g., various technological inputs used in production (excluding data on pesticide use)
   b. Food balance sheets — accounting of production, imports, exports and end uses of various products, as well as composition of diet and energetic value of each item, per each social system considered
   c. Data on agricultural production
   d. Data on population and land use.
   Data on pesticides have been estimated starting from Pimentel (1997).

2. A common set of energy conversion factors, which have been applied to these statistical data sets, to provide a common reference value. These conversion factors tend to apply generalized values, but at the same time reflect peculiar characteristics of various socioeconomic contexts in which agricultural production occurs (reflecting the system of aggregation provided by FAO statistics).

Factors of production are given in Table 10.1 (from the FAO data set). Assessments of pesticide consumption have been rearranged from Pimentel (1997) to fit the FAO system of aggregation. Assessments of fossil energy used in primary production of crops, based on the inputs reported in Table 10.1, are given in Table 10.2. Fossil energy conversion factors are discussed below:

1. **Machinery** — To assess the energy equivalent of machinery from FAO statistics, I adopted basic conversion factors suggested by Stout (1991), since they refer directly to the FAO system of accounting: a standard weight of 15 metric tons (MT) per piece (both for tractors and for harvester and thresher (H&T)) for the U.S., Canada and Australia; a common value of 8 MT for pieces in Argentina and Europe; and a common value of 6 MT for pieces in Africa and Asia. For the resulting machinery weight Stout suggests an energy equivalent of 143.2 GJ/MT of machinery. This value (which includes maintenance, spare parts and repairs) is quite high, but it has to be discounted for the life span of machinery. It is the selection of the discount time that will define, in ultimate analysis, the energy equivalent of a metric ton of machinery. Looking at other assessments, made following a different logic, it is possible to find in the literature values between 60 MJ/kg for H&T and 80 MJ/kg for tractors, but only for the making of the machinery. The range of 100 to 200 MJ/kg found in Leach's (1976) analysis also includes the depreciation and repair. Pimentel and Pimentel (1996) suggest an overhead of 25 to 30% for maintenance and repairing to be added to the energy cost of making. In general, a 10-year life span discounting is applied to these assessments. The original value of 143.2 GJ/MT of machinery suggested by Stout can be imagined for a longer life span than 10 years (the higher the cost of maintenance and spare parts, the longer should be the life span). Depending on different types of machinery, the range can be 12 to 15 years. Therefore, in this assessment a flat discount of 14 years has been applied to the tons of machinery, providing an energy equivalent of 10 GJ/MT/year.

2. **Oil consumption per piece of machinery** — The conversion factors are from Stout (1991); again, these factors refer directly to data found in FAO statistics. The estimates of consumption of fuel per piece are the following: 5 MT/year for the U.S., Canada and Australia; 3.5 MT/year

TABLE 10.1

Means of Production in Primary Vegetal Production in World Agriculture

	Arable Land (Million ha)	Work Supply (Billion hours)	Irrigation (Million ha)	H&T (1000)	Tractors (1000)	Nitrogen (Million MT)	Phosphorous (Million MT)	Potassium (Million MT)	Pesticides (Million MT)
World	1379	2385	268	4193	26,335	85.1	32.9	23.3	2.7
Developed countries	632	85	66	3372	19,879	31.5	12.5	12.3	1.5
Developing countries	747	2300	201	820	6455	53.6	20.4	11.0	1.2
Asia	499	1886	187	1917	7079	43.9	12.6	2.5	0.9
Africa	175	352	12	39	557	2.2	0.9	0.5	0.1
Europe	294	56	25	1222	11,198	20.7	6.3	12.0	0.8
North America	222	6	22	794	5511	12.9	4.9	5.2	0.6
Latin America	134	80	18	159	1587	4.5	3.2	3.1	0.2
Oceania	55	5	3	60	401	1.0	1.5	0.5	0.1

Note: Work supply = labor force in agriculture (number of workers) × 1800 h/year (a flat workload).

Source: Giampietro, M., (2001), Fossil energy in world agriculture, in *Encyclopedia of Life Sciences*, Macmillan Reference Limited, available at http://www.els.net/.

TABLE 10.2

Fossil Energy Input in Primary Vegetal Production (Million GJ) in World Agriculture

	Machinery	Fuel	Nitrogen	Phosphorous	Potassium	Irrigation	Pesticides	Other Input	Total
World	2730	5000	6650	550	325	1130	1130	650	18,200
Developed countries	2150	3800	2460	200	170	280	630	650	10,350
Developing countries	580	1200	4190	350	155	850	500	0	7850
Asia	540	1200	3450	240	34	780	380	50	6700
Africa	35	70	170	20	7	60	40	0	400
Europe	995	1960	1600	110	164	100	340	300	5600
North America	950	1400	1000	90	71	90	250	250	4100
Latin America	140	270	350	60	42	80	80	0	1000
Oceania	70	100	80	30	7	20	40	50	400

Source: Giampietro, M., (2001), Fossil energy in world agriculture, in *Encyclopedia of Life Sciences*, Macmillan Reference Limited, available at http://www.els.net/.

for Argentina and Europe; and 3 MT/year for Africa and Asia. The energy equivalent suggested by Stout is quite low (42.2 GJ/MT of fuel — typical for gasoline, without considering the cost of making and handling it). A quite conservative value of 45 GJ/MT as average fossil energy cost of fuel has been adopted.

3. **Fertilizers** — The conversion factors are from Hesel (1992), within the encyclopedia edited by Stout (1992). These assessments include the packaging, transportation and handling of the fertilizers to the shop. Values are:

- For nitrogen, 78.06 MJ/kg — This is higher than the average value of 60 to 63 MJ/kg for production (Smil, 1987; Pimentel and Pimentel, 1996) and lower than the value estimated for production of nitrogen in inefficient plants powered by coal (e.g., in China), which can reach the 85 MJ/kg reported by Smil (1987).

- For phosphorous, 17.39 MJ/kg — This is higher than the standard value of 12.5 MJ/kg reported for the process of production (Pimentel and Pimentel, 1996), but still in the range reported by various authors: 10 to 25 MJ/kg by Smil (1987), 12.5 to 26.0 MJ/kg by Pimentel and Pimentel (1996). The packaging and the handling can explain the movement toward the upper value in the range.

- For potassium, 13.69 MJ/kg — Also in this case, the value is higher than the standard value of 6.7 MJ/kg reported for production. Ranges are 4 to 9 MJ/kg given by Smil (1987) and 6.5 to 10.5 MJ/kg given by Pimentel and Pimentel (1996). Clearly, the energy related to the packaging and handling in this case influences in a more evident form the increase in the overall cost per kilogram.

4. **Irrigation** — The conversion factors suggested by Stout (1991) are 8.37 GJ/ha/year for Argentina, Europe, Canada, the U.S. and Asia; and 9.62 GJ/ha/year for Africa and Australia. These values refer to full fossil energy-based irrigation. However, when looking at FAO statistics on irrigation, one can assume that only 50% of it is machine irrigated, so this conversion factor has been applied only to 50% of the area indicated as irrigated (except in Australia).

5. **Pesticides** — A flat value of 420 MJ/kg has been used for both developed and developing countries. This includes packaging and handling (Hesel, 1992). Values in the literature vary between 293 MJ/kg for low-quality pesticides in developing countries and 400 MJ/kg in developed countries (without including packaging and handling).

6. **Other energy inputs** — At the agricultural level there are other technical inputs that are required for primary production, for example, infrastructures (commercial buildings, fences), electricity for on-farm operations (e.g., drying crops), energy for heating, embodied energy in vehicles and fuels used for transportation. For this reason, a flat 5% of the sum of previous

TABLE 10.3A

Primary Vegetal Production (Million MT) in World Agriculture: Major Items

	Cereal	Starchy Roots	Pulse	Oil Crops	Vegetables	Fruits	Forage
World	2030	631	55	447	532	502	5000
Developed countries	976	183	15	137	142	137	2000
Developing countries	1055	448	40	310	390	365	3000

Source: Giampietro, M., (2001), Fossil energy in world agriculture, in *Encyclopedia of Life Sciences*, Macmillan, available at http://www.els.net/.

TABLE 10.3B

Total Amounts Consumed (Kilograms p.c. per Year) per Production Item

	Cereal	Starchy Roots	Pulse	Oil Crops	Vegetables	Fruits	Forage
World	350 (46%)	109 (56%)	10 (66%)	77 (n.a.)	91 (99%)	86 (80%)	860 (0%)
Developed countries	647 (21%)	153 (50%)	12 (25%)	102 (n.a.)	113 (89%)	127 (71%)	1550 (0%)
Developing countries	260 (65%)	97 (59%)	9 (78%)	70 (n.a.)	85 (91%)	74 (83%)	666 (0%)

Note: Includes crops used for feeding animals, making beer and processed by industry. The percentages in parentheses indicate the amount consumed directly as food.

Source: Giampietro, M., (2001), Fossil energy in world agriculture, in *Encyclopedia of Life Sciences*, Macmillan, available at http://www.els.net/

TABLE 10.3C

Animal Production (Million MT) in World Agriculture: Major Items

	Meat	Beef	Sheep	Pig	Poultry	Milk	Eggs
World	212	56	11	81.0	60	545	53
Developed countries	101	30	3	35.4	30	340	18
Developing countries	111	26	8	45.5	30	205	35

Note: These values include amount of products processed.

Source: Giampietro, M., (2001), Fossil energy in world agriculture, in *Encyclopedia of Life Sciences*, Macmillan, available at http://www.els.net/

TABLE 10.3D

Animal Products (Kilograms p.c. per Year) in the Diet: Major Items

	Meat	Beef	Sheep	Pig	Poultry	Milk	Eggs
World	36	10	2	14	10	93	9
Developed countries	76	23	2	27	22	246	14
Developing countries	25	6	2	10	7	50	8

Note: These values include amount of product processed (e.g., milk for cheese).

Source: Giampietro, M., (2001), Fossil energy in world agriculture, in *Encyclopedia of Life Sciences*, Macmillan, available at http://www.els.net/

energy inputs has been adopted in this analysis. This has been applied only to agricultural production in developed countries. It should be noted that this assessment of other inputs does not include infrastructure and machinery related to animal production, which are reported in Table 10.5.

An overview of world agricultural production includes (1) production of plant products (Table 10.3a), (2) estimate of food energy intake from consumption of plant products (Table 10.3b), (3) production of animal products (Table 10.3c) and (4) estimate of energy intake from consumption of animal products (Table 10.3d) (all data from the FAO database). It should be noted that the effect of trade (imports and exports) is almost negligible, when using this level of aggregation. The only relevant item is 10% of the total production of cereal, which is moving from developed to developing countries. The trade of animal products across these groups is not significant.

Data reported in Table 10.3 are from the FAO Food Balance Sheet (excluding forage, which is estimated; more detail is given in the notes to Table 10.4). Values reported in Table 10.3b are total amounts, which are consumed at the level of the social system considered (either directly or indirectly as feed for animals, seeds, processed in the food industry or lost in the food system). Also in this case, the data on forage are estimated. Values related to animal production do not include animal fat and offal.

Starting from these data, the assessment of how much fossil energy is invested in plant production and in animal production is not an easy task. In fact:

1. A certain fraction of food energy produced in the form of plants (e.g., 37% of cereal at the world level) is used as feed for animal production.
2. A fraction of plant products (e.g., vegetable oil) implies the existence of by-products, which are then used as feed. This implies a problem of accounting — joint production dilemma (same fossil energy input used for producing both vegetable oils and feed).
3. A large fraction of animal feed comes from forage, and it is difficult to attribute fossil energy costs to forage production (in relation to different continents, developed and developing countries), when following the aggregation used in FAO statistics.
4. Additional inputs of fossil energy, which are not included in the assessment reported in Table 10.2, are required for animal production (e.g., for making and heating buildings, technology and energy used for running feedlots, producing milk and eggs, etc.).

Therefore, additional information is needed to assess global fossil energy investments in plant and animal production, both in developed and developing countries. This information is given in Table 10.4 and Table 10.5. An overview of inputs used for feed in animal production is given in Table 10.4.

Values reported in Table 10.4a are mainly based on the assessment provided by Tamminga et al. (1999). This study has been used, since it provides an assessment of requirements of both concentrates and forage for animal products, divided per item and continent, following the organization of the FAO statistics. Minor adjustments have been made to the original data in relation to the consumption of concentrates (following the need of congruence between quantities of concentrates produced and consumed within developed and developing countries). The same check has been applied to estimates of forage for the world, developed and developing countries, as well as per continent. Estimates of forage reported in Table 10.4 are consistent with (1) assessments of consumption for various animal productions and (2) data on estimated yields from permanent pasture reported by FAO statistics.

- An overview of fossil energy input in animal production is given in Table 10.5. Conversion factors adopted for the making of concentrates in this table are:

- For cereal — 4.7 MJ/kg from Sainz (1998), this reflects an output/input in cereal production of 2.5/1, which is more or less common in developed and developing countries.
- For starchy roots — 1.8 MJ/kg in developed countries and 0.8 MJ/kg in developing countries (assuming output/input ratios of 1.5/1 and 5/1).

TABLE 10.4A

Types of Animal Production and Feed Use (Million MT) in World Agriculture

	Large Ruminants (Beef and Milk)			Small Ruminants (Sheep)			Monogastric (Pigs)			Fowl (Includes Eggs)		
	Meat	Concentrate	Forage	Meat	Concentrate	Forage	Meat	Concentrate	Forage	Meat	Concentrate	Forage
World	56.0	340.0	3500.0	11.0	N	900.0	81.0	400.0	500.0	59.0	400.0	N
Developed countries	29.0	250.0	1500.0	2.8	N	230.0	36.5	180.0	225.0	29.5	200.0	N
Developing countries	27.0	90.0	2000.0	8.2	N	670.0	44.5	220.0	275.0	29.5	200.0	N
Asia	12.5	46.0	860.0	5.8	N	480.0	42.5	210.0	262.0	19.8	134.0	N
Africa	3.8	13.0	280.0	1.9	N	160.0	0.8	4.0	5.0	2.4	16.0	N
Europe	13.0	112.0	600.0	1.6	N	130.0	24.4	121.0	151.0	11.0	75.0	N
North America	12.8	112.0	600.0	N	N	N	9.0	45.0	56.0	15.9	108.0	N
Latin America	11.5	37.0	860.0	0.5	N	40.0	3.5	17.0	22.0	9.0	61.0	N
Oceania	2.5	20.0	300.0	1.1	N	90.0	0.4	2.0	2.0	0.6	4.0	N

Note: N = Negligible

Source: Giampietro, M., (2001), Fossil energy in world agriculture, in *Encyclopedia of Life Sciences*, Macmillan, available at http://www.els.net/

TABLE 10.4B

Animal Products, Feed Ingredients and Forage Used in Animal Production (Million MT)

	Total			Ingredients for Making Concentrate and Feeds							Forage No Fossil Energy Input	Forage With Fossil Energy Input
	Meat	Milk	Eggs	Cereal (%)	Starchy Root (%)	Pulses (%)	Oil Seed By-product	Other Industries By-Product	By-Product No Fossil	Total Concentrate		
World	212.0	545.0	52.4	(34)–688	(21)–134	(25)–14	120	40	140	1140	2600	2300
Developed countries	101.0	340.0	17.7	(54)–452	(23)–46	(63)–10	80	40	Nt	630	450	1500
Developing countries	111.0	205.0	34.7	(20)–235	(20)–88	(10)–4	40	N	140	510	2350	600

Note: Percent of cereal, starchy roots and pulse in parentheses; refer to percentage of total production of these crops. Oil seed and other industry by-products refer to concentrates requiring fossil energy input for their processing. Other by-product no fossil refers to high-quality feed obtained from residues and by-products in developing countries. N = Negligible.

Source: Giampietro, M., (2001), Fossil energy in world agriculture, in *Encyclopedia of Life Sciences*, Macmillan, available at http://www.els.net/

- For pulses — 8 MJ/kg in developed countries and 4.7 MJ/kg in developing countries (assuming output/input ratios of 1.8/1 and 3/1).

- A 10% overhead of fossil energy for making feed from the harvested crops has been applied to the production of these crops, following the estimate of Sainz (1998).

- For oil seed by-products — No fossil energy charge for their production is considered. In fact, oil seeds represent an important item in the diet of both developed and developing countries. A fossil energy cost of 1.2 MJ/kg has been calculated for processing the by-product into feed.

- For other feeds and industrial by-products — These include vegetables, fruits, residues from sugar and beer production; a value of 10 MJ/kg, following Sainz (1998).

- Forage — It is almost impossible to imagine a unique conversion factor for the production of forage in developed and developing countries on marginal land at different levels of productivity. Such an assessment is therefore necessarily approximated. A conversion factor of 1.7 MJ/kg has been used for forage production in developed countries (including the production and handling of harvested biomass), whereas a factor of 0.8 MJ/kg has been used for production of forage in developing countries (referring only to the production of biomass). However, a very conservative estimate of the amount of forage produced by using fossil energy input has been used for developing countries. In this regard, there is an item of Table 10.4b (140 million MT of by-products are used to make concentrates in developing countries without the use of fossil energy for their handling) that reflects the completely different logic in animal production between developed and developing countries. In developing countries a large amount of feed is obtained by recycling by-products rather than by producing *ad hoc* crops or forage.

10.3.2.1 Overall Assessment of Fossil Energy Used in Animal Production — According to the conversion factors listed above, the average fossil energy cost of concentrates in 1997 can be estimated in 3.86 MJ/kg at the world level, 4.76 MJ/kg for developed countries and 2.75 MJ/kg for developing countries. These values include (1) the production of crops (cereal, starchy roots, pulses) (reported in Table 10.5b), (2) an extra 10% of the previous amount of energy for the preparation of feed from these crops (reported in Table 10.5b), and (3) the processing of oil seed by-products and the processing of other industrial by-products (reported in Table 10.5b). The fossil energy used in running the infrastructure used for the production of animal products has been assessed following Sainz's (1998) indications that the fossil energy embodied in the feed is a fixed percentage of total production costs. He suggests the values of 90% for beef, 88% for milk, 65% for pig, 50% for poultry and 58% for eggs. At this point, the item additional input for buildings, technical devices and energy for running feedlots and producing milk and eggs has been calculated starting from the assessment of energy in feed. That is, the figures in this column represent complementing percentages to 100% of total production, using the estimates of Sainz. The production of sheep meat was not considered a relevant item for the assessment of fossil energy consumption. Estimates of fossil energy consumption in animal production are given per type of production (Table 10.5a) and per type of input (Table 10.5b).

An overall assessment of fossil energy used in both plant and animal production at the world level, in developed and developing countries (resulting from the set of data reported in previous tables), is given in Table 10.6. They are calculated using data from Table 10.2 and Table 10.5. Data on food energy supply (consumed in the diet) are from the FAO Food Balance Sheet (average of daily consumption times the relative population).

Various indicators related to fossil energy use in world agriculture, derived from the assessments presented in this study, are given in Table 10.7.

10.3.3   Looking Ahead: What Can We say about the Future of Agricultural Production?

Some of the indicators calculated so far are listed in Table 10.8 together with relevant characteristics of the various socioeconomic aggregates considered in this study. An integrated analysis of these data can be useful for discussing possible future trends of fossil energy use in agriculture.

TABLE 10.5A

Fossil Energy Input in Animal Production (Million GJ) per Type of Production

	Concentrate Beef	Forage Beef	Concentrate Milk	Forage Milk	Concentrate Pig	Forage Pig	Concentrate Poultry	Concentrate Eggs	Additional Input	Total Fossil Energy
World	950	2000	500	600	1500	400	1300	250	2000	9500
Developed countries	800	1700	400	400	900	300	800	100	1550	7100
Developing countries	150	300	100	200	600	100	500	150	450	2400

Source: Giampietro, M., (2001), Fossil energy in world agriculture, in *Encyclopedia of Life Sciences*, Macmillan, available at http://www.els.net/

TABLE 10.5B

Fossil Energy Input in Animal Production (Million GJ) per Type of Input

	Crop Production	Forage Production	Extra Feed from crop	Extra Feed from By-product	Other Inputs Beef and Milk	Other Inputs Poultry and Eggs	Other Inputs Pigs	Total Animal Products
World	3500	3000	400	600	500	900	600	9500
Developed countries	2300	2500	250	500	300	750	500	7100
Developing countries	1200	500	150	100	200	150	100	2400

Source: Giampietro, M., (2001), Fossil energy in world agriculture, in *Encyclopedia of Life Sciences*, Macmillan, available at http://www.els.net/

TABLE 10.6

Overall Analysis of Food Production at the World Level

	Total Input Primary Product (Million GJ)	Fraction for Feed (Million GJ)	Total Input Plant Product (Million GJ)	Total Input Animal Product (Million GJ)	Total Input for Food (Million GJ)	Plant Food Consumed (Million GJ)	Animal Food Consumed (Million GJ)	Total Food Consumed (Million GJ)	Plant Food Produced (Million GJ)	Population (Millions)
World	18,200	6500	11,700	9500	21,200	20,700	3800	24,500	34,700	5801
Developed countries	10,300	4800	5550	7350	12,900	4700	1700	6400	13,200	1293
Developing countries	7850	1700	6150	2150	8300	16,000	2100	18,100	21,500	4507

Source: Giampietro, M., (2001), Fossil energy in world agriculture, in *Encyclopedia of Life Sciences*, Macmillan, available at http://www.els.net/

TABLE 10.7

Different Indicators about Fossil Energy Use in Agriculture

	O/I Ratio				
	Food Consumption	**Plant Consumption**	**Animal Consumption**	**Plant Products**	**Total Food Consumption p.c. MJ/day**
	(Fossil Energy)	(Fossil Energy)	(Fossil Energy)	(Fossil Energy)	
World	1.2/1	1.8/1.0	1/2.5	1.9/1	11.6
Developed countries	1/2	1.0/1.2	1/4.3	1.3/1	13.6
Developing countries	2.2/1	2.6/1.0	1/1.0	2.7/1	11.1

	Animal Food Consumption p.c. (MJ/day)	**Total Plant Product p.c. (MJ/day)**	**Fossil Energy Arable Land (GJ/ha)**	**Fossil Energy Agricultural Work (MJ/h)**	**Food Diet Arable Land (GJ/ha)**	**Food Diet Agricultural Work (MJ/h)**
World	1.8	16.4	15.4	9	17.7	10
Developed countries	3.6	28.0	20.4	152	10.1	75
Developing countries	1.3	13.0	11.1	4	24.2	8

First, an important point is that only for the step of agricultural production, humankind is using an amount of fossil energy that is very close to the amount of food energy intake in the diet. That is, 0.83 MJ of fossil energy is spent in world agriculture for each megajoule of food energy consumed at the household level. To make things worse, it should be noted that the direct consumption of fossil energy input in agricultural production (the assessment given in Table 10.6) is just a fraction of the total amount of fossil energy spent in the food system. This is especially true in developed countries. Additional sources of fossil energy consumption in the food system are:

1. Energy required to move around the various food items imported and exported in world agriculture.
2. Energy required by the food industry for processing and packaging of food products. To give an example, a 455-g can of sweet corn requires 2.2 times more fossil energy for the making of the can than for the production of the corn (Pimentel and Pimentel, 1996).
3. Energy required for distribution, handling and refrigeration of food items to and within the shops.
4. Energy required for preparing, cooking and consuming food at the household level.

Pimentel and Pimentel (1996) estimate that in the U.S., for home refrigeration, cooking and heating of food, dishwashing and so forth, there is a consumption of about 38 MJ/day/person (more than the amount of fossil energy needed to produce the food consumed in the diet).

A second observation should be related to the large differences found when considering indices describing the energetic performance of agriculture in developed and developing countries. Even though agriculture is doing the same thing in these two systems (producing food for the rest of society), the logic and strategy for achieving such a goal are totally different. That is, assessments of indices referring to fossil energy consumption as world averages are totally useless for a discussion of trends and future scenarios. This different logic is shown by the differences in the output/input energy ratio (in this study the output is food energy in food products disappearing at the household level, whereas the input is fossil energy spent in the production of the various food items consumed). Developed countries spend twice as much fossil energy in the step of agricultural production than the energy food they consume (1/2), whereas developing countries get more energy food from the diet than the fossil energy they consume in agricultural production (2.2/1). Particularly relevant is the difference in relation to the output/input ratios of animal production. Developed countries have a large investment for the production of animal product (4.3 MJ of fossil energy in production per megajoule of food energy intake of animal products). On the contrary, developing countries get a very good output/input energy ratio for animal production (they are producing 1 MJ of animal product per megajoule of fossil energy invested in production). This ratio is even higher than the ratio achieved by developed countries in relation to plant food. Developed countries spend 1.2 MJ of fossil energy in production per megajoule of food energy intake of vegetal products. This difference in fossil energy use in the production of animal products between developed and developing countries indicates that poor societies are mainly using spare feed (e.g., by-products and forage obtained on marginal land), rather than investing valuable crops in animal production.

However, this lower investment in producing feed is reflected by the lower consumption of animal products in the diet (1.3 MJ/day in developing countries vs. 3.6 MJ/day in developed countries). This reflects the general rule always valid when dealing with agricultural conversions: a better output/input for a given flow is linked to a lower throughput.

A third important observation can be related to the fact that the output/input energy ratios, in both developed and developing countries, do not depend much on technical coefficients in agricultural production. Rather, the composition of the diet, especially the double conversion of crops (mainly cereal) into meat, is the major determinant of such a ratio. We can look at a different output/input energy ratio by considering (1) total output of plant production, which in 1997 was 34,700 million GJ, including uses of cereal, starchy roots, pulses, oil from oil seed crops, sugar, alcohol, vegetables and fruits and (2) fossil energy input in primary plant production (the 18,200 million GJ reported in Table 10.2). When considering this ratio, both developed and developing countries have an output/input energy ratio in plant production larger than 1 (1.3/1 and 2.7/1, respectively). However, this huge supply of plant production is not all used for direct human consumption. Besides the making of feed for animals there are other important alternative uses of crops — especially cereal — in the food system. We can recall the example of the embodied consumption of cereal discussed in Figure 3.1. In 1997, in the U.S., of the 1015 kg of cereal consumed per capita (using the chain of physiological conversions as the mechanism of mapping), more than 600 kg was used for feeding animals, more than 100 kg was consumed in the form of bread, pasta, pizza or other simple products, and an additional 280 kg was used for making beer, for seeds and in industrial processes, plus losses in processing and distribution. To give a comparison, in developing countries, only 260 kg per capita of cereal was consumed in the same year; only 50 kg of it went in feed, 10 kg in beer, and only minor quantities in other uses. The larger investment of plant products in alternative uses explains why the index of total plant food energy produced per capita is more than double in developed (28 MJ/day) than in developing (13 MJ/day) countries, even though this difference disappears when considering the amount of food energy intake from vegetables directly consumed in the diet (10 MJ/day in developed and 9.8 MJ/day in developing countries). That is, consumers of developed countries consume more than double the plant products consumed per capita in developing countries, but in an indirect way.

The larger use of fossil energy in developed countries is reflected by the index of consumption both per hectare (20 vs. 11 GJ/ha in developing countries) and per hour of agricultural work (152 vs. 4 MJ/h in developing countries). When analyzing the pattern of food consumption in relation to available arable land and available workforce in agriculture, we obtain a first important observation. The arable land used to generate the same amount of food energy consumed in the diet is much higher in developed countries (0.1 ha/GJ/year) than in developing countries (0.04 ha/GJ/year). (Note that these values are the inverse of the values 10.1 and 24.2 GJ/ha reported in Table 10.7.) This larger availability/use of land can be related to the better quality of the diet of developed countries. In spite of the fact that the agriculture of developed countries is using more fossil energy per hectare of arable land (twice as much than developing countries), it is also using more than twice as much land in production per capita. Finally, looking at the ratio of energy of food consumed in the diet/hour of labor invested in agriculture, we find another large difference. Developed countries have a value (75 MJ/h) that is almost 10 times larger than that of developing countries (8 MJ/h).

TABLE 10.8

Indicators for Studying Trends of Fossil Energy in World Agriculture

	Arable Land p.c. Hectare	Agriculture Workforce Fraction TOTwforc	Arable Land per Worker Hectare	Fossil Energy per Worker GJ/year	Fertilizer and Irrigation per Hectare GJ/year	Machinery and Fuel per Hour MJ	Fossil Energy in Agriculture Fraction Total Fossil Energy
World	0.24	0.5	1	16	6.3	3.2	0.06
Developed countries	0.49	0.1	12	273	4.9	70.0	0.05
Developing countries	0.17	0.6	1	6	7.4	0.8	0.08

Source: Giampietro, M., (2001), Fossil energy in world agriculture, in *Encyclopedia of Life Sciences*, Macmillan, available at http://www.els.net/

To better understand the causes and implications of these differences, we can analyze the different patterns of use of technical inputs (Table 10.8). The larger use of fossil energy in developed countries is mainly due to the larger use of machinery and fuel. In fact, when assessing the fossil energy per worker, we obtain an investment of 273 GJ/year of fossil energy in the agriculture of developed countries vs. an investment that is 45 times smaller in developing countries (6 GJ/worker/year). When fossil energy-based inputs are separated into two classes, then their different roles become more evident. When accounting only for fertilizers and irrigation, we can see that developing countries are already using more fossil energy per hectare than developed countries (7.4 vs. 4.9 GJ/year). The situation is totally reversed when looking at machinery and fuel. These inputs are virtually absent in many poor crowded countries. The difference between developed and developing countries on this index (70 vs. 0.8 MJ/h) is about 90 times.

Another general observation is that, all things considered, fossil energy consumption in agricultural production, even when assessed by including indirect amounts of fossil energy (production of fertilizers and machinery generally recorded in industrial sectors), is not particularly relevant in relation to the total fossil energy consumption of modern societies. The total assessment of 21,200 million GJ in 1997 is only 6% of the total fossil energy consumption at the world level (about 350×10^{18} J according to BP-Amoco (1999)). This percentage does not change much for developed (5% — over an estimated total of 245×10^{18} J) and developing (8% — over an estimated total of 105×10^{18} J) countries.

Last but not least, it should be observed that the differences in indices reported in Table 10.7 and Table 10.8, between developed and developing countries, are quite significant, even when considering possible adjustments in the set of conversion factors adopted to assess fossil energy consumption or different estimates for various inputs.

Getting into an analysis of possible scenarios, this set of data for sure generates concern. At the moment, fossil energy consumption in developing countries is mainly due to high demographic pressure (0.17 ha of land in production per capita vs. 0.49 ha in production p.c. in developed countries). Such pressure can only get worse. Almost the totality of population growth expected in the next decades (we are talking of another 2 or 3 billion) will occur basically within developing countries. The other big force driving agricultural changes (the reduction of workforce in agriculture and the improvement of farmer income) still did not enter in play in developing countries, or did so only very marginally.

Looking at developed countries, the fancy excess of animal products in the diet of the rich is reflected by a consumption of cereal per capita of 650 kg/year (see Table 10.3b). Probably, using a double conversion of cereal into animal products and beer is the only known solution for making humans consume so much. In this way, in fact, it is possible to maintain a high economic demand for cereal (the easiest crop to produce in a mechanized way). That is, in this way, it is possible to keep high the income of farmers in developed countries. This has been especially important in Europe, where a relative shortage of land per farmer has pushed toward more and more animal products in the diet and cereal in the beer (the only accessible option for boosting the economic labor productivity of farmers). For example, in the Netherlands the consumption of cereal for beer (106 kg/year p.c.) is higher than the direct consumption of cereal products in the diet (75 kg/year p.c.).

As noted earlier, it is evident that the strategy of fossil energy use in developed countries has not been driven by the need of increasing land productivity. The strategy of technical changes in agriculture of developed countries has been aimed at:

1. Improving the nutritional quality of the diet, rather than optimizing the energy supply of the diet. This implies a dramatic elimination of the dominant role of a cereal staple food (e.g., by increasing the availability of fresh vegetables and fruits, the diversity of products available on the market, and the fraction of animal proteins). The opposite is true for typical diets of developing countries.
2. Reducing, as much as possible, the fraction of workforce absorbed by food production, trying to preserve at the same time the economic viability of farmers.

Developed countries can still look for improvement in relation to the first goal (a better quality of diet), whereas we cannot expect more changes toward the second goal (further reduction of the number of farmers). Statistics reported in Table 10.8 are, in a way, misleading, since the system of accounting of the FAO includes within the aggregated developed countries the countries of the ex-Soviet Union and the countries of East Europe (where the fraction of the workforce in agriculture is still relatively high). In reality, the fraction of the workforce in agriculture in countries such as the U.S., Canada and Australia is already at 2%, and in the European Union it is already smaller than 5%. In addition to that, a growing fraction of farmers in developed countries are engaged in part-time, off-farm economic activities to sustain their income (that is, they are working less than the flat amount of 1800 h used in the calculations of this study). A further reduction in the number of farmers in the most developed countries would generate major social problems at the level of rural communities, which are already under stress because of too low a density of population and serious problems of aging.

A reduction of the percentage of farmers in the economically active population will also be quite difficult for developing countries, but for completely different reasons:

1. The amount of available land, both per capita and per worker, is much smaller in developing countries than in developed countries, so that mechanization is almost impossible due to the very small plots.
2. A massive process of mechanization (to increase by 90 times the level of fossil energy investment per hour of work) would require huge economic resources. These resources are simply not there.
3. Demographic growth keeps occurring in rural areas, generating many unemployed with little education. This makes impossible a large-scale implementation of techniques aimed at a reduction of jobs in agriculture.
4. An acceleration of the movement of rural population into cities is impossible to achieve in the short or medium horizon.

The speed of this move is already overwhelming the capability of building and handling functional cities in many developing countries. Looking at the existing number of farmers in developing countries (plus the billions of expected new arrivals), the possibility of achieving a quick reduction seems to be out of question. On the other hand, the traditional solution of farming, without support of machines, due to its intrinsic low productivity of labor, tends to spell poverty for the large number of farmers in developing countries. To make things worse, they will be, more and more, forced to produce on a shrinking area of arable land. That is, the more agricultural production spreads into marginal land, the higher will be the demand for agricultural inputs in production, the higher the risk of critical environmental impact and the lower the return for farmers.

10.3.4 The Sustainability Issue: What Are the Future Biophysical Constraints in Agriculture?

When dealing with the issue of sustainability, we can say that existing heavy reliance on fossil energy is bad not only when considering as a relevant criterion of performance the degree of dependency of food security on a nonrenewable resource, but also (and especially) when considering as a relevant criterion of performance the environmental impact that high-input agriculture implies. Getting into an analysis of environmental impact of fossil energy-based inputs, we can make a distinction between (Giampietro, 1997b):

1. Technical inputs aimed at producing more per hour of labor, which imply:
   - Synchronization of farming activities in time and concentration of farming activities in space
   - Homogeneity in the patterns of application of inputs and harvesting of outputs (to get economies of scale in the development and operation of new technologies)

   This translates into heavy reliance on monocultures, a large dimension of crop fields (loss of edge effect, good for preservation of biodiversity in agroecosystems), erosion of genetic diversity of crops (due to the use of commercial seeds linked to technological packages) and more in general to the spread of technology of extensive adaptation (not tailored on location-specific characteristics of different agroecosystems).

2. Technical inputs aimed at producing more per hectares of land, which imply:
   - Unnatural concentration of fertilizers and toxic substances in the soil, in the agroecosystem and, more in general, in the food and the environment
   - Local alteration of the water cycle (e.g., salinization in crop fields and depletion of underground water reservoirs)

   This translates into the depletion of natural resources that usually are considered renewable (fertile soil and freshwater stored under ground) due to the rate of their exploitation being much higher than the rate of their replenishment, and a dramatic reduction of biodiversity at the agroecosystem level (due to the effects of these alterations).

10.3.4.1 Can We Forecast an Inversion in the Medium Term of Existing Negative Trends? — The data set presented in this study, especially data in Table 10.6 through Table 10.8, does not give reasons for much optimism. Given the existing trends of population growth and the strong aspiration for a dramatic improvement in economic and general living conditions in the rural areas of developing countries, a reduction of fossil energy consumption in agriculture, in the next decades, seems to be a very unlikely event indeed. On the positive side, it is important to observe that the total consumption of fossil energy in agriculture is still a small fraction of the total energy consumption in the rest of the society. In a way, the concern for the ecological impact of high-input agriculture should be, for the moment, much more worrisome than the concern for the limitation of existing reserves of fossil energy.

Boundary conditions and physical laws do impose biophysical constraints on what is feasible for human societies in spite of our wishes and aspirations. Therefore, it is important to have a sense of what are the most likely biophysical constraints faced by humankind in the future in terms of food security. In addition to shortages of arable land, soil erosion, and a lack of alternatives to high-input agriculture to guarantee elevated productivity both per hour and per hectare, there are two other crucial constraints:

1. Water shortages — Presently, 40% of the world's people live in regions that compete for small water supplies. Related to these growing shortages is the decline in per capita availability of freshwater for food production and other purposes in the arid regions of the world. There is very little that technology can do when human development clashes against limits in the water cycle. As noted earlier, the amount of energy required to sustain the water cycle in the biosphere is almost 4000 times the entire amount of energy controlled by humankind worldwide. The

idea that human technology will substitute for the services provided today by nature in this field is simply based on ignorance of biophysical realities.

2. Biodiversity for the long-term stability of the biosphere — The dramatic reduction of species caused by the conversion of natural ecosystems into agroecosystems has been discussed earlier. Since we need biodiversity to stabilize the structure and functions of the biosphere, we cannot transform all terrestrial ecosystems into agricultural fields and cement for housing and infrastructure. A large diversity of species is vital to agriculture and forestry and plays an essential role in maintaining a quality environment and recycling the vital elements such as water, carbon, nitrogen, and phosphorus. This is another clear example of a form of natural capital that cannot be replaced by human technology.

Other well-known global problems can be added to this list: changes in the composition of the atmosphere (greenhouse effect, ozone layer depletion), the cumulative effects of pollution and the intensification of the occurrence of contagious diseases (e.g., AIDS, viruses moving to humans from livestock kept at a too high concentration). I am mentioning them here not to get into the very well known argument between cornucopians and neo-malthusians, but only to make the point that in spite of current disagreements about the evaluation of the seriousness of these problems, we have to acknowledge the obvious fact that there are biophysical limits to the expansion of human activity. These limits can be avoided only by adequately reacting to feedback signals coming from disturbed ecosystems. Whenever humans are not able to obtain reliable indications about the room left for expansion (or when they are not able to understand those signals), they should be reluctant to further expand their disturbance, when it is not absolutely needed.

To end this discussion on a positive note, we can say that in the past two centuries humans have shown an almost magic ability to adapt to fast changes as soon as they detect and acknowledge the existence of a major reason for doing so. For sure, the challenge of sustainability of food security will soon represent one of those reasons. Therefore, we can expect that some major readjustments — the genuine emergence of new patterns — will occur in how humans produce and consume their food in the new millennium. Even though we cannot guess today what they will be, we can — the sooner the better — clarify as much as possible the terms of reference of the sustainability predicament.

References

BP-Amoco, (1999), World-Energy Statistics, available from http://www.bpamoco.com/worldenergy/.

Conforti, P. and Giampietro, M., (1997), Fossil energy use in agriculture: an international comparison, *Agric. Ecosyst. Environ.*, 65, 231–243.

FAO Agricultural Statistics, available from http://apps.fao.org/cgi-bin/nph-db.pl?subset = agriculture.

Fluck, R.C., (1992), Energy in human labor, in *Energy in Farm Production*, Fluck, R.C., Ed., Elsevier, Amsterdam, pp. 31–37 (Vol. 6 of *Energy in World Agriculture*, Stout, B.A., Editor in Chief).

Giampietro, M., Pimentel, D., and Cerretelli, G., (1992), Energy analysis of agricultural ecosystem management: human return and sustainability, *Agric. Ecosyst. Environ.*, 38, 219–244.

Giampietro, M., Bukkens, S.G.F., and Pimentel, D., (1994), Models of energy analysis to assess the performance of food systems, *Agric. Syst.*, 45, 19–41.

Giampietro, M., (1997a), Socioeconomic pressure, demographic pressure, environmental loading and technological changes in agriculture, *Agric. Ecosyst. Environ.*, 65, 219–229.

Giampietro, M., (1997b), Socioeconomic constraints to farming with biodiversity, *Agric. Ecosyst. Environ.*, 62, 145–167.

Giampietro, M., (1998), Energy budget and demographic changes in socioeconomic systems, in *Life Science Dimensions of Ecological Economics and Sustainable Use*, Ganslösser, U. and O'Connor, M., Eds., Filander Verlag, Fürth, Germany, pp. 327–354.

Giampietro, M., Bukkens, S., Pimentel, D., (1999), General trends of technological change in agriculture, *Crit. Rev. Plant Sci.*, 18, 261–282.

Giampietro, M., (2001), Fossil energy in world agriculture, in *Encyclopedia of Life Sciences*, Macmillan Reference Limited, available at http://www.els.net/.

Gomiero, T., Giampietro, M., Bukkens, S.G.F., and Paoletti, M.G., (1997), Biodiversity use and technical performance of freshwater fish aquaculture in different socioeconomic contexts: China and Italy, *Agric. Ecosyst. Environ.*, 62, 169–185.

Hall, C.A.S., Cleveland, C.J., and Kaufmann, R., (1986), *Energy and Resource Quality*, John Wiley & Sons, New York.

Hannon, B.M. and Ruth, M., (1994), *Dynamic Modeling*, Springer-Verlag, New York.

Hayami, Y. and Ruttan, V., (1985), *Agricultural Development and International Perspective*, 2nd ed., John Hopkins University Press, Baltimore.

Helsel, Z.R., (1992), Energy and alternatives for fertilizers and pesticide use, in *Energy In Farm Production*, Fluck, R.C., Ed., Elsevier, Amsterdam (Vol. 6 of *Energy in World Agriculture*, Stout, B.A., Ed.) 177–201.

Lascaux, R., (1921), *La Production et la Population*, Payot & Co, Paris.

Leach, G., (1976), *Energy and Food Production*, IPC Science and Technology Press Ltd., Guilford, Surrey, U.K.

Lotka, A.J., (1956), *Elements of Mathematical Biology*, Dover Publications, New York.

Odum, H.T. (1983), *Systems Ecology*. John Wiley, New York.

Margalef, R., (1968), *Perspectives in Ecological Theory*, University of Chicago Press, Chicago.

Pimentel, D., Harvey, C., Resosudarmo, P., Sinclair, K., Kurz, D., McNair, M., Crist, S., Shpritz, L., Fitton, L., Saffouri, R., and Blair, R., (1995), Environmental and economic costs of soil erosion and conservation benefits, *Science*, 267, 1117–1123.

Pimentel, D. and Pimentel, M., (1996), *Food, Energy and Society*, rev. ed., University Press of Colorado, Niwot.

Pimentel, D., Ed., (1997), *Techniques for Reducing Pesticide Use: Environmental and Economic Benefits*, John Wiley & Sons, Chichester, U.K., 444 pp.

Sainz, R., (1998), Framework for Calculating Fossil Fuel in Livestock Systems, Technical Report, University of California, Davis.

Smil, V., (1987), *Energy, Food, Environment*, Clarendon Press, Oxford.

Smil, V., (1991), Population growth and nitrogen: an exploration of a critical existential link, *Popul. Dev. Rev.*, 17, 569–601.

Stout, B.A., (1991), *Handbook of Energy for World Agriculture*, Elsevier, New York.

Stout, B.A., Ed., (1992), *Energy in World Agriculture*, 6 vols., Elsevier, Amsterdam (Vol. 1, Singh, R.P., Ed., *Energy in Food Processing*; Vol. 2, Hesel, Z.R., Ed., *Energy in Plant Nutrition and Pest Control*; Vol. 3, McFate, K.L., Ed., *Electrical Energy in Agriculture*; Vol. 4, Parker, B.F., Ed., *Solar Energy in Agriculture*; Vol. 5, Peart, R.M. and Brooks, R.C., Eds., *Analysis of Agricultural Systems*; Vol. 6, Fluck, R.C., Ed., *Energy in Farm Production*).

Tamminga, S., Zemmelink G., Verdegem M.C.J., and Udo, H.J.M., (1999), The role of domesticated farm animals in the human food chain, in Quantitative Approaches in System Analysis 21, Technical Report, DLO Research Institute for Agro-biology and Soil Fertility, C.T. de Wit Graduate School for Production Ecology, Wageningen University, Wageningen, Netherlands.

USBC (United States Bureau of the Census), (1994), *Statistical Abstract of the United States, 1990*, U.S. Department of Commerce, Washington, D.C.

WRI (World Resources Institute), (1994), *World Resources 1994–95*, Oxford University Press, New York.

Zipf, G.K., (1941), *National Unity and Disunity: The Nation as a Bio-social Organism*, Principia Press, Bloomington, IN.

11

Multi-Scale Integrated Analysis of Farming Systems: Benchmarking and Tailoring Representations across Levels*

This chapter deals with farming system analysis, a topic that entails dealing with all the typologies of epistemological problems associated with complexity discussed so far. A useful knowledge of farming systems, in fact, has to be based on a repertoire of typologies of farming systems. On the other hand, all farming systems are special, in the sense that their representations must include the specificity of their history and the specificity of local constraints. To make things more difficult, the very concept of farming systems implies dealing with a system that is operating within two nonequivalent contexts: a socioeconomic context and an ecological context. That is, any real farm is operating within a given typology of socioeconomic system and within a given typology of ecosystem. The two identities of these two contexts are very important when selecting an analytical representation of a farming system. In fact, a typology of farming system has to be related, by definition, to an expected associative context. This is the step where concepts such as impredicative loop analysis (ILA) and multi-criteria performance space (MCPS) become crucial. In fact, they make it possible to characterize the reciprocal constraints associated with the dynamic budget of the farming system considered, which is interacting with its two contexts exchanging flows of energy, matter and added value. A given selection of typologies used to represent its identity (system, typical size, metabolic flows considered) has to be compatible with the set of typologies used to represent the identities of its socioeconomic and ecological context.

This chapter is organized in three sections. Section 11.1 introduces in general terms basic concepts related to farming system analysis found in literature. These concepts are translated into a narrative compatible with the theoretical concepts and analytical tools presented in Part 2. Section 11.2 presents an approach (land–time budget) useful for applying ILA to farming systems. This approach can be used to (1) individuate useful types across levels for a multi-scale integrated analysis (MSIA) and (2) establish a link between socioeconomic types used to represent farming systems across levels. Section 11.3 illustrates the possibility of linking a multi-level analysis of farming systems based on typologies across levels to a multi-level characterization of land uses associated with these types. In this way, a multi-level multi-criteria analysis of farming systems can be tailored to the various strategy matrices used by relevant agents. This section ends by providing an overview on how the heterogeneous information space built by adopting the analytical tool kit suggested in this chapter (different ILAs based on land–time budget and multi-criteria performance space associated with land use maps over multiple hierarchical levels) can be handled when discussing possible policies and scenario analysis.

11.1 Farming System Analysis

11.1.1 Defining Farming System Analysis and Its Goals

An overview of literature about the challenges implied by an integrated analysis of farming systems provides a list that is very similar to that discussed so far about the challenges implied by sustainability analysis:

---

* Tiziano Gomiero is co-author of this chapter.

1. Agricultural systems are complex systems operating on several hierarchical levels (with parallel processes definable only on different spatio-temporal scales). This makes impossible an exhaustive description of them with a set of assumptions typical of a single scientific discipline (Hart, 1984; Conway, 1987; Lowrance et al., 1987; Ikerd, 1993; Giampietro, 1994a, 1994b; Wolf and Allen, 1995).

2. Any substantive comparison of farming options would require the simultaneous consideration of (1) a large variety of different production processes, strategies, techniques and technologies that can be found all around the world; (2) the need to use agronomic, ecological and socioeconomic analyses in parallel to verify the compatibility of farming techniques with different sets of constraints coming from both the biophysical and socioeconomic characteristic of the system; and (3) the need to expand the range of assessments of the farming system over multiple and alternative views of it, to check the feasibility of proposed solutions in ecological, economic and social terms (Altieri, 1987; Brown et al., 1987; Lockeretz, 1988; Brklacich et al., 1991; Allen et al., 1991; Schaller, 1993).

3. Specific policies or technological changes are unlikely to generate absolute improvements (when considering all possible hierarchical levels of organization and every possible perception found among stakeholders). We can only expect to obtain trade-offs, when assessing the effect of changes on different scales and in relation to different descriptive domains.

   Recent enthusiasm regarding win–win scenarios in many cases is buoyed by scaling error. Explicit recognition of the implications of necessary trade-offs, both positive and negative, promotes the development of mechanisms to support losers. Failure to confront the fact that losers are consistently produced exaggerates the negative impact they have on system performance. (Wolf and Allen, 1995, p. 5)

4. The trade-offs faced when comparing the effects of different options are not always commensurable (when facing cases of sustainability dialectics). Costs and benefits generated by a particular change in relation to a given criterion and a relative indicator of performance can be measured indeed. However, this can be done only by mapping changes in an observable quality associated with a given descriptive domain (at a given scale) at the time. As soon as we deal with problems of sustainability (when different relevant scales have to be considered simultaneously) and when our selection of relevant stakeholders includes several social groups (when the existence of legitimate but contrasting views is unavoidable), the various assessments of heterogeneous perceptions of costs and benefits become nonreducible and incommensurable (Martinez-Alier et al., 1998; Munda, 1995). A perfect example of this scientific impasse is found when scientists are asked to quantify costs–benefits related to the dilemma of fighting hunger in the present generation vs. preserving biodiversity for future generations.

5. When dealing with the issue of sustainability, a substantive definition of rationality cannot be adopted (Simon, 1976, 1983). After accepting that conflicting effects on different levels, when evaluated from different perspectives and values, cannot be quantitatively evaluated by a reduction and aggregation into a single indicator of costs–benefits, we are forced to admit that an optimum strategy of development for farming systems cannot be selected by experts once and for all. The very definition and perception of sustainability is inherently sensitive to changes in the analytic context (Wolf and Allen, 1995; Allen et al., 2001). Sustainability in agriculture has to do with conflict management and an adequate support for decision making in the context of complexity (e.g., participatory techniques and multi-criteria methods as discussed in Chapter 5). These methods require analyses able to link actions at one scale to consequences generated at other scales. Moreover, the choices made to represent these consequences have to reflect the variety of perceptions found among the stakeholders.

In conclusion, an integrated analysis of agroecosystems requires the ability to describe farming systems simultaneously on different space–time scales (e.g., biosphere, regional and local ecosystems; macro-economic, community, micro-economic, farmer levels) and by adopting nonequivalent descriptive domains (when considering the economic, ecological, technical, social and cultural dimensions). In particular, it requires the ability to tailor the selection of an integrated package of indicators of performance on the set of system characteristics that are relevant for the agents that are making relevant decisions within the farming system considered.

When translating this set of challenges in the narrative proposed so far in this book, we can say that farming system analysis is about selecting a finite set of useful perceptions and representations of the performance of agroecosystems in relation to events occurring within a local space–time domain. This entails that within such a representation the farming system is assumed to be interacting with a context that is made of both socioeconomic and ecological systems. Both of these self-organizing systems, in turn, do have (and have to be characterized by using) a given set of identities.

According to the discussion presented in Chapter 5, the challenge for scientists willing to perform an integrated analysis of farming systems becomes that of finding a useful problem structuring for framing in formal terms the specific problem of sustainability considered. Such a framing has to be able to cover relevant scales and dimensions of analysis. General principles and disciplinary knowledge are certainly necessary for this task. However, at the same time, they are not enough. Crucial disciplinary knowledge has to be tailored to the specificity of a given situation found at a given point in space and time.

As noted in Chapter 5 any multi-criteria analysis of sustainability requires starting with a preanalytical definition of:

1. Relevant stakeholders to be considered when deciding what are the relevant perspectives to be addressed by the problem structuring (the set of goals and fears to be considered relevant in the analysis to be able to reflect the relative set of legitimate but nonequivalent perceptions of costs and benefits for relevant agents).

2. A performance space used for the evaluation (a set of indicators of performance able to characterize the effects of changes in relation to the set of relevant criteria of performance selected in the previous steps).

3. A package of models able to generate a multi-scale integrated analysis of possible changes. This requires the individuation of:

    a. Key attributes and observable qualities determining the particular set of identities used to represent the investigated system.

    b. Key mechanisms generating and maintaining the various forms (and relative perceptions) of the metabolism of the system that we want to sustain. The analysis has to deal with the ability to stabilize key flows such as endosomatic energy, exosomatic energy, added value and other critical matter flows (e.g., water, nitrogen) and with the ability to reduce the emission of harmful flows (e.g., pollutants). This implies addressing the problem of how to characterize the identity of the metabolic system as a whole (at the level n) and, in relation to this, whole how to characterize the relevant identities of its lower-level components controlling the various metabolic flows considered relevant (at the level $n - 1$). These two set of identities within the requirement of sustainability, in turn, have to be compatible with the characteristics of the larger context (level $n + 1$) and lower-lower-level characteristics associated with the definition of input and wastes (level $n - 2$).

    c. The set of existing constraints on the possible actions (policies, choices) to be adopted. The individuation of constraints is related to the existence of nonequivalent dimensions of feasibility (e.g., biophysical, technical, socioeconomic, cultural, ecological).

    d. Existing drivers that are determining current evolutionary trends.

Only at this point does it become possible to gather data and set up experimental designs to operate such an integrated package of models and indicators useful to discuss scenarios and options. The generation of this scientific input has been called in Chapter 5 the development of a discussion support system, and this should be considered a crucial starting point for a sound process of integrated analysis and decision making.

From what was discussed in Parts 1 and 2, we can say that any farming system is organized — as any other complex adaptive system made up of humans — in a hierarchy of nested typologies. This entails that when analyzing these systems, we should expect to find several agents operating at different levels. In turn, this requires the consideration of several sets of relevant identities to be studied in a multi-scale analysis. These agents can be individual households (composed of individual human beings) that are organized in larger units, villages and communities (composed of households) that are organized in larger units, provinces and regional administrative units (composed of villages) that are operating within larger socioeconomic contexts, and countries and macro-economic areas. These socioeconomic systems (perceived at various levels) in turn are embedded in ecological entities that are also organized in nested hierarchies (which are perceived in terms of different identities at different levels of organization).

According to what was said in Part 1, we cannot expect to find a standard set of perceptions and representations of performance (associated with the building of a single descriptive model) that can be used once and for all to deal with such a multi-scale integrated analysis of the performance of farming systems. Any individual model used to assess the performance of farming systems will reflect just a given selection of relevant criteria, key mechanisms and contiguous hierarchical levels. Therefore, no matter how elaborated, mathematical models will be necessarily referring to a single descriptive domain at a time (a given definition of identity for the modeled system), which is associated with a particular point of view. What is needed to get out from this predicament is a characterization, based on a parallel reading of agroecosystems at different hierarchical levels. Such a characterization must be rich enough to be useful for the discussion and negotiation of policies among relevant stakeholders. This is the criterion to be used for controlling the quality of a given characterization of a farming system.

This last requirement implies an additional problem. Scientific information has to be packaged in a way that will be useful for the various agents that are in charge of decision making at different levels. As noted in Chapter 8, this implies the ability to consider in parallel the characteristics of different observed–observer complexes. In the case of farming system analysis, the observed are (1) terrestrial ecosystems managed by humans and (2) relevant human agents. The observers are the various interacting agents, which are both acting and deciding how to act. Within this frame, humans making relevant decisions about land use are included in the complex observed–observer two times: as observers and as observed.

Decisions in agriculture can refer to the particular mix of crops to be produced and the selection of related techniques of production. The various complexes observed–observer, however, are operating in parallel at different scales, and they do decide, act and change their characteristics at different paces. For example, in a market economy governments can only implement their choices about the adoption of a given set of production techniques using policies and regulations. On the contrary, farmers can decide directly to adopt one given technique rather than another. In general, we can say that human agents operating within a given farming system base their decisions on:

1. An option space (perception and representation of the severity of constraints coming from both the ecological and the socioeconomic interfaces)
2. A strategy matrix (the perceived or expected profile of nonequivalent costs and benefits associated with the various options, which is weighted and evaluated in relation to a given set of goals/wants and fears reflecting cultural values)

The couplets of option spaces and strategy matrices adopted by agents operating at different hierarchical levels (e.g., governments vs. farmers) are nonequivalent. As noted in Part 1, the combined use of nonequivalent couplets of option space and strategy matrix often result in the adoption of different strategies (recall the example of more taxes, which is good for the governments and bad for farmers). This is another way to say that a generalization of a standard problem structuring (an optimizing model) providing a substantive definition of optimal performance within a farming system is impossible.

To make things more difficult, not only should we expect differences in the definitions of both option space and strategy matrix when dealing with agents operating at different hierarchical levels, but also it is normal to expect important differences in the characteristics of both option space and strategy matrix for agents that are operating in different typologies of context (meat producers in Sahel and in the Netherlands) and have a different cultural background (Amish and high-tech farmers in Canada). The existence of unavoidable differences in the definition of both option space and strategy matrix for nonequivalent observers/agents will obviously be reflected in the existence of legitimate, but contrasting *optimizing strategies* adopted by these agents.

For example, pastoralists operating in marginal areas tend to minimize their risks by keeping a certain redundancy (safety buffers) in their farming system even though this implies not taking full advantage of momentarily favorable situations (a suboptimal level of exploitation of their resources on a short time horizon). Often traditional techniques imply choosing or accepting to operate in conditions that provide a return that is lower than the maximum that would be achievable at any particular moment. In this case, pastoralists are not considering the short-term maximization of technical efficiency as a valid optimizing criterion. Actually, the solution of keeping a low profile, so to speak, can increase the resilience of this system over the long run. In the long term, in fact, shocks and fluctuations in boundary conditions are unavoidable for any dissipative system. Therefore, the bad performance of pastoralists — perceived when representing their performance in terms of limited productivity on the short term, when compared with beef lots — can be explained, when expecting future changes still unknown at the moment, by the greater ability of a redundant system to cope with uncertainty.

On the contrary, meat producers of developed countries are mainly focused on the maximization of the economic return of their activity (maximizing efficiency in relation to short-term assessment). This is equivalent to granting an absolute trust in the current definition (perception or representation) of optimization for the performance of the system of production (maximization of output/input under present conditions). This trust is justified by the fact that when deciding about technical and economic choices, the physical survival of individual members of the household is not at stake. In developed countries, in fact, the responsibility for guaranteeing the life of individual citizens against perturbations, shocks and unexpected events has been transferred to functions provided by structures operating in the society at a higher hierarchical level (e.g., in the indirect compartment where, at the country level, one can find organizations in charge of health care and emergency relief). This is another example of how changes in the indirect compartment (more services) can affect changes in the direct compartment (more short-term efficiency).

The process of selection of techniques and related technologies is also affected by the extreme variability of the characteristics of the context. That is, after deciding what to produce and the how to produce it (basic strategies), farmers have to implement these choices, at the farming system level, in the form of a set of procedures that are linked to the operation of a set of specific technologies. Again also in this case, subsistence farming is affected in this step by the existence of location-specific constraints (e.g., techniques of food processing in the Sahel areas are not feasible in Siberia and vice versa), whereas farmers operating in developed countries can afford to use extensive adaptation technologies (e.g., fertilizers, pumps and machinery used in the U.S. can also be used in Australia or the Netherlands).

These examples show again that deciding about the advisability and feasibility of choices made by farmers at a given point in space and time is not a task that can be formalized in an established protocol to be applied to whatever farming system. Concepts such as feasibility and advisability have to be checked each time at different levels and in relation to different criteria and different dimensions of performance. This multiple check is required for every step of the chain of choices going from the definition of basic strategies for socioeconomic systems (a definition that is obtained at the level of the whole society) to the final step of adoption of production technologies in a given day, at the field level. Different typologies of constraints can only be studied in relation to cultural identity, sociopolitical organization, characteristics of the institutional context, macro-economic variables, availability of adequate know-how in the area, available knowledge about local ecological processes and micro-economic variables affected by short-term fluctuations.

11.1.2 Farming System Analysis Implies a Search for Useful Metaphors

After accepting the point that farming systems belong to the class of nested metabolic systems organized in holarchies, we should expect that they are affected by the epistemological paradox discussed in Parts 1 and 2 of this book:

1. Holons are organized according to types. This is what makes it possible to make models of them.
2. At the same time, individual elements of holarchies are all special, since they are particular realizations of a type. Because of this, they have their own special history that makes them unique.

In this regard we can recall the example of Gina and Bertha discussed in Chapter 8. The four pictures given in Figure 8.1 and Figure 8.3 can be seen as either the same set of four types (girl, adult woman, mature woman, old woman) realized by two distinct individualities or two given individualities getting through a set of expected types. This distinction is important to understand the difference between basic disciplinary knowledge and applied knowledge for sustainability. For example, medicine is interested in knowing as much as possible about typologies of diseases. Relating this to the two sets of pictures shown in Figure 8.1 and Figure 8.3, we can say that the four typologies are the information that matters for the development of disciplinary knowledge. On the other hand, a doctor facing an emergency has to take care of patients one at a time. That is, the general knowledge about types given by medicine is required to provide the physician with a certain power of prediction. However, when coming to a specific serious case, it is always the special situation of a particular patient that counts. As discussed in Chapter 4, this situation found often in medicine can be associated with the typical situations faced in the field of science for governance (postnormal science). In these cases standard protocols cannot be applied by default. Even the best physician in the world cannot decide a therapy that implies a certain level of hazard without first interacting with the patient to get an agreement on the criteria to be adopted in the choice. Another useful metaphor that can be used to illustrate the difference of relevance of (scientific) information based on typologies vs. information that is tailored on the special characteristics of an individual realization is given in Figure 11.1.

In the upper part of the figure there is a graph reporting the trend of suicides in Italy over the period 1980–1992. Using this set of data, it is possible to gain a certain predicting power on the characteristics of the class. For example, it is possible to guess the number of suicides in a given year (e.g., 1987) even when this information is missing from the original set. On the other hand, by looking at the poem given in the lower part of Figure 11.1 — the last words written by Mayakovsky before his suicide — it is easy to notice that the information given in the upper part of the figure is completely irrelevant when dealing with actions of individuals. That is, a data set useful for dealing with the characteristics of an equivalence class has limited usefulness when dealing with the actions of individual realizations. Statistical information about the suicides of a given country is no good for (1) predicting whether a particular person will commit suicide at a given point in space and time, or (2) preventing the suicide of that person.

The limited usefulness of information related to typologies for policy making is directly related to the challenge found when dealing with the analysis of farming systems. In fact, it is possible and useful to define typologies of farming systems. These typologies could be subsistence farming system in arid areas based on millet, paddy rice farming system in densely populated areas, high-input corn monoculture on large farms and shifting cultivation in tropical forests. Starting from a set of typologies, we can also get into even more specific typologies by adding additional characteristics (categories) to be included in the definition of the identity of the particular farming system — e.g., Chinese farming system based on a mix of subsistence and cash crops, characterized by paddy rice and a rotation based on a mix of vegetables sold to the urban market. However, no matter how many additional categories and specifications we use for defining an identity in terms of a typology for the farming system under analysis, it is unavoidable to discover that as soon as one gets into a specific place, doing fieldwork, each person, each farm, each field, each tribe, each town, each watershed is special. Moreover, to this special individuality special events are happening all the time. That is, no matter how elaborated is the label that we use to describe a given farming system in general terms, it is always necessary to deal with the unavoidable existence of special characteristics associated with a given situation. As discussed at length in Chapter 2, this is an unavoidable predicament associated with the perception and representation of holons, which can only be obtained, by humans, in terms of types and epistemic categories. Any characterization based on a finite selection of types, however, will cover only a part of the relevant characteristics of a real learning holarchy operating in the real world to which it refers. To make things more difficult, the validity of this coverage is bound to expire.

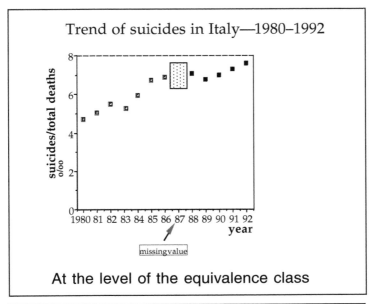

FIGURE 11.1 Information about suicides.

The consequent dilemma for the analysts is to look for a representation that should be able to achieve a sound balance between (1) the need of adopting general types (what makes it possible to learn, compress and transfer knowledge from one situation to another) and (2) addressing the peculiarity of individualities (e.g., tuning the analysis to a point that it can include the feelings of individual human systems found in the study, taking into account the special history of the investigated system).

A theoretical discussion of this dilemma can be related to the distinction, proposed by Robert Rosen, between models and metaphors when dealing with the representation of complex systems (Rosen, 1985, 1991; Mayumi and Giampietro, 2001):

Model — A process of abstraction that has the goal of representing within a formal system of inference causal relations perceived in a subset of relevant functional properties of a natural system. This subset represents only a small fraction of the potential perceptions of observable qualities of the modeled natural system. A model, to be valid, requires a syntactic tuning between (1) the relation among values taken by encoding variables (used to represent changes in relevant system qualities) according to the mathematical operations imposed on them by the inferential system and (2) the causal relation perceived by the observer among changes in the finite set of observable qualities of the natural system included in the model. That is, after having performed the calibration of a given model to a specific situation, it is possible to check the validity of such a model by checking its ability to simulate and provide predictive power (a congruence between 1 and 2) to those using it. As noted in Chapter 8, when dealing with the evolution (sustainability) of complex adaptive systems organized in nested hierarchies, all models are wrong by definition and tend to become obsolete in time. The seriousness of this predicament depends on the number of legitimate but nonequivalent perspectives that should be considered in the problem structuring and by the speed of becoming of (1) the observed system, (2) the observer and (3) the complex observed–observer.

Metaphor — The use of a basic relational structure of an existing modeling relation, which was useful in previous applications, to perform a decoding step (to guess a modeling relation) applied to a situation in which the step of encoding is not possible. That is, we are using the semantic power of the structure of relations of a class of models, without having first calibrated a given individual model on a specific situation and without having measured any observable quality of the natural system about which we are willing to make an inference.

Translating the technical definition of a metaphor into plainer words, we can say that a metaphor makes it possible, when studying a given system at a given point in space and time, to infer conclusions, guess relations and gain insights only by taking advantage of analogies with other systems about which we have preliminary knowledge. Therefore, metaphors make it possible to use previous experience or knowledge to deal with a new situation. A metaphor, to be valid, must be useful when looking at a given natural system for the first time in our life, to guesstimate relations among characteristics of parts and wholes that can be associated with systemic properties, even before interacting with the particular investigated system through direct measurements. From what has been said so far, we can say that to generate a useful metaphor, we have to be able to share the meaning assigned to a set of standard relations among typologies and expected associative contexts. According to this definition, the four-angle figures given in Figure 11.2 and Figure 11.3 are examples of metaphors.

When coming to farming system analysis, the use of metaphor should make it possible to apply lessons learned from studying a farming system producing millet in Africa to the solution of a problem of corn production in Mexico. A metaphor can be used to define the performance of a given system in relation to a given criterion of performance (e.g., when assessing the trade-off of efficiency vs. adaptability), but using a set of variables (a definition of indicators) that is different from the set adopted in a previous study (e.g., when applying general principles learned about milk production to aquaculture). To be able to do that, however, the analysts have to frame their analysis in a way that generates relational patterns within a system that share a certain similarity with other relational patterns found in other systems. When looking for useful metaphors, the local validation of individual models obtained using sophisticated statistical test ($p = .01$) is beside the point. The accuracy of prediction associated with a given model in a given situation does not guarantee the possibility of exporting the validity of the relative basic metaphor (within which the model has been generated) to other situations. When moved to another situation, the same model can lose either relevance or predictive power, or both. Therefore, the real test of usefulness is whether a given set of functional relations among system qualities — indicated as relevant within the metaphor — will actually be useful in increasing our understanding of other situations.

In this sense, impredicative loop analysis provides a common relational analogy (a typology) of self-entailment among the values taken by parameters and variables — used to characterize parts and the whole — within a standardized representation of autocatalytic loops (see Figure 9.1). This relational analogy over autocatalytic loops can be applied to the analysis of the metabolism of different systems (see examples in Chapter 7) using different choices of representation of such a metabolism. By adopting the approach proposed so far, this translates into a selection of (1) variables used to characterize flows (e.g., the characterization of size as perceived from the context — selection of extensive variable 2 — e.g., food energy, solar energy, added value, water, exosomatic energy) and (2) variables used to characterize the black box (e.g., the characterization of size as perceived from within — selection of extensive variable 1 — e.g., human activity, land area, kilograms of biomass).

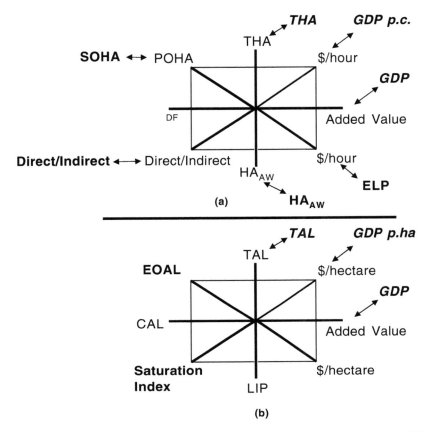

FIGURE 11.2 (a) A look at the impredicative loop of added value (EV2) in relation to human activity (EV1). (b) A look at the impredicative loop of added value (EV2) in relation to land area (EV1).

In conclusion, to face the challenge associated with an integrated analysis of agroecosystems across hierarchical levels, we should base our representation of performance of farming systems on useful metaphors (classes of meta-models), rather than on specific models. This requires developing a tool kit made up of a repertoire of tentative problem structurings that have to be selected and validated in relation to a specific situation before getting into a more elaborate analysis of empirical data. This preliminary selection of useful typologies, relevant indicators and benchmarking of expected ranges of values for the variables will represent the basic structure of the information space used in the analysis. After having validated this basic structure in relation to the specificity of the given situation, the analyst can finally get into the second phase (based on empirical data) of a more detailed investigation.

11.1.3 A Holarchic View of Farming Systems (Using Throughputs for Benchmarking)

The viability and vitality of holarchic metabolic systems can be checked in relation to two nonequivalent categories of constraint:

1. **Internal constraints** — Constraints associated with the characteristics of the identities of lower-level components of the black box. Internal constraints do limit the ability of the system to increase the pace of the throughput (the value taken by intensive variable 3). This limitation can be associated with (1) human values expressed at lower levels and (2) the (in)capability of providing the required amount of controls for handling and processing a larger throughput. The presence of these constraints translates into a set of limitations of the value that can be

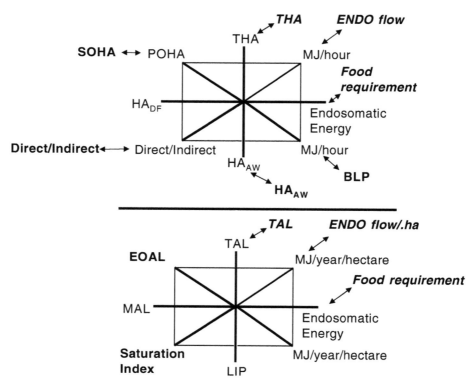

FIGURE 11.3 (a) A look at the impredicative loop of food energy (EV2) in relation to human activity (EV1). (b) A look at the impredicative loop of food energy (EV2) in relation to land area (EV1).

taken by the different variables used as IV3, when characterizing the throughput at the level *n*. Besides the existence of cultural curtaining on human expansion into the environment due to ethical reasoning (e.g., as in the case of Buddhists or Amish), technical bottlenecks (shortage of technical devices) can prevent a socioeconomic element from handling more power (e.g., reaching higher values of EMR_i). We can describe this internal technical limitation as the (in)ability to generate more goods and services in the working compartments (reaching higher values of BPL_i) even when additional input and sink capacity would be available. In economic terms, we can describe an internal constraint as the (in)ability to generate more added value per unit of labor (reaching higher values of ELP_i). Within an economic discourse, internal constraints are in general related to shortages of various forms of human-made capital (e.g., technology or know-how).

2. **External constraints** — Constraints associated with the characteristics (the weak identity) of the environment (level *n* + 1). As noted in the technical section of Chapter 7, the required admissibility of boundary conditions for the black box can be seen as a weak identity assigned to the environment, which is supposed to supply — by default — a certain flow of inputs and absorb the relative flow of wastes to the metabolic system. External constraints are those limiting the value that can be taken by the extensive variables used to characterize the size of the metabolic system (the carrying capacity of the context, so to speak). In terms of input and output, this refers to a limit on (1) the available input that can be appropriated from the context over a given period of time and (2) the sinking capacity of the context (a limited capability of absorbing the wastes associated with the given metabolism of the black box). Put another way, external constraints entail a limit — referring to the selection of the extensive variable 2 and in relation to extensive variable 1 — on how big the metabolism of the black box can be in relation to what is going on in its context. Within this representation, external constraints can

be characterized — after having selecting EV1 and IV3 —in terms of the resulting availability and adequate supply of inputs (EV2), and to the possibility to safely dispose of the relative flow of wastes.

The existence of these constraints is often ignored by neoclassical economists who assume that economic systems will always be able to replace or substitute limiting resources or limiting environmental services thanks to their ingenuity. Put another way, an economic narrative tends to neglect the existence of external constraints. Some economists admit that external constraints might exist, but because of the assumption of moderate scarcity (the economic version of the biophysical default assumption of admissibility of boundary conditions), they never consider them as ultimate constraints. In economics, the potential role of external constraints is accounted for in terms of availability and quality of natural resources and environmental services. Also, when coming to the definition of external constraints in biophysical terms, we get into slippery territory. In fact, the very concept of weak identity for the environment implies acknowledging an unavoidable level of ignorance in the real definition of these constraints (it is impossible to predict all the possible mechanisms of incompatibility with ecological processes that should be included in such an evaluation).

It should be noted that any assessment of external constraints (the limit that ecological processes operating at the level $n + 2$ can imply on the stability of favorable boundary conditions of socioeconomic processes considered at the level $n + 1$) would require the ability to compare (1) the biophysical size of the metabolism of the socioeconomic system (using a set of variables for EV2) and (2) the biophysical size of the metabolism of the ecological system in which the metabolism of socioeconomic systems is occurring (using the same set of variables for EV2). Such an analysis would require the study of the nature of the interaction of these two processes of self-organization and the relative interference that the metabolism of socioeconomic processes implies over the metabolism of ecological systems. That is, the analysis should study how the value taken by EV2 used to characterize the size of the socioeconomic system affects the value taken by EV2 used to characterize the size of the ecological processes guaranteeing the stability of boundary conditions. Coming to the possible selection of a mechanism of mapping useful for comparing the relative size of these two self-organizing systems, it becomes obvious that such an assessment cannot be done from within a descriptive domain provided by economics (e.g., when adopting added value as extensive variable 2). In fact, a monetary variable reflects the representation of the perceptions of usefulness for human observers/agents that are interacting within a structured economic process (a view from within). That is, assessments of flows of added value (the relative variable is a proxy of the monetary value associated with perceptions of the utility of exchanged goods and services) refer only to the equivalence class of transactions occurring within a given economic process characterized in terms of a specified market, preferences and institutions. This mechanism of mapping of exchange values within a given market (the system quality measured by monetary variables) does not and cannot account — when considered as an extensive variable 2 — for the perception of the size of the black box (the socioeconomic system) from the outside by those observers/agents determining the identity of ecological systems (Giampietro and Mayumi, 2001; Mayumi and Giampietro, 2001). Put another way, monetary variables are crucial for checking the compatibility of the characteristics of a socioeconomic system at different levels with human aspirations expressed by elements operating within human holarchies (the economic and socioeconomic viability of solutions for the various human elements making up a farming system). However, monetary variables are not relevant for assessing the ecological compatibility. What humans are willing to pay for preserving trees has nothing to do with the determination of the threshold of interference on the natural mechanisms of control of a terrestrial ecosystem, which can imply a collapse of its integrity (recall the discussion about Figure 10.15).

The impossibility of using a descriptive domain based on an economic narrative for performing an analysis of ecological compatibility should not be considered a problem. As discussed in previous chapters, when adopting a multi-criteria framework, there is no need to collapse nonequivalent descriptive domains into a unique cost–benefit analysis to deal with integrated analysis of sustainability. .

In conclusion, the analysis of internal constraints of a socioeconomic system deals with the perception and representation of feasibility of humans operating from within the black box. Such an analysis refers to processes and mechanisms about which the only relevant observers and agents are humans. On the contrary, the analysis of external constraints should necessarily deal with perceptions and representations obtained from outside the black box. In this case, there are also perceptions and representations adopted by nonhuman observers and nonhuman agents (those operating within ecological processes) that count. These perceptions and representations are used by ecological agents that are interacting when generating the process of self-organization of ecosystems. This can seem a trivial consideration. However, this implies that a substantive representation of the metabolism of human societies from the outside cannot be obtained. Humans can only perceive and observe themselves from within their cultures and their social structures. Processes that are operating outside the black box are determined by decisions taken by nonequivalent observers and agents (elements of ecosystems), which are not sharing their meaning with us. Therefore, the mechanisms regulating these processes can only be partially known by humans. To make things worse, many human agents, which are relevant because of their decisions, always represent an unknown context of other human agents.

Within this frame, farming system analysis is location specific by default. Therefore, it is unavoidable to expect a large dose of uncertainty and ignorance about how to perceive and represent the role of external constraints, when deciding how to structure sustainability problems. Using as an example the well-known scientific debate over the sustainability of human progress, there are scientists/observers that do not see any future problem of sustainability for humans. Some of them (the so-called cornucopians) imagine as possible an unlimited adjustment of boundary conditions (the characteristics of ecological processes and the cultures of other human systems) on their own cultural definition of what human systems should be. This is at the basis of the myth of technological fix. Technology will give immortality to humans both to current individuals (e.g., through clonation) and to current civilization (e.g., through a continuous supply of silver bullets). On the contrary, other scientists/observers (the so-called prophets of doom) see no hope for sustainability since at the moment there is no known solution to accommodate existing trends of evolution of the characteristics of socioeconomic systems (expected changes in population and expected standard of living) within the room allowed by the compatibility with ecological processes. A third group of scientists/observers (that we could call the hyperadaptationists) imagine as perfectly acceptable and feasible a heavy and dramatic readjustment of the characteristics of human systems to less favorable boundary conditions — e.g., back to caves and human muscles, when environmental services will be in shortage, with or without much trouble. Obviously, nobody can decide, in a substantive way, who is right and who is wrong, in this debate. In reality, all these positions just reflect nonequivalent definitions of the original problem structuring used for characterizing sustainability.

For this reason, when dealing with integrated analysis of sustainability, it is important to characterize and handle in an integrated way and simultaneously the various pieces of the puzzle. It is crucial to focus the discussion on a shared meaning given to the problem structuring. Sustainability, when framed within the metaphor of holarchic metabolic systems, has to do with the ability to maintain the compatibility across levels of different sets of identities associated with equivalence classes of organized structures (which in turn can be associated with characteristic metabolic patterns) making up the various nested holons. This requires a chain of compatibility between the established mechanisms operating inside the black box (the metabolism of structural lower-level elements) and the set of identities associated with the maintenance of stable boundary conditions on the interface black box–environment (the essences associated with the validity of relational functions expressed at higher levels) across levels. As noted in Chapter 7, the simultaneous validity of identities of elements of metabolic holarchies requires forced congruence between the characteristics of processes occurring within the black box and the characteristics of processes occurring outside the black box, which are required to guarantee a stable associative context to those metabolic patterns. This reciprocal constraining of characteristics between the lower level and higher level is at the basis of impredicative loop analysis. An analysis of these characteristics can be very useful in farming system analysis.

11.1.4 Benchmarking to Define Farming System Typologies

The concept of benchmarking in the context of ILA translates into the characterization of a given typology of farming system in relation to the selection of (1) a set of extensive and intensive variables and (2) an integrated set of typologies of activities (investments of human activity or land in relation to densities of flows) that can be used to establish a relation between the characteristics of parts and the whole. An ILA implies looking for a forced relation over the loop associated with a dynamic budget and the pace of a given throughput. An example of the rationale of this approach is given in Figure 10.1. There the two nonequivalent definitions of throughput (assessed using food as EV2) are (1) productivity per hectare of investment of total land (EV1) in the typology land in production and (2) productivity per hour of investment of total human activity (THA) (EV1) in the typology agricultural labor. These two definitions of throughput are applied at two hierarchical levels: (1) at the field level (this translates into an assessment that reflects technical coefficients) and (2) at the level of the whole country (this translates into an assessment that refers to demographic and bioeconomic pressure perceived and represented at the level of the socioeconomic context in which the farm is operating). This comparison of the value taken by the two nonequivalent definitions of IV3 at different hierarchical levels makes it possible to compare the relative compatibility of the typologies (defined on two hierarchical levels: the field level and the country level), both associated with this throughput (Figure 10.2). The rationale of this approach has been discussed using theoretical examples of analysis of metabolic systems (the relation between the characteristics of organs and the whole body) in Figure 6.1 and Figure 6.2 and in theoretical terms in Chapter 7.

Getting more into the details of impredicative loop analysis, we can say that this approach makes it possible to establish a relation between the values taken by a set of intensive variables 3 used to characterize a given farming system at different levels (e.g., at the level of individual households or at the level of a village) and the values taken by the same set of intensive variables 3 used to characterize the socioeconomic context within which this farming system is operating. The same can be done in biophysical terms when comparing densities of matter flows against land area. This has been discussed in Figure 9.1 through Figure 9.3. Here we want to discuss more in detail this feature in relation to benchmarking, that is, the characterization of a range of values expected for indicators, which can be associated with the identity of a typology of a farming system. To introduce the basic rationale, we can use two generic ILAs referring to a generic socioeconomic entity (either a household or a village) belonging to a given farming system (Figure 11.2 and Figure 11.3).

The analysis of Figure 11.2 is based on the adoption of assessments of flows of added value used as EV2, which are represented in the loop in relation to a given set of typologies of possible human activities (activities assumed to be typical of the considered farming system):

Figure 11.2a: EV1 = profile of investments of human activity — This analysis of the impredicative loop starts with a given size of the entity expressed in terms of human activity, EV1 (the size of either the household or the village is expressed by this variable in terms of hours; recall that THA reflects the number of persons making up such an entity). Then, by using added value as EV2, the four-angle figure represents on the axis on the right the level of economic interaction of this entity with its context (the total added value generated and consumed in a year by the household or the village). This represents, at this level of analysis, the equivalent of the total gross domestic product (GDP) assessed at the level of the country. The peculiarity of this system of accounting should be recalled here once again. Even though we are calling this an extensive variable determining the size of the system, this variable indicates the amount of added value produced and consumed per year by this entity when interacting with its context (it indicates an amount of added value per year). At this point we can calculate an intensive variable 3, which is characterizing the throughput of the metabolic holon/household (or holon/village) in relation to its higher-level holon, country. This IV3 represents the flow of added value per hour of human activity (or per capita (p.c.) per year), which can be associated with the particular typology of household or village characterized with this ILA. An assessment of an $IV3_m$ makes it possible to compare the characteristics of the farm/household (at the level m) with those of the larger holon within which it is operating. For example, we can define an $IV3_{m+1}$ for the village to which the household type belongs, an $IV3_{m+2}$ for the province to which the village belongs and an $IV3_{m+3}$ for the country to which the province belongs. Assume that the village is the level $m + 1$, the province is the level $m + 2$ and the country the level $m + 3$. In this way, the value of the intensive variable dollars per

/hour can be used to characterize the performance of the economic metabolism of an element defined at the level m (in this case a household) against the average value of dollars per hour of the village in which the household is embedded at the level $m + 1$. In turn, the village can be assessed against the average dollars per hour of the province at the level $m + 2$. In the same way, the province can be benchmarked against the values found in the country (recall here the discussion about Figure 6.1 and Figure 6.2) at the level $m + 3$. In this manner, we can characterize a special household type as being richer than the average found in its village (a local optimum). But at the same time, the analyst can be aware that this local optimum represents a very bad performance when compared with the average of the country. Thus, such an analysis can provide in parallel the big picture (the existence of huge differences related to differences in boundary conditions at the village level) and local fine-grain resolution (the ability to deal with small differences that still count — at the local level — within the village). Depending on the goal of the analysis, the analyst has to select an opportune set of benchmark values to be used in the problem structuring, when reading the performance at different scales and in relation to nonequivalent indicators. The right selection of a benchmark value is crucial. If we were to adopt an indicator of economic labor productivity tailored (in the step of representation) on average values obtained by Dutch farmers (with ELP in the order of tens of U.S. dollars per hour), we would not be able to detect even differences of 100% in the ELP in a farming system in Laos. In fact, such a change would occur in a range of values expressed in cents of U.S. dollars.

Figure 11.2b: EV1 = profile of investments of land area — Another impredicative loop analysis is provided in the lower four-angle figure, but this time the pace of flows (EV2 is always added value) is mapped against land area. Also in this example, it is possible to establish a bridge between different dimensions of the analysis. On the vertical axis, we can characterize how the demographic pressure is determining the total available land of this entity. Then, in the upper-left quadrant, we can characterize the decision made — imagine describing the system at the village level — in relation to the ecological overhead on available land (EOAL). That is, the angle EOAL can be defined as the difference between total available land (TAL) and colonized available land (CAL). This difference is determined not only by the existence of a biophysical overhead (the fraction of available land that cannot be colonized by humans because of severe biophysical limitations), but also because in general there is a fraction of disposable available land area, so to speak, that is not used for production. This fraction is in general set aside to preserve the diversity of habitats and ecological processes (e.g., natural parks, religious sites). Finally, the lower-right quadrant characterizes the saturation index of CAL, that is, what fraction of colonized land is used for land uses alternative to agricultural production. The drive toward higher levels of economic activity (associated with an increase in bioeconomic pressure) tends to increase the saturation index. This means that to assess this saturation index, we have to aggregate typologies of land use within two categories: (1) alternative to agriculture and (2) associated with agriculture. They both are competing for the same fraction of TAL that is invested into CAL. Therefore, the profile of investments of hectares of CAL over this set of possible land use typologies will determine a complex set of trade-offs. For example, using a larger fraction of CAL for supporting industrial activities can increase the density of added value per unit of area, but it can also generate a higher environmental impact through pollution and reduce the internal supply of food. The same analysis can be applied to the mix of typologies of land use adopted within the direct sector. When considering the flow of produced food as EV2, the direct sector becomes land in production (LIP). For example, a large investment in the typology high-input monoculture — associated with larger yields per hectare — implies a reduction of the requirement of land in production per unit of throughput (the number of hectares required to generate a given internal supply of food input). In terms of trade-offs, high-input monocultures can imply a higher level of interference with terrestrial ecosystems and a larger dependence on fossil energy for the internal supply.

Again, this is just an overview of this approach, and it is no time to get into specific analyses. The main point to be driven home in this section is that an integrated use of nonequivalent ILAs can make it easier to deal with multifunctional land uses. Recall here the example of multifunctional land use analysis described in the lower part of Figure 6.2. If a Japanese farmer decides to invest a few hectares of her or his farmland to establish a driving range for practicing golf, then, with this approach, we can characterize how 1 ha of the typology of land use "golf-driving range" generates a density of added value that is much higher (by several times) than that of 1 ha of intensive production of rice. On the other hand, such a choice will not provide any internal supply of rice for Japan. The resulting overall set of trade-offs will depend on the indicators chosen to characterize this choice (the problem structuring chosen by the analyst to characterize and compare the two options).

For this reason, it is important to generate an integrated mix of ILAs able to track changes in relation to nonequivalent indicators, which can be linked to a multi-criteria analysis. For example, the two four-angle figures given in Figure 11.3 are perfectly similar to the two four-angle figures illustrated in Figure 11.2. The only difference is related to the selection of EV2, which in this case is food. Human activity is used as extensive variable 1 in Figure 11.3a, whereas land area is used as extensive variable 1 in Figure 11.3b. As observed before, the intensity of the throughput assessed by IV3 at the level of the household or the village can be compared to the average value found in the society in which the farm is operating.

By assessing differences and similarities among (1) the characteristics of lower-level elements (e.g., technical coefficients and economic characteristics, which can be associated with the set of activities used to represent the profile of investments of human activity and the set of land uses used to represent the profile of investment of land area) and (2) the characteristics of the system as a whole (the value taken by EV1 and EV2 at the level n), it becomes possible to generate indicators characterizing the performance of different types of households and villages.

The common metaphor shared by the four-angle figures shown in Figure 11.2 and Figure 11.3 makes it possible to use the average characteristics of the system under analysis — defined at the focal level m — as indicators. When considering a farm in this analysis, a useful indicator can be obtained by the value of the variable income per capita — dollars per hour ($IV3_m$) — which is associated with the angle in the upper-right quadrant. This indicator is comparable with the indicator used to assess the performance (in relation to the same criterion) of the larger socioeconomic element to which the farm belongs (the larger context, which can be used as a benchmark to characterize the performance of the household) — the average GDP p.c. ($IV3_{m+1}$). At the same time, lower-level characteristics — the economic labor productivity of the direct compartment of human activity (related to the mix of activities in which the household invests labor to generate added value) — can be analyzed and characterized using the same variable $IV3_{m-1}$.

All these indicators can be related to the characteristics of the same impredicative loop applied at various levels. That is, the pace of throughput of added value per unit of human activity (the level of dollars per hour in the upper-right quadrant, $IV3_m$) can be related to (1) the fraction of human activity lost to physiological overhead on human activity (how much THA cannot be considered disposable human activity, because of age structure, sleep and other activity dedicated to personal maintenance (the upper-left quadrant)) and (2) how much of the disposable human activity is invested in activities generating added value vs. alternative activities (the angle in the lower-left quadrant). Finally, within the direct compartment dealing with the internal generation of EV2, we can still focus on the characterization of (1) the set of possible options and tasks (in this case, possible investments of human activities in performing different tasks associated with the generation of added value) and (2) the actual profile of investments of human activity, within the direct compartment, over this set of possible options and tasks. That is, when adopting this approach, the average economic labor productivity of labor in the direct compartment ($IV3_{m-1}$) can be expressed as a function of the average economic return of each of the possible tasks defined at the lower level ($IV3_{m-2}$) and the profile of investment of human activity chosen by the farmer over this set of tasks.

The values taken by the variables used to characterize the loop in the various quadrants reflect key characteristics of the system on different hierarchical levels. These characteristics in turn are associated with the identity of typologies that can be compared with other typologies found in different farming systems in different contexts. To make things more interesting, with ILA it is possible to look at the congruence (or lack of congruence) between total consumption and internal production. The resulting assessment (an internal supply that is larger than, equal to or smaller than the total demand) can be used as an additional indicator. In this way, after having determined how an individual household is doing, in terms of metabolism, in relation to this larger context (an assessment related to the total consumption of EV2), we can also look at the various processes generating the internal supply in the compartment defined as direct (according to the choice of variable for the assessment of EV2). In this way, we can individuate limiting factors (different bottlenecks in relation to different dimensions) on the internal supply in relation to different definitions of flows (e.g., added value, food).

In the case of economic reading, it is possible to do a benchmarking by comparing the level of ELP_m achieved in the element considered in this analysis (the average return of labor of the farm under analysis) against the average value found in the socioeconomic system in which such an element is operating (ELP_{m+1}, that is, the average return of labor of the society within which the farm is operating). This value can also be used to compare the average return of labor of the farm under analysis with the average economic return of other farms belonging to the same typology of farm or rural village (remaining within the level ELP_m). This will indicate how special this farm is in relation to the typology to which it is supposed to belong.

By moving at the level $m - 1$, we can characterize the average economic return of labor referring to individual techniques of production (or tasks) over which working hours are invested in this farming system (ELP_{m-1}). That is, we can explain the value ELP_m using our knowledge of lower-level characteristics $(ELP_{m-1})_i$. In the same way, we can also compare the various economic returns of individual tasks (e.g., the added value generated in working hours in producing rice, aquaculture, flower production) within the average values found for the whole farm, at the level $m - 1$, that is, with the average return of labor obtained for the same set of tasks in other farms belonging to the same farming systems or even to farms belonging to different farming systems. In this way, the effects and constraints associated with changes in bioeconomic pressure (from Figure 11.3a) and demographic pressure (from Figure 11.3b) at the level of the whole country can be understood and linked to changes in boundary conditions perceived and represented at the level of the farm.

In this regard, we want to recall and comment again on three examples of ILA, which have been briefly discussed in Chapter 7. These three examples are applications of ILA aimed at investigating the existence of bottlenecks and biophysical (external) limits for a particular characterization of a dynamic budget associated with the metabolism of farming systems.

Figure 7.14: an ILA of the dynamic budget of net disposable cash (NDC), at the household level, based on land area as EV1 and added value as EV2 — This example points to the critical shortage of available land in relation to the relative required flow of added value for the rural household type considered. That is, according to the identities of land use types found in this farming system and the profile of distribution of CAL over these land use types, this rural household type does not have enough land in production (size of LIP) to cover even a significant fraction of the required flow of net disposable cash.

Figure 7.15: an ILA of the dynamic budget of net disposable cash, at the household level, based this time on human activity as EV1 and added value as EV2 — This ILA is based on a characterization of the set of lower-level typologies of investment of human activity that are included in the direct compartment. In this analysis, the direct compartment includes investments of human activity in tasks generating added value. In this case, not all the activities generating flows of added value require the availability of a relative amount of land in production. It is only because of the option to perform this additional set of tasks (independent from land) that it is possible for this type of rural household to generate an adequate internal supply of net disposable cash. At this point, it is the average economic labor productivity of this household type that is the most relevant parameter. In presence of very severe shortages of land, the average ELP is mainly determined by the average economic return of tasks performed in the off-farm compartment (e.g., off-farm wages). Actually, the particular profile of investments of the resource human activity over the set of options considered in the working compartment can be used to label this typology of rural household. Off-farm rural households are those households that are investing in the category off-farm activities a fraction of their investment of human activity in working, which is higher than the fraction invested in on-farm activities.

Figure 7.16: an ILA of the dynamic budget of food, at the household level, applied to shifting cultivation in Laos at different speeds of rotation (i.e., 3, 5 and 10 years), which is based on land area as EV1 and food as EV2 — This ILA has been used to detect a constraint of a different nature (a nonbiophysical one), which enters into play because of an increase in demographic pressure. In this system, a higher demographic pressure makes it more difficult to maintain the coherence in the reciprocal entailment of identities (of parts and the whole) that would be required to maintain coherence over the pattern of shifting cultivation operating over a 10-year time window (for more details, see the discussion given in Chapter 7).

In general terms, we can say that different mixes of ILAs based on two choices of EV1 (both human activity and land area) and two choices of EV2 (both added value and food) can be used to characterize in agricultural systems the (lack of) congruence between (1) total requirement and (2) internal supply at which the particular entity is operating. This check can be used to develop indicators and to characterize the particular role (e.g., related to a particular definition of EV2) that the entity plays within the food system. Households producing much more food than that consumed are households of farmers, whereas households producing less food than that consumed are just rural households. By using this distinction, we can find either rich farmers or poor farmers depending on the level of added value consumption that the household manages to stabilize in time. In the same way, we can find well-nourished rural households and malnourished rural households by looking at the flow of nutrients that a household manages to stabilize in relation to the requirement.

An example of several ILAs performed in parallel is shown in Figure 11.4. The one indicated in Figure 11.4a can be used to check the internal supply in relation to total demand of food. This analysis is useful for determining the degree of coverage of food in terms of subsistence. In the same way, another ILA (Figure 11.4b) can be used to check the internal supply of net disposable cash in relation to the constraint represented by land. This analysis provides an indication of the existence of a bottleneck associated with the requirement of land (given the existing characterization of the possible set of activities generating added value in crop production). The ILA represented in Figure 11.4c indicates that even if the bottleneck of land were not there, the considered set of farming activities (determining the average return of labor — dollars per hour — associated with the given mix of produced crops, $IV3_{m-1}$) would be, in any case, not enough to support the flow of net disposable cash required. The requirement of NDC is determined by the typology of consumptions associated with the characterization of household lifestyle obtained in the upper-right quadrant (the flow of net disposable cash per capita per year, $IV3_m$). In this example, the set of ELP_i associated with the set of agricultural activities would be too low to make these farmers rich. That is, even if an adequate amount of land would be available to saturate all the available internal supply of working time (no external constraints), this particular selection of lower-level activities (mix of crops produced and relative technical coefficients and economic variables) implies the existence of an internal constraint on the flow of added value that can be produced in this way.

When we consider these three ILAs in parallel, we can appreciate the existence of a clear trade-off, which is reflected by the relative changes in the two indicators — (1) degree of internal coverage of food security with subsistence crops vs. (2) level of internal generation of net disposable cash from agricultural activities. In fact, a larger fraction of the land in production, which is allocated to subsistence crops (to obtain a better coverage of subsistence need), will be reflected in a smaller fraction of LIP that can be allocated in generating net disposable cash. The terms of this trade-off can be analyzed (using the trick discussed in Figure 7.14), by including in the analysis (as an additional reduction of LIP) the amount of land lost to buy technical input to boost the production of crops, for both subsistence and cash. At this point, to analyze the technical aspects of this trade-off, one can analyze how lower-level characteristics (e.g., technical coefficients) related to the productivity of land and labor for both subsistence and cash crops will affect each other, when considering different options. An example of this analysis is illustrated in Figure 11.4d. This relation will be discussed more in detail later on (Figure 11.9).

An integrated set of indicators to characterize the performance in relation to nonequivalent descriptive domains (using these multiple parallel nonequivalent readings) is given in Figure 11.5. This figure is

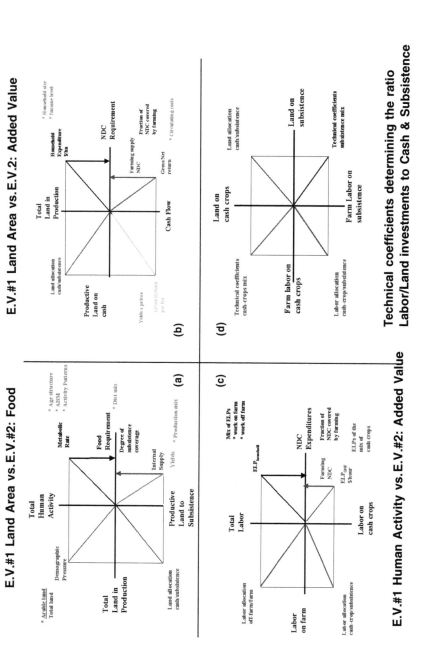

FIGURE 11.4 Mosaic of constraints affecting the characteristics of the throughputs. (a) EV1 of land area *vs.* EV2 of food. (b) EV1 of land area *vs.* EV2 of added value. (c) EV1 of human activity *vs.* EV2 of added value. (d) Technical coefficients determining the ratio of labor/land investments to cash and subsistence.

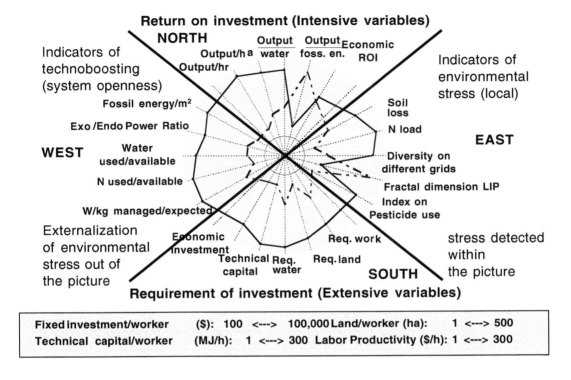

FIGURE 11.5 Integrated representation at the farm level.

based on a radar diagram containing different axes. The various indicators associated with these axes are aggregated in four quadrants over four categories:

1. North — intensive variables (return on investment)

    This set of indicators of performance is based on a list of output/input ratios reflecting a choice of variables relevant for characterizing the performance of the system. In semantic terms, this section deals with the return (output) on an investment (input). These inputs are called within the economic narrative production factors. According to the approach presented so far, assessments framed in terms of return on investment can be considered members of the semantic class of IV3 variables. In particular, in the selection given in Figure 11.5, we included:

    - Produced output per hour (e.g., kilograms per hour) — biophysical narrative: biomass output per unit of investment of human activity in agricultural labor
    - Produced output per hectare (e.g., kilograms per hectare) — biophysical narrative: biomass output per unit of investment of land in production
    - Dollars of output per hour — economic narrative: added value generated per hour of agricultural labor
    - Dollars of output per hectare — economic narrative: added value generated per hectare of land in production
    - Produced output per unit of consumed freshwater (e.g., kilograms per kilogram) — biophysical narrative: biomass output per unit of consumption of available freshwater
    - Produced output per unit of consumed freshwater (e.g., dollars per kilogram) — economic narrative: added value generated per unit of consumption of available freshwater
    - Produced output per unit of consumed fossil energy (e.g., kilograms per Exo megajoules) — biophysical narrative: biomass output per unit of consumption of fossil energy input

- Economic return on investment (e.g., dollars per dollar) — economic narrative: economic return per unit of economic investment

2. South — extensive variables (total requirement of investment)

These indicators represent the extensive variables associated with the intensive variables considered in the previous list:

- Total work supply (hours of human activity invested in agricultural work per year) — This is the total amount of working hours required to run the given entity (e.g., a farm or a village) at a viable productivity level.

- Total land in production (hectares of land area controlled by the entity invested in production) — This is the total amount of hectares required to run the given entity (e.g., a farm or a village) at a viable productivity level.

- Total freshwater consumption (cubic meters of freshwater consumed by the entity in a given year to obtain the given biomass production) — This is the total amount of freshwater consumption required to run the given entity at a viable productivity level.

- Total fossil energy consumption (based on one of the possible assessments of gigajoules of fossil energy embodied in the technical inputs consumed by the entity in a given year to obtain the given biomass production) — This is the total amount of fossil energy required to run the given entity at a viable productivity level. This can be assumed to be a proxy of the technical capital requirement.

- Total economic investment (based on one of the possible assessments of the requirement of capital, fixed and circulating) — This is the flow of capital required to run the given entity at a viable productivity level.

Before getting into an analysis of the two sets of nonequivalent indicators included in the other two quadrants, it is important to pause a moment for some considerations. The values taken by the two sets of indicators in the north quadrant and in the south quadrant can be interpreted using the metaphor of the four-angle figures. Within that frame, they are assessments that refer to (1) angles (those belonging to the family of returns on the investment) and (2) lengths of segments defined on axes (those belonging to the family of total requirements of investment). Because of this fact, the two sets of values taken by these two sets of variables are not and cannot be considered as independent. This trivial observation is particularly relevant when considering, as indicated in the box at the bottom of the figure, the huge differences that can be found when characterizing in this way, in terms of benchmarks, different farming systems in the world. Examples of ranges of values of IV3 — level of capital requirement per worker — are given on the left (expressed by adopting both an economic and a biophysical narrative). The two assessments on the right are related to the different degrees of conditioning of the context in terms of existing levels of (1) demographic pressure (societal average of land in production per worker) and (2) bioeconomic pressure (societal average of labor productivity). The relative differences between possible values found in the feasibility range are in the order of hundreds. Actually, when coming to the economic narrative, which is more sensitive at human perceptions of gradients of usefulness, we arrive at a range of differences that is in the order of thousands. In relation to the analysis of biophysical constraints (both external and internal), we can see that a level of labor productivity of hundreds of kilograms of grain per hour (a threshold value that within developed countries translates into rich farmers and a workforce that is mainly allocated to the operation of the industrial and service sectors) requires an amount of land per workers that must be at least in the two-digit range in terms of number of hectares. To make things worse, the heavy mechanization of agriculture associated with Western models of production requires not only a large amount of LIP per worker, but also the possibility to invest huge amounts of financial resources in the agricultural sector. On the other hand, in spite of this high requirement of capital per worker, the agricultural sector is not always able to reach the same level of economic return on the investment reached by other economic sectors (especially in an era of globalization).

These are well-known considerations in the field of farming systems; however, the link between the value taken by the variables represented in the north quadrant and the value that can be taken by the variables represented in the south quadrant is often neglected by those working in technical development of agriculture. A few analysts seem to not be aware that whenever we start with less than 1 ha of land in production per worker, there is very little that can be done to increase the labor productivity of farmers (that is, their economic performance), especially if these farmers are also required to produce (at a high level of demographic pressure) food for themselves and to pay taxes. When facing a major biophysical constraint (a crucial shortage of LIP), and when experiencing a growing gap in the consumption level of the household in comparison with the socioeconomic context (when farmers feel that they are remaining behind in the process of economic development occurring in the socioeconomic system to which they belong), farmers will stop investing a large fraction of their resources to the optimization of agricultural techniques. Rather, they will start looking around, that is, outside their farm, to diversify their investments of human activity and land area, looking for a mix of economic activities and land uses that includes also agricultural production. In this case, the decision about how to use available resources to farm is determined according to a multi-criteria evaluation of performance, in which the agricultural production is evaluated in relation to nonequivalent definitions of costs (and cost opportunities) and benefits (and benefit opportunities). In this situation, keeping the focus of the analysis on a standard agronomic definition of performance (the optimizing goal is linked to a continuous increase of the productivity of production factors), it is not always a wise choice. On the other hand, this optimizing goal still represents the basic assumption used to justify the transfer of production technologies developed within developed countries (high-input, high-capital agriculture) to farming systems operating in socioeconomic and ecological contexts in which these technologies do not make any sense.

Let us now get back to the analysis of Figure 11.5. On the two quadrants indicated east and west, we included, in this example, two sets of indicators referring to a characterization of such a system in relation to the ecological dimension. Obviously, we could have included in these two quadrants other indicators related to the compatibility with the socioeconomic dimension, as will be illustrated later on in different examples.

3. East — indicators assessing local environmental stress

Such a list, in this example, must necessarily be very generic. As discussed several times so far, it is not possible to indicate a sound list of indicators of local environmental stress in general terms. Therefore, since the special characteristics of the entity considered in this analysis have not been specified, we cannot indicate what should be considered a valid selection of indicators. Just to make it possible to indicate in the graph a set of generic indicators, we are assuming that this integrated analysis is related to a farming system producing grain. The list in this quadrant, for the moment, has the only goal of preserving the general overview obtained with this approach (an example of a real selection of indicators, referring to real systems, is given in Figure 11.6).

- Soil loss
- Nitrogen load into the water table
- Indices based on spatial analysis — e.g., vegetal diversity assessed on grids at different scales, an assessment obtained through remote sensing
- Indices based on spatial analysis — e.g., fractal dimension (ratio perimeter/area) of LIP, an assessment obtained through remote sensing
- Index related to pesticide use

4. West — indicators of technoboosting (system openness)

Indicators considered in this quadrant refer to the lack of congruence between the total requirement of a natural production factor and its internal supply (the amount of this production factor that would be available according to naturally generated boundary conditions). Put another way, this indicator assesses what fraction of the flows (EV2) that are consumed by a

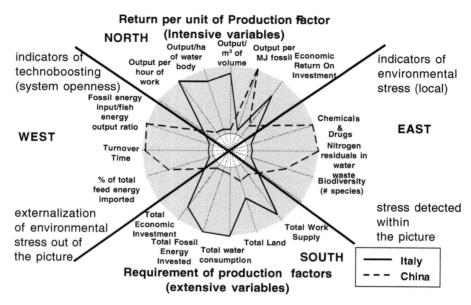

FIGURE 11.6 Comparing freshwater aquaculture systems of China and Italy.

farming system is not made available by natural processes included in the definition of boundary given to the system. In this example, the selection includes:

- Fossil energy per unit of area — This energy form is related to a depletion of stock. Therefore, we can associate the typologies of farming system metabolism heavily dependent on fossil energy with sustenance by a stock depletion within the system. When considering stocks of fossil energy as resources stored under human control, we can conclude that the system is not in a quasi-steady-state situation, but rather is eating itself.

- Exo/Endo power ratio — Also in this case, this indicator points to the existence of a hypercycle of energy forms (humans control machines that, eating fossil energy, make available more machines and more fossil energy to humans) that is logically independent from the autocatalytic loop of energy forms, which are used to sustain the human species when operating in a full ecological mode (within the natural essence that ecological systems negotiated in the past for the species *Homo sapiens*).

- Water consumed/available — The flow of freshwater consumed in LIP is often boosted through irrigation based on stock depletion and imports. In this way, humans manage to use for the production of useful biomass more water than the amount that would be naturally available. Mining freshwater (pumping out irrigation water at a pace noncompatible with the pace of recharge of the water table) should be considered an analogous to mining fossil energy or mining the soil.

- Nitrogen consumed/available — The flow of nitrogen input consumed in LIP is often boosted by importing fertilizers. The ratio between the actual consumption of nitrogen vs. the amount of nitrogen that would be available in production according to natural processes of supply represents an indicator of technoboosting on the cycling of nutrients.

- The ratio level of dissipation per unit of standing biomass assessed for the altered ecosystem compared with the level associated with the expected typology of ecosystem in the area. This is an indicator that has been discussed in Chapter 10 (Figure 10.16).

The discussion of the various indicators used in the integrated representation of Figure 11.5 (whether this particular selection is adequate, how to calculate and measure individual indicators) is obviously not relevant here. The only relevant point here is the emergent property represented by the shape obtained when considering the profile of values over the various sets of indicators arranged in the different quadrants (the overall message obtained when using this type of representation). To discuss this point, let us imagine comparing two general typologies of systems of production using this graphic representation: the two shapes illustrated in the graph in Figure 11.5. The first system is characterized using the solid line, and we can associate this expected shape to a high-input Western-type farm. The second system is characterized using a dotted line, and we can associate this expected shape to a low-input farm typical of a poor developing country. When adopting this integrated analysis, we characterize the performance of these farming systems in terms of intensive and extensive variables. In this case, the values taken by the productivity of land and labor for the high-input farm are higher than the values relative to the low-input farm. However, this difference can be easily explained by the larger size of the farm obtained when characterizing its size using a set of extensive variables (the scale of the system is bigger in terms of land, fossil energy, capital and freshwater).

But this is only part of the story. In fact, not only the size of the farms belonging to these two generic typologies is very different when considering extensive variables (the total economic investment associated with the typology of farm, the total amount of exosomatic devices associated with the typology of farm, the total amount of land associated with the typology of farm, etc.), but also the lack of congruence between what is consumed by the metabolism of such a system at the local level and what is generated by environmental services in terms of local favorable boundary conditions is widely different. The high-input Western-type farm is not only bigger in size, but also uses flows of inputs that are boosted by technology at a pace that would be unthinkable according to the natural associative context implied by internal constraints (the physiological conversion of food into power) and external constraints (the input supply and the sink capability imposed by ecological boundary conditions). As noted earlier, a farmer driving a 100-hp tractor is delivering an amount of technical power equivalent to that of 1000 human workers. In a preindustrial society, 1000 human workers (and their dependents) could not have worked and been sustained by that single farm. Moreover, the continuous harvesting of a few tons of biomass per hectare cannot be sustained in a normal terrestrial agroecosystem without the external supply of fertilizers.

An expected consequence of the massive effect of technoboosting (very high values for all the indicators included on the east quadrant) is that the indicators of local environmental stress (on the east quadrant) should also indicate a higher level of stress. That is, when characterizing in this way the performance of a high-input Western-type farm vs. a low-input farm operating in a poor developing country, we should also expect the existence of certain relations between the values taken by the indicators on the left and the values taken by the indicators on the right.

When looking at the graphical pattern generated by the two lines (solid and dotted) over this selection of indicators, we can observe that the pattern — higher values for the solid line and lower values for the dotted line — is reversed only for two indicators: (1) return on investment of fossil energy (in the north quadrant), according to the maximum power principle effect already discussed in Chapter 6 and (2) fractal dimension of LIP, since the crop fields in low-input farms are organized in small, scattered plots, whereas they tend to be organized in large plots in mechanized agriculture. This confusion in the visual pattern is generated by the particular procedure of representation adopted in this graph (higher values for the variables considered for the various axes/indicators are positioned far away from the origin of the axis). To avoid this problem, one should discuss how to handle representations of this type in a way that makes it possible to generate more evident systemic patterns on the integrated representation. This can be obtained by (1) normalizing the values in relation to a given range for each indicator and (2) giving a common orientation to the various indicators in relation to the preliminary definition of a criterion of performance. With this organization, within the feasibility domain, far away from the origin is good and close to the origin is bad.

Coming back to the analysis provided in Figure 11.5, which implies expected patterns in the shapes, we can say that at this point, the two shapes indicated on the graph should be considered another example of metaphorical knowledge. In fact, the profile of relative values over the various axes indicated in Figure 11.5 is not reflecting experimental data. Rather, the two shapes represent a typical pattern of expected differences (reflecting the particular characterization indicated in that figure), which can be associated with the typologies of the farming system considered. The question at this point becomes, Is the metaphor indicated in Figure 11.5 useful? As noted in the previous section, the only way to answer this question is to look to a totally different situation, applying this tentative problem structuring for organizing the information space. An example of this is given in Figure 11.6, which presents a comparison of two typologies of aquaculture (in rural China and in rural Italy). A detailed presentation of this comparison is available in Gomiero et al. (1997). What is important about this figure is the clear similarity of the pattern found in Figure 11.6 (when comparing two systems producing and cultivating fish in China and Italy using a real data set) and the pattern associated with the metaphor suggested in Figure 11.5 (when comparing two hypothetical systems producing cereal in a developing and a developed country using typical expected values for these systems). The west and east quadrants in these two figures are only sharing the same semantic message (openness of the system and generation of local environmental stress), whereas in terms of the formalization of these concepts in the form of indicators (proxy variables and measurement schemes), the two analyses are totally different.

11.2 Individuating Useful Types across Levels

11.2.1 The Land–Time Budget of a Farming System

The examples of ILA provided so far should have clarified to the reader the meaning and usefulness of the rationale of the four-angle figures. So now we can finally move on to practical applications of this method, no longer based on the use of this class of figures. In particular, we start by introducing a method that has been named the land–time budget, which is used, in the next example, to characterize the chain of choices faced by a given household in relation to its livelihood. With this method, it is possible to characterize a relevant set of choices made by a household in terms of two profiles of investments of the original endowments of (1) human time (EV1, human activity) and (2) land (EV1, land area). A graphic view of this approach is given in Figure 11.7. In this way, it is possible to characterize the decisions made by the household according to the rationale of ILA as discussed using the four-angle figures. The set of choices made by farmers when deciding how to use their production factors is translated into a graphical representation of a chain of reductions applied in series to the initial budgets.

The analysis given in Figure 11.7 is tailored to a farming system operating in China; therefore, the selection neglects financial capital among the relevant production factors to be considered. The supply of the two production factors (tracked as EV1) in this figure is represented using solid arrows along compartments indicated by ellipsoids. The consequences associated with a given profile of choices (e.g., the level of internal supply of EV2 achieved or the profile of EV1) are illustrated by the value taken by the indicators associated with gray rectangles. In this example, the three gray rectangles coincide with three variables that can be used as indicators of performance for the household in this farming system.

In the example given in Figure 11.7, the chain of decisions of a given household is represented starting from the left and right with:

1. A definition of an amount of disposable human activity for the household — EV1, labeled "budget of disposable human activity." This represents the amount of investment of this resource that can be allocated according to the decisions of the household within its option space. In this example, such a budget is 14,000 h/year (the meaning of this value is explained in Figure 7.15)
2. A nonequivalent definition of disposable investment (EV1), which is related to the amount of land area for the household (labeled "budget of colonized available land"). In the example given in Figure 11.7, such a budget is 0.5 ha (this value is related to a study of a farming system in populated areas of China, which is discussed in detail in the rest of this section).

With this approach, different choices made by different households can be characterized in terms of different profiles of investments (expressed in terms of fractions of the two available budgets) of (1) disposable human activity, over the set of possible typologies of activities and (2) colonized available land, over the set of possible typologies of land uses.

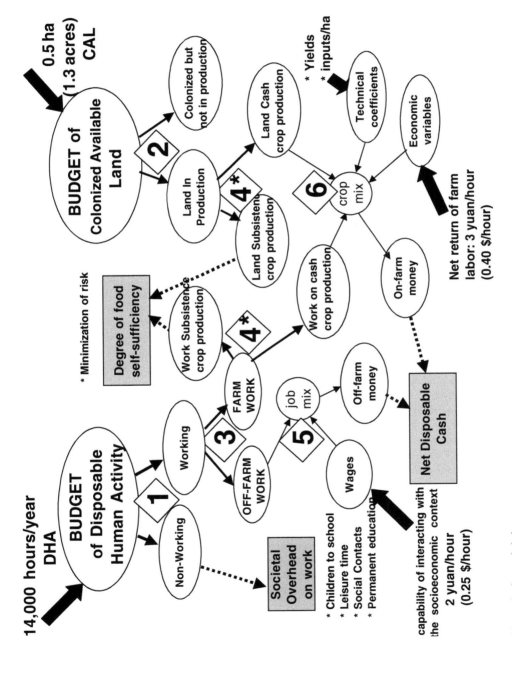

FIGURE 11.7 Looking at the farmers' choices.

The chain of choices that a given household can perform when deciding how to use (and invest) these two budgets over activities dedicated to either production or consumption of a flow of EV2 (e.g., food or added value) is indicated in Figure 11.7 by the set of diamonds with black numbers. Recall here the theoretical discussions about impredicative loops and complex time. In spite of the ordinal sequence suggested by the numbers written within diamonds, it is important to avoid thinking that these various choices are occurring one after another in simple time.

Let us start with the diamond marked with a 1. This refers to a decision determining what fraction of the disposable human activity is invested in the compartment labeled "working" vs. the investment of human activity in the compartment labeled "nonworking." This first decision has two consequences:

1. It determines the value taken by an indicator of quality of life (called in this example societal overhead on work). In fact, an increase in the fraction of disposable human activity invested in nonworking translates into more education for children and youngsters, and more social interaction and more leisure for adults in the household.
2. It determines the availability of the resource working human activity required to perform the tasks associated with the stabilization of the metabolism of the household (a crucial production factor for food and net disposable cash).

In parallel with choice 1 is choice 2. Choice 2 deals with the decision of what to do with (how to use) the available colonized land. In general, the investment of a part of TAL for preservation of ecological processes (the reduction associated with EOAL discussed in Figure 11.2b) is decided at a hierarchical level higher than the level at which the household is operating. Therefore, in general, at the household level the option space for the farmers is related only to how to use their CAL for practical tasks — that is, how to choose among land uses associated with crop production (LIP) and other land uses not associated with crop production. This decision is indicated by the diamond labeled 2. Another relevant choice made by the household is that indicated by the diamond 3. This is the decision related to how to split the available amount of working human activity over the two compartments labeled "off-farm work" and "farm work." This choice forces us to deal with the complexity implied by such an analysis. In fact, at this point, the two choices related to how to use the available budgets of (1) human activity invested in farm work and (2) hectares of colonized land invested in land in production are no longer independent of each other. When deciding how to invest a certain amount of hours of working activities (the first diamond 4* on the left) and a certain fraction of the hectares of land in production (the other diamond 4* on the right) over the two compartments — subsistence crop production and cash crop production — we have two sets of choices conditioned by a reciprocal entailment. These choices are affecting each other in relation to the characteristics of other lower-level elements determining the ILA (see, for example, Figure 11.4d). That is, depending on (1) the mix of crops produced (both in subsistence crop production and cash crop production) and (2) the set of technical coefficients characterizing the various production (e.g., productivity of land and productivity of labor), we can determine the existence of a link between the effects of the choices indicated by the two diamonds 4* (the two choices must be congruent with each other).

After having defined the lower-level characterization (profile of investment of labor and land on the given mix of crops and technical coefficients for each crop), the amount of land and labor invested in subsistence crop production will directly define one of the three indicators of performance selected — the degree of food self-sufficiency. As noted before, this criterion can be totally irrelevant for a farmer operating in the U.S. or Europe, but it can be very relevant for a farmer operating in a marginal rural area in China or Africa.

To also characterize the economic performance of this household, we have to include the effect of additional choices — in particular, a choice related on how to invest the amount of hours of human activity of adults, which are invested in the compartment off-farm work (the choice indicated as 5). This amount of hours of off-farm work is allocated on a mix of jobs according to a set of criteria, considered as relevant by the household. In the same way, the farmer will choose, according to the decision indicated by the diamond 6, the special mix of crops cultivated over the hectares of LIP invested in cash crop production. This will imply supplying the relative hours of human activity (determined by technical coefficients) that have to be invested in the compartment cash crop production. According to the scheme indicated in Figure 11.7, choices 5 and 6 will determine the overall value of IV3 — economic labor productivity — for the hours of human activity invested in dollars of work (off-farm work plus farm work). The combination of the two EV1 (the hours of human activity invested and the hectares of land invested) and the two IV3 (the economic labor productivity and the economic land productivity) make it possible to (1) define another crucial indicator of performance for the household, in this case the level of net disposable cash and (2) individuate the existence of internal constraints (bottlenecks) over such a throughput in relation to the available budget of land and human time. An example of such an indication has been discussed in Figure 7.14. When analyzing that ILA, in fact, it becomes quite clear that the typology of household considered in that example has a very limited option (given the relative ELP) when using the available LIP to sustain the actual requirement of net disposable cash.

The analysis of the land–time budget presented in Figure 11.7 — in relation to the household level — can be standardized for a given farming system as illustrated in Figure 11.8. We say standardized, since this second overview can be applied also to a hierarchical level higher than the household. For example, this makes it possible to use large-scale analyses of land use to define ecological indicators of performance. Starting with a given budget of hours of human activity (in the upper white box on the left) for a given socioeconomic entity and with a given budget of hectares of land (in the lower white box on the right), we can represent the two chains of reductions as a chain of decisions splitting the available budget into two lower-level compartments. Depending on the selection of EV1 (THA or TAL), different sets of lower-level typologies have to be used to define the size of the two lower-level compartments generated by the splitting of the higher compartment. After the split, only one of the two compartments is considered for the supply of EV1 to the direct compartment in the next splitting (see Figure 11.8).

We can go quickly, once again, through the list of acronyms/labels used to characterize and standardize the chains of reductions.

Starting with a total budget THA (indicated in gray in the upper-left box), this budget is split into (1) physiological overhead on human activity (POHA), the amount of hours invested in sleeping and personal care and the hours of human activity of persons that do not belong to the working force and (2) human activity disposable fraction (HADF), the maximum amount of human activity that could be invested in working.

The HADF is the relevant compartment in terms of the fraction of resource (EV1) that can be invested in the direct compartment. The size of HADF is then split into two other compartments: (1) leisure and education (L&E) and (2) human activity in work (HAWork). This is the amount of hours of human activity that is invested in working. The fraction of HADF that is not used in HAWork can be used as an indicator of social performance (recall the ratio THA/HA_{PS} discussed in Chapter 9, which is a good indicator of development also at the level of the whole country). At this point, it is the compartment HAWork that becomes the relevant compartment determining the supply of hours of human activity for the direct compartment. The compartment HAWork is split into two lower-level compartments:

1. Wsub — Work in subsistence. This includes chores, which are required for the production of goods and services contributing to the material standard of living of the household (the monetary value of these services can be included in the assessment of the income of the household). The food produced and consumed in subsistence, however, does not generate market transactions and therefore does not generate monetary flows of added value to be included in the assessment of net disposable cash.

2. W$ — Work in cash generation. This includes the various activities associated with monetary flows. The compartment W$ is now the relevant compartment for the supply of hours of human activity invested in generating NDC. This compartment is split between:

   a. W$-off-farm — Work for money in off-farm activities. This includes the various activities aimed at the generation of flows of money, which do not require a land investment, or at least a demand of space negligible.

   b. W$-land – Work for money on-farm. This includes the various activities aimed at the generation of flows of money, which requires an associated given amount of investment of farmland.

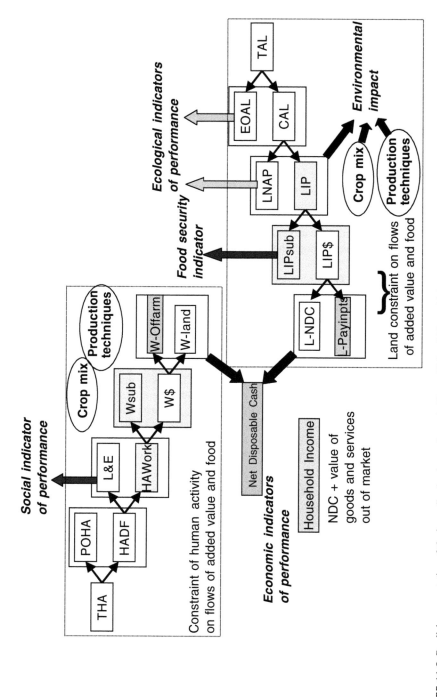

FIGURE 11.8 Parallel representation of the impredicative loops shown in Figure 11.2 and Figure 11.3.

The representation of the chain of decisions (preserving the closure of the various compartments at each step) in relation to land area (as EV1) is given in the white lower box on the right of Figure 11.8. This time, the chain of decisions considered is generating a profile of investments expressed in terms of fraction of the total available land. Starting with a given amount of total available land (TAL), we have to split this amount of hectares into two lower-level compartments: (1) EOAL, the amount of hectares that are left not managed by humans and (2) CAL, the hectares that are controlled (managed) directly by humans. This compartment is then split into two other compartments: (1) land not in agricultural production (LNAP) and (2) LIP. The names of these two compartments are self-explanatory. Then the compartment LIP is split into two lower-level compartments:

1. Land in production allocated to subsistence (LIPsub). This includes all productions contributing to the material standard of living of the household (the value of these services can be included in the assessment of the income). But this production does not generate market transactions and therefore monetary flows of added value accounted in the assessment of net disposable cash.

2. Land in production used for cash generation (LIP$). This includes the various productions of the farm associated with monetary flows. Then the compartment LIP$ can be split between: (1) land providing net disposable cash (L-NDC), the fraction of land that is generating a net flow of added value (after discounting a fraction of land lost to generate cash crops, whose revenue is used to pay for input) and (2) land allocated to cash crops that is subtracted from the total to account for the loss of land associated with the cost of production (to pay for inputs used in agricultural production) (L-payinputs).

In the overview provided by Figure 11.8, it is possible to appreciate that the resulting analysis can be used for:

1. An integrated representation of the performance of the system. For example, we can imagine using this structuring of the information space to obtain several nonequivalent indicators of performance of this farming system (as done in Figure 11.6). For example, the level of leisure and education of the household can be used as an indicator of performance when addressing the social dimension (an indicator of material standard of living). The level of self-sufficiency of the farming system can be used as an indicator of food security (in those systems in which such an indicator is relevant). In economic terms we can calculate both (1) the income of the household and (2) the net disposable cash, which are two nonequivalent indicators of economic performance. The analysis of land use related to the density of flows of input and output can be used to develop indicators of environmental impact.

2. An analysis looking for internal constraints that are reducing the option space of different typologies of farmers. Depending on the expected values of the variable used as indicators of performance, we can study the possible limiting effects of the forced relation between the value taken by extensive variables (e.g., availability of natural resources) and the intensity of through-puts (e.g., internal supply vs. total requirement of food or added value). This can be studied by considering the relative technical coefficients and economic variables.

3. Verify the relevance of a particular representation of the farming system in relation to the strategy matrix adopted by various agents in the farming system considered.

In relation to this last point, it should be noted that the two chains of decisions indicated as linear tree choices in Figure 11.8 (using a representation that preserves the closure across levels and compartments at each choice), in reality, are not either sequential or linear at all. On the contrary, the various agents deciding at different levels within a given farming system are choosing simultaneously a given profile of investments in relation to both the budget of land and the budget of human activity. This choice is based on (1) an expected set of costs and benefits that are associated with the selected profiles of investments and (2) the existing perception of a set of biophysical constraints. Put another way, the various agents when deciding what to do with their budgets of land and time are not dealing with a chain of binary decisions that can be handled one at a time (as represented by Figure 11.8). Rather, they have to go for a selection of a given profile of values (depending on the agent considered) in relation to a set of choices that must be: (1) congruent over the various nonequivalent definitions of constraints and (2) effective in relation to the goals. To make the life of agents more difficult, nonequivalent definitions of constraints can only be studied by adopting different ILAs, that is, by representing in nonequivalent ways different dynamic budgets of extensive and intensive variables for the same metabolic system.

In practical terms, the only way out for real agents operating in a real situation (in a finite time with imperfect information) is to go for a validation of a tentative set of choices determining the profile of investments of THA and CAL. This validation has to be obtained in relation to (1) the acceptability of the resulting indicators of performance on the socioeconomic side, (2) the compatibility of the resulting indicators of performance in relation to ecological processes and (3) the feasibility according to the existing technical coefficients determining the relation of biophysical flows across parts and the whole. In this way, we are back to basic concepts that have been discussed in Part 1: a Peircean semiotic triad that has to be applied in relation to the three incommensurable and nonreducible dimensions of sustainability, when looking for satisficing solutions.

11.2.2   Looking for the Mosaic Effect across Descriptive Domains

When discussing of the various diamonds representing choices in Figure 11.7, we observed that a few of these choice are not independent from each other (e.g., the two choices indicated by the two diamonds marked 4*). The nature of this link can be explored using the concept of mosaic effect across scales (Chapter 6). That is, the density of relevant flows (e.g., nonequivalent definitions of IV3) at level m (e.g., the farm) can be linked to the density of flows (e.g., nonequivalent definitions of IV3) at level $m - 1$ (e.g., technical coefficients for individual activities) using the same mechanism illustrated in Chapter 6. In our case, this requires applying to the farming system a system of accounting of the type illustrated in Figure 11.7 and then characterizing the various lower-level activities (e.g., producing rice, producing piglets) in terms of (1) technical coefficients (requirement of hectares and hours of work per unit of biophysical flow) and (2) economic variables (economic return of both labor and land). This information refers to the perception and representation of events at the level $m - 1$. The profile of investment of EV1 (either human activity or land area) over the various activities considered in the lower-level compartment (e.g., working in agriculture) will then define the characteristics of the relative intensive variables at the level m. Put another way, we can use the specific mix of crops produced (profile of investment over the set of options) and the characteristics of individual crops to estimate aggregate values of flows referring to the agricultural compartment as a whole.

Examples of the mosaic of relations are given below. Starting with IV3, ELP (economic labor productivity), assessed over the compartment working (at the level m), we can write:

$$[\text{Level } m] \; \text{ELP}_W = [\text{Level } m - 1] \; (X_{OFF} \times \text{ELP}_{OFF}) + (X_{ONF} \times \text{ELP}_{ONF}) \tag{11.1}$$

where:

ELP_{OFF} and ELP_{ONF} = the characteristics of the two lower-level compartments. These are two IV3 — that is, the two levels of economic labor productivity (assessed in dollars per hour) of the compartments working off-farm and working on-farm.

X_{OFF} and X_{ONF} = the fractions of the total amount of hours of human activity of the compartment working that are invested in the two lower-level compartments working off-farm and working on-farm. Since we can write $X_{OFF} + X_{ONF} = 1$, these two values represent the profile of investment of the fraction of resource THA invested in the compartment working (at the level m) over the possible set of lower-level types.

We can express (at the level $m - 1$) the two values of ELP_{OFF} and ELP_{ONF} in relation to lower-lower-level characteristics — to identities referring to the level $m - 2$. For example,

$$[\text{Level } m - 1] \; \text{ELP}_{OFF} = [\text{Level } m - 2] \; (X_{job1} \times \text{wage}_1) + (X_{job2} \times \text{wage}_2)$$
$$+ (X_{job3} \times \text{wage}_3) \tag{11.2}$$

where:

wage_i = the characteristics of lower-level compartments (IV3 — economic labor productivities — characterizing the various off-farm tasks, labeled job_i). We assume that in this case there are three types of off-farm jobs accessible to this household (job_i) and that they can be characterized by a variable that we call wage_i.

X_{jobi} = the fraction of the total amount of hours of human activity of the compartment working off-farm that are invested in the off-farm task job_i. Since $\Sigma\, X_{\text{jobi}} = 1$, these three values represent the profile of investment of the fraction of the resource THA invested in working off-farm over the set of lower-level types of off-farm tasks: job_i.

The same reasoning can be applied to the characterization of the other IV3 — level $m - 1$:

$$[\text{Level } m - 1]\; \text{ELP}_{\text{ONF}} = [\text{Level } m - 2]\, (X_{\text{crop1}} \times \text{ELP}_{\text{cr1}}) + (X_{\text{crop2}} \times \text{ELP}_{\text{cr2}})$$
$$+ (X_{\text{crop3}} \times \text{ELP}_{\text{cr3}}) \tag{11.3}$$

where:

ELP_i = the characteristics of lower-level compartments (IV3 — the economic labor productivity of the various on-farm tasks, labeled as crop_i). We assume that in this case there are three types of crops produced in this system (crop_i) and that they can be characterized by a variable that we call ELP_i.

X_{cri} = the fractions of the total amount of hours of human activity of the compartment working on-farm that are invested in the on-farm task, labeled crop_i. Also in this case, $\Sigma X_{\text{cri}} = 1$; therefore, these three values represent the profile of investment of the fraction of the resource THA invested in working on-farm over the set of lower-level types on-farm tasks: crop_i.

We can establish a bridge between economic and biophysical variables when defining the lower IV3. In fact, the gross economic labor productivity of each of these three crops (the i crops considered in Equation 11.3) can be written as

$$\text{GELP}_i = [(\text{Yield}_{\text{cropi}} \times \text{Price}_{\text{cropi}}) - (\text{Yield}_{\text{cropi}} \times \text{Cost}_{\text{cropi}})]/\text{Work-hours}_{\text{cropi}} \tag{11.4}$$

The cost of a crop (crop i) can be related to the level of consumption of inputs. Imagining three types of inputs (e.g., A = fertilizer, B = pesticides, C = irrigation), the total requirement of each of these three inputs can be written as

$$\text{Tot. Req. Input A} = \Sigma(\text{kg input A/hectares})_{\text{cropi}} \times (\text{hectares})_{\text{cropi}} \tag{11.5}$$

$$\text{Tot. Req. Input B} = \Sigma(\text{kg input B/hectares})_{\text{cropi}} \times (\text{hectares})_{\text{cropi}} \tag{11.6}$$

$$\text{Tot. Req. Input C} = \Sigma(\text{kg input C/hectares})_{\text{cropi}} \times (\text{hectares})_{\text{cropi}} \tag{11.7}$$

The information given by Equations 11.5 through 11.7 is not only useful for the determination of costs, but also useful for the direct calculation of indicators of environmental impact (e.g., amount of pesticides, consumption of freshwater in irrigation, leakage of nitrogen in the water table) and indices of efficiency in relation to the use of inputs.

The total cost of crop i, at this point, can be written as the combination of the costs related to the inputs used in production. In this simplified example, this can be written as

$$\text{Cost}_{\text{cropi}} = (\text{Input A} \times \text{cost}_{\text{inputA}}) + (\text{Input B} \times \text{cost}_{\text{inputB}}) + (\text{Input C} \times \text{cost}_{\text{inputC}}) \tag{11.8}$$

Technical coefficients can also be used to calculate the biophysical labor productivity per different types of crops (for the assessment of subsistence coverage):

$$\text{BLP}_{\text{cropi}} = (\text{Yield}_{\text{cropi}} \times \text{hectares}_{\text{cropi}})/\text{Work-hours}_{\text{cropi}} \tag{11.9}$$

At this point, we have all the ingredients required to calculate economic labor productivity by mixing together the information provided by Equation 11.1 through Equation 11.9 ($\text{ELP}_i = \text{GELP}_i - \text{Cost}_i$).

However, it would be unwise to continue to write down these semantic relations with the goal of obtaining a full formalization. As already discussed, the series of equations from Equation 11.1 to Equation 11.9 should be considered a set of equations obtained through a combination of intensive and extensive variables over an impredicative loop. The numerical values assigned to the various labels making up these relations do affect each other. In fact, agents operating at different levels are using the relative values taken by these variables as relevant signals for action. Therefore, depending on the time differential that is relevant for a particular goal of the analysis, some of these labels have to be considered variables, others parameters and others constant. Moreover, the predicament of an arbitrary definition of categories (the choice of the set of formal identities to be used in the model) is also in play. Put another way, the particular procedure that has to be used to formalize the structure of these equations in practical situations has to be decided according to the circumstances.

For example, Equation 11.2 includes the assessment of three different wages associated with three different typologies of job, in this case, how to deal with commuting time — that is, how to account for the time spent by the worker to move from the house to the workplace. This can be accounted as human activity invested in the compartment working off-farm. In this case, this choice would result in a reduction of the value assigned to the variable wage (dollars per hour). That is, let us assume that the wage actually paid in job1 is $1 per hour, and that commuting requires the addition 10% to the actual working hours in job1. Then the wage relative to job1 should be reduced, in this system of accounting, by 10%. On the other hand, the time spent commuting can be accounted for as human activity that must be invested in the compartment chores (activities necessary for stabilizing the metabolism, but not generating a direct return of added value). A third alternative choice of accounting could be, if the commuting is done by bus where the worker has a pleasant social interaction or some leisure time (e.g., reading a book), to account for that investment of human activity in the category leisure.

It should be noticed that nowadays the challenge of keeping coherence and congruence in a system of accounting of this nature has been greatly simplified by the availability of powerful software that can be run on every PC. Actually, the very popular Microsoft Excel makes it possible to establish an interface between the mechanism of accounting applied to a database and a graphic form of representation of multiple indicators (e.g., the radar diagram illustrated in Figure 11.6). Another useful and popular software that can be use when structuring the analysis of systems organized in different hierarchical levels is STELLA, which makes it possible to (1) visualize in the form of graphs the set of relations across levels and (2) keep separated classes of variables belonging to different descriptive domains (e.g., economic reading vs. biophysical reading). An example of the analysis of the set of relations discussed before is given in Figure 11.9. Details are not relevant now. What is relevant is the clear distinction that can be made between the parameters defined within economic domains (selling prices and costs of inputs), visualized in a the box labeled economic variables, and the parameters defined within biophysical domains (productivity of production factors), visualized in the box labeled technical coefficients. Then, when moving up to a higher hierarchical level and when characterizing the productivity of labor at the level of the farm, we can notice that this economic characteristic (at the level m) in reality is affected by biophysical characteristics of the system (perceived and represented at the level $m - 1$). In the same way, biophysical characteristics such as the productivity of land — assessed in terms of biophysical output per hectare — are affected by economic characteristics (e.g., the possibility of affording the purchase of a lot of technical inputs per hectare). Because of this, we believe that it is important to develop an integrated analytical approach that explicitly addresses the reciprocal entailment of economic and biophysical characteristics at different levels within a given farming system.

There is another important point to be made about the existence of mosaic effects across levels and dimensions. When looking simultaneously at Figure 11.5, Figure 11.7 and Figure 11.9, we can appreciate even more the case already made about the severe challenge faced by individual agents when selecting a given profile of investments (making a multiple and simultaneous choice in relation to all the diamonds illustrated in Figure 11.7). When doing that, an agent has to (1) consider the existing set of constraints (the set of existing relations between economic and biophysical characteristics illustrated in Figure 11.9) and (2) evaluate the performance of the farming system considered (the actual shape over the package of indicators used to characterize such a performance vs. the expected shape, as illustrated in Figure 11.5) in relation to the existing goals.

The reciprocal entailment across levels of characteristics and the requirement of congruence over those choices that imply the sharing of the same pool of production factors (e.g., the two diamonds 4* in Figure 11.7) indicate that individual farmers, for example, cannot modulate their profiles of investments

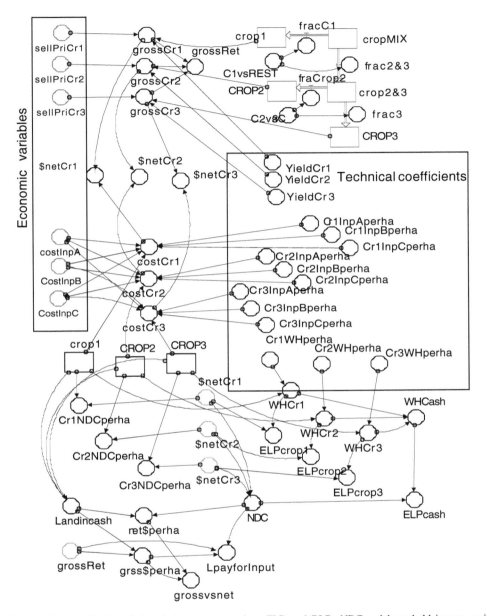

FIGURE 11.9 Characterization of the relations among various ELP$_i$ and BLP$_i$, NDC and household income, using a STELLA diagram.

of both human activity and land area in a continuous way. On the contrary, we can imagine that the reciprocal entailment among characteristics of lower-level elements (feasibility on the biophysical/structural side) and the reciprocal entailment among characteristics of higher-level elements (desirability on the cultural/functional side) should imply a sort of quantization of the option space for farmers. That is, we can hypothesize that agents tend to follow attractors/typologies/packages of profiles of choices in terms of investments of production factors and tend to adopt attractors/typologies/packages of profiles of weighting factors when dealing with the selection of a strategy in face of the unavoidable presence of incommensurable sustainability trade-offs (e.g., minimization of risk vs. maximization of return, preservation of cultural values vs. integration in a fast-changing socioeconomic context) and uncertainty.

The analytical tools presented so far make it possible to study this phenomenon both (1) on the side of the analysis of different profiles of investment of production factors and (2) on the side of the analysis of the different profiles of weighting factors adopted when selecting an overall strategy in terms of achievements on the multi-criteria performance space. A given profile of investments of production factors, in fact, can be associated with a given shape of the characterization of the performance in relation to a given multi-criteria performance space. An analysis of this type is briefly discussed below, applied to the hierarchical levels of both household and village.

Again, one has to be very careful about imagining that it is possible to formalize this type of analysis. In fact, it is important to always keep in mind that the data set used for such an analysis is derived from an impredicative loop analysis. That is, the data already reflect a representation of the reality that is biased by the set of preliminary choices made by analyst. When trying to apply the analytical approach presented so far, we are confronted by questions such as:

- What should be considered managed land and natural land?
- What should be considered the size of the household in terms of human activity when dealing with hired work? (For example, should we stick with the household THA and consider hired work an economic input accounted for only by a reduction of ELP and L-NDC?)
- What should be considered the size of the household in terms of land when dealing with inputs such as feed that have embedded a land requirement (the land used to produced the feed elsewhere)?
- How should we deal with the assessment of fixed capital (e.g., actual value and depreciation) when assessing gross and net economic return?
- How should we calculate with accuracy the value of goods and services obtained by the household outside market transactions?

As soon as one accepts (or acknowledges) that this type of analysis requires working with useful metaphors rather than with substantive models, the unavoidable arbitrariness in the original data set is no longer a problem. The ambiguity entailed by these questions is only lethal when attempting the construction of exact substantive models.

On the contrary, analysts looking to increase the quality of their integrated analysis of complex systems should consider the challenge represented by ambiguous questions (such as those listed above) an opportunity. Ambiguous questions that pop up as soon as one tries to do an integrated assessment of a real farming system should be viewed as opportunities to check with other nonequivalent observers the quality of the problem structuring. In this way, it becomes possible to explicitly discuss the implications of the assumptions and the selection of identities and categories used in the problem structuring. Such a discussion has to be done not only with other scientists dealing with the same system but operating in different disciplinary fields, but also with nonscientists who are relevant stakeholders in the problem to be tackled.

11.2.3 An Example of the Selection of Useful Typologies

The following example of an analysis of useful typologies in a multi-scale integrated analysis of the farming system is based on the results of a 4-year research project in China entitled "Impacts of Agricultural Intensification on Resources Use Sustainability and Food Safety and Measures for Its Solution in Highly-Populated Subtropical Rural Areas in China." This project included a farming system analysis group; an overview of the research activity of this group has been published in a special issue of *Critical Reviews in Plant Sciences*, Vol. 18, Issue 3, 1999.

The goal of this section is to provide a general overview and a qualitative presentation of the nature of the analysis. Therefore, graphs and figures presented below do not have the ultimate goal of explaining in detail the choices adopted when formalizing the analysis in this case study. Interested readers can get a more exhaustive explanation of the procedures adopted in this study by referring to Li et al. (1999), Giampietro and Pastore (1999) and Pastore et al. (1999).

In fact, it was during the processing and analysis of the data gathered during this project (working in parallel at the household and village levels) that the group of farming system analysis realized that the most important findings of this analysis were linked to the ability of characterizing the farming system using an integrated selection of useful typologies. By an integrated selection of useful typologies we mean a set of tasks and a profile of investments to be used to characterize household types, a set of household types and a profile of investments to be used to characterize villages, and a set of village types and a profile of investments to characterize the aggregate performance of the farming system at a larger scale.

11.2.3.1 *The Frame Used to Compare Household and Village Types* — In this section

we present an example of characterization of a set of typologies of households (at level m) that can be linked to a set of typologies of villages (at level $m + 1$). Such a characterization is based on a multi-objective integrated representation of performance — the profile of values taken by a package of indicators over a radar diagram Figure 11.10 — that can be associated with a given profile of choices over the set of diamonds illustrated in Figure 11.7.

Let us start with an example of characterization of two typologies of households that are compared in Figure 11.10 (types 4 and 6). These are just two of the six typologies of households individuated in the farming systems considered. This representation and characterization, obviously, reflects the preliminary selection of a set of relevant criteria associated with the indicators included in the radar diagram, which is in common for all the typologies.

The radar diagram is divided into four quadrants organized over two axes of symmetry. A vertical axis divides the two quadrants on the left, which refer to indicators of socioeconomic performance, from the two quadrants on the right, which refer to indicators of ecological impact. The horizontal axis divides the two upper quadrants, which refer to a local perception and representation (indicators) of performance, from the two quadrants on the lower part, which refer to information characterizing the farm in relation to its larger context. The two sets of indicators included in the two lower quadrants, however, provide information of a different nature. The selection of indicators dealing with the socioeconomic dimension, on the left, reflects the perspective of agents operating in the larger socioeconomic context (e.g., administrators of the province, the government of China) about the performance of the farm. That is, these are systems qualities of the household that are relevant for the higher-level holon (the village or the country) within which that household is operating.

When coming to the representation of the effects that the characteristics of the farm can have on the larger ecological context, it is not possible to use, at this level, indicators of environmental impact for large-scale ecological systems. In fact, there is a mismatch of scale between the local disturbance generated by the specific characteristics of a given household and the large scale that would be required by an ecological analysis of sustainability. Moreover, the possibility given by the exosomatic metabolism to stabilize useful flows in the short period by relying on (1) stock depletion and abuse of sink capacity and (2) trade of environmental services and limiting natural resources makes it impossible to associate the effect of local patterns with events occurring elsewhere or on a different scale. For this reason, the lower-right quadrant includes a set of indicators that are labeled "technoboosting." This is a set of indicators that is useful for defining how bad is our representation of the metabolic system as a system that is in steady state. A high level of technoboosting required and used to keep the pace/density of relevant flows much higher than natural rates implies (1) a high dependency on external inputs (which generally is associated with stocks depletion somewhere else and (2) a high probability of generating harmful interferences with the natural mechanisms of regulation of flows within terrestrial ecosystems (see Chapter 10).

The idea of using in parallel different sets of indicators to reflect the different perspectives of nonequivalent agents can be related to the rationale discussed in relation to Figure 5.3. A multi-criteria performance space entails characterizing the same system using a set of different indicators that reflect nonequivalent perceptions and representations of what is good and bad in relation to different dimensions of sustainability and in relation to different agents operating at different levels. In this example, referring to the two sets of indicators of socioeconomic performance (upper and lower parts on the left side of the radar diagram), we have:

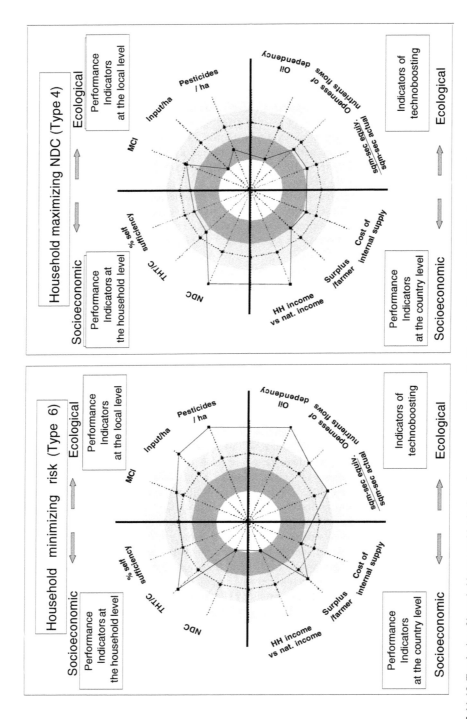

FIGURE 11.10 Typologies of households on a multi-criteria performance space. (Pastore, G., Giampietro, M., and Li, J., (1999), Conventional and land-time budget analysis of rural villages in Hubei Province, China, *Crit. Rev. Plant Sci.*, 18, 331–358.)

- **Indicators reflecting the perceptions of performance at the level of the household**
- The three selected criteria are:
  1. Minimization of risk (the goal here is to make individual members of the household as safe as possible against external fluctuations in market variables and from external perturbations such as climatic and political disturbances).
  2. Maximization of the level of social interaction of the household with the socioeconomic context, in relation to the exchange of goods and services within the economic process. This criterion becomes very important in those cases in which the set of activities expressed by a particular household in the compartment leisure and education is different and less diversified than the set of activities expressed in the same compartment by the average household in the socioeconomic context.
  3. Maximization of the fraction of total human activity that is invested outside the compartment working. As noted in Chapter 9, this criterion (which is useful also at the level of the whole society) can be related to the possibility of expanding the potentiality of human beings — e.g., better education, better social interactions and processing of information, traveling.
- Coming to the numerical assessment the relative selected indicators are:
  1. Minimization of risk in terms of food security — The degree of coverage of food requirement through subsistence production. This indicator is related to the difference between the requirement of food products associated with a given diet of the household and the mix of food products available through subsistence to the household.
  2. Maximization of socioeconomic interaction — Maximization of the flow of net disposable cash (as noted earlier, this is different from an assessment of income, since it does not include (1) the monetary value of subsistence food consumed and additional goods and services obtained outside the market and (2) consumption of added value associated with the depreciation (discount) of fixed investment and the direct loss on circulating investments). This indicator is related to the ability of the household to interact with the socioeconomic context in relation to those activities requiring monetary transactions.
  3. Maximization of the ratio total human activity/working time. In practical terms, this reflects the fraction of disposable human activity of the household not invested in working activities. This indicator is affected by the (1) level of education of children, (2) leisure time for workers, (3) workload for children and (d) workload for elderly.
- **Indicators reflecting the perceptions of performance of the government of China**
- In this study, the government of China was considered a relevant agent when deciding the selection of indicators. In relation to this choice, the three relevant criteria selected were:
  1. Minimization of negative gradients between the income of the rural household and the average household income of the country. This is a very sensitive issue in China, where big gradients of wealth between rural and urban population already generated in the past social tension and even revolutions.
  2. Maximization of the surplus of food per household. This is a crucial indicator in relation to the double goal of guaranteeing a good coverage of internal food security and, at the same time, reducing the workforce engaged in agriculture. Since this workforce in agriculture is facing, in any case, a severe biophysical constraint in terms of shortage of LIP, it is important for China to move an increasing fraction of the human activity invested in the working compartment to other economic activities to foster the economic development of the country. On the other hand, such a process of development implies that the internal food security of an increasingly urban population depends on the availability of food surplus produced in rural areas. An analogous criterion could be the maximization of the surplus of rice produced per unit of area of a given household typology.
  3. Minimization of the cost of food security for the Chinese economy. This is an obvious economic criterion that does not need explanation.

- In relation to this selection of criteria, the three indicators used for the characterization of the farming system are self-explanatory.

For the moment we will skip a discussion of the criteria and indicators used to deal with the ecological dimension. This will be done later on, when discussing the integrated representation over different hierarchical levels. In fact, ecological stress can be defined at different levels and scales, and this requires the adoption of different criteria of analysis and relative indicators.

The same approach used in Figure 11.10 to characterize two typologies of households can also be used to characterize typologies of villages. This is illustrated in Figure 11.11. The structure of the two radar diagrams (including the selection of the four sets of indicators for the various quadrants) is exactly the same. The only difference is a different hierarchical level, which implies a larger size, when considering the selection of extensive variables EV1 (human activity and land area) and EV2 (added value and food) for the entity. However, when the representation of this farming system at two different levels (household and village) is based on the same selection of extensive and intensive variables, it becomes possible to apply the mechanism of mosaic effects across levels to bridge different levels of analysis. According to what was said in Chapter 6, we can obtain the representation or characterization given in Figure 11.11 in two nonequivalent ways: (1) by a direct measurement of the characteristics of a village (e.g., through a direct survey), at the level $m + 1$ and (2) by a simulation of the characteristics of a virtual village based on our knowledge of its lower-level components and their organization into a whole, at the level m. To estimate the characteristics of a virtual village made up of typologies of households, we need to specify (1) a set of household types making up the village and (2) the curve of distribution of the population of households over the set of typologies, and complement (3) the information about parts of the villages that are not made up of households (e.g., communal land, roads, societal infrastructures).

This possibility of double-checking the empirical analysis of a given farming system is very interesting. Unfortunately, in this specific case study, this option was discovered only after the fieldwork for this project was complete. Because of this, the final selection of typologies used for an MSIA (required to get a full closure of EV1 across the two levels — household and village) was not totally compatible with the original choices of sampling procedures for the empirical study of households and villages.

Before getting into a more detailed analysis of the integrated analysis over the two levels in this study, it is necessary to briefly explain the type of fieldwork used to gather data. The first step was an analysis of general trends of agricultural development (to obtain benchmark values useful for characterizing the situation of Chinese agriculture compared with those of other countries). The second step was an analysis of historical trends within Chinese agriculture (to obtain benchmark values useful for characterizing the situation of the province under analysis compared with those of other provinces). The third step was an institutional analysis and overview based on interaction with local stakeholders of the targeted area to characterize relevant aspects of the farming system's focus of the study. In this way, five villages were selected to represent special local situations (close to a big city; very severe shortage of land; two generics, with various crop mixes; and in the middle of nowhere). Then a random sampling of 250 households was selected over the five villages. Data collected provided:

1. A general overview of socioeconomic parameters (household size, household composition, age structure, life span, average income, other conventional indicators of development)
2. Assessments of land availability and profiles of land use
3. Assessments of time allocation, with a particular focus on profiles of working time use
4. An assessment of cash flows by source and profiles of cash expenditures
5. The degree of food self-sufficiency (based on assessment of internal production vs. levels of food consumption)

This information was gathered using questionnaires and discussed by the households with Chinese Ph.D. students involved in the research. In addition to the direct interviews, a parallel analysis of technical coefficients for the various cropping systems, and an analysis of land use, made it possible to generate a mosaic effect across levels in relation to the congruence of different profiles of land–time budgets

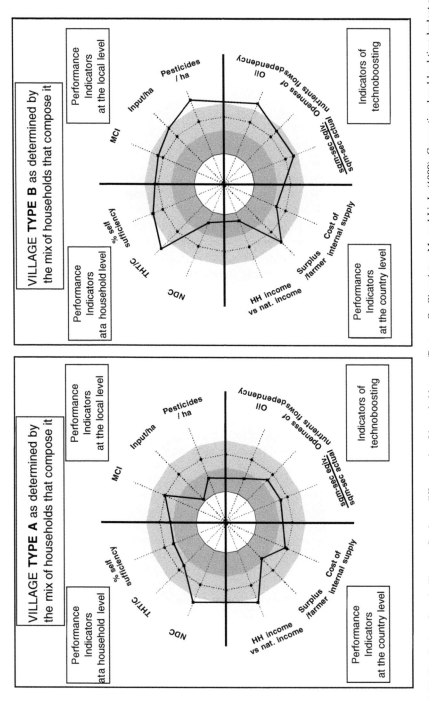

FIGURE 11.11. Typologies of rural villages reflecting a different mix of household types. (Pastore, G., Giampietro, M., and Li, J., (1999), Conventional and land-time budget analysis of rural villages in Hubei Province, China, *Crit. Rev. Plant Sci.*, 18, 331–358.)

declared in the questionnaires. At this point, it was possible to check whether the various solutions were able to balance a given supply and consumption of (1) goods and services out of the market and (2) net disposable cash.

11.2.3.2   *Characterization of Useful Household Typologies* — The characterization of household types was done in relation to the profiles of investments of human time and land area.

The six types of households individuated by that analysis were:

Type 1, almost totally off-farm — Households that allocate more than 80% of their total working time to off-farm activities

Type 2, mainly off-farm — Households that allocate between 50 and 80% of their working time to off-farm activities

Type 3, mainly cotton cropping — Households that allocate more than 50% of their working time to farming activities and that have the largest investment of their harvested land in cotton (special typology related to a special local situation)

Type 4, mainly vegetable cropping — Households that allocate more than 50% of their working time to farming activities and have the largest investment of their harvested land in vegetables

Type 5, mainly cereal cropping — Households that allocate more than 50% of their working time to farming activities and have the largest investment of their harvested land in cereal

Type 6, traditional farming — Households that do not perform any off-farm productive activities, but allocate 100% of their working time to agriculture

The individuation of types 1 and 2 refers to a single choice illustrated by the diamond 3 in Figure 11.7, whereas the individuation of types 3 to 5 is related to choices 4* and 6 in Figure 11.7. In this analysis, it was concluded that choices 1, 2 and 5 are less relevant when determining the formulation of strategy of these agents. A few comments taken by the text of the original paper can be used to explain this point.

When opportunities for investments of net disposable human activity into the compartment off-farm work are available — whenever there is a supply of off-farm jobs — we are in a situation in which the amount of working activity that can be allocated to the generation of net disposable cash is not affected by an external biophysical constraint (e.g., demand of land area per hour of work investment). Put another way, a farmer working in producing rice cannot work 2000 h/year when cultivating only 0.1 ha of land (the available land is the bottleneck for the possible investments of human activity in the working compartment). When the same farmer has the option of working part-time in a factory making tennis shoes, she or he can decide to also work overtime (above the basic workload) if the wage perceived is judged economically convenient. As noted by Georgescu-Roegen (1971), the number of labor hours that can be allocated to agricultural work per hectare per year are determined by a set of biophysical constraints and lag times determined by the speed of natural processes (e.g., the lag time between transplanting and harvesting rice).

This makes it possible to explain the peculiar fact that household types that are better off in economic terms (higher NDC) have a lower average return of labor in terms of net disposable cash generated per hour. The maximum power principle is still at work here. A larger throughput of EV2 for the worker (a higher NDC over the year) is often paid by a lower output/input (a lower IV3 — the average ELP). In fact, after having saturated all those available tasks that provide a very high ELP_i, to gain more money, the farmers are forced to accept work in activities with a lower ELP_i. In the farming system considered, we found that very small investments of human activity in raising small livestock, aquaculture and cash crops were characterized by levels of ELP_i larger than the wages obtained in off-farm activities. The problem with this set of tasks was linked to the existence of external biophysical constraints that were limiting the amount of hours of human activity that could be invested in raising piglets and in producing fish using organic by-products.

Any profile of investments that satisfies the conditions of (1) saturating the existing budgets of land, labor time and capital and (2) operating within the feasibility domain of the selected set of indicators of performance, should be considered a viable technical option for farmers, and therefore it represents a possible state (a potential type) for the farm. Each state defined in this way implies a certain combination of trade-offs for the environment and the national economy. This implies that different shapes over the multi-criteria performance space (Figure 11.10) can be used to study different strategies adopted by farmers (e.g., maximization of economic return vs. minimization of risk, different choices about how to use production factors).

Two general points about the analysis performed in this case study are:

1. The average value of the size of farm (in this case, the most relevant is EV1 — land area) can be used for benchmarking. In this case, such a value is 0.53 ha per farm (0.12 ha per capita with an average size of 4.4 persons per household) over the sample of 250 households. The range of values found in the sample include (1) smallest, 0.33 ha per farm for type 3, cotton cropping (0.08 ha per capita with an average size of 4.1 persons per household) and (2) largest, 0.70 ha per farm for type 5, cereal cropping (0.16 ha per capita with an average size of 4.4 persons per household). The largest endowment of land — 0.16 ha per capita of arable land invested in production — would be considered a very small amount of land under any international standard (against an international benchmarking of possible contexts). However, within our local context, a value of 0.16 ha per capita of arable land is *double* that available to the household belonging to the cotton cropping type.

   At the local level, we can see that the mechanism associated with the bottleneck of land can explain the choices made by households belonging to the totally off-farm type 1 (0.10 ha per capita) and the partially off-farm type 2 (0.12 ha per capita). In the same way, households belonging to the traditional farmers' type (type 6) also have a limited budget of land (0.11 ha per capita), and this can also be one of the factors determining their low-profile, low-risk strategy. The only difference with typologies 1 and 2 is that because they are located in the middle of nowhere or for cultural reasons (older age), these households do not have the option of off-farm working available, and therefore, they simply reduce both risks and workloads. What is relevant in this example from our discussion of an MSIA of a farming system is the ability of this analysis to provide in parallel general and local benchmarking. That is, we can say that the situation of this farming system is characterized by a very severe demographic pressure, when comparing the level of LIP in China with those available to farmers operating in other countries. This means that a difference in the budget of land of 0.2 ha in land in production for two farms is totally irrelevant in explaining the choices of a farmer operating within the U.S. On the contrary, this difference can be crucial in the farming system considered in this case study.

2. In this particular case study, the farming system considered is based on the adoption of multiple croppings per year and on forced rotation patterns of crop mixes. This implies that different crops are (1) cultivated in "packages" rather than individually and (2) over cycles operating over a time period longer than individual years. At the beginning, the handling of the analysis of different rotation patterns, which generates multiple croppings per year in relation to cycles expressed over periods longer than a year, appeared to be a major complication. On the contrary, when useful typologies of households were individuated, this characteristic resulted in a major simplification for ILA. The existence of naturally occurring packages of rotation and crop mixes can be associated with the existence of mosaic effects across dimensions and levels, as illustrated in Figure 11.9. In fact, among the bound, but still large, set of possible combinations of crop mixes and crop rotations that are feasible according to the reciprocal constraints imposed by internal mosaic effects, farmers tend to converge on those solutions that provide a larger overall satisfaction according to their particular perception of performance (which is affected by culture, religion and personal feelings). In other words, some farm types can be associated with profiles of allocation of production factors, the ones giving the most satisfying shapes of the multi-criteria performance space. These standard profiles of allocation of production factors can be seen as a sort of attractor for farmers belonging to a given farming system when deciding how to organize their farm (e.g., farmers will imitate those neighbors doing better). Because of social interactions and societal regulations based on patterns, these attractor types will be

amplified over a larger scale due to establishment of systemic mechanisms of stabilization — e.g., economies of scale in the generation of supply and processing of output. This, in turn, will increase the predictability of the mechanisms generating these patterns, a major plus for an adaptive system.

Because of this mechanism of lock-in, we can expect that a consistent fraction of farmers in a rural area, sharing some internal characteristics and the same typology of boundary conditions, will tend to settle down into the same household typology. In turn, the characteristics of these amplified farm types will affect with their specific pros and cons other hierarchical levels (changes in values of indicators of performance belonging to different quadrants). Even in a uniform ecological and socioeconomic context and under a common set of constraints operating on households (same technology, prices and credit access), heterogeneity in farmers' characteristics (e.g., age, aspirations, fears) will guarantee the existence of several distinct farm types, unless the socioeconomic or ecological context is overwhelmingly powerful in dictating farmers' choices (Arctic households and European farmers in the 1980s were forced to converge on very few uniform farm types that were more or less imposed on them by the context). In any case, the study of the distribution of farm households over possible farm types always requires significant input from the social sciences.

11.2.3.3   *Analysis of the Strategies behind Household Typologies* — The convenience
of adopting a particular profile of investments of production factors, and later on a particular cropping pattern rather than another, depends on four classes of factors:

External factors:
1. Demographic pressure (CAL per capita)
2. Socioeconomic pressure (minimum acceptable fraction of NDHA in nonworking activities and minimum acceptable flow of NDC)

Internal factors:
3. Technical coefficients (yields, labor demand, input demand)
4. Economic variables such as revenue of produced crops, cost of inputs (real cost and subsidies), taxes and off-farm job opportunities (wages)

Classes 1, 3 and 4 are completely outside the control of households, whereas a little room for free decision can be found in class 2 variables. Households can accept living with a flow of NDC that is lower than average and compensate with a higher net disposable human activity in nonworking. However, this room for decision is increasingly reduced as the difference in standard of living between farmers and the rest of China increases. When the gap in material standard of living grows too wide, it becomes imperative for farmers to adjust their lifestyle to get closer to average values of the socioeconomic context within which the farming system is operating. Moreover, a dramatic reduction in the budget of available land (increase in demographic pressure) has the effect of reducing the options on how to use land and working time in the LIP. This increasingly forces the household to obliged choices.

At this point, it should be noted that factors belonging to class 3, technical coefficients, are slow to change; factors associated with class 1, demographic pressure, are slow and difficult to control (China has a vast experience in this field); and finally, factors belonging to class 2, fast economic growth in all developing countries, are affected by large global dynamics that are difficult to control. This makes factors belonging to class 4, economic variables, the only ones by which it is possible to operate in the short term through policy making. This, however, requires the availability of financial resources for those willing to implement the relative policies.

The formulation of policies affecting economic variables is made at the provincial or central government level. Therefore, it is important that at the moment these policies are formulated, due consideration is given to possible side effects that changes in economic variables will induce on the various levels of the system and in relation to different dimensions of analysis. For example, in this case, the relatively high convenience of a farm type based on total cotton production (which has to be both feasible and advisable for local farmers) was mainly generated by government policies. In this case, the goal of the country (getting hard currency with exportable goods, cottons, generated at the rural level) drove special policies helping cotton growers in terms of lower taxes, lower costs and therefore higher revenues. Ecological side effects or long-term effects of such a choice on Chinese food security were obviously ignored. This can be done without prejudice if such a solution is limited in size (it affects only a limited fraction of the land endowment of this farming system). To be able to verify this point, in any case, it is important to have an integrated analysis of this solution (a local solution over a part in relation to the effect on the whole).

In general terms, the fast economic development of China implies a quick change in the characteristics of the socioeconomic context of this farming system. This translates into the need for a dramatic increase in the cash flow of farmers, which is coinciding with a dramatic decrease of available land per household. We are in a case in which both bioeconomic and demographic pressure are pushing this farming system into a difficult situation. A general solution to this double challenge — in this investigated farming system — has been obtained through a dramatic shift of rural households to off-farm work. Actually, a remarkable finding of this study was that the set of activities generating added value, without a direct requirement of arable land (livestock plus aquaculture plus off-farm wages), were the most important source of NDC for this farming system. It is important to keep in mind that this is the rice basket of China. On the other hand, as soon as one performs an ILA associated with a land–time budget analysis, such a choice seems to be quite unavoidable under present conditions. Moreover, this is a choice that will be very difficult to reverse in the near future.

When setting two goals (formalized with variables referring to 1997, when the study took place) — (1) disposable cash over 2000 Yuan/year/capita, plus (2) a good level of food security at the household level — it is easy to prove that these two goals cannot be achieved by allocating the available budget of working time only to farming activities, when at the same time the budget of available land is less than 0.10 ha per capita. One important consequence of this fact is that in spite of the common definition of agricultural area, activities not related to the compartment LIP are the most important economic activities. This means that the challenge posed by the development of rural China cannot be faced by considering only agricultural factors and by implementing agricultural policies. Ignoring the consequences of industrial policies, which are determining the availability of off-farm opportunities, where to invest the abundant surplus of disposable human activity, will cause us to miss a crucial piece of information. More and more in the future, we can expect that the availability of off-farm jobs and their characteristics will determine the agricultural choices of the majority of households in this province.

In general terms, one can guess that the clustering of choices over profiles of investments of human activity should map onto the adoption of (1) a given integrated representation of costs, opportunity and constraints and (2) a given strategy matrix that is used to structure and validate later on the particular choice of typology. For example, we can generalize and aggregate according to common goals the six types considered before:

- Farmers that maximize the net disposable cash NDC through cultivation of cash crops and off-farm labor, even though this means taking risks (neglecting subsistence crops and reliance on the market) and a heavy workload (a low value for the indicator net disposable human activity invested in nonworking).
- Farmers that minimize their risk by growing mainly subsistence crops and that maximize their leisure time (a high value for the indicator net disposable human activity invested in nonworking) by avoiding off-farm jobs, even if this implies remaining behind in the fast process of the modernization of China (low NDC).
- Farmers that operate combining the minimization of risk (relying on subsistence crops) and the maximization of net disposable cash (off-farm jobs and cultivating whenever possible cash crops). This choice is paid for by heavy workloads (a low value for the indicator net disposable human activity invested in nonworking).

Getting back to the analysis of the case study, farm type 1 implies higher income for farmers but at the same time a larger environmental load and a total lack of rice surplus to feed the urban population of China (actually, these farmers are net consumers of rice). If farm type 1 were the only one practiced in all of rural China, the country would no longer be able to feed its urban (and rural) population without heavily relying on imports. In addition, the large amount of labor time invested in off-farm activities implies that farm type 1 is not based on traditional environmentally friendly techniques. Indeed, when assessed at the country level, a massive switch of rural households to farm type 1 can have negative implications not only for food self-sufficiency but also for ecological processes (e.g., they are stopping all environmental friendly tasks).

A detailed discussion of other farm types is graphically illustrated and discussed in Pastore et al. (1999). We just want to emphasize here that each of these farming types — defined at the household level — can be linked to a certain pattern of landscape use (defined on the space scale of the farm) and to certain socioeconomic effects (when aggregated on a large scale) on the national economy. This means that by scaling up potential effects of choices made by individual farmers (given a spatial distribution of rural villages in a determined area and assuming several different distributions of the population of rural households over the set of possible farming types), it is possible to link this type of analysis to (1) changes in landscape use (and therefore related environmental impact indicators) and (2) effects on the national economy generated by the (simulated) changes in the area. In this way, we can also study the effect of government policies or technological changes by simulating the effect that they will have on the distribution of households over possible farm types. Clearly, dramatic changes in technology, farmers' feelings, environmental settings and governmental policies can scramble the existing picture by introducing new possible farm types, making existing ones obsolete and generating dramatic changes in the distribution of individual households over the accessible set of farm types. These types of nonlinear events cannot be predicted, and this is the subject of the last section of this chapter.

Trade-offs faced at the household level by the various household types can be discussed as done in Figure 11.12. Such a graph has on its two axes the variables (1) ratio of NDHA/working, that is, hours invested in disposable human activity divided by hours invested in working and (2) net disposable cash. For those household types facing external constraints that prevent an increase in NDC (shortage of land or no off-farm job opportunities — types 5 and 6), it becomes reasonable to focus on the other parameter determining their satisfaction — that is, increasing the coverage of food security with subsistence and reducing the fraction of NDHA invested in working time (going for the maximization of NDHA/working). Traditional farmers (household type 6) therefore focus on a profile of investments of production factors that maximizes the return of labor and food security (self-sufficiency), even though this is paid for in terms of a lower NDC. This implies a lower integration with the process of evolution of the socioeconomic context. A side effect of this choice is that these household types are frozen in their present situation. By sticking with this choice, they have little hope to change in a short time their present status by participating in the process of the fast modernization of China. On the other hand, this household type minimizes the risk of food security (in the short term) should market perturbations or big economic crises hit the economy of China.

On the opposite side, we find household types that maximize NDC with a low NDHA/working ratio (e.g., type 1). Household types off-farm (type 1), mainly off-farm (type 2), mainly vegetables (type 4) and mainly cereal (type 5) accept working for more hours at a lower return than do households belonging to the traditional farmer type. In this way, however, they are able to remain in touch with the pace of changes of their socioeconomic context. This choice points to a strong aspiration (personal feelings) for a dramatic improvement in the near future. Peculiar is the situation of households belonging to the farm type cotton cropping (type 3), with a high NDC and a high NDHA/working ratio. This win–win situation in relation to these two criteria, however, is due not only to the bold choice made by these households (which abandoned completely the goal of risk minimization through self-production of food), but also to the existing government policy regarding cotton (no taxes and higher revenues for these cash crops). Moreover, it should be noted that when the ecological dimension is also considered, type 3 is the one with the largest emissions of pesticide per hectare.

11.2.3.4 *Characterization of Useful Village Typologies* — The original goal of this study was the characterization of five villages considered to be relevant examples of typologies of rural villages of this area. However, the decision to record empirical data about a population of 250 households operating under different combinations of socioeconomic and ecological constraints made it possible

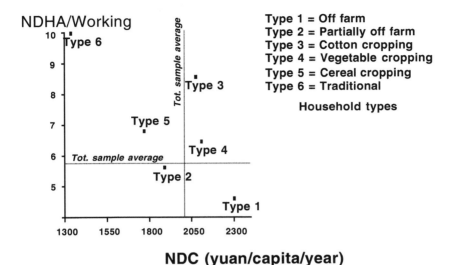

NDHA/Working

Type 1 = Off farm
Type 2 = Partially off farm
Type 3 = Cotton cropping
Type 4 = Vegetable cropping
Type 5 = Cereal cropping
Type 6 = Traditional

Household types

NDC (yuan/capita/year)

FIGURE 11.12 Representing the choices of household types in relation to two different criteria. (Pastore, G., Giampietro, M., and Li, J., (1999), Conventional and land-time budget analysis of rural villages in Hubei Province, China, *Crit. Rev. Plant Sci.*, 18, 331–358.)

later on to also analyze the same data set at the hierarchical level of households. That is, the total sample of households was analyzed, without any reference to the village of origin, looking for clusters of profiles of working time investment over the set of various productive activities. Unfortunately, at this point, it was found that the choice of a random sampling of 50 households per village was not compatible with this nonequivalent use of the original data set. After individuating the six relevant household types listed before as useful for an ILA, it was discovered that the sample of household types presented a very skewed distribution within the population of households of the original five villages. Obviously, cotton farmers were mainly present in the cotton village, and villages close to the city had a different profile of typologies from villages in the middle of nowhere. Unfortunately, households belonging to the six typologies were not adequately sampled. The problem in this case was that the empirical analysis should have adopted a stratified sampling procedure, at the beginning, according to the typologies selected at the end (another example of the chicken–egg problem).

Demographic pressure (defined as the ratio between total population and available land) was found to be very high in all five villages — 0.12 ha per capita — and the major constraint to the development of this rural population. As noted earlier, this value is less than half of the world average. Available LIP ranges from 0.06 ha per capita in Qun Lian (Village 5) to a maximum of only 0.16 ha per capita in Zhuang Chang (Village 2). Differences in land availability are only partially compensated by different intensities of land exploitation, as measured by the multiple-cropping index (MCI), harvested area/available land, and other indices of technoboosting.

The five villages are:

Village 1 — Located near the main town and is representative of villages open to the market and with many off-farm activities. Production of cash crops is important as a source of income but less relevant than off-farm work.

Village 2 — Located in the lowland and relatively far away from the town market. It is representative of farming systems focused on intensive cultivation of cereal, although off-farm activities are an important source of income.

Village 3 — Located in between the high- and lowlands is representative of traditional subsistence agricultural patterns with few off-farm activities. The main cropping pattern is rice–wheat rotation. Cotton is the favorite cash crop, although its cultivation is restricted to only a small fraction of the land.

Village 4 — Located in the lowland relatively near to a large market (as is Village 1). It is representative of farming systems open to the market with a lot of off-farm employment. In contrast to Village 1, cash crops are more important than off-farm work as a source of income.

Village 5 — Located in the highlands and has been selected for its specialization in cotton production. A large fraction of land is allocated to cotton production at the expense of cereal cultivation. Besides cotton cultivation, a large fraction of working time is allocated to off-farm employment. This village is therefore the most dependent on market variables (revenues from off-farm activities and from cotton, cost of inputs, and prices of food commodities) for its food security.

A characterization of these villages in terms of types implies selecting some criteria that can be used to define the types we want to describe. Actually, this operation was done, at the beginning of this project, when selecting these five villages according to the advice obtained by the experts. Such a characterization, however, was not formalized in terms of a set of parameters or variables to be considered in the definition of typologies.

In this example, the characterization of typologies of villages has been based on the same set of indicators and profiles of production factors used for the analysis of household typologies. This choice makes it possible to use mosaic effects, and therefore to express the characteristics of villages in relation to two nonequivalent external referents: (1) empirical data gathered in relation to the villages (e.g., when adopting a survey at the village level, as done in this case study) and (2) simulating villages using information gathered about household types. This makes it possible to use theoretical and empirical analyses in parallel to (1) validate the assumptions adopted in the simulation (in terms of both definition of types and shape of distribution curves), or (2) validate the selection of types in the lower-level analysis used to do the scaling from one level to the other.

For example, this study found that the geographic location of villages (implying different access to markets and off-farm job opportunities for households belonging to different villages) was a significant factor affecting the distribution of farmers over the possible farm types. Similar hypotheses could have been tested when considering the physiological and social characteristics (age and sex structure, ethnic origin, level of education) of households as possible factors affecting the distribution over the existing set of farm types. The problem was that this double analysis was not planned at the beginning.

The metaphorical knowledge generated by this type of analysis makes it possible to compensate, in part, the impossible mission associated with the existence of chicken–egg processes in ILA. In fact, if it is true that by applying this method, one can find out how to do a better sampling in step 2, only after having done a suboptimal sampling procedure in step 1, then it is also true that a good understanding of the set of basic mechanisms (parallel ILAs) determining the definition of useful typologies within a given farming system represents in any case a useful result, independent from the statistical significance of the relative data set. In fact, such an understanding will make it possible, in a successive empirical study, to start the fieldwork with a much better problem structuring and a much better set of hypotheses to be tested. Probably, a new MSIA done in the same area (based on the findings of this first study) will end up with a new set of questions associated with a more refined problem structuring, rather than with a definitive set of answers proved with statistical tests. However, what is important for the quality of the research is that each new generation of questions be associated with a higher level of understanding of the investigated problem and a higher level of communication of nonequivalent observers (scientists and stakeholders) about the common problem structuring. An additional problem is represented by the fact that, when getting back to the same villages after 7 years, one is at risk of not recognizing the places anymore. The pace of change in rural areas undergoing a quick process of transition is so high that it is important to rely as much as possible on metaphorical knowledge rather than on sound statistical data. The lag time required to understand "what" should be sampled and "how" guarantees that the original problem structuring already lost its validity.

We suggested in Figure 11.11 the possibility of generating virtual villages belonging to a given farming system, by simulating their focal-level characteristics from our knowledge of lower-level farm types found in the same farming system. For example, we can imagine that the village described in Figure

11.11 on the left is characterized by a majority of farmers that optimize the net disposable cash (this simulation is based on a distribution of 80% of farmers belonging to farm type 1, 10% to farm type 2, and 10% to farm type 3). The village described in Figure 11.11 on the right is characterized by a majority of farm households practicing traditional agriculture, hence minimizing risks and time allocated to work (simulation is based on a distribution of 80% of farm households belonging to farm type 2, 10% to farm type 1, and 10% to farm type 3).

Thus, there are two types of differences between the two graphs in Figure 11.10 and those in Figure 11.11. In Figure 11.10 we have two characterizations: (1) referring to the household level, (2) which are based on empirical data. In Figure 11.11 we also have two characterizations: (1) referring to the village level, (2) which are simulated using lower-level information. Obviously, the space–time domain of the two characterizations is different: the village in Figure 11.11 is larger in terms of both human activity (1.75 million h), associated with about 200 people, and land in production, about 20 ha. Villages are also slower in reacting to changes.

Considering the shape of the profile of values taken by the various indicators selected in the multi-criteria performance space in Figure 11.11, we see that the first virtual village — Village A on the left (market-driven choice based on off-farm work and intensive production of cash crops) — is the one that generates by far the higher environmental loading and is to the larger extent dependent on coal and oil for food production. From the national perspective, this village does not produce any surplus of rice; on the contrary, it erodes the rice surplus produced by nearby villages. As expected, however, what is detrimental to the environment and the food self-sufficiency of the country also has its positive side: a high net disposable cash for farmers. The productive pattern adopted by Village A is therefore benign to the villagers and to the people of the closeby town, who have access to a cheap supply of fresh vegetables and other food. On the contrary, the second virtual village — Village B on the right —provides a high surplus of rice (good for self-sufficiency of China) and generates a moderate environmental impact (good for the environment). This environmental benign solution is paid for in terms of low net disposable cash from agriculture. People living in Village B are at risk of losing contact with the dramatic socioeconomic transformation that is taking place in China. A general amplification of the Village B type will imply locking a large part of the Chinese rural population into a situation of poverty and lack of modernization.

We believe that the analysis of these links is useful from a policy-oriented perspective. This can be used to study general trends, which could be useful for understanding the evolutionary behavior of agricultural systems in different geographic areas or socioeconomic conditions. In fact, the amplification of village types and farm types can be directly related to the spread of relative patterns of landscape use and to changes in vertical power relations across levels. Since some of the possible types are more benign to a level perspective than others, we can expect that the shape of the distribution curve of individuals over types will reflect power relations among levels' perspectives. For example, if the Chinese government wants to slow down spreading of the land use pattern typical of the Village A type (dramatic reduction of rice cultivation, abandonment of environmentally friendly farming techniques), it has to negotiate alternative solutions with farmers, for example, guaranteeing higher incomes to young farmers attracted to farm typology 1. This could be achieved by changing the combination of options available to farmers at the local scale, for instance, by offering off-farm jobs only to those farmers that plant rice on their land, or by taxing cash crops and rice alike. Clearly, each of these interventions will induce as side effects a rearrangement of the profile of the values taken by the various indicators on different levels and dimensions. Hence, each potential change will produce food for further thought.

As done with the example of household type, it is possible to also perform a trade-off analysis at the village level, as shown in Figure 11.13. In this example, we get back to the original data of the study of Pastore et al. (1999), which are based on real characterizations of the five villages described before. The two axes selected for this analysis are net surplus of grain per unit of area of farming system (assessed in kilograms per hectare) and net disposable cash from agriculture per unit of area of farming system (assessed in Yuan per hectare per year). With this choice, we check the trade-off between the two criteria (1) the ability of China as a country to feed the urban population with the available arable land used by existing farming systems and (2) the stability of agricultural activities in rural areas (in fact, patterns of crop production that provide a level of NDC too low will be opted against by households' choices).

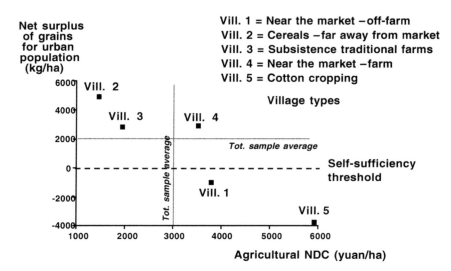

FIGURE 11.13 Representing the performance of village types in relation to two different criteria. (Pastore, G., Giampietro, M., and Li, J., (1999), Conventional and land-time budget analysis of rural villages in Hubei Province, China, *Crit. Rev. Plant Sci.*, 18, 331–358.)

A quick look at this graph indicates an evident problem. An excessive industrialization of rural areas carries the risk of eroding rice surpluses. In fact, Village 1, which is operating in a full market mode (high rate of off-farm and conventional cash crops), and Village 5 (high rate of off-farm and cotton production) do not generate any surplus of rice to feed China's growing urban population. Actually, they are already net consumers (they are both below the threshold of self-sufficiency). The village with the highest production of rice surplus (Village 2) is the one far away from the town market and with the lowest NDC from agriculture. However, off-farm activities in Village 2 are able to sustain the income of the villagers. Village 3 is in a worse situation; its low NDC from agriculture is not compensated by large revenue from off-farm activities. Therefore, from the perspective of self-sufficiency of the whole country (availability of enough surplus of rice produced in rural areas to support the urban population moved into the cities by the industrialization process), China should be careful to sustain those combinations of investments of production factors (ILAs) that generate surpluses of rice at the rural level. However, due to the shortage of land, the only way to make such an option appealing to farmers is either (1) to change the economic variables in a way that increases the overall performance of a profile of choices generating rice surpluses (pay more to the rice farmers) or (2) to make available to villages specializing in cereal production off-farm job opportunities with high wages in the off-farm compartment. This could compensate the lower ELP of the choice of producing rice. Clearly, coming to the perspective of China as a country, solution 1 would increase the cost of surpluses for Chinese economy, whereas solution 2 would increase the control of the planners on both industrial choices (e.g., by imposing a rule that only part-time farmers planting rice can be assumed by rural factories) and farmers' choices (to get access to well-paid jobs, they have to plant rice). On the other hand, when considering the dynamic of demographic and socioeconomic pressure and the actual critical condition, it is also important to keep tight control over the directions taken by economic growth in relation to food security and environmental security.

However, to include in this multi-scale integrated analysis the ecological dimension of sustainability, it is necessary to bridge the typologies used to represent household and village types in terms of parallel investments of human activity and CAL to a representation of energy and matter flows at the landscape level. In this way, different typologies of farming systems can be related to the level of interference they generate on the system of control expressed by the terrestrial ecosystems embedding them. An example of how to establish such a bridge is given in Section 11.3.

To conclude this section, it is important to note that the basic rationale adopted in the analysis of this case study is not that of telling the Chinese people how to run their agricultural sector or how to plan strategies of economic growth. This analytical approach does not individuate what is the best solution or what should be considered the optimal use of production factors. In our view, the goal of multi-scale integrated analysis is that of providing a richer understanding of the complexity of relations found in an evolving farming system. Therefore, this analysis is only descriptive and — on purpose — tends to avoid any normative tone. As observed in Chapter 5, an MSIA of agroecosystems should be considered an input related to a discussion support system. This means that a parallel process of societal multi-criteria evaluation is required with the goal of translating such rich understanding into action.

11.3 An Overview of the MSIA Tool Kit and the Impossibility of Multi-Agent Simulations

11.3.1 Linking Alfa-Numerical Assessments to Spatial Analysis of Land Uses across Levels

The previous section illustrates the possibility of using mosaic effects within an alfa-numerical data set, which can be used to generate an integrated characterization of a farming system on different levels. Mosaic effects can be used to select a set of useful typologies that can be adopted in an impredicative loop analysis useful for studying the socioeconomic sustainability of farming systems at different levels. On the other hand, a multi-level analysis of ecological compatibility of an agroecosystem requires information of a different nature. Alfa-numerical data sets can be used for the analysis of matter and energy flows; however, to be useful in relation to the ecological dimension, they have to be integrated in spatial maps whose identities are based on land use typologies. For this reason, it is important to establish a link between a characterization across levels of household typologies based on alfa-numerical assessments (in the form of packages of indicators over a feasibility domain) and spatial analysis of land use.

This section provides an example of how it is possible to link an MSIA of a farming system (the socioeconomic analysis described in the previous section) to an analysis of land use at different levels. The goal of this section is to illustrate an example of application of this method. That is, it describes the general structural organization of the information space adopted in a different case study (in Vietnam), which is based on tables, graphs and maps. At this point, it is not necessary to again get into a discussion of technical aspects of the analysis, characteristics of the given farming system or a narrative regarding the implications of the analysis (i.e., what the farming types individuated mean in relation to the objectives of the analysis). Data and figures of this section are taken from Gomiero and Giampietro (2001).

In this case study, the MSIA method has been applied in an *ex post* evaluation of a project aimed at rural development of marginal areas in Upland Vietnam (on the border with Laos). The project, whose effects were evaluated, had the goal of involving ethnic minorities in a program of reforestation. In relation to this goal the project was not particularly successful. Against this background, an MSIA of this farming system had the goal to check whether the actions suggested and supported by the program (with the proposed development policy) were compatible with the option space as seen and perceived by the various household typologies found in this farming system. The conclusion of the study was that especially in relation to the household typologies representing the ethnic minority, the options proposed by the program were both not feasible and not advisable according to the strategy matrices and option spaces associated with these household types.

11.3.1.1 *Definition of Lower-Level Characteristics and Household Types* — By assessing the profile of investments of human time and land in production related to the tracking of flows of added value and food (total consumption and internal supply), it is possible to identify a finite set of tasks that can be used to describe the activities of the various households. This set of task has to be defined to obtain (1) closure over the two budgets of human activity and colonized available land and (2) a finite set of possible compartments. In particular, it is important to have a fine resolution in relation to the identities referring to the mix of tasks of the direct compartments. The direct compartments are those that are related to the stabilization of the metabolic flows (food and added value). The result of

TABLE 11.1

Time Allocation per Household Type

Indicators	Total Sample	Type 1 Off-farm Crop$_{mix}$	Type 2 Husbandry Crop$_{mix}$	Type 3 S&B Crop$_{mix}$	Type 4 NTFP Crop$_{mix}$
Total Time Allocation					
Total worked time per household (hours/year)	3630	4016	4550	3236	2066
% worked time/disposable working time	27	32	32	24	19
Worked time per capita (hours/capita/year)	706	854	820	619	428
Worked time (hours/worker/year)	932	1107	1059	815	612
Chores					
% total available working time	17	18	18	16	16
Worked Time Allocation (% of Total Time Worked)					
Home garden	13	10	9	14	49
Paddy	5	3	4	7	8
Cropland	23	7	25	30	13
Husbandry	21	13	37	10	0
S&B	17	4	10	33	5
NTFP	4	0	2	5	23
Off-farm	15	62	12	0	0

Note: S&B = slash and burn; NTFP = non-timber forest products.

Source: Gomiero, T. and Giampietro, M., (2001), Multiple-scale integrated analysis of farming systems: the Thuong Lo Commune (Vietnamese Uplands) case study, *Popul. Environ.*, 22, 315–352.

this analysis is a set of tables. An example of this type of information is given in Table 11.1, Table 11.2, Table 11.3 and Table 11.4.

11.3.1.2 Definition of Household Types — At this point, it is possible to use different profiles of investment of disposable human activity and land in production over this set of possible activities to characterize household types. In this case study, four typologies of households were selected in this way. The set of possible typologies of investments of hours of human activity in agricultural work were based on packages of techniques (which reflect established patterns found in this typology of farming system in Southeast Asia).

The four typologies individuated can be divided into two different categories:

1. Typologies that do not depend heavily on the forest for the stabilization of their metabolism

   Type 1 — Based on a mix of off-farm and conventional crops

   Type 2 — Based on husbandry and conventional crops.

2. Typologies that heavily depend on the exploitation of tropical forest for the stabilization of their metabolism (mainly adopted by the ethnic minority)

   Type 3 — Has a large investment of land in slash and burn

   Type 4 — Has a large investment of working time in the extraction of non-timber forest products

After selecting a set of indicators to be used to build a multi-criteria performance space, it is possible to characterize these various typologies of households in two nonequivalent ways, as done in Figure 11.14 and Figure 11.15.

Looking at Figure 11.14 and Figure 11.15, we have (1) on the left alfa-numerical data organized within the MCPS and (2) on the right a characterization of the typologies of land use (mapped in terms of extensive variable — square meters — per typology of land use). A simple comparison of the two

TABLE 11.2

Land Use Pattern per Household Type

Land Use (ha)	Total Sample	Type 1 Off-Farm $Crop_{mix}$	Type 2 Husbandry $Crop_{mix}$	Type 3 S&B $Crop_{mix}$	Type 4 NTFP $Crop_{mix}$
Total land:					
TOT	85.3	2.4	25.0	52.1	3.3
per HH	2.19	0.34	2.10	3.57	0.66
per capita	0.32	0.13	0.26	0.52	0.14
%	100	100	100	100	100
Land outside of the commune:					
S&B land	64.7	0.7	16.3	44.6	1.0
TOT	1.66	0.10	1.36	2.97	0.20
per HH	0.24	0.02	0.18	0.41	0.04
per capita	76	29	16	86	31
%					
Land within the county:					
TOT	20.6	1.7	8.7	7.5	2.3
per HH	2.04	0.24	0.73	0.50	0.46
per capita	0.08	0.05	0.09	0.07	0.09
%	24	71	84	14	69
Home garden:					
TOT	8.3	1.0	2.8	2.9	1.6
per HH	0.21	0.14	0.23	0.19	0.32
per capita	0.03	0.03	0.03	0.03	0.07
%	10	41	11	6	50
Paddy:					
TOT	4.1	0.4	1.3	2.0	0.4
per HH	0.11	0.06	0.11	0.13	0.08
per capita	0.02	0.01	0.01	0.02	0.02
%	5	16	5	4	12
Crop:					
TOT	7.8	0.3	4.6	2.6	0.2
per HH	0.20	0.05	0.39	0.18	0.05
per capita	0.03	0.01	0.05	0.03	0.01
%	9	14	18	7	7
Pasture land[a]	5.8	1	12	5	0
Forestland for NTFP[b]	230	0	166	300	500

[a] Assuming 8 ha pasture per cow per year feeding in low-quality pasture (Vu and Nguyen, 1995). Pasture land in Thuong Lo is quite degraded, most of it on sloping fallow land that surrounds the commune.

[b] NTFP, mainly rattan climbing palms (e.g., *Calamus* genus) and cap leaves. Forest surrounding Thong Lo Commune is secondary forest, impoverished by the collecting pressure exerted in the last decades. For that reason, rattan collection requires a long time to be spent in the forest, a very harsh and risky activity.

Source: Gomiero, T. and Giampietro, M., (2001), Multiple-scale integrated analysis of farming systems: the Thuong Lo Commune (Vietnamese Uplands) case study, *Popul. Environ.*, 22, 315–352.

typologies of land use shows the huge difference in terms of requirement of forestland between these two household types.

But also in this case, it is important to make a distinction between a local perspective and a larger-scale perspective to be able to appreciate the real effect that the various household typologies play in the characterization of a given farming system. In fact, in general, households performing slash and burn are those usually accused of being the ecological villains in farming systems operating on the edge of tropical forests. On the other hand, when an analysis of land use is performed at a larger scale (at a

TABLE 11.3

Technical Coefficients of Activities

Activities	Type 1 Off-farm Crop$_{mix}$ (kg/ha/year)	Type 2 Husbandry Crop$_{mix}$ (kg/ha/year)	Type 3 S&B Crop$_{mix}$ (kg/ha/year)	Type 4 NTFP Crop$_{mix}$ (kg/ha/year)
Home garden:				
Starchy roots	—	—	2000	2000
Corn	—	400	400	400
Beans	200	200	200	200
Vegetable	500	500	500	500
Fruits	1000	1000	200	200
Paddy field:				
Rice^{wr}	2200	2200	2200	2200
Cropland:				
Rice^{dr}	—	1000	1000	1000
Cassava	—	4000	4000	4000
Corn	400	400	400	400
Beans	200	200	200	200
Vegetable	1000	—	—	—
Husbandry	—	40^H	40^H	—
Slash and burn^{SB}:				
Rice	—	1000	1000	—
Cassava	—	3500	3500	—
NTFP^{NT}:	—	—		
Rattan			300	300
Honey			3	3

Note: wr = wet rice, generally with two crops per year; dr = dry rice, one crop per year; SB = a cycle of 2 years of cultivation (rice cassava) and 4 to 5 years of fallow; H = assuming a cow feeding on 8 ha of pasture land; NT = this activity is carried on over several hundred square kilometers of forest. Rattan is obtained by climbing the palms of the *Calamus* genus, and it is used to make forniture.

Source: Gomiero, T. and Giampietro, M., (2001), Multiple-scale integrated analysis of farming systems: the Thuong Lo Commune (Vietnamese Uplands) case study, *Popul. Environ.*, 22, 315–352.

level higher than the one of individual villages), it is easy to discover that households belonging to this typology were using less than 10% of the CAL — when defined at the level of the Thu Lo Commune (an administrative unit including the three villages considered in this study). Put another way, when looking at the big picture, we can appreciate that the decisions of individual households, in this situation, affect only 10% of the forest area considered in this study. The remaining 90% was managed by a nationally owned company. This means that the activities related to this land use were decided by agents operating far away (in the capital of the province or in the capital of the country). When perceiving the various relations of cause and effect on multiple levels, it becomes evident that it is unfair to blame the ecological problems of this forest on the actions of local households, since the most important driver of environmental impact is associated with the activities of management expressed by the national company.

11.3.1.3 *Moving to Higher Hierarchical Levels* — This example of mismatch between the picture obtained at the local level and the picture obtained at a larger scale is important. It focuses on the need to scale up in terms of MSIA. Especially when dealing with the issue of sustainability (long-term perspective) and the ecological dimension (ecosystem perspective), it is important to reach a level of scale, in the analysis of land uses, at which it becomes possible to have the right level of analysis to address the environmental problem of concern. In fact, it is only when the various effects of policies and techniques of production are considered within the appropriate frame (perception and representation referring to the right scale) that it becomes possible to detect which choices are more relevant for the sustainability of these forests. Obviously, different categories of land use require different indices of environmental impact. Again, this points to the need for having a multiple reading on different levels.

TABLE 11.4

Economic Throughput per Hour of Work (ELP) and per Hectare of Tasks

Economic Performance	Type 1 Off-farm Crop$_{mix}$	Type 2 Husbandry Crop$_{mix}$	Type 3 S&B Crop$_{mix}$	Type 4 NTFP Crop$_{mix}$
Home garden:				
ELP$_{HG}$ (VND /h)	4000	2000	500	500
Dollar supply$_L$ (VND/ha)	7,000,000	5,000,000	1,000,000	1,000,000
Cropland:				
ELP$_C$ (VND/h)	1000	1000	800	800
Dollar supply$_L$ (VND/ha)	5,000,000	5,000,000	4,000,000	4,000,000
Husbandry:				
ELP$_H$ (VND/h)	—	1000[a]	1000[a]	—
Dollar supply$_L$ (VND/ha)	—	125,000	125,000	—
Slash and burn:				
ELP$_{SB}$ (VND/h)	—	2080	2080	—
Dollar supply$_L$ (VND/ha)	—	1,000,000	1,000,000	—
NTFP:				
ELP$_{NTFP}$ (VND/h)	—	—	1750	1750
Dollar supply$_L$ (VND/ha)	—	—	1000	1000
Off-farm:				
ELP$_{OFF}$ (VND/h)	3000	1500	2000	—

[a] Considering two cows.

Note: VND = Vietnamese dong (US\$1 = 12,300 VND in 1997); NTFP = non-timber forest products; HG = home garden; S&B = slash and burn; C = crop; H = husbandry; OFF = off-farm

Source: Gomiero, T. and Giampietro, M., (2001), Multiple-scale integrated analysis of farming systems: the Thuong Lo Commune (Vietnamese Uplands) case study, *Popul. Environ.*, 22, 315–352.

At the level of the village, as shown in Figure 11.16, it becomes possible to associate alfa-numeric indicators of performance referring to the socioeconomic dimension of sustainability with an overview of land use associated with the identity of the village. In fact, at this scale it becomes possible to apply to this land use analysis a representation based on remote sensing. This implies that it also becomes possible to apply techniques of elaboration of data generating indicators of environmental stress (e.g., diversity of vegetal species over a given grid). In the socioeconomic analysis of Figure 11.16, the characteristics of lower-level households (specified before) and the profile of distribution of actual households over this set (illustrated in the lower-right quadrant in the radar diagram on the right side of the figure) are used to infer some of the characteristics of the village. Obviously, additional information is required to fill in gaps of information about land use. These gaps refer to land uses decided at the level of the village, which are not included in the analysis performed at the household level. This can include, for example, schools, communal buildings, roads and infrastructure, which have to be inserted, in the scaling up, using additional sources of information (referring to nonequivalent measurement schemes applied at the village level).

The same approach can be used to move the integrated analysis to a higher hierarchical level. In this case study (Figure 11.17), this higher hierarchical level was represented by the Thuo Lo Commune, made up of three villages. Also in this case study, the geographic location of villages in relation to the city and the internal characteristics of households (ethnicity, age, level of education) resulted in significant factors to explain different profiles of distribution of household types in the villages.

11.3.1.4 An Overview of the Organizational Structure of the Information Space —

Finally, an overview of the relations established among different typologies of information within such a multi-scale integrated analysis is provided in Figure 11.18. The process starts with the characterization of the set of possible activities at the lower level of the household (using the land–time budget). In this way, it is possible to individuate useful typologies of households (using the ILA on different metabolic flows to be stabilized) as discussed in Section 11.2. At that point, a profile of distribution of the population of households over the given set of household typologies generates different typologies of villages. Each

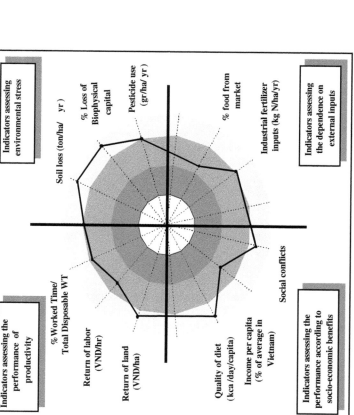

FIGURE 11.14 Household type 1, Upland Vietnam. (Gomiero, T. and Giampietro, M., (2001), Multiple-scale integrated analysis of farming systems: the Thuong Lo Commune (Vietnamese Uplands) case study. *Popul. Environ.*, 22, 315–352.)

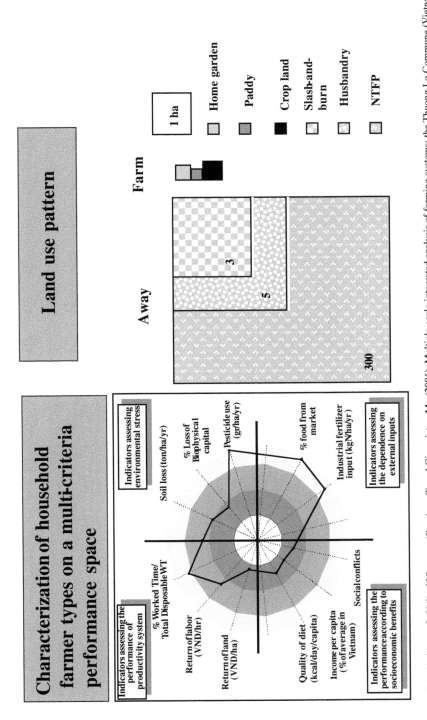

FIGURE 11.15 Household type 3, Upland Vietnam. (Gomiero, T. and Giampietro, M., (2001), Multiple-scale integrated analysis of farming systems: the Thuong Lo Commune (Vietnamese Uplands) case study, *Popul. Environ.*, 22, 315–352.)

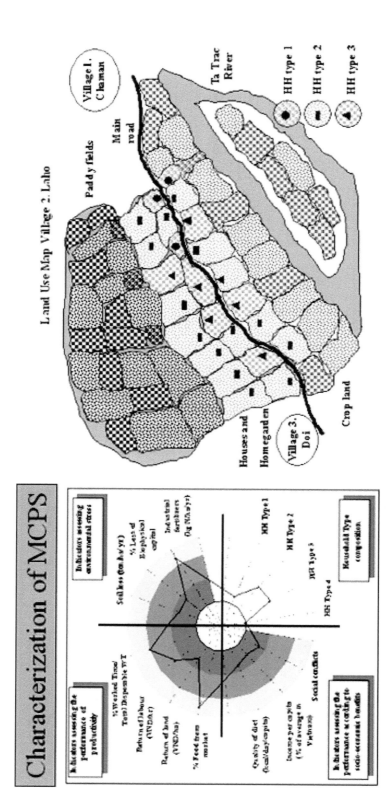

FIGURE 11.16 Village 2 (Laho), Upland Vietnam. (Gomiero, T. and Giampietro, M., (2001), Multiple-scale integrated analysis of farming systems: the Thuong Lo Commune (Vietnamese Uplands) case study, *Popul. Environ.*, 22, 315–352.)

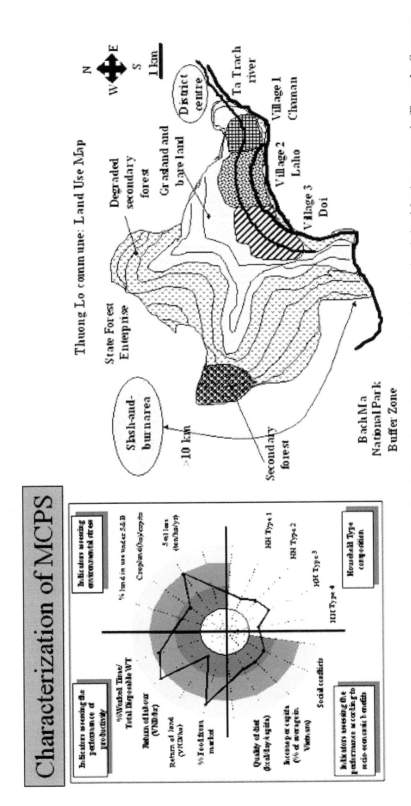

FIGURE 11.17 Thuong Lo Commune, Upland Vietnam. (Gomiero, T. and Giampietro, M., (2001), Multiple-scale integrated analysis of farming systems: the Thuong Lo Commune (Vietnamese Uplands) case study, *Popul. Environ.*, 22, 315–352.)

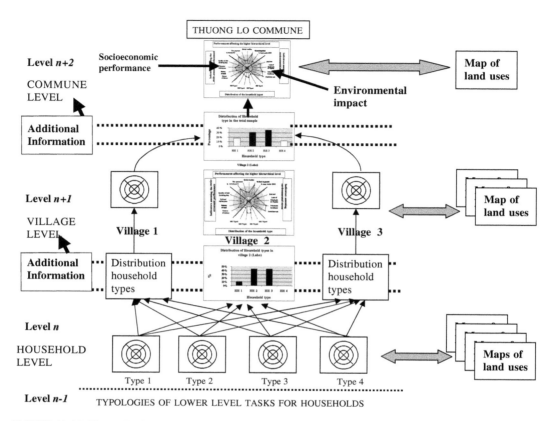

FIGURE 11.18 Thuong Lo Commune. Typologies of lower-level tasks for households. (Gomiero, T. and Giampietro, M., (2001), Multiple-scale integrated analysis of farming systems: the Thuong Lo Commune (Vietnamese Uplands) case study, *Popul. Environ.*, 22, 315–352.)

typology of villages can be associated with a map of land uses, which is related to the typologies of land use within the compartment CAL defined at the level of households. In turn, these typologies of villages can be related to a higher-level typology of entity belonging to the same farming system, but defined on a larger space–time domain (in this case, the Thuo Lo Commune).

It is important to acknowledge the need for additional sources of external information whenever one wants to perform a scaling from lower-level characteristics (at a level x) to focal-level characteristics (at a level $x + 1$). This is required by the simple consideration that the holon considered at the higher level is expressing emergent properties and therefore requires the use of additional categories not included in the characterization of lower-level entities.

The set of relations illustrated in Figure 11.18 provides an integrated vision of how it is possible to keep a certain level of coherence in a multi-scale integrated representation of the performance of a farming system in relation to various dimensions and hierarchical levels. However, also in this case one should not expect too much from this approach in terms of the building of formal models able to handle such a multi-scale integrated analysis. This approach provides some coherence, but cannot provide a full predictability of changes, as discussed in the next two sections.

11.3.2 Multi-Objective Characterization of Performance for Different Agents

The analytical tools presented so far can be used to describe the effect of changes in a farming system in parallel on different hierarchical levels (space–time scales) and according to a given selection of perspectives (those selected in terms of indicators in the four quadrants). However, this multidimensional representation that is required to provide useful input to the discussion of pros and cons related to a given decision making in agriculture (e.g., to evaluate possible scenarios) implies a few steps that make a substantive (formal) analysis impossible:

1. The selection of a finite set of indicators to describe the effects of a particular change (the representation of the system in terms of states on an MCPS) on different levels and according to different perspectives is just one of a virtually infinite set of possible options. This implies that this is always an arbitrary operation.

2. The various perceptions and representations of the states of the system on different scales and in relation to different dimensions of analysis are not reducible to each other in formal terms. They refer to nonequivalent descriptive domains (incompatible definitions of simple time and space).

3. Models used to link changes perceived and represented at one level to changes perceived and represented at a different level work only in one direction. They define a directon of causality according to the assumptions required for the triadic filtering to get out of hierarchical complexity. Therefore, their usefulness is local and depends on the credibility of the relative assumptions.

Very often these basic epistemological problems are amplified by the fact that those looking for a substantive, optimal solution for the sustainability of a farming system (in relation to all possible dimensions of analysis, all possible hierarchical levels at which relevant identities are expressed, and the various legitimate perspectives of all possible agents) get themselves into an impossible mission in epistemological terms. On the other hand, if one attempts to analyze only the perspective of one agent at a time, things can become a little bit easier. At least the unavoidable epistemological challenge to be faced is reduced within the window of reality of competence of that agent.

In fact, the various analytical tools presented so far can be used to study the behavior of households seen as agents. An overview of this analysis is given in Figure 11.19. Depending on how we decide to characterize a given farming system using extensive and intensive variables over an impredicative loop, we will end up determining an option space for the households, which is determined by the information contained in the three boxes on the left of the figure. In this example, the upper box includes information that has been labeled under the category parameters. Then the center box, referring to the available budget of production factors considered in the ILA, has been labeled using the category constraints. Finally, the lower box includes system qualities that can be changed according to the decisions made by the household/agent; therefore, this box has been labeled using the category variables. We have to repeat again that what is relevant here is the semantics of the categories used for these boxes and not the selection of items included in them. Depending on the situation and the hierarchical level of analysis, the list of items to be included in the various boxes can change.

Depending on a given strategy matrix (what are the goals of the agent and the perceived priority among them), the agent will select a possible state over the option space. This choice will generate a result that can be characterized in terms of (1) a given shape on a multi-criteria performance space (which is relevant in relation to a socioeconomic dimension of sustainability) and (2) a given pattern of land use (which is relevant in relation to an ecological dimension of sustainability). In a way, this integrated characterization (MOIR stands for multi-objective integrated representation) can be related to the concept of payoff matrix (the effect associated with a given choice made by the agent).

As discussed in the previous section, decisions made at the household level by individual agents when aggregated on a larger scale (e.g., village level) will determine the pattern of land use at that scale. That is, the choices made by individual farmers about farming techniques (crops selected in the mix, techniques and technologies adopted, which will determine the intensity of cultivation) and about the amount of work allocated to farming and off-farming activities will determine the characteristics of a given farming system at a higher level. An overview of the mechanism through which it is possible to aggregate the effects of individual farmers' choices — using useful typologies of households — is given in Figure 11.20.

The various characterizations of the effects generated at the local level by the decisions of the various agents can be related to the effects generated at the level of the village. At this point, however, the peculiar nature of complex systems organized in nested hierarchies enters into play, whereas decisions of individual agents (e.g., households), when considered one at a time, are not relevant for higher-level

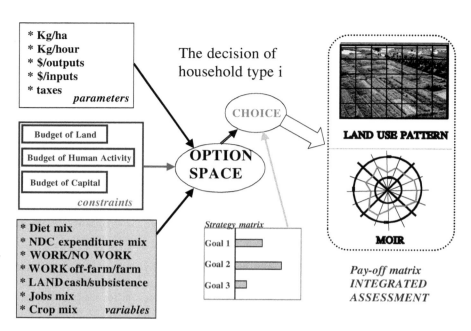

FIGURE 11.19 Households seen as agents.

agents (e.g., the government). When lower-level agents converge on a given set of typologies and these typologies are amplified by some mechanisms operating across levels, the situation changes dramatically. A lower-level pattern (associated with a given agent behavior) can be amplified to a scale large enough to become relevant for higher-level agents. At this point, holons of different levels interacting within a healthy holarchy are able to interact in a way that is impossible to model using a conventional formal system of inference (we are dealing with complex time and an open and expanding information space).

11.3.3   The Impossible Simulation of the Interaction of Agents across Levels

An overview of the information given by Figure 11.19 and Figure 11.20 is given in Figure 11.21.

If we assume that the behavior of lower-level agents (households) is organized over typologies and that the mechanisms of self-entailment across levels are able to express identities at the level n determined by (1) the set of characteristics of lower-level states accessible for the element at the level $n - 1$ and (2) the profile of distribution of lower-level elements over the set of possible states, then we have a system of interaction that is impossible to formalize. In fact, the definition of what should be considered a possible state for a household will depend on different types of information (what is socially acceptable, what is ecologically compatible, what is economically viable and what is technically feasible), which can only be defined on nonequivalent hierarchical levels. That is, if we try to define such an option space in formal terms, we face again the well-known problem of identities and perceptions referring to nonequivalent descriptive domains.

In particular, there is a dimension of sustainability that is related to human feelings and aspiration (cultural and economic variables), and there is a dimension of sustainability that is related to ecological processes and biophysical constraints. Everything within this information space is changing in time, but with different paces. The formation and evolution of typologies within the human side (associated with changes in institutions, laws, cultural rules, ethical principles) has a different speed from the formation and evolution of typologies within the ecological side (recall the discussion about the definition of ecological essences in Chapters 8 and 10). To make things more difficult, the strategy matrix and the perception of constraints, costs and opportunities of agents operating at different levels also change at different paces. Technical coefficients and economic characteristics (such as prices and costs) that are parameters for households are variables for national governments. Agents at the national level can decide

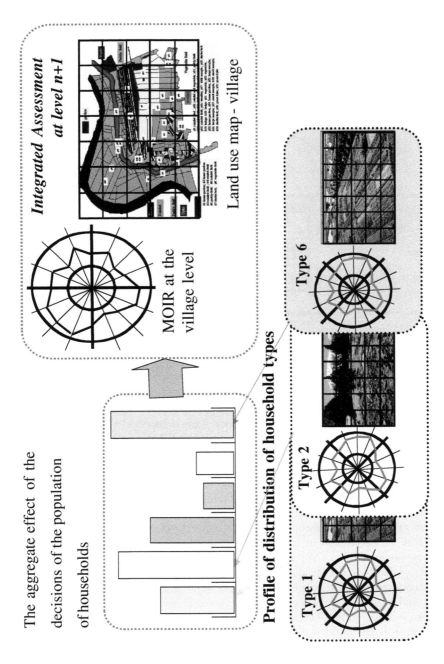

FIGURE 11.20 Scaling up the effect of household choices.

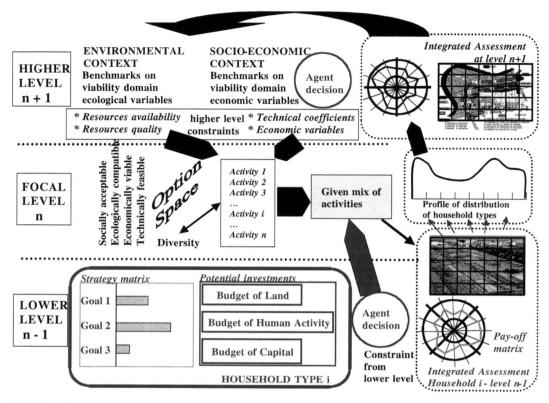

FIGURE 11.21 The reciprocal representation of cause–effect among agents operating at different hierarchical levels.

programs of technological innovation and economic policies of subsidies. In a situation of parallel decision making across scales, it is impossible to guess from the beginning whether, in the long run, the aggregate behavior of farmers will affect the decisions made by the government or vice versa. But even imagining that at a particular point in time we can guess a direction of causality in the short term (a given policy of the government will affect the choices of the farmers) will have, in any case, a limited validity in time. The change imposed from the higher level will induce a readjustment of lower-level characteristics. This can generate temporary feedback in terms of economic variables in relation to the choices made at the government level. For example, subsidies to produce a given crop can generate an excessive move toward the production of that crop that implies an excessive drop in the relative price. On the other hand, the change induced by government policy can be more drastic, such as the eradication of a given technique or technology from a farming system (e.g., elimination of animal power in rural transport). Put another way, the continuous interaction across levels can imply either a simple readjustment of the profile of distribution of lower-level elements over a give set of possible states or the emergence of new possible states and the elimination of obsolete states from the original set.

We can relate this discussion of the analysis of the evolution of the information space to the metaphor of the lazy 8 proposed by Holling (in Chapter 8). The introduction of new states (a phenomenon of emergence) can induce a new definition of useful typologies within a given ILA (recall the general overview of ILA given in Figure 9.1). A major readjustment over the loop can lead to the individuation of and a definition of a new metabolic system in a population of interacting observers. Then this new metabolic system will change in time during its evolution. This will be obtained by changing the profile of distribution of its extensive variables 1 over lower-level types, deleting obsolete types and amplifying those types that are more useful for the whole according to boundary conditions. In this process some typologies are improved in terms of both a better-organized structure and a better definition of functions (recall the description of the evolution of car types given in Chapter 8). Finally, the accumulation of changes will become so relevant that for the population of observers, it will be easier to deal with this system by changing its initial definition of identity. The cycle of creative destruction reached a point that for the observers, it becomes more useful to adopt another basic definition of what this metabolic system is — a new definition that is based on a different integrated set of useful typologies.

If we accept this explanation of the metaphor proposed by Holling about the evolution of metabolic systems in relation to the issue of sustainability, we have to also accept that when dealing with the analysis of this process applied to farming system analysis, metaphorical knowledge is much more relevant than formalized characterizations. Put another way, in farming system analysis it is better to use a mix of high-tech tool kits, gut intuitions, political skill and common sense (to be a 007 researcher) than to base the research activity on a mix of formalisms, statistical tests and complicated mathematical models (to be a $p = .01$ researcher).

References

Allen, P., van Dusen, D., Lundy, J., and Gliessman, S., (1991), Expanding the definition of sustainable agriculture, *Am. J. Sustain. Agric.*, 6, 34–39.

Allen, T.F.H., Tainter, J.A., Pires, J.C., and Hoekstra, T.W., (2001), Dragnet ecology, "just the facts ma'am": the privilege of science in a postmodern world, *Bioscience*, 51, 475–485.

Altieri, M., (1987), *Agroecology: The Scientific Basis for Alternative Agriculture*, Westview Press, Boulder, CO.

Brklacich, M., Bryant, C., and Smit, B., (1991), Review and appraisal of concept of sustainable food production system, *Environ. Manage.*, 15, 1–14.

Brown, B.J., Hanson, M.E., and Merideth, R.W., Jr., (1987), Global sustainability: toward a definition, *Environ. Manage.*, 11, 713–719.

Conway, G.R., (1987), The properties of agroecosystems, *Agric. Syst.*, 24, 95–117.

Georgescu-Roegen, N., (1971), *The Entropy Law and the Economic Process*, Harvard University Press, Cambridge, MA.

Giampietro, M., (1994a), Using hierarchy theory to explore the concept of sustainable development, *Futures*, 26, 616–625.

Giampietro, M., (1994b), Sustainability and technological development in agriculture: a critical appraisal of genetic engineering, *Bioscience*, 44, 677–689.

Giampietro, M. and Pastore, G., (1999), Multidimensional reading of the dynamics of rural intensification in China: the AMOEBA approach, *Crit. Rev. Plant Sci.*, 18, 299–330.

Giampietro, M. and Mayumi, K., (2001), Integrated Assessment of Sustainability Trade-Offs: The Challenge for Ecological Economics, paper prepared for the EC High Level Scientific Conference, ESEE Frontiers 1: Fundamental Issues of Ecological Economics, available at http://www.euroecolecon.org/frontiers, paper 112.

Gomiero, T. and Giampietro, M., (2001), Multiple-scale integrated analysis of farming systems: the Thuong Lo Commune (Vietnamese Uplands) case study, *Popul. Environ.*, 22, 315–352.

Hart, R.D. (1984), The effect of inter-level hierarchical system communication on agricultural system input-output relationships. In: *Workshop Agroecology Paris: CIHEAM*, 166 p. (Options Mediterraneennes: Serie Etudes; n. 1984-I — Limits to Agricultural Production, 1984/01 Zaragoza, Spain). http://ressources.ciheam.org/om/pdf/s07/CI010840.pdf.

Ikerd, J.E., (1993), The need for a system approach to sustainable agriculture, *Agric. Ecosyst. Environ.*, 46, 147–160.

Li, J., Giampietro, M., Pastore, G., Cai L., and Luo H, (1999), Factors affecting technical changes in rice-based farming systems in southern China: case study of Qianjiang municipality, *Crit. Rev. Plant Sci.*, 18, 283–298.

Lockeretz, W., (1988), Open questions in sustainable agriculture, *Am. J. Sustain. Agric.*, 3, 174–181.

Lowrance, R., Hendrix, P., and Odum, E.P., (1986), A hierarchical approach to sustainable agriculture, *Am. J. Alternative Agric.*, 1, 169–173.

Martinez-Alier, J., Munda, G., and O'Neill, J., (1998) Weak comparability of values as a foundation for ecological economics, *Ecol. Econ.*, 26, 277–286.

Mayumi, K. and Giampietro, M., (2001), Epistemological Challenge for Modelling Sustainability: Exploring the Difference in Meaning of Risk, Uncertainty and Ignorance, paper prepared for the EC High Level Scientific Conference, ESEE Frontiers 1: Fundamental Issues of Ecological Economics, available at http://www.euroecolecon.org/frontiers, paper 119.

Munda, G., (1995), *Multi-criteria Evaluation in a Fuzzy Environment*, Physica Verlag, Heidelberg.

Pastore, G., Giampietro, M., and Li, J., (1999), Conventional and land-time budget analysis of rural villages in Hubei Province, China, *Crit. Rev. Plant Sci.*, 18, 331–358.

Rosen, R., (1985), *Anticipatory Systems: Philosophical, Mathematical and Methodological Foundations*, Pergamon Press, New York.

Rosen, R., (1991), *Life Itself: A Comprehensive Inquiry into the Nature, Origin and Fabrication of Life*, Columbia University Press, New York, 285 pp.

Schaller, N., (1993), The concept of agricultural sustainability, *Agric. Ecosyst. Environ.*, 46, 89–97.

Simon, H.A., (1976), From substantive to procedural rationality, in *Methods and Appraisal in Economics*, Latsis, J.S., Ed., Cambridge University Press, Cambridge, U.K.

Simon, H.A., (1983), *Reason in Human Affairs*, Stanford University Press, Stanford, CA.

Vu Biet Linh, and Nguyen Ngoc Binh (1995), *Agroforestry Systems in Vietnam* (Vietnam APAN) Agriculture Publishing House, Hanoi. 160 pp.

Wolf, S.A. and Allen, T.F.H., (1995), Recasting alternative agriculture as a management model: the value of adapt scaling, *Ecol. Econ.*, 12, 5–12.

Index

9 780367 394813